T0271874

Discrete Mathematics with Coding

For a first undergraduate course in discrete mathematics, this book systematically explores the relationship between discrete mathematics and computer programming. Unlike most discrete mathematics texts focusing on one or the other, *Discrete Mathematics with Coding* investigates the rich and important connection between these two disciplines and shows how each reinforces and enhances the other.

The mathematics in the book is self-contained, requiring only a good background in precalculus and some mathematical maturity. New mathematical topics are introduced as needed.

The coding language used is VBA Excel. The language is easy to learn, has intuitive commands, and the reader can develop interesting programs from the outset. Additionally, the spreadsheet platform in Excel makes for convenient and transparent data input and output and provides a powerful venue for complex data manipulation. Manipulating data is greatly simplified using spreadsheet features, and visualizing the data can make programming and debugging easier.

The VBA language is seamlessly integrated into the spreadsheet environment with no other resources required. Furthermore, as some of the modules in the book show, intricate patterns, graphs, and animation in the form of moving cells is possible.

Features

- Introduces coding in VBA Excel, assuming no previous coding experience.
- Develops programs in linear analysis, logic, combinatorics, probability, and number theory.
- Contains over 90 fully tested and debugged programs. The code for these, as well as the exercises, are available on the author's website.
- Encompasses numerous examples that gradually introduce the reader to coding techniques.
- Includes programs that solve systems of linear equations, linear programming problems, combinatorial problems, Venn diagram problems, programs that produce truth tables from logic statements, logic statements from switching and gate circuits, encrypt and decrypt messages, and simulate probability experiments.

Hugo D. Junghenn is a professor of mathematics at The George Washington University. He has published numerous journal articles and is the author of several books, including *A Course in Real Analysis* and *Principles of Analysis: Measure, Integration, Functional Analysis, and Applications*. His research interests include functional analysis, semigroups, and probability.

Textbooks in Mathematics
Series editors:
Al Boggess, Kenneth H. Rosen

Transition to Advanced Mathematics
Danilo R. Diedrichs and Stephen Lovett

Modeling Change and Uncertainty
Machine Learning and Other Techniques
William P. Fox and Robert E. Burks

Abstract Algebra
A First Course, Second Edition
Stephen Lovett

Multiplicative Differential Calculus
Svetlin Georgiev and Khaled Zennir

Applied Differential Equations
The Primary Course
Vladimir A. Dobrushkin

Introduction to Computational Mathematics: An Outline
William C. Bauldry

Mathematical Modeling the Life Sciences
Numerical Recipes in Python and MATLAB™
N. G. Cogan

Classical Analysis
An Approach through Problems
Hongwei Chen

Classical Vector Algebra
Vladimir Lepetic

Introduction to Number Theory
Mark Hunacek

Probability and Statistics for Engineering and the Sciences with Modeling using R
William P. Fox and Rodney X. Sturdivant

Computational Optimization: Success in Practice
Vladislav Bukshtynov

Computational Linear Algebra: with Applications and MATLAB® Computations
Robert E. White

Linear Algebra With Machine Learning and Data
Crista Arangala

Discrete Mathematics with Coding
Hugo D. Junghenn

https://www.routledge.com/Textbooks-in-Mathematics/book-series/CANDHTEXBOOMTH

Discrete Mathematics with Coding

Hugo D. Junghenn

CRC Press
Taylor & Francis Group
Boca Raton London New York

CRC Press is an imprint of the
Taylor & Francis Group, an **informa** business

A CHAPMAN & HALL BOOK

First edition published 2024
by CRC Press
6000 Broken Sound Parkway NW, Suite 300, Boca Raton, FL 33487-2742

and by CRC Press
4 Park Square, Milton Park, Abingdon, Oxon, OX14 4RN

CRC Press is an imprint of Taylor & Francis Group, LLC

© 2024 Hugo D. Junghenn

Library of Congress Cataloging-in-Publication Data

Names: Junghenn, Hugo D. (Hugo Dietrich), 1939- author.
Title: Discrete mathematics with coding / Hugo D. Junghenn.
Description: Boca Raton : CRC Press, 2024. | Series: Textbooks in
  mathematics | Includes bibliographical references and index. | Summary:
  "The book is an attempt to integrate coding with discrete mathematics in
  the hopes that each discipline will complement the other and allow for a
  fuller treatment of concrete mathematical computations. Mathematics
  provides fertile ground for interesting, useful, and certainly
  nontrivial programs"-- Provided by publisher.
Identifiers: LCCN 2023004242 (print) | LCCN 2023004243 (ebook) | ISBN
  9781032398525 (hardback) | ISBN 9781032398563 (paperback) | ISBN
  9781003351689 (ebook)
Subjects: LCSH: Discrete mathematics--Textbooks. | Coding
  theory--Textbooks.
Classification: LCC QA297.4 .J86 2024 (print) | LCC QA297.4 (ebook) | DDC
  511/.1--dc23/eng/20230322
LC record available at https://lccn.loc.gov/2023004242
LC ebook record available at https://lccn.loc.gov/2023004243

ISBN: 978-1-032-39852-5 (hbk)
ISBN: 978-1-032-39856-3 (pbk)
ISBN: 978-1-003-35168-9 (ebk)

DOI: 10.1201/9781003351689

Typeset in CMR10
by KnowledgeWorks Global Ltd.

Publisher's note: This book has been prepared from camera-ready copy provided by the authors.

TO MY FAMILY

Companions
in the
Great Adventure

Contents

Preface

This book is an attempt to explore the fertile connection between discrete mathematics and computer programming and to show how each discipline reinforces and enhances the other. The connection between the two subjects is, of course, well-known and also not surprising given the algorithmic nature of much of mathematics. But while there are many excellent texts that treat the mathematical topics in this book and many other texts that teach coding, few make a systematic effort to exploit in a detailed way the relationship between the two disciplines. It is hoped that this book can make a contribution to this endeavor.

The mathematics in the book is self-contained, requiring only a background in standard precalculus and some mathematical maturity. The book treats only discrete mathematics; calculus is not covered. New mathematical topics are introduced as needed.

The main feature of the book is the use of Excel's spreadsheet to display in an engaging and informative way the output of a program. For example, the reader will discover programs that show the evolution of a maze construction and its solutions; the initial, intermediate and final tableaus of a linear programming problem; simulation of the spread of a disease; random walks of cell "organisms" that produce "offspring;" an animated solution to the Tower of Hanoi puzzle; Venn diagram solutions for three and four sets; particle swarm optimization; dynamic simulation of probability experiments including Markov processes; and the encryption and decryption of cipher algorithms, including a working spreadsheet dynamic model of the Enigma Machine.

The spreadsheet platform in Excel also makes for convenient and transparent data input and output and provides a powerful venue for complex data calculations. Manipulating data is greatly simplified using spreadsheet features, and visualizing the data in a transparent and convenient form can make programming and debugging easier. Furthermore, Excel is readily available and spreadsheet navigation is easy to learn.

The coding language used to produce the programs in the book is Excel VBA. VBA is seamlessly and naturally integrated into the spreadsheet environment of Excel and requires no additional resources. The language is easy to learn, has intuitive commands, and the reader can develop interesting programs from the start. Indeed, Dartmouth mathematics professors Kemeny and Kurtz developed BASIC precisely to introduce beginners to coding with a simple, all purpose language. Furthermore, Excel VBA is designed to make the transition from algorithm to code easy and transparent, unhindered by complex or obscure commands.

The book falls into six parts:

  I. Essentials of VBA

 II. Linear Analysis

III. Logic

IV. Combinatorics

 V. Probability

VI. Properties of Numbers

Part I develops fundamental coding principles such as conditionals, loops, and arrays, and integrates these with spreadsheet features. The chapters here emphasize those aspects of programming that are aligned with the mathematical goals of the book. A variety of examples are included to illustrate standard coding techniques.

Part II develops methods and programs that solve systems of linear equations and linear programming problems as well as evaluate matrix algebraic expressions and determinants. Applications include input-output analysis and curve fitting.

Part III explores propositional logic and constructs programs that calculate truth tables from logical statements and vice versa. Applications are made to switching circuits and logic gate circuits. Programs here simulate circuits on the spreadsheet, including current flow, and generate logical expressions from circuit diagrams.

Part IV develops set theory and describes various methods of enumeration. There are programs that evaluate set expressions, solve Venn diagram problems, generate permutations and combinations, and evaluate algebraic expressions involving permutations.

Part V develops basic probability theory and constructs programs that simulate probability experiments and visually demonstrate the evolution of various Markov processes.

Part VI discusses basic number theory and its applications. Programs include finding the greatest common divisor of a pair of integers, displaying the division algorithm, and generating the prime factorization of integers. A variety of applications are made to cryptography with programs that encrypt and decrypt messages. Programs are also developed that carry out calculations with numbers represented by stings. This allows computations using numbers of virtually unlimited size, opening up the possibility for additional cipher programs.

There are over 90 modules in the book and more in the exercises. These have been fully tested and debugged. While the vast majority of the modules actually carry out useful tasks, several of the examples, such as the random walk of multiple organisms, the random growth of a plant, or the nonsense generator, are admittedly somewhat whimsical. But each example has been chosen to illustrate particular coding constructs or techniques, and if a program is visually engaging, so much the better. The modules developed in the book are open-ended in the sense that the reader may tinker with the code to produce modifications and extensions. Additionally, many of the coding exercises explore new ideas. The code for the examples and exercises is available on my blog blogs.gwu.edu/hdj.

The book has benefitted from my fortunate experience over the years in teaching courses on discrete mathematics and math foundations. An effort was made in these courses to emphasize the algorithmic nature of mathematics. The book attempts, via code, to give full expression to this philosophy.

Finally, many thanks go to my editor Bob Ross for his invaluable guidance and suggestions and to the staff at Taylor & Francis/CRC Press for their professionalism in the development of this book.

Hugo D. Junghenn
George Washington University
Washington, D.C.

Part I

Essentials of VBA

Chapter 1

Introduction

In this chapter we describe some general features of Excel VBA, leaving coding constructs such as conditionals and loops to later chapters. The reader is assumed to have some familiarity with entering data into a spreadsheet and using the arrow keys to navigate among the cells.

1.1 Modules, Subs, and Functions

A VBA *module* or *program* is a text file consisting of a collection of VBA statements that are executed by Excel. Modules are written in the VBA editor window. To access the editor you need to carry out the following steps, the first four of which enable the Developer tab.

- Pull up a blank sheet in Excel.
- Right-click any of the existing tabs, such as File.
- Select Customize The Ribbon.
- Select the Developer checkbox and click OK.
- Click the Developer tab and then click on View Code to launch the VBA editor.

The VBA editor works like most editors but with some additional features that help catch errors and make the process of coding more efficient. Once in the editor window you can begin writing the module. Click "View Code" under the Developers tab in the spreadsheet to get to the editor. Use ALT + F11 to switch between editor and spreadsheet.

A module consists of one or more building blocks called *procedures*. These are collections of statements executed together as a unit. Breaking up a lengthy program into smaller procedures can improve efficiency and greatly enhance clarity, allowing the overall logic of the program to emerge unfettered by details. Such partitioning can also reduce code and simplify the process of debugging, an unfortunate but inevitable and all too frequent aspect of programming.[1]

---

[1] On the other hand, debugging can sometimes be an interesting logical challenge and can lead to a firmer grasp of coding principles.

There are two types of procedures, *sub procedures* and *function procedures*. They have the following similar forms:

```
Sub name()                      Function name()
    Declaration of variables        Declaration of variables
    Statements                      Statements
End Sub                          End Function
```

The essential distinction between the two types of procedures is that a function can return data while a sub cannot. However, a sub can always return data through an *argument*, discussed later.

The reader may see in some sources the word Private preceding the declaration Sub or Function. With this modifier the procedure may be accessed only from another procedure in the same module. Without the modifier, the procedure can be accessed from other modules. Since all of our programs are single modules, the prefix Private will usually be omitted. The exception is in the case of user defined data types, discussed in Section 1.6.

1.2   Data Output and Input

Many programs process data in one form or another and so must have ways to access and display the data. We illustrate with some simple examples.

Outputting Data

The following module consists of a single sub procedure. The program does nothing more than output text in cell A2 of Sheet1 using the *Range object*.

```
Sub OutputExample()
    Sheet1.Range("A2").Value = "GREETINGS, PLANET EARTH!"
End Sub
```

Alternate code that accomplishes the same thing uses the *Cells object*, which has the form Cells(row,column). For example, replacing the middle line of the above code by

```
Sheet1.Cells(2,1).Value = "GREETINGS, PLANET EARTH!"
```

produces exactly the same output. Note that the row-column order is reversed in the two notations: A2 and $(2,1)$ both refer to the cell located in column 1 and row 2 of Sheet1. The Cells object is particularly useful when one needs to treat the spreadsheet as a coordinate system or when many data items need to be read or printed.

The above project may be saved as a Macro-Enabled Workbook. To run the program go to the text editor and click "run." The output will then appear in

Sheet1. Using the prefix "Sheet2" or "Sheet3" outputs the data accordingly. This is occasionally useful if, for example, data needs to be "hidden" or stored for another program run. If all output is to go to Sheet1, then the prefix may be omitted, as it is the default.

Inputting Data

The Range and Cells objects mentioned above may also be used to input data, that is, to read data from the spreadsheet into *variables*. These are user-chosen names that refer to locations in the computer's memory. VBA allows many types of variables. The list below describes the most common of these and the ones that we shall use throughout the book. Detailed descriptions of these and other data types may be found online.

- `Integer`: Holds negative, zero, and positive integers.
- `Long`: Similar to Integer, but holds integers within a larger range.
- `Single`: Stores negative or positive decimal values.
- `Double`: Similar to Single, but holds larger decimals.
- `Currency`: Holds positive and negative monetary values.
- `String`: Stores text.
- `Boolean`: Holds two values, True or False, the so-called *Boolean values* used in conditional statements and loops.
- `Variant`: The VBA default data type. An undeclared variable is always taken to be Variant. The data type can contain string, Boolean, or numeric values, and can convert the value it contains from one type to another. The downside is that it takes more memory than other variables.

Here's simple example that illustrates some of the features of data assignment. Other features appear in examples throughout the book.

```
Sub InputExample()
    Dim v1 As Double, v2 As Integer, v3 As Currency, v4 As String
    v1 = Range("A3").Value          'read data from spreadsheet into v1
    v2 = v1                          'assign the value of v1 to v2
    v3 = v1                                  'and to v3
    v4 = v1                                  'and to v4
    Range("B3").Value = v2          'output the values in cells B3, C3, D3
    Range("C3").Value = v3
    Range("D3").Value = v4
End Sub
```

Notice how the data types of the variables `v1,v2,v3,v4` are declared in the first line of the code using the keyword `Dim`. In the second line the program reads a numerical quantity that was presumably entered by the user in cell A3. Running the program causes the value to be assigned to the variable (memory location) `v1` and subsequently to the variables `v2,v3,v4`. The results are then printed in cells B3, C3, and D3.

The text following the apostrophes in the above procedure is meant to elucidate the code; it is ignored by the Excel compiler. While the code here is largely self-explanatory, programs can be quite complex, and liberal use of comments greatly enhances clarity. It is not uncommon for a programmer to return to past code and wonder what the program is supposed to be doing. (The author admits to being guilty of this on more than one occasion.)

The spreadsheet corresponding to the above procedure is depicted below. Notice how the data output in cells B3, C3, and D3 conforms to the data type: the original value in A3 is printed as a rounded integer in B3, as a currency in C3, and as a string (text) in D3. VBA performs these conversions automatically.

	A	B	C	D
3	1234.567	1235	$1,234.57	1234.567

We have used short names for the variables, but it is frequently the case that longer names are needed to clarify the code. For example, instead of `v1` we could have used any one of the following standard forms `varOne`, `VarOne`, or `var_one`. Here are the rules for naming a variable:

- It must start with a letter. Numbers may be used elsewhere.

- It must not be an Excel key word.

- It cannot contain spaces, periods, or special characters (except underscore).

Also, keep in mind that VBA variables are *not* case-sensitive.

1.3 Cell Characteristics

The Value property of the Range object introduced in the previous section is one of many attributes that the object possesses. Others include font size and style, color, and number formats. Attributes may be viewed by right clicking on any cell and then clicking Format Cells. As an illustration, the procedure `CellAttributes` below prints a string with various color and font features added.

```
Sub CellAttributes()
    Cells.RowHeight = 15                    'reset to standard value
    Cells.ColumnWidth = 8.43                'reset to standard value
    Columns("A:C").ColumnWidth = 16            'expand column width
```

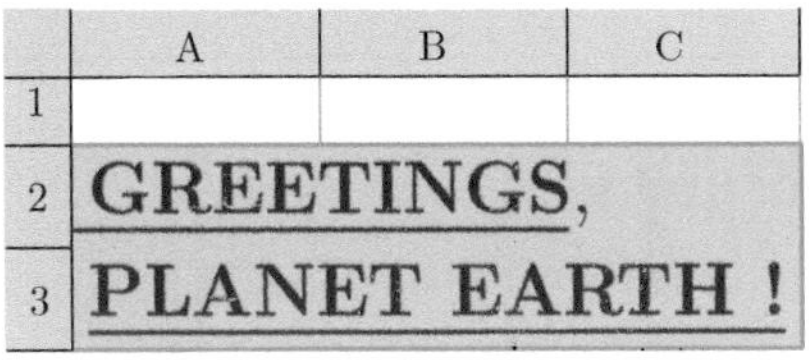

```
        Rows("2:3").RowHeight = 54                      'expand row height
        Range("A2:C3").Font.Name = "Times New Roman"    'set the font type
        Range("A2:C3").Font.FontStyle = "Bold"          'set its style
        Range("A2:C3").Font.Size = 30                   'and size
        Range("A2:C3").Font.Underline = True            'underline the text
        Range("A2:C3").Font.ColorIndex = 3              'color the letters red
        Range("A2:C3").Interior.ColorIndex = 6          'make cell interior yellow
        Range("A2:C3").Borders.ColorIndex = 4           'outline cells green
        Range("A2").Value = "GREETINGS,"                'print the text
        Range("A3").Value = "PLANET EARTH!"
    End Sub
```

The first two lines in the procedure expand the column width and row height of the spreadsheet from their standard values (8.34 points and 15 points, respectively). The code `Range("A2:C3")` is an example of the *block feature* of the Range object. One may also use `Range(Cells(2,1),Cells(3,3))`. The font "Times New Roman" is one of many that are available. The font styles are "Regular", "Italic", "Bold", and "Bold Italic". In addition to `Font.Underline` there are the attributes `Font.Strikethrough`, `Font.Subscript`, and `Font.Superscript`.

The above procedure uses the VBA `ColorIndex` property to set the colors of the interior and border of the range A2:C3. Color indices have values 1–56.

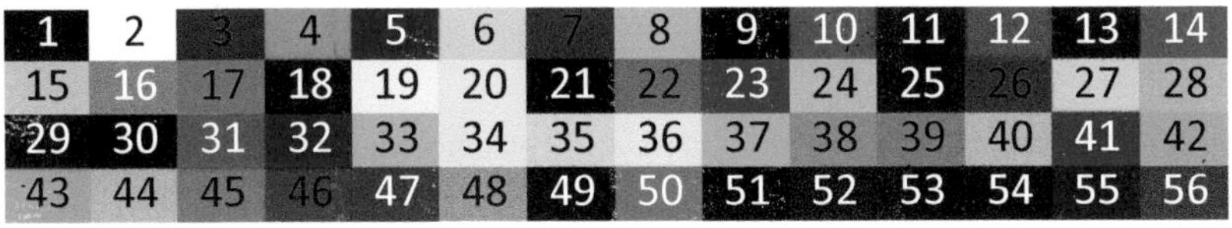

FIGURE 1.1: VBA color indices.

To eliminate repetition of the code fragment `Range("A2:C3")` one can use the VBA `With...End With` statement:

```
With Range("A2:C3").Font
                .Name = "Times New Roman"
                .FontStyle = "Bold"
                .Size = 30
                .Underline = True
                .ColorIndex = 3
                .Interior.ColorIndex = 6
                .Borders.ColorIndex= 4
    End With
```

1.4   Number Formats

The `Range` or `Cells` object may also be used to set the number format of cells, as shown in the procedure `NumberFormat` below. The program first copies

	A	B	C	D
1	1234.000	1,234	$1,234.00	1234

FIGURE 1.2: Spreadsheet output for `NumberFormat`.

the number 1234 into cells A1 to D1. The cells are then formatted in various ways with output shown in Figure 1.2.

```
Sub NumberFormat()
    Range("A1:D1").Value = 1234              'output 1234 to cells in block
    Range("A1").NumberFormat = ".000"                'produces 1234.000
    Range("B1").NumberFormat = "#,###"                 'produces 1,234
    Range("C1").NumberFormat = "$#,##0.00"           'produces $1,234
    Range("D1").NumberFormat = "General"                'produces 1234
End Sub
```

The reader may wish to experiment with other formats and cell attributes. The ones described in this section are sufficient for our purposes.

1.5   Passing Arguments

Both subs and functions can accept and return data via arguments (variable names) placed within the parentheses of the procedure. There are two ways to do this: either by *reference*, using the keyword `ByRef`, or by *value*, using the keyword `ByVal`. If you pass an argument by reference, the memory address of the variable is passed, enabling the procedure to modify the original value located at that address. On the other hand, if you pass an argument by value, only a copy of the argument's value is passed, so that the procedure may modify the copy without changing the original value. By default, VBA passes all arguments by reference. So if you wish the original value of the variable to be preserved, you must pass the argument `ByVal`.

The module `SwitchIntegers` illustrates the distinction between the two methods of passing arguments. It takes a pair of integer variables and calls two procedures, `SwitchInt(var1,var2)` and `NoSwitchInt(var1,var2)`, each of which attempts to switch the values of the arguments by introducing an auxiliary variable `temp` to store the first while the switch is made. The reader may check by running the program that only `SwitchInt` was successful.

```
Sub SwitchIntegers()
    Dim m As Integer, n As Integer
    m = 3: n = 7
    Call SwitchInt(m, n)
    Range("A4").Value = m: Range("B4").Value = n    'prints 7,3: switched
    Call NoSwitchInt(m, n)
    Range("A3").Value = m: Range("B3").Value = n     'prints 3,7: failed
End Sub

Sub SwitchInt(var1 As Integer, var2 As Integer)                    'ByRef default
    Dim temp As Integer
    temp = var1: var1 = var2:  var2 = temp  'switch values of var1, var2
End Sub

Sub NoSwitchInt(ByVal var1 As Integer, ByVal var2 As Integer)
    Dim temp As Integer
    temp = var1: var1 = var2:  var2 = temp             'values not switched
End Sub
```

In the above code we have used a feature of VBA Excel that allows several statements, separated by colons, to be placed on the same line. While this is a handy space-saving device, it should be used only when it does not compromise readability. (The author admits to occasionally ignoring his own advice in this regard.)

1.6 User-Defined Data Types

A *user defined data type* (UDT) is a custom data structure that consists of multiple *fields*, each of which represents some data type. It allows the user to group together within a single object a set of conveniently accessed variables of different data types.

The following code constructs a UDT TDog that summarizes data for the user's dogs. We have used the letter T in the name of the type as a shorthand way of distinguishing its special role as a UDT. We shall apply this convention throughout the book.

```
Private Type TDog
    Breed       As String
    Age         As Integer
    Weight      As Double
    RabiesVac   As Boolean
End Type
```

The variable TDog may be assigned values as illustrated by the code

```
Tdog.Breed = "pug": Tdog.Age = 7: TDog.Weight = 15: TDog.RabiesVac = True
```

A UDT declaration must be placed at the top of the module and declared `Private`. Any procedure passing a UDT variable must also be declared `Private`. UDT's will be used frequently in later chapters.

1.7 Command Buttons

A command button allows the user to run a program directly from the spreadsheet rather than from the VBAProject window. To install the button, carry out the following steps:

- In the Developer tab, click Insert and then under ActiveX Controls click Command Button. Click the worksheet location where you want the button to be placed. The box may be dragged to other locations and expanded or contracted.

- Assign code to the button as follows: Right-click on the button and then click 'view code'. In the resulting sub procedure that appears in the VBA editor window, type the name of the procedures you wish to run from the button and any additional code.

- To assign a caption to the button, right-click the button, and then click CommandButtonObject, Edit.

- To change the properties of the button, right-click the button and then click Format Control or Properties.

- To delete a command button, select Visual Basic from the Developer tool bar, click Design Mode, move to the button, and hit the delete key.

To illustrate how a command button works, let's add two buttons to the procedure `NumberFormat`, one to run the program and the other to restore the format settings to their defaults and clear the numeric content of the spreadsheet. Here is the additional code needed:

```
Sub CommandButton1_Click()                    'click to run program
    Call NumberFormat
End Sub

Sub CommandButton2_Click()                       'click to clear
    Range("A1:D1").ClearContents
    Range("A1:D1").ClearFormats
End Sub
```

Here are command buttons for the procedure `CellAttributes`:

```
Sub CommandButton1_Click()                    'click to run program
    Call CellAttributes
End Sub

Sub CommandButton2_Click()                        'click to clear
    Range("A2:C3").ClearContents
    Range("A2:C3").Interior.ColorIndex = 0          'remove color
    Range("A2:C3").Borders.LineStyle = xlLineStyleNone   'remove borders
    Cells.RowHeight = 15                    'restore to standard value
    Cells.ColumnWidth = 8.43                              'ditto
End Sub
```

1.8   Spin Buttons

Spin buttons consist of two arrows that change a number in an associated cell. Clicking one arrow increases the number, clicking the other decreases it. To install the button, carry out the following steps:

- In the Developer tab, click Insert and then under ActiveX Controls click Spin Button. Next, click the worksheet location where you want the button to be placed. The box may be dragged to other locations and expanded or contracted.

- Assign code to the button as follows: Right-click on the button and then click 'view code'. In the resulting code that appears in the VBA editor window, type the name of the procedures you wish to run from the button and any additional instructions.

- To delete a spin button click Design Mode in the Developers tab, move to the button and hit the delete key (gently).

The module `MixColors` below uses a spin button to display the colors available with the VBA function `RGB`. This function, together with the VBA Color property, allows one to create hundreds of colors and hues. The syntax of the function is `RGB(red,green,blue)`, where the arguments `red,green,blue` are integers between 0 and 255 that determine the proportions of red, green, and blue in the color. For example, `RGB(255,0,0)` yields red, `RGB(0,255,0)` yields green, and `RGB(0,0,255)` yields blue. Varying the values of the arguments `red,green,blue` produces mixtures of these colors. For example, the code snippet `Range("B2").Interior.Color = RGB(249,251,219)` colors the interior of cell B2 a pale yellow.

To run the program the user enters integers from 0 to 255 in each of the cells B4, C4, and D4, these holding values of the arguments `red,green,blue` of the function `RGB`. Depressing the arrow keys changes the numbers and places the corresponding color in cell B4. Figure 1.3 depicts part of the spreadsheet showing values that result in a lavender.

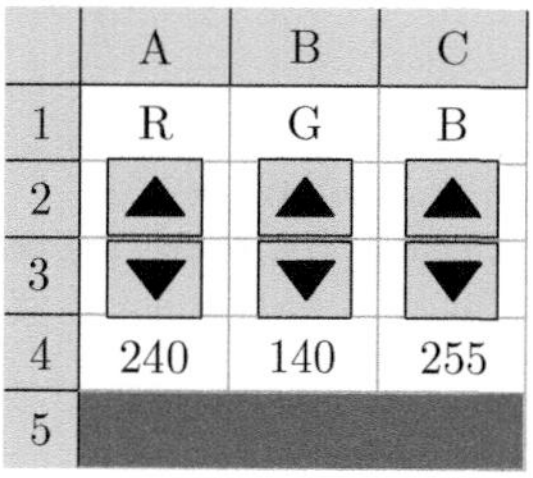

	A	B	C
1	R	G	B
2	▲	▲	▲
3	▼	▼	▼
4	240	140	255
5			

FIGURE 1.3: Spreadsheet for Mixcolors.

Here is the code for the module. The spin buttons are linked to the cells A4, B4, and C4; the range of values is 0–255 in steps of 1. Linking is accomplished by right clicking on the spin button in Design mode and then filling in the appropriate entries in the pop-up Properties box.

```
Sub SpinButton1_Change() 'linked to A4, 0 - 255 in steps of 1
    Dim i As integer, j As integer, k As Integer
    i = Range("A4").Value: j = Range("B4").Value: k = Range("C4").Value
    Range("A5:C5").Interior.Color = RGB(i,j,k)           'print the color
End Sub

Sub SpinButton2_Change() 'linked to B4, 0 - 255 in steps of 1
    Dim i As integer, j As integer, k As Integer
    i = Range("A4").Value: j = Range("B4").Value: k = Range("C4").Value
    Range("A5:C5").Interior.Color = RGB(i,j,k)
End Sub

Sub SpinButton3_Change() 'linked to C4, 0 - 255 in steps of 1
    Dim i As integer, j As integer, k As Integer
    i = Range("A4").Value: j = Range("B4").Value: k = Range("C4").Value
    Range("A5:C5").Interior.Color = RGB(i,j,k)
End Sub
```

1.9 Exercises

1. Experiment with the procedure `CellAttributes` by altering values, changing colors, fonts, styles, etc. Use the With-End statement.

2. Write a procedure `ElbowPath(row1 As Int, col1 As Int, row2 As Int, col2 As Int)` with a command button that takes as arguments user-entered positions of two cells, `row1,col1` and `row2,col2`, and draws an

L-shaped yellow path from one cell to the other (see Figure 1.4). Use the block feature of the Range object to draw and then erase the path, using the color index `xlNone` for the latter. Insert the statement `MsgBox"clear"` after the drawing part of the code and before the erasing part and see what is does.

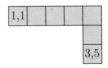

FIGURE 1.4: Elbow path.

3. Write a program `FontColorText` that takes as input three things: text, fontstyle, and a color index, entered in cells B3, B4, and B5, respectively, and outputs the text in the chosen color and font as in Figure 1.5. Use the With End With statement and a command button. Some interesting fonts to try are Algerian, AR BERKLEY, Broadway, Comic Sans MS, Copperplate Gothic Bold, Forte, Goudy Stout, Harlow Solid Italic, Jokerman, Kunstler Script, Magneto, Matura MT Script Capitals, Modern Love, Niagara Engraved, Papyrus, Script MT Bold, Snap ITC, and Wide Latin.

	A	B
1	run program	
2	fiddlefaddle	
3	text:	fiddlefaddle
4	font:	Times New Roman
5	color index:	3

FIGURE 1.5: Spreadsheet for `FontColorText`.

Chapter 2

VBA Operators

In this chapter we discuss three types of operators: arithmetic, used in computations, and comparison and logical, used largely in conditional statements and loops.

2.1 Arithmetic Operators

The most common arithmetic operators are

$$
\begin{array}{ll}
+ & addition \\
- & subtraction \\
* & multiplication \\
/ & division \\
\char`^ & exponentiation
\end{array}
$$

The operators follow the usual order of precedence in arithmetic. For example, in the code $2 * 3\char`^4 + 5/(6 - 7 * 8)$, exponentiation is carried out before the multiplication, parentheses before the division, and multiplication and division before addition and subtraction. It should be noted that Excel has a feature that calculates such expressions directly in a spreadsheet cell. However, this feature may be impractical if complicated arithmetic calculations must be performed on large data sets. Coding these operations is then a more reasonable approach.

There are four other arithmetic operators that are of importance:

$$
\begin{array}{ll}
\text{Int} & greatest\ integer \\
\text{Mod} & modulus \\
\backslash & integer\ division \\
\text{Abs} & absolute\ value
\end{array}
$$

The first operator rounds down a number to the next lowest integer value. For example $\text{Int}(2.99) = 2$ and $\text{Int}(-2.01) = -3$. The operation $a \bmod b$ produces the remainder on division of the integer a by the integer b. For example, the statement 14 Mod 3 returns the integer 2. The operation $a \backslash b$ divides the integer

DOI: 10.1201/9781003351689-2

a by the integer b, discards the remainder, and returns the integer part. For example, $14 \backslash 3 = 4$. The operators are related by the equation

$$a \text{ Mod } b = b * (a/b - a \backslash b),$$

as illustrated by

$$3 * (14/3 - 14 \backslash 3) = 3 * (2/3) = 2 = 14 \text{ Mod } 3.$$

Finally, the operator Abs returns a number without the sign. For example,

$$\text{Abs}(-2) = 2 = \text{Abs}(2).$$

Recall that equality is used to assign values to variables. For example, the code x = 5 assigns the value 5 to the memory location called x. One can take this a step further by having the right side of the assignment contain the variable x. For example, the code x = x+2 tells the computer to add 2 to the current value of x, which was 5, and place the resulting value 7 in the memory location called x.

2.2   Comparison Operators

As the name suggests, comparison operators compare two numbers. Here is the list.

a = b	*equal to*
a <> b	*unequal to*
a < b	*strictly less than*
a > b	*strictly greater than*
a <= b	*less than or equal*
a >= b	*greater than or equal*

Expressions such as these have values True or False. For example, both expressions $1 < 2$ and $1 <= 2$ evaluate to True, while $1 = 2$ evaluates to False. Note how the use of equality in this context differs from its use in assigning values to variables.

2.3   Logical Operators

Boolean expressions are statements that are either true or false. They may be combined by the *logical operators* And, Or, and Not. The use of these operators is governed by the following rules, where P and Q denote Boolean expressions:

P And Q   True if and only if both P and Q are True.
P Or Q   False if and only if both P and Q are False.
Not P   True if and only if P is False.

The first two connectives extend to arbitrarily many statements. For example, the code Range("A1").Value = 2 < 3 And Not 3 < 4 Or 5 < 6 places the word True in cell A1. The order of precedence of the operations in this statement is as follows: After the inequalities are assigned truth values, the statement becomes True And Not True Or True. The Not statement is then evaluated, producing True And False Or True. Next in precedence is the And statement, yielding False Or True. Finally, the Or statement is evaluated, producing True.

Parentheses may be used to change operator precedence. For example, the code 2 < 3 And Not (3 < 4 Or 5 < 6) returns the value False.

If operators from more than one category appear in a line of code, they are evaluated in the order (1) arithmetic operators, (2) comparison operators, and (3) logical operators.

2.4   Exercises

1. Write a function Triangle that takes three lengths a, b, and c and returns True if these can be the lengths of a triangle and False otherwise. (Hint: The sum of two sides must be greater than the third.) Test the function by writing a program with a command button that reads the values of a, b, and c entered in the spreadsheet and prints out True or False accordingly.

2. Write a function NumSeconds(h,m,s) that takes the elapsed time in hours h, minutes m, and seconds s, and returns the number of seconds. For example, NumSeconds(1,25,43) should return 5143. Test the function with a program that prints out the answer. Use a command button.

3. Write a procedure TestArithOps that reads user-entered numbers a, b from the spreadsheet and prints the difference a Mod $b - b * (a/b - a \backslash b)$, (which should always be zero). Use a command button.

4. Write a procedure SpinArith that uses a Spin Button to change the values of an arithmetic expression entered in the spreadsheet. For example, you could link spin buttons to A3, A4, and A5 and enter expressions like = A2^2 + B2^2 + C2^2 in cell A4 (note the required equals sign).

5. Describe the sequence generated by each of the expressions
   (a) $n \backslash 3$ (b) $((-1)^{(n \backslash 3)} + 1)/2$,   where $n = 1, 2, 3, \ldots$.

6. Let a, b, and c denote positive integers. Describe the sequence generated by the expression $b + ((-1)^{(n\backslash 3)} + 1) * (a - b)/2, \quad n = 0, 1, 2, \ldots$. What is 1,000,000th term of this sequence? (To find this, long divide; do not write out 1,000,000 terms unless you're really fond of the letters a and b.)

7. A *prime number* is a positive integer larger than 1 that has no positive factors other than 1 and itself. The first 100 primes are

$2, 3, 5, 7, 11, 13, 17, 19, 23, 29, 31, 37, 41, 43, 47, 53, 59, 61, 67, 71, 73, 79, 83, 89,$

$97, 101, 103, 107, 109, 113, 127, 131, 137, 139, 149, 151, 157, 163, 167, 173, 179,$

$181, 191, 193, 197, 199, 211, 223, 227, 229, 233, 239, 241, 251, 257, 263, 269, 271,$

$277, 281, 283, 293, 307, 311, 313, 317, 331, 337, 347, 349, 353, 359, 367, 373, 379,$

$383, 389, 397, 401, 409, 419, 421, 431, 433, 439, 443, 449, 457, 461, 463, 467, 479,$

$487, 491, 499, 503, 509, 521, 523, 541.$

Write a program `EulerPoly` that uses spin buttons to calculate values of the polynomial $n^2 - n + 41$ (Euler, Legendre) for $n = 1, 2, \ldots$. What do you notice?

8. Write a procedure `Arithxy` that reads values x and y from the spreadsheet and prints the values

$$\frac{1}{2}(x + y + |x - y|) \quad \text{and} \quad \frac{1}{2}(x + y - |x - y|).$$

What do you notice? Can you prove your conclusions?

9. It may be shown that $n^7 - n$ is always divisible by 42. Write a program `Div42(n)` that prints the other factor for any positive integer n. Use a spin button that increments n.

Chapter 3

Conditional Statements

Conditional statements allow the path of a program to branch to other statements depending on conditions that arise during run time. These conditions are usually in the form of Boolean expressions (see Section 2.3) and typically involve the VBA comparison and logical operators discussed in the previous chapter. In the current chapter we describe two conditional statements, the If Then Else statement and the Select Case statement. We begin with the former.

3.1 The If Then Else Statement

This is the most common conditional statement. It has the following general form:

```
If condition1 Then
    statements1
ElseIf condition2 Then
    statements2
    ...
ElseIf conditionk Then
    statementsk
Else
    statements
End If
```

There may be many `ElseIf` statements or none at all. The code executes as follows: if `condition1` is true, then `statements1` are executed and control passes to the statement following `End If`. On the other hand, if `condition1` is false, then `condition2` is tested, etc. The `Else` statement, which is optional, is executed if conditions 1 to k are false. Figure 3.1 is a schematic of the logic for the case $k = 3$. It shows how the flow of the program depends on the truth values of the conditions in the diamond-shaped boxes.

For a simple example we consider the procedure `Grade` below, which uses an If Then Else statement to determine the letter grade of a test score. A score from 0 to 100 is entered in cell B3 and the program prints the corresponding letter grade in cell B4.

DOI: 10.1201/9781003351689-3

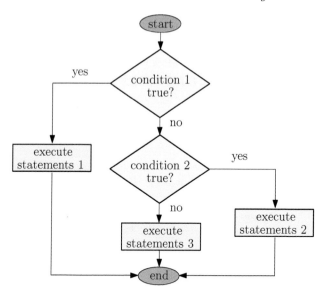

FIGURE 3.1: If Then flowchart.

```
Sub Grade()
    Dim score As Single, grade as String              'declare variables
    score = Range("B3").Value        'read score 0 to 100 from spreadsheet
    If score >= 90 Then
        grade = "A"                  'exit the statement and output grade A
    ElseIf score >= 80 Then
        grade = "B"                  'exit the statement and output grade B
    ElseIf score >= 70 Then
        grade = "C"                                             'etc.
    ElseIf score >= 60 Then
        grade = "D"
    Else
        grade = "F"
    End If
    Range("B4").Value = grade                         'output letter grade
End Sub
```

One could introduce additional ElseIf statements to include grades of A+, A−, etc. However, this would make the code cumbersome and less readable. A better way is to use the Select Case statement, discussed next.

3.2 The Select Case Statement

The Select Case statement is a version of the If Then Else statement. It is useful when many ElseIf statements are needed. Here is the general form:

```
Select Case  TestThisQuantity
       Case  condition1
             statements1
                . . .
       Case  conditionk
             statementsk
       Case  Else
             statements
End Select
```

The code works like this: The numeric quantity `TestThisQuantity` is tested against conditions 1 to k in that order. If `condition1` is found to be true, then the code `statements1` is executed and control passes to the statement following `End Select`. If `condition1` is false, then the next condition is tested, etc. If none of the conditions is true, then the statements under `Case Else` (which is optional) are executed.

We illustrate the Select Case statement with a spin button procedure that adds edges to a cell using the `Cells.Borders` feature of VBA. A variety of such edges is available as well as a facility to remove them. For example, the following code snippet places and then removes the left edge of a cell:

```
Cells(row, col).Borders(xlEdgeLeft).LineStyle = xlContinuous
Cells(row, col).Borders(xlEdgeLeft).LineStyle = xlNone
```

Other edges that may be added are `xlEdgeTop`, `xlEdgeBottom`, `xlEdgeRight`, `xlDiagonalDown`, and `xlDiagonalUp`. A border may be made thicker or thinner by using `xlThick` or `xlThin`. In place of `xlContinuous` one can use broken borders, for example, `xlDashDotDot`. Color may be added using the color index property.

The procedure's spin button causes the program to cycle through the various edges. The button assigns the integers 1 – 6 to cell A3. These values are set in the properties box. The program reads the number in the cell and the Select Case statement assigns the corresponding edge.

```
Sub SpinButton1_Change()
    Dim row As Integer, col As Integer, edge As Integer, n As Integer
    row = 2: col = 2                     'cell B2 displays the edges
    n = Range("A3").Value               'print spin button value in A3
          'delete any previous edges and diagonals
    Cells(row, col).Borders.LineStyle = xlNone
    Cells(row, col).Borders(xlDiagonalDown).LineStyle = xlNone
    Cells(row, col).Borders(xlDiagonalUp).LineStyle = xlNone
```

```
Select Case n                'select an edge corresponding to number in A3
    Case Is = 1:  edge = xlEdgeTop
    Case Is = 2:  edge = xlEdgeRight
    Case Is = 3:  edge = xlEdgeBottom
    Case Is = 4:  edge = xlEdgeLeft
    Case Is = 5:  edge = xlDiagonalDown
    Case Is = 6:  edge = xlDiagonalUp
End Select
With Cells(row, col).Borders(edge)
                    .LineStyle = xlContinuous
                    .Weight = xlThick
                    .ColorIndex = 3                          'red edge
End With
End Sub
```

Output in cell B2 is shown in the Figure 3.2.

FIGURE 3.2: Output of CycleEdges.

3.3   Exercises

1. Write a function `EvenOdd` that takes a positive integer and returns the string "even" if the number is even and "odd" otherwise. Write code that tests the function using a command button.

2. Write a procedure `Arithmetic` that reads Double values x, y entered in the spreadsheet and returns $x + y$, $x - y$, $x * y$, or x/y depending on the instruction add, subtract, multiply, or divide entered (as strings) in the spreadsheet. Use a Select Case statement.

3. Use the If Then Else statement in a function `Max(a,b,c)` that returns the maximum (largest) of the (Double) numbers `a,b,c`. Do the same for the minimum. Incorporate the function into a module with a command button that takes numbers `a,b,c` entered into the spreadsheet by the user and prints out the maximum and minimum values as well as the value `Min(a,b,c) + Max(-a,-b,-c)`, which should always be zero (why?).

4. Write a function `SecondLargest(a,b,c)` that takes three *distinct* Double values and returns the second largest. Incorporate the function into a module with a command button.

5. Write a procedure `ConvertTemp` that reads a temperature from one of the two cells B3 or B4 with the other empty. The first cell is reserved for a Celsius temperature, the second for a Fahrenheit temperature. If B3 is not empty then the program converts the number there to Fahrenheit and prints it in B4. If B4 is not empty then the program converts the number there to Celsius and prints it in B3. Use the VBA function `IsEmpty`; for example, `IsEmpty(Range("B3"))` returns True if there is nothing in the cell, and False otherwise. Use the formula $C = 9(F - 32)/5$.

6. Use a Select Case statement to expand the grading procedure `Grade` to include plus and minus grades such as C- and B+. Use the following values: A: 93 – 100; A-: 90 – 92; B+: 87 – 89; B: 83 – 86; B-: 80 – 82; C+: 77 – 79; C: 73 – 76; C-: 70 – 72 (C-); D+: 67 – 69; D: 63 – 66; D-: 60 – 62; F: below 60. Incorporate the function into a program with a command that reads a user-entered score and prints the letter grade.

7. Write a function `Digit2String(d As Integer)` that takes an integer $d = 0, 1, \ldots, 9$ and returns the corresponding string "zero", "one", ..., "nine". Use a spin button to change the values of d.

8. A *complex number* is an expression of the form $\alpha + \beta i$, where α and β are real numbers called, respectively, the *real* and *imaginary parts* of $\alpha + \beta i$, and i, a symbol with the property that $i^2 = -1$. Write a program `Quadratic` that solves a quadratic equation $ax^2 + bx + c = 0$ $(a \neq 0)$. The solutions x take two forms depending on the sign of the *discriminant* $d = b^2 - 4ac$:

$$x = \frac{-b}{2a} \pm \frac{\sqrt{d}}{2a} \text{ if } d \geq 0, \quad x = \frac{-b}{2a} \pm \frac{\sqrt{-d}}{2a} i \text{ if } d < 0.$$

You will need a conditional statement to distinguish between the cases. The program should read the coefficients a, b, c entered by the user and print out the real and imaginary parts of the solutions, as in Figure 3.3. Use the VBA function `SQR(number)` to calculate square roots.

	A	B	C
4	$a =$	1	
5	$b =$	2	
6	$c =$	3	
7		real part	imaginary part
8	solutions:	-1	1.41421354
9		-1	-1.41421354

FIGURE 3.3: Spreadsheet for QuadraticFormula.

Chapter 4

Loops

In this chapter we consider statements that cause a sequence of instructions to execute repeatedly in a loop until a given condition changes, causing the loop to exit. Such statements are frequently used in conjunction with the VBA operators discussed in Chapter 2. We discuss the three main loops: For Next, Do While, and Do Until.

4.1 The For Next Statement

This statement has the general form

```
For counter = firstval To lastval   Step x
        statements
   Next counter
```

Figure 4.1 illustrates the flow of logic. The variable `counter` is given the initial

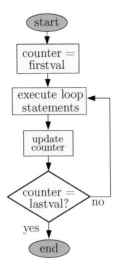

FIGURE 4.1: For Next flowchart.

DOI: 10.1201/9781003351689-4

value `firstval`, the statements in the body of the loop are exercised, and `counter` is incremented by the amount `x`, which can be negative. The process is repeated, the final iteration occurring when `counter` has the value `lastval`. If the code `Step x` is omitted, then the step is taken to be 1.

4.2 The Do While Loop

This loop has the two forms

```
Do While condition          Do
    statements                  statements
Loop                        Loop While condition
```

In the version on the left, the Boolean expression `condition` is tested *before* `statements` are executed; in the version on the right it is tested *afterwards*. The looping continues as long as `condition` is `True`. Figure 4.2 illustrates the logical flow.

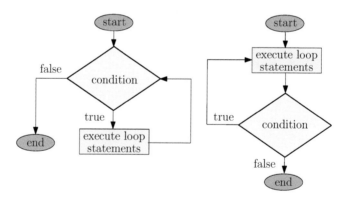

FIGURE 4.2: Do While flowcharts.

The example below uses both the For Next and the Do While statements. The program reads two Long integers, `N1`, `D1`, and reduces the fraction `N1/D1` to its lowest terms by canceling common factors.

```
Private Sub CommandButton1_Click()
    Dim num As Long, den As Long               'numerator and denominator
    num = Range("B4").Value: den = Range("B5").Value    'entered by user
    Call ReduceFraction(num, den)                       'reduce it
    Range("C4").Value = num: Range("C5").Value = den    'print reductions
End Sub
```

```
Sub ReduceFraction(N1 As Long, D1 As Long)
    Dim i As Long
    For i = 2 To N1 / 2              'a divisor cannot be bigger than N1/2
        Do While N1 Mod i = 0 And D1 Mod i = 0 'if i divides both N1, D1
            N1 = N1 / i: D1 = D1 / i              'then cancel the i
        Loop
    Next i
End Sub
```

Our second example, NumberOfTerms, is passed an exponent $p \leq 1$ and an upper bound U and returns the smallest integer n needed for the sum

$$1 + \frac{1}{2^p} + \frac{1}{3^p} + \cdots + \frac{1}{n^p}$$

to exceed U. These sums grow larger and larger without bound but they do so very slowly. For example, the reader may check that when $p = 1$, $272,400,600$ terms are needed for the sum to exceed the value 20, and an additional 468,061,001 terms are needed to exceed 21. (This may take several minutes; higher values are not recommended unless you're ready for a long break.) As p is decreased so is the number of terms required to exceed a limit. For example, when $p = .9$, only $48,744$ terms are needed for the sum to exceed the value 20, and $p = .7$ reduces this to 606.

```
Function NumberTerms(exp As Double, UpperBound As Double) As Long
    Dim sum As Double, NumTerms As Long  'automatically initialized to 0
    If exp > 1 Then
        MsgBox "exp must be <= 1"  'sums don't get arbitrarily large then
        Exit Function
    End If
    Do While sum < UpperBound
        NumTerms = NumTerms + 1                      'increment count
        sum = sum + 1 / NumTerms ^ exp               'add next term
    Loop
    NumberTerms = NumTerms
End Function
```

The program uses the VBA message feature MsgBox, which displays a message until the icon OK is clicked.

4.3 The Do Until Loop

Like the Do While loop, the Do Until loop has two forms:

```
Do Until condition        Do
    statements                statements
Loop                      Loop Until condition
```

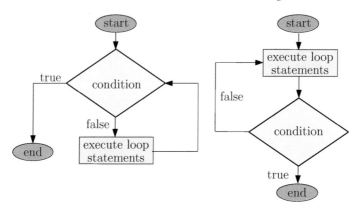

FIGURE 4.3: Do Until flowchart.

In the version on the left, `condition` is tested before `statements` are executed; in the version on the right it is tested after. Unlike Do While loops, Do Until looping continues as long as `condition` is *false*. We give an example in Section 4.5.

4.4   Exit Statements

An exit statement causes a program to immediately depart from the code block which contains it and go to the statement that follows the block. If used in a nested loop, the program exits the loop and resumes at the next higher loop. Here are the most common types of exit statements.

`Exit For`	Exits a For Next loop.
`Exit Do`	Exits a Do loop.
`Exit Sub`	Exits a Sub procedure.
`Exit Function`	Exits a Function procedure.

We give an example in the next section.

4.5   Finding the Zeros of a Function

A *zero* or *root* of a function f is a number z such that $f(z) = 0$. For example, the quadratic function $x^2 - 5x + 6$ has zeros $z = 2$ and $z = 3$. The zeros in this example may be found exactly by factoring or applying the quadratic formula. However, in many cases one must rely on algorithms that produce only approximations. The simplest such algorithm is the *interval halving* or *bisection* method, which works for *continuous functions*, that is, those that have no gaps or jumps in their graphs.

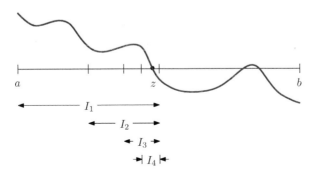

FIGURE 4.4: Interval Halving method.

The interval halving method is illustrated in Figure 4.4. The function f is defined on an interval $[a, b]$ and is assumed to have opposite signs at the endpoints, that is, $f(a)f(b) < 0$. Since f is continuous, its graph must cross the x-axis at some point in between. To home in on the point we proceed as follows: Let $c = (a + b)/2$, the midpoint of the interval. It must then be the case that either $f(a)$ and $f(c)$ have opposite signs or $f(b)$ and $f(c)$ have opposite signs. If the former then f must have a zero in $[a, c]$; if the latter then f has a zero in $[c, b]$. We then apply the same reasoning to the appropriate halved interval. Continuing in this manner, we obtain a sequence of shrinking intervals I_n containing a zero. The process stops when the width of the interval is small enough to satisfy a predetermined accuracy condition. The desired zero may then be approximated by either of the endpoints. The figure shows that the interval halving method may miss some zeros. To capture these one needs to adjust the initial interval $[a, b]$.

The program below uses the Do Until statement and several Exit statements. The command button code launches the function `FindZero`, the main procedure in the module. It assumes that the user has entered the endpoints of the interval and the desired degree of accuracy into cells B3, B4, and B4, respectively, and has also entered the function expression with variable x into cell B6. For example, to find where in the interval $(0, \pi/2)$ the graphs of functions $f(x) = x^2 \sin(x)$ and $g(x) = \cos(x)$ cross, enter .1, 1.5, .0001, and the expression `=x^2*Sin(x) - Cos(x)` into the aforementioned cells (note the required equals sign). Running the program yields the x value .8953 (rounded). Using a smaller accuracy value results in greater precision.

```
Sub CommandButton1_Click()
    Dim left As Double, right As Double, z As Double
    Dim accuracy As Double, error As Boolean, expr As String
```

```
      left = Range("B3").Value: right = Range("B4").Value        'endpoints
      accuracy = Range("B5").Value                  'desired level of accuracy
      expr = Range("B6").Value                      'function expression
      z = FindZero(expr, left, right, accuracy, error)          'calculate!
      If error Then
         Range("B7").Value = "error"
      Else
         Range("B7").Value = z
      End If
   End Sub
```

The function `FindZero` takes the input data and returns an approximate zero of the function, if possible. A Do Until loop carries out the bisections. For each new interval, the procedure `FunctionVal` calculates the values of the function at the left endpoint, right endpoint, and midpoint. A new interval is determined from these by testing signs at these points, as mentioned earlier.

```
Function FindZero(expr As String, left As Double, right As Double, _
               accuracy As Double, error As Boolean) As Double
   Dim mid As Double, lValue As Double, rValue As Double
   Dim mValue As Double, z As Double
   Do
      mid = (left + right) / 2         'get midpoint of current interval
      lValue = FunctionVal(expr, left)      'function values at left,
      rValue = FunctionVal(expr, right)            'right, and
      mValue = FunctionVal(expr, mid)              'mid points
      If lValue * rValue > 0 Then error = True: Exit Do 'check interval
      If lValue = 0 Then z = left: Exit Do          'exit if zero found
      If rValue = 0 Then z = right: Exit Do
      If mValue = 0 Then z = mid: Exit Do
      If lValue * mValue < 0 Then              'if different signs, then
         right = mid                           'new interval is left half
      ElseIf rValue * mValue < 0 Then          'if different signs, then
         left = mid                            'new interval is right half
      End If
      z = mid                                  'tentative zero
   Loop Until right - left < accuracy    'keep getting better approx.
   FindZero = z
End Function
```

The procedure `FunctionVal(expr,u)` uses the VBA concatenation operator & to splice together text (in this case to surround u with parentheses), and uses the VBA function `Replace` to substitute numerical value u for every occurrence of the character x in `expr`. Both of these concepts are discussed in detail in Chapter 6. The VBA function `Evaluate` then calculates the function value.

```
Function FunctionVal(expr As String, u As Double) As Double
   Dim str As String
   str = Replace(expr, "x", "(" & u & ")")         'put parens around u
   FunctionVal = Evaluate(str)              'find value of expression
End Function
```

4.6 Exercises

1. Use a For Next loop in a function that returns the *factorial* of a nonnegative integer n, defined by

$$n! = n \cdot (n-1) \cdot (n-2) \cdots 3 \cdot 2 \cdot 1, \quad 1! = 1, \ 0! = 1.$$

2. Write a function PrintBetween(m,n) that takes a pair of integers $m < n$ and prints all integers strictly between them.

3. Write a program ExcludeMultiples(m,N) that prints in a column all positive integers less than N that are not multiples of m.

4. Write a program that generates the sequence $((-1)^{n \backslash c} + 1)(a - b)/2$ for user-entered integers a, b, c and $n = 0, 1, 2, 3, \ldots, M$. What do you notice?

5. Write a program ExtractDigits that takes a Long positive integer n with at most 9 digits and prints its digits in a column. (Hint: Use integer division by powers of 10, starting with the exponent 9 and keep reducing.)

6. Write a function LargestDigit that takes a Long positive integer and returns its largest digit. For example, LargestDigit(357671) = 7. (Use the idea in ExtractDigits.) Test the function with a command button procedure.

7. Write a function AddDigits that takes a Long positive integer with at most 9 digits and returns the sum of the digits. Test the function with a command button procedure.

8. Write a procedure KeepAddingDigits that takes a Long positive integer and keeps calculating the sum of the digits until a single digit appears. Display each number in the process in a column. For example, KeepAddingDigits(456789) prints the numbers 39, 12, 3. (Use the function AddDigits.) Test the function with a command button procedure.

9. A *Pythagorean triple* consists of three positive integers (a, b, c) that satisfy $a^2 + b^2 = c^2$, for example, $(3, 4, 5)$ and $(5, 12, 13)$. Write a program PythagTriples that generates Pythagorean triples. Use the fact that for positive integers n, x, y with $x > y$ and x, y odd,

$$\left(n(x^2 - y^2)/2, nxy, n(x^2 + y^2)/2\right)$$

is a Pythagorean triple, as the reader may check. (It may be shown that all Pythagorean triples are of this form.)

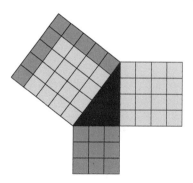

FIGURE 4.5: Pythagorean triple $(3, 4, 5)$.

10. It may be shown that every positive integer can be written as the sum of 4 perfect squares (some of which may be 0). For example,

$$2 = 1^2 + 1^2 + 0^2 + 0^2, \quad 5 = 2^2 + 1^2 + 0^2 + 0^2, \quad 79 = 1^2 + 2^2 + 5^2 + 7^2.$$

Write a procedure `Sum4Squares` that takes a positive integer N and prints four numbers whose squares add up to the number. (Use nested For Next loops.)

11. Write a program `SinCosInterval` that finds the percentage of times out of N trials that the values $\sin(n)$ and $\cos(n)$, n a positive integer, fall inside a user-entered interval (a, b). Use the VBA functions `Sin` and `Cos`.

	A	B	C
3	num trials	50,000	
4	lower limit	0	
5	upper limit	.7	
6		sin	cos
7	percentage hits	24.676	24.682

FIGURE 4.6: SinCosInterval spreadsheet.

12. Write a program `Collatz` that starts with a user-entered positive integer n and prints in a column the results of the following steps:

(1) If n is even, print $n/2$; if n is odd print $3n + 1$.

(2) Repeat (1) with the number obtained.

For example, for $n = 17$ the steps are

$$17 \to 52 \to 26 \to 13 \to 40 \to 20 \to 10 \to 5 \to 16 \to 8 \to 4 \to 2 \to 1$$

The *Collatz conjecture* asserts that the process will eventually produce the cycle 4,2,1 forever. (As of 2020, this has been found to be the case for all initial values up to 2^{68}.)

13. Every positive integer n of the form $8k + 1$, $8k + 3$, or $8k + 5$ may be written as the sum of 3 perfect squares. For example,

$$8 \cdot 2 + 1 = 4^2 + 1^2 + 0^2, \ 8 \cdot 2 + 3 = 3^2 + 3^2 + 1^2, \ 8 \cdot 2 + 5 = 4^2 + 2^2 + 1^2.$$

Write a procedure `Sum3Squares` that takes numbers $r = 1, 2, 3$ and an upper bound U entered in the spreadsheet and generates the columns of the form suggested by the following sample spreadsheet, where $a \le b \le c$, $a^2 + b^2 + c^2 = n$, and the numbers generated are $\le U$.

	A	B	C	D	E
3	r	3			
4	U	50			
5	k	$8k + r$	a	b	c
6	0	3	1	1	1
7	1	11	1	1	3
8	2	19	1	3	3
9	3	27	1	1	5
10	3	27	3	3	3
11	4	35	1	3	5
12	5	43	3	3	5

FIGURE 4.7: Sum 3 squares spreadsheet.

14. Each of the expressions

$$6^{1/2} \left(1 + \frac{1}{4} + \frac{1}{9} + \frac{1}{16} + \cdots + \frac{1}{n^2} \right)^{1/2}$$

$$4 \left(1 - \frac{1}{3} + \frac{1}{5} - \frac{1}{7} + \cdots + (-1)^n \frac{1}{2n + 1} \right)$$

$$12^{1/2} \left(1 - \frac{1}{4} + \frac{1}{9} - \frac{1}{16} + \cdots + (-1)^{n+1} \frac{1}{n^2} \right)^{1/2}$$

gets closer to π as n gets larger and so may be used to approximate π. Write a program `ApproxPi` that calculates the values of these expressions for a user-entered value of n. Format the output cells to fourteen decimal places by using `Range("B4:B9").NumberFormat = "0.00000000000000"`.

Discrete Mathematics and Coding

Calculate the absolute value of the difference between each approximation and the value $\pi \approx 3.141592653589793$ (accurate to fourteen decimal places) and use this to measure which value appears to give the best approximation. Here is π to 63 decimals:

3.141592653589793238462643383279502884197169399375105820974944

	A	B
3	num terms	
4	pi to 14	
5	approximations	errors
6	3.14157355512957	0.00001909846022
7	3.14161265318978	0.00001999959999
8	3.14159265320780	0.00000000038199

FIGURE 4.8: ApproxPi spreadsheet.

Chapter 5

Arrays

An array is a named list of variables of the same type whose entries may be accessed by indices. Arrays greatly enhance the flexibility and power of programs and are indispensable for certain tasks. In this chapter we describe these data structures and illustrate their use in various examples.

5.1 Declaring Arrays

Arrays are declared in a manner similar to individual variables. Here are some typical examples:

- `Dim arr(10) As String`
  Declares `arr` as a one-dimensional list of 11 String data types starting (by default) at `arr(0)` and ending at `arr(10)`. The array is automatically initialized with null strings.

- `Dim arr(1 To 100) As Integer`
  Declares `arr` as a one-dimensional list of 100 Integer data types starting at `arr(1)` and ending at `arr(100)`. The array is automatically initialized with zeros.

- `Dim arr(3,20) As Single`
  Declares `arr` as a two-dimensional list of $4 \times 21 = 84$ Single data types starting (by default) at `arr(0,0)` and ending at `arr(3,20)`. The array is automatically initialized with zeros.

- `Dim arr(7 To 15, 11 To 20) As Double`
  Declares `arr` as a two-dimensional list of $9 \times 10 = 90$ Double data types starting at `arr(7,11)` and ending at `arr(15,20)`. The array is automatically initialized with zeros.

- `Dim arr() As Long`
  Dimensions here are unspecified. Memory allocated later in program using `ReDim` (see discussion below).

Members of an array are accessed by indices and may be used exactly as variables, for example, `arr1(5) = 3*arr2(1)`.

Arrays may be allocated during run time. For example, the code snippet `ReDim arr(1 To size)` uses the `ReDim` statement to allocate memory after the variable `size` has been determined by the program. We illustrate this idea in later modules.

5.2 First and Last Indices of an Array

There are two important VBA functions associated with arrays, `UBound` (upper bound) and `LBound` (lower bound). The former returns the last index of an array and the latter the first. Here's an example.

```
Dim Arr1(10) As Double, Arr2(7 To 15,11 To 20) As String
Dim v As Integer
v = LBound(Arr1)            'assigns to v the default lower index 0
v = UBound(Arr2,2)          'assigns to v upper index 20 of 2nd dimension
```

The number of entries in an array can be calculated by taking the difference of `UBound` and `LBound` and adding 1. For example, the number of entries in the first dimension of `Arr2` above is `UBound(Arr2,1)-LBound(Arr2,1) + 1`= 9.

5.3 Passing and Returning Arrays

An array is always passed to a procedure by reference and as such may be modified by the procedure. Arrays may also be returned by a function. The module that follows demonstrates this. The command button procedure declares an array with unspecified dimensions. The function `ReturnArray` fills the array and returns it to the command button procedure.

```
Sub CommandButton1_Click()
    Dim Arr() As Integer
    Arr = ReturnArray()                  'function returns array
    Range("A4").Value = Arr(3)                   'print value
End Sub

Function ReturnArray() As Integer()
    Dim Arr(1 To 3) As Integer, i As Integer
    For i = 1 To 3                        'assign values to array
        Arr(i) = i ^ 2
    Next i
    ReturnArray = Arr                    'return array location
End Function
```

5.4 Variants and the Array Function

Recall that a variable of Variant type can contain string, Boolean, or numeric values. With the use of the VBA function `Array`, a Variant variable can also be made to hold an array. For example, running the following code snippet causes 2.3, dog and False to be placed in column 1.

```
Dim Arr As Variant, k As Integer
Arr = Array(2.3, "dog", 1 = 2)                 'place these in Arr
For k = LBound(Arr) To UBound(Arr)
    Cells(2 + k, 1).Value = Arr(k)             'print Arr entries
Next k
```

The same result is achieved by using the For Each Next statement to select the items in `Arr`:

```
Dim Arr As Variant, item As Variant, k As Integer
Arr = Array(2.3, "dog", 1 = 2)     'generate Arr with Array function
For Each item In Arr
    Cells(3 + k, 1).Value = item               'print members of Arr
    k = k + 1
Next item
```

The general syntax for the For Each Next statement is

```
For Each element in var
    statements
Next element
```

Both `element` and `var` must be Variant types.

5.5 Sorting Numeric Data

In this and the remaining sections of the chapter we develop programs that make essential use of arrays. The sorting procedure in the current section reads numeric data entered by user in column A starting at row 3 and sorts the data from smallest to largest. The main procedure `NumericSort` initially determines how many data items have been entered by using the Excel row count feature

```
Cells(Rows.count, col).End(xlUp).row,
```

which returns the row number of the last nonempty cell of column `col`. The array `data` is then allocated dynamically using the `ReDim` statement. The procedure reads the data into the array and passes it to the sorting procedure `BubbleSort`. The latter sorts the data, which is then printed in the column next to the original data.

```
Sub NumericSort()
    Dim data() As Double, i As Integer
    Dim numData As Integer, firstRow As Integer, lastRow As Integer
    firstRow = 3                                    'first row of data
    lastRow = Cells(Rows.count, 1).End(xlUp).row       'last row of data
    numData = lastRow - firstRow + 1
    ReDim data(1 To numData)        'allocate memory dynamically for array
    For i = 1 To numData                    'read data starting at firstrow
        data(i) = Cells(i - 1 + firstRow, 1).Value
    Next i
    Call BubbleSort(data)                              'sort the array
    For i = 1 To numData
        Cells(i - 1 + firstRow, 2).Value = data(i)     'print sorted data
    Next i
End Sub
```

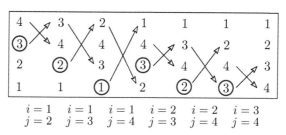

FIGURE 5.1: Bubble sort of 4321.

The workhorse of the module is the procedure `BubbleSort`, so named because its sorting mechanism causes smaller ("lighter") numbers to "float" to the top. Figure 5.1 illustrates how the array $\{4, 3, 2, 1\}$ is sorted. In column one, the number 3, being less than 4, is selected to move to the top and so is swapped with 4, producing column two. In column two, the number 2, being less than the top number 3, is selected to move to the top and so is swapped with 3, producing column three, etc. The numbers that "float up" are circled. It should be mentioned that the bubble sort algorithm is one of several sorting techniques used in computer science.

```
Sub BubbleSort(data() As Double)
    Dim i As Integer, j As Integer, last As Integer, swap As Double
    last = UBound(data)                         'last index of data()
    For i = 1 To last                           'run through the data entries
        For j = i + 1 To last                   'move smaller entries up list
            If data(j) < data(i) Then
                swap = data(i)          'found entry smaller than data(i) so:
                data(i) = data(j)         'move data(j) up to position i and
                data(j) = swap            'move data(i) down to position j
            End If
        Next j
    Next i
End Sub
```

We remark that the Sort Range function in Excel VBA can sort multiple columns on a key. For example, the code

```
Range("A8":"D17").Sort Key1:=Range("B8"), Order1:=xlAscending,Header:=xlNo
```

sorts in increasing order the data in B8:B17 as well as (in tandem) the rest of the data in the block A8:D17. The procedure `BubbleSort` is nonetheless occasionally useful in programs requiring sorted arrays.

5.6   Finding Nearest Numbers

The program `NearestNumbers` takes a column of numbers entered by the user and finds k numbers closest to each column entry, where k is specified by the user. It illustrates the technique of storing the indices of one array in a second array to reference data in the first. The numeric data is entered in column B starting at row 3 and the value of k is entered in cell C2. Figure 5.2 illustrates input and output with $k = 3$. Three nearest numbers to each of these are printed to the right of the number in columns C–E.

	B	C	D	E
2	$k =$	3		
3	1.4	0.7	2.3	3.5
4	0.7	1.4	2.3	3.5
5	3.5	4.1	2.3	1.4
6	2.3	1.4	3.5	0.7
7	4.1	3.5	2.3	1.4

FIGURE 5.2: Spreadsheet input and output for NearestNumbers.

The main procedure of the module first determines the number of data items by using the rows count feature described in the last section. It then reads the data into the array `Data` with a For Next loop. A second loop repeatedly calls the procedures `FindNear` and `PrintNear`. For each i, the former finds the indices of `Data` corresponding to k numbers nearest `Data(i)` and stores these in the index array `idxNear`. The procedure `PrintNear` prints the k numbers next to data item `Data(i)` in column B.

```
Sub NearestNumbers()
    Dim Data() As Double, numData As Integer, i As Integer, k As Integer
    Dim  idxNear() As Integer, firstRow As Integer, lastRow As Integer
    firstRow = 4                              'first row of data
    lastRow = Cells(Rows.count, 2).End(xlUp).row    'last row of data
    numData = lastRow - firstRow + 1
```

```
        k = Range("C2").Value                'desired number of nearest numbers
        If k > numData - 1 Then
            k = numData - 1                 'enforce the inequality k <= NumData - 1
        End If
        ReDim Data(1 To numData)                    'storage for data
        For i = 1 To numData                    'read the data in column B
            Data(i) = Cells(i + firstRow - 1, 2).Value
        Next i
        For i = 1 To numData  'for each i store indices of k nearest numbers
            idxNear = FindNear(i, Data, k)      'to Data(i) in array idxNear
            Call PrintNear(i, Data, idxNear)              'and print them
        Next i
    End Sub
```

The procedure `FindNear` is passed the current index i, the array `Data` and the integer k. The outer For Next loop finds k numbers in `Data` nearest to the data item `Data(i)` (excluding `Data(i)`) and stores the k indices of the data in `idxNear`. This is accomplished by two For Next loops, the first to get a candidate for an available nearest number and the second to find an actual nearest number. The VBA absolute value function `Abs` is used to test to measure nearness of the numbers.

```
Function FindNear(i, Data() As Double, k As Integer) As Integer()
    Dim n As Integer, j As Integer, nearest As Double
    Dim idxNear() As Integer, numData As Integer
    ReDim idxNear(1 To k)                   'storage for indices of Data array
    numData = UBound(Data)
    For n = 1 To k                      'get k nearest numbers to Data(i)
        For j = 1 To numData            'get a candidate for a nearest no.
            If j <> i And IsAvailable(j, idxNear) Then
                nearest = Data(j)       'found a candidate for a nearest no.
                idxNear(n) = j                      'store its index
                Exit For
            End If
        Next j
        For j = 1 To numData            'get nearer no. from unused Data
            If j <> i And IsAvailable(j, idxNear) Then
                If Abs(Data(i) - Data(j)) < Abs(Data(i) - nearest) Then
                    nearest = Data(j)               'found a nearer number
                    idxNear(n) = j                  'store its index
                End If
            End If
        Next j
    Next n
    FindNear = idxNear
End Function
```

The function `IsAvailable` used in the two inner For Next loops of `FindNear` determines whether an index j is already in `idxNear`, that is, whether a data item is already in the list of numbers closest to the current entry of `Data`. If it is, then the function returns `False`, resulting in no action for that iteration.

```
Function IsAvailable(j, idxNear() As Integer) As Boolean
    Dim i As Integer, idxAvailable As Boolean
    idxAvailable = True                            'default value
    For i = 1 To UBound(idxNear)
        If idxNear(i) = j Then idxAvailable = False        'already used
    Next i
    IsAvailable = idxAvailable                      'return availability
End Function
```

The remaining procedure prints k nearest numbers to `Data(i)`.

```
Sub PrintNear(i, Data() As Double, idxNear() As Integer)
    Dim n As Integer, k As Integer
    k = UBound(idxNear)
    For n = 1 To k              'print k nearest numbers next to Data(i)
        Cells(3 + i, 2 + n).Value = Data(idxNear(n))
    Next n
End Sub
```

The program runs fairly quickly. For example, the reader may test its speed by entering the numbers 1 to 100 in column A and taking $k = 20$. More complicated data may be entered automatically using a random number generator (Section 10.1).

5.7   Stacks

A *stack* is an abstract data type that may viewed as a column of data entries. There are two main operations on a stack, *push* and *pop*. The former places a data item on the top of the stack if the stack is not full, and the latter removes a data item from the top of the stack if the stack is not empty. (See Figure 5.5.) In this section we model stacks and the push and pop operations.

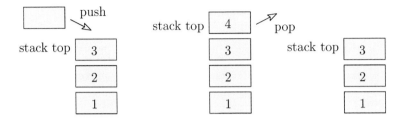

FIGURE 5.3: Push and pop at work.

For our stack we take a UDT that incorporates the data in the stack, the size of the stack, and the index of the top of the stack.

```
Private Type TStack
    data()      As String
    size        As Integer
    top         As Integer
End Type
```

The stack memory is the String data array in the UDT. The variable `size` refers to the desired maximum memory of the data array, that is, `Ubound(stack.data)`. The variable `top` is the current top of the stack, that is, the largest index for which the data array has items.

The procedure `CreateStack`, as the name implies, creates an instance `stack` of the UDT. It is passed the variable `size` and returns a stack of that size.

```
Private Function CreateStack(size As Integer) As TStack
    Dim stack As TStack
    stack.size = size                   'maximum memory for the stack
    stack.top = 0                                      'no data yet
    ReDim stack.data(1 To size)                   'allocate memory
    CreateStack = stack                           'return the stack
End Function
```

The procedure `Push` takes a data item and, if there is room, places it on top of the last one, incrementing the top of the stack by 1.

```
Private Sub Push(stack As TStack, DataItem As String)
    If stack.top = stack.size Then
        MsgBox "stack full"                      'no more room for data
    Else
        stack.top = stack.top + 1
        stack.data(stack.top) = DataItem    'put data item at the new top
    End If
End Sub
```

The function `Pop` takes a stack and returns the top item, decrementing the top of the stack by 1.

```
Private Function Pop(stack As TStack) As String
    Dim DataItem As String
    If stack.top = 0 Then
        MsgBox "stack empty"                          'no data items left
    Else
        DataItem = stack.data(stack.top)              'retrieve the item
        stack.top = stack.top - 1          'item gone (but not forgotten)
    End If
    1Pop = DataItem
End Function
```

Here's a sample spreadsheet for the code below that shows the functions in action. The spreadsheet illustrates the basic principle of stacks: last in, first out.

	A	B
3	push	Aaron
4	push	Betty
5	push	Sally
6	pop	Sally
7	pop	Betty
8	pop	Aaron

FIGURE 5.4: Pushing and popping.

```
Sub TestPushPop()
    Dim DataItem As String, stack As TStack
    stack = CreateStack(3)
    For n = 3 To Cells(Rows.Count, 1).End(xlUp).Row
        If Cells(n, 1).Value = "push" Then       'push item in Cell(n,2)
            Call Push(stack, Cells(n, 2).Value)          'onto stack
        ElseIf Cells(n, 1).Value = "pop" Then    'pop item in Cell(n,2)
            Cells(n, 2).Value = Pop(stack)               'from stack
        End If
    Next n
End Sub
```

5.8 Exercises

1. Write a function `ArrayIndex(arr() As Integer, k As Integer)` that takes a positive integer array `arr` starting with index 0, and a positive integer `k` and returns the index of the first instance of the number `k` in `arr` or returns -1 if `k` is not in the array.

2. Write a function `LargestEntry` that takes an array of integers starting at index 1 and returns its largest entry.

3. Write a function `LargestDiff` that takes an array of numbers and returns the absolute value of the largest difference. For example, $\{4, 2, 3, 5, 8\} \rightarrow 6$.

4. Write a function `SumOdd` that takes an array of positive integers and returns the sum of the odd entries. For example, $\{7, 2, 3, 5, 6\} \rightarrow 16$.

5. Write a function `SumOddIdx(arr As Integer)` that takes an array of positive integers and returns the sum of those entries in odd positions.

6. Write a function `OddEntries` that takes an array of positive integers and returns an array of the odd entries.

7. Write a function `Average` that takes an array of Double numbers and returns the average.

8. The *median* of ordered data is the middlemost entry. For example, the median of the nine numbers 1, 1, 2, 2, 2, 3, 4, 5, 5 is 2, since there are 4 to the left and 4 to the right. The median of 1, 1, 2, 2, 2, 3, 4, 5, 5, 6 is the average $(2 + 3)/2 = 2.5$ of the two middle numbers, 2,3. Write a function `Median` that takes an array of Double numbers and returns the median (use `BubbleSort`).

9. The *mode* of data is the most frequent value of the data. For example, the mode of each data set in Exercise 8 is 2. There can be more than one mode. Write a program `Mode` that takes an array of Double numbers and returns a mode.

10. Write a function `TwoDim2OneDim(arr)` that take an $m \times n$ array and returns it as a $1 \times mn$ array by concatenating the rows.

11. (Concatenation of arrays.) Write a function `ConcatArrays(arr1,arr2)` that takes two integer arrays `arr1` and `arr2` starting at indices 0 and joins them in a single array with `arr1` first. For example,

$$\{1, 3, 2\}, \{1, 5, 8\} \rightarrow \{1, 3, 2, 1, 5, 8\}$$

12. (Complement of arrays.) Write a function `ArrComplement` that takes two integer arrays `arr1,arr2` with indices starting at 1 and returns an array that contains those members of `arr1` that are not members of `arr2`.

13. Write a function `Balance` that takes an integer array `arr(1 To n)` and returns an index k with the property that

```
arr(1) + arr(2) + . . . + arr(k) = arr(k+1) + arr(k+2) + . . . + arr(n).
```

If there is no such index the function should return -1.

14. Write a function `DistinctEntries` that takes an array of integers with index starting at 1 and returns an array consisting of the *distinct* entries. For example, if the input array is $\{1, 2, 3, 1, 2\}$ then the output should be the array $\{1, 2, 3\}$.

15. (a) Write a program `Jam(stack,item,m)` that inserts `item` into the stack at position m in the stack

(b) Write a function `Jerk(stack,item,m)` that removes and returns the `item` at position m.

(c) Test the functions in a program.

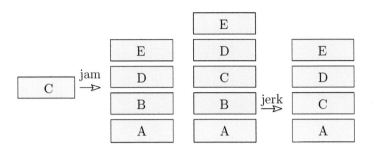

FIGURE 5.5: Jam and Jerk at work.

Chapter 6

String Functions

VBA has a number of functions that manipulate strings (text). In this chapter we describe the most common of these and the ones that will be needed in the remainder of the book. We begin with the simplest operator, concatenation.

6.1 The VBA Concatenation Operator

This operator joins several strings into a single string. For example, the code `"con" & "cat" & "e" & "nate"` returns the string "concatenate". The For Next loop

```
For j = 1 To 9
    Cells(j, 1).Value = "A" & j
Next j
```

places the labels A1, A2, ..., A9 in column one. Note that there are no quotes around the integer j. Indeed, inserting quotes would result in the unintended labels Aj, ..., Aj.

Duplicating Strings

Here's a simple example that takes a string and returns a string of duplicates separated by a delimiter (which could be the *null string* "").

```
Function DupString(str As String, n As Integer, dlm As String) As String
    Dim k As Integer, dup As String
    If n = 0 Then Exit Function
    dup = str                               initial string
    For k = 1 To n - 1                      'make duplicates
        dup = str & dlm & dup               'with delimiters
    Next k
    DupString = dup
End Function
```

For example the expression `DupString("baa",2,",")` & `"black sheep"` evaluates to `"baa,baa black sheep"`. We remark that VBA has a function

```
String(numChar As Long, Char As String)
```

that returns `Char` repeated `numChar` times. For example, `String(4, "A")` returns `AAAA`, as does `String(4, "ABC")`.

6.2 The VBA Extraction Function Mid

This function takes a string and returns a substring at a specified position and of a specified length. The syntax for the function is

$$\text{Mid(text,start,[length])}$$

where `text` is the given string, `start` is the position of the beginning of the substring to be extracted, and `length` is the length of the substring. The last argument is optional; if omitted the Mid function returns all characters from the start position to the end of the string. Thus `Mid("abcdef",2,3)` returns the string `"bcd"`, while `Mid("abcdef",2)` returns `"bcdef"`. Here are a few applications of `Mid`.

Matching Symbols

The function `IsMatch(expr,left,right)` returns `True` if there are the same number of occurrences of the symbol designated by the String variable `left` as there are of the symbol `right`. This is useful for error-checking an arithmetic expression containing various types of brackets. etc. For example the statement `IsMatch("(2+(3(-4)))","(",")")` returns True since the number of left and right parentheses are the same. The procedure uses the VBA function `Len`, which returns the length of an input string.

```
Function IsMatch(expr As String, left As String, right As String) _
   As Boolean
   Dim i As Integer, numleft As Integer, numright As Integer
   For i = 1 To Len(expr)                           'length of expr
      If Mid(expr,i,1) = left Then numleft = numleft + 1
      If Mid(expr,i,1) = right Then numright  = numright + 1
   Next i
   IsMatch = (numleft = numright)                   'returns True if equal
End Function
```

Inserting, Replacing a String

The function `InsertString` is passed a string `expr`, a position `pos` in the string, and another string `addstr`, and outputs a string with `addstr` inserted into `expr` immediately after position `pos`, which can be zero. For example, `InsertString("abcde",3,"xy")` returns the string `"abcxyde"`, and `InsertString("abcde",0,"xy")` returns `"xyabcde"`.

```
Function InsertString(expr As String, pos As Integer, addstr As String) _
          As String
    Dim left As String, right As String, out As String, L As Integer
    L = Len(expr)                                   'length of expr
    If pos >= L Then
        out = expr & addstr                         'addstr goes at end
    ElseIf pos <= 0 Then
        out = addstr & expr                         'put addstr at beginning
    Else
        left = Mid(expr, 1, pos)                     'left part of expr
        right = Mid(expr, pos + 1, L - pos)          'right part after pos
        out = left & addstr & right                 'insert addstr in between
    End If
    InsertString = out                              'return expanded string
End Function
```

The function `RemoveString` is passed a string `expr`, and two positions, `startpos` and `endpos`, in the string and returns `expr` with the portion of the string removed starting from the position `startpos` and ending with position `endpos`, inclusive. For example, the code `RemoveString("abcde",2,4)` returns `"ae"`.

```
Function RemoveString(expr As String, startpos As Integer, _
                      endpos As Integer) As String
    Dim left As String, right As String, out As String
    If startpos > endpos Then Exit Function
    If startpos > 1 Then left = Mid(expr, 1, startpos - 1)
    If endpos < Len(expr) Then right = Mid(expr, endpos + 1)
    out = left & right              'remove portion from startpos to endpos
    RemoveString = out                      'return collapsed string
End Function
```

The following function combines `InsertString` and `RemoveString`, returning the input expression with the portion starting at `startpos` removed and replaced by the string `addstr`. For example, the code `RemoveInsertString("abcde",2,"xyz")` returns `"axyzd"`.

```
Function RemoveInsertString(str As String, startpos As Integer, _
                            addstr As String) As String
    Dim outstr As String, endpos As Integer
    If addstr = "" Then RemoveInsertString = str: Exit Function
    endpos = startpos + Len(addstr) - 1
    outstr = RemoveString(str, startpos, endpos)
    outstr = InsertString(outstr, startpos - 1, addstr)
    RemoveInsertString = outstr                     'return expanded string
End Function
```

A Variation of Mid

VBA returns an error if the number `start` in `Mid(text,start,length)` is less than 1 or bigger than `length`. As this is sometimes inconvenient, we give the following easy fix, which returns the single-space string " " in these cases. The reader may wish instead to make the choice of the return character an argument, say `defaultreturn`, passed to the function.

```
Function Midd(text As String, start As Integer, length As Integer) _
         As String
   Dim str As String
   str = " "                                          'default return
   If start + length <= Len(text) + 1 And start > 0 Then
      str = Mid(text, start, length)                        'no error
   End If
   Midd = str
End Function
```

6.3   The VBA ASCII Functions Asc and Chr

ASCII is an abbreviation for "American Standard Code for Information Interchange." The code provides a numerical representation of English letters and other characters. For example, the capital letters A–Z have code 65–90, the small letters a–z have code 97–122, and the digits 0–9 have code 48–57. We shall not need other codes, but the interested reader will find the complete code online in various sources.

The function `Asc(inputstr)` returns the ASCII code for the first character in the input string. The function `Chr(code)` does the reverse, taking as input an ASCII code number and returning the corresponding character. Thus `Asc("Zebra")` returns 90, the ASCII code for Z, and `Chr(90)` returns the character Z.

Boolean ASCII Functions

The following functions are occasionally helpful in detecting various types of characters.

```
Function IsUpper(char As String) As Boolean
     IsUpper = 65 <= Asc(char) And  Asc(char) <= 90
End Function

Function IsLower(char As String) As Boolean
     IsLower = 97 <= Asc(char) And  Asc(char) <= 122
End Function
```

```
Function IsLetter(char As String) As Boolean
    IsLetter = (97 <= Asc(char) And  Asc(char) <= 122) Or _
               (65 <= Asc(char) And  Asc(char) <= 90)
End Function

Function IsDigit(char As String) As Boolean
    IsDigit = 48 <= Asc(char) And  Asc(char) <= 57
End Function
```

6.4 VBA Data Conversion Functions CStr, CInt, CLng, CDbl

VBA Excel has a number of functions that convert from one data type to another. We consider four of these, the ones used in the book.

- `CStr`: Converts a numerical value into a string, facilitating access to each digit. For example, `CStr(1.2345)` returns the string "1.2345".

- `CInt`: Converts a string into a rounded integer. For example, `CInt("2.5001")` and `CInt("-2.5001")` return 3 and -3, respectively, and `CInt("2.4999")` and `CStr("-2.4999")` return 2 and -2, respectively, as do `CInt("2.5000")` and `CStr("-2.5000")`.

- `CLng`: Similar to `CInt` but converts the argument into a Long integer.

- `CDbl`: Converts arithmetic strings and numeric values into Double numbers. For example, `CDbl(4*(2.3)^2 + 3.7)` and `CDbl("24.86")` both return the Double value 24.86.

6.5 The VBA Function Replace

This function replaces parts of a string with another string. For example, the code `Replace("abracadabra","ra","roo")` replaces each occurrence of the string "ra" by the string "roo", returning the well-known incantation "abroocadabroo." In general the instruction

`Replace(expression, find, replacement)`

takes every occurrence of the string `find` in `expression` and replaces it by the string `replacement`. We give two applications of the function.

Removing Spaces From a String

The following function takes a string as argument and returns the string devoid of all spaces.

```
Function RemoveWhiteSpace(inputstr As String) As String
    Dim i As Integer, outputstr As String
    outputstr = inputstr                        'copy string
    For i = 1 To Len(outputstr)        'keep replacing single white space
        outputstr = Replace(outputstr," ","")        'by null string
    Next i
    RemoveWhiteSpace = outputstr            'return compressed string
End Function
```

Removing Duplicate Characters

The following function removes duplicate characters from an input string. For example `RemoveDupChar("aaabbbbcccc")` returns the string `"abc"`.

```
Function RemoveDupChar(inputstr As String) As String
    Dim char As String, temp As String, outputstr As String
    temp = inputstr                            'copy to temp
    Do While Len(temp) > 0
        char = Mid(temp, 1, 1)        'get first character of current temp
        outputstr = outputstr & char            'append it to outputStr
        temp = Mid(temp, 2, Len(temp) - 1)        'get remainder of string
        temp = Replace(temp, char, "")        'kill char throughout temp
    Loop
    RemoveDupChar = outputstr
End Function
```

To see how this works, consider the string "aabbacb". In the first iteration of the loop result in

char = "a", outputstr = "a", temp = "abbacb", temp = "bbcb"

and second iteration in

char = "b", outputstr = "ab", temp = "bcb", temp = "c".

The third iteration reduces `temp` to the null string `""` and produces the desired output `"abc"`.

6.6   The VBA Function InStr

The function `InStr` takes as arguments a string and a substring and checks if the latter is contained in the former. If so, the function returns the position of the substring in the input string; otherwise it returns 0. The function performs exact matches. Here is the general syntax for the function:

```
InStr([start], inputstr, substr, [compare])
```

The optional argument `start` is the starting position of the search, with 1 as the default. For a non case-sensitive search, set the optional argument `compare` to the VBA constant `vbTextCompare`. Here are some examples based on the string `str = "cat in a hat"`:

- `InStr(str, "hat")` returns 10, the position of `"hat"` in `str`.

- `InStr(str, "HAT")` returns 0: not found.

- `InStr(9, str, "HAT", vbTextCompare)` returns 10; case insensitive.

- `InStr(11, str, "hat")` returns 0: not found after position 10.

- `InStr(str, "at")` returns 2, position of first `"at"`.

- `InStr(3, str, "at")` returns 11, position of second `"at"`.

Here's an example that will be useful later. The function `ExtractString` takes an expression `expr`, a position `pos` in the expression, and a string of characters `allow`, and returns the sequence of letters starting at `pos` that match the characters in `allow`, stopping when a match fails. The procedure also modifies `pos` to point to the position of the character immediately after the last match. For example, if `expr = "ax12.03yz2.7"`, `pos = 3`, and `allow = "0123456789."`, then the function returns `"12.03"` and sets `pos` to 8, the position of the y.

```
Function ExtractString(str As String, pos As Integer, allow As String) _
        As String
    Dim k As Integer, char As String, L As Integer
    L = Len(str)
    If pos <= 1 Then ExtractString = str: Exit Function
    If pos > L Then ExtractString = "": Exit Function
    k = pos                              'remember starting position
    char = Mid(str, pos, 1)          'get the character at position pos
    Do While pos <= L And InStr(allow, char) > 0        'get the rest
        pos = pos + 1
        If pos <= L Then char = Mid(str, pos, 1)    'get next character
    Loop                    'stop when match fails; pos now after last match
    ExtractString = Mid(str, k, pos - k)        'return string of matches
End Function
```

6.7 The VBA Operator Like

This operator determines whether a given string satisfies a given pattern, returning `True` if so and `False` otherwise. Thus the operator may considered as a complement to the function `InStr`. The general syntax for the operator is

```
string Like pattern
```

Its use is best explained with some examples:

- `"A"` Like `"A"` returns `True`. `"A"` Like `"AA"` and `"A"` Like `"a"` each return `False`.

- `"Ab2De"` Like `"A*e"` returns `True` and so detects strings that begin with `"A"` and end with `"e"`. The asterisk is a wild card matching any character (including `""`) or sequence of characters.

- `"A5e"` Like `"A?e"` returns `True` and `"Abde"` Like `"A?e"` returns `False`. The question mark is a wild card matching any *single* character.

- `"A5e"` Like `"A#e"` returns `True` and `"Ade"` Like `"A#e"` returns `False`. The hash sign is a wild card matching any single digit 0-9.

- `"R"` Like `"[A-Z]"` returns `True` but `"R"` Like `"[!A-Z]"` (note the exclamation sign), `"R"` Like `"[A-Q]"`, and `"r"` Like `"[A-Z]"` all return `False`. The pattern `"[A-Z]"` matches any capital letter, while `"[A-Q]"` matches any capital letter up to and including `"Q"`. The analogous remarks applies to the pattern `"[a-z]"`

- `"bExf"` Like `"b[D-P]?[c-m]"` and `"bE5f"` Like `"b[D-P]#[!a-e]"` both return `True`.

- `"brJx3z"` Like `"b?J*"` returns `True`. Thus the statement detects strings with first character b, any second character, third character J, and zero or more arbitrary remaining characters.

6.8 The VBA Functions Join and Split

The VBA function `Join` takes an array of strings and an optional delimiter and returns the string obtained by concatenating the members of the array separated by the delimiter. Here is the description:

```
Join(arr[,delimiter])
```

In the absence of a delimiter the function will separate the elements of the array by spaces. If the delimiter is the null string "" then no separation occurs. For example, if `LBound(arr)=1` and `UBound(arr)=3`, then `Join(arr,"*")` returns the string `arr(1)*arr(2)*arr(3)`. If `"*"` is replaced by `""`, then the function returns `arr(1)arr(2)arr(3)`.

The function `Split` is the reverse of `Join`. It takes as input a string and returns a one-dimensional array of substrings (with index starting at 0) based on a specified *delimiter*, which tells the function where the input string should be split. Here is a simplified version of the syntax, sufficient for our purposes:

$$\text{Split(expression [,delimiter] [,limit])}$$

The argument `expression` is the input string that is to be split. The argument `delimiter`, which is optional, specifies where `expression` is to be split. If omitted the delimiter defaults to a space character. The argument `limit`, which is also optional, specifies the maximum number of substrings split from `expression`; left unspecified the function will split out all the substrings. Here are some examples:

- `Split("cat*in*a*hat", "*")` returns the array {"cat", "in", "a", "hat"}, splitting the string at the asterisks.

- `Split("cat in a hat")` returns the same array using the space character as the default delimiter.

- `Split("cat*in*a*hat", "*", 2)` returns {"cat", "in*a*hat"}.

It is sometimes inconvenient that the lower bound of the array returned by `Split` is zero. The following version has `LBound = 1`. Also, unlike `Split`, if `delim = ""` then the function returns an array of the individual characters of the expression

```
Function Split1(expr, delim) As String()
    Dim i As Integer, A() As String, B() As String
    If delim = "" Then                       'if null string delimiter
        ReDim B(1 To Len(expr))                  'then create an array
        For i = 1 To Len(expr)            'with single characters of expr
            B(i) = Mid(expr, i, 1)           'and index starting at 1
        Next i
    Else
        A = Split(expr, delim)                   'index starts at 0 so
        ReDim B(1 To UBound(A) + 1)              'increase indices by 1
        For i = 1 To UBound(A) + 1
            B(i) = A(i - 1)
        Next i
    End If
    Split1 = B
End Function
```

Deleting Duplicate Strings

The function `DeleteDupStr` is an application of both `Split` and `InStr`. It takes as input a string consisting of substrings separated by a delimiter, splits the string into substrings, and then reassembles it, omitting duplicate substrings. For example, applying the function to `"aaa,bbb,AAA,bbb,aa"` produces the string `"aaa,bbb,AAA,aa"`. The function employs a For Each statement (Section 5.4) to extract substrings and uses the case sensitive version of `InStr` to compare with previous substrings, discarding duplicates.

```
Function DeleteDupStr(str As String, dlm As String) As String
    Dim item As Variant, temp As String        'temp initialized as ""
    For Each item In Split(str, dlm)        'check if item already in temp
        If InStr(1, dlm & temp & dlm, dlm & item & dlm) = 0 Then
            temp = temp & dlm & item        'if item not in temp, attach it
        End If
    Next item
    DeleteDupStr = Mid(temp, Len(dlm) + 1)      'remove delim at beginning
End Function
```

Here's how the function works for the input string `str = "aa,bbb,bbb,cc"` with the comma as delimiter. Initially, the string `"aa"` is extracted from `str`. The function `InStr` then determines that `"aa"` is not already a part of `temp`, which at this stage is the null string, and so updates `temp` to `",aa"`. The next item checked in `str` is the string `"bbb"`. The function `InStr` determines that this string is not part of `temp` and so the string is attached to `temp`, which is now `",aa,bbb"`. In the next iteration of the loop, `"bbb"` is extracted. `InStr` determines that the string is already part of the current value of `temp` and so the latter is not updated. In the last iteration `InStr` determines that the string `"cc"` is not already a part of `temp` and so is attached to `temp`. The loop is then exited. The delimiter attached to `temp` during the first iteration (when `temp` was still the null string) is deleted by `Mid` and the resulting string `"aa,bbb,cc"` is returned by the function.

6.9   Exercises

1. Write a function `NumDigits(x)` that returns the number of digits in a Long integer x.

2. Write a function `LastDigit(x)` that returns the last digit of a Long integer x.

3. Write a function `RemoveLastDigit(x)` that removes the last digit of a Long integer x.

4. Write a function `CircularShift` that takes a string, removes the last letter and attaches it at the beginning. For example, `CircularShift("abcde")` should return `eabcd`.

5. Write a procedure `NumUpper(str)` that takes a string and returns the number of upper case letters. Write the analogous function `NumLower` as well. (Use `IsUpper,IsLower`.)

6. Write a function `IsVowel(char)` that takes a character and returns True if the character is a vowel and False otherwise. Write the analogous function `IsConsonant`.

7. Write a function `RemoveVowels(str)` that removes the vowels from `str`, collapsing the leftover letters into a single string in the same order that they appear in `str`.

8. Write a function `Lower2Upper(char)` that takes a lower case letter `char` and returns the corresponding upper case letter. Write a function `Upper2Lower(char)` that does the reverse.

9. A *palindrome* is a sequence of symbols that is unchanged when reversed, for example, radar, kook, level, rotor, 1001, and nurses run. Write a function `IsPalindrome(str)` that returns `True` if `str` is a palindrome and `False` otherwise.

10. Write a function `SwitchUpperLowerCase(str)` that takes a string of letters and changes every upper case to lower case and vice versa. For example, `abZadXy` becomes `ABzADxY`.

11. Write a function `OddConcat` that takes a string and returns the characters in the odd positions, concatenated in the same order that they appear in the string. For example, `OddConcat(sweet)` returns `set`.

12. Write a function `Interlace` that takes two strings of equal length and interlaces the letters. For example, `Interlace("abcde", "uvxyz")` returns `aubvcxdyez`.

13. Write a function `DecimalPart` that takes a decimal number and returns the fractional part, if any. Thus `DecimalPart(12.345)` should return the Double number .345. Write the analogous function `WholePart`.

14. Write a function `ExpandDate` that takes a date string such as "8/4/2022" and returns the string "August 4, 2022."

15. Write a function `DupCharacters` that takes a string of letters interspersed with digits 1–9 and returns a string that duplicates each letter as indicated by the numbers. For example, `DupChar("3a2b4c")` should return `aaabbcccc`.

16. Write a function `FindDoubleChars` that takes a string and returns the first pair of consecutive duplicate characters if there is such a pair and returns the null string `""` if not. For example, `DoubleLetters("scoop")` returns oo while `DoubleLetters("scope")` returns the null string.

17. The *Hamming distance* between two strings of equal length is the number of character positions where the strings differ. For example, the Hamming distance between the strings "below" and "belie" is 2, since the strings differ in two positions. Write a function `HammingDistance` that calculates Hamming distance between two strings. The Hamming distance is used in information theory to measure the minimum number of errors that might have occurred in changing a string.

18. Write a function `StringIntersect(Str1,Str2)` that takes a two strings and returns the common portion of the strings (which could be null) in the order in which the letters appear. For example, `StringIntersect("impossible","implausible")` returns the string "impsible".

19. Write a function `ReverseWords(inputstr)` that reverses the words in a sentence, keeping the letters of the words in the correct order.

20. Write a function `ExtractSymbols(inputstr, mode)` that extracts from an input string either all letters (mode="letters"), all small letters (mode="lower"), all capital letters (mode="upper"), or all digits (mode="digits"). For example,

    - `ExtractSymbols("X12aBc","lower")` returns ac
    - `ExtractSymbols("X12aBc","upper")` returns XB
    - `ExtractSymbols("X12aBc","letters")` returns XaBc
    - `ExtractSymbols("X12aBc","digits")` returns 12

21. (Fractions). Write a function `AddFrac(frac1,frac2)` that takes a pair of String fractions `frac1,frac2` and returns the string obtained by adding the fractions. Each fraction should be a string of the form `num/den`, where `num,den` are digit strings. For example, `AddFrac("1/2","3/4")` should return `"5/4"`.

22. Write a function `MidDelim` that takes as input a string consisting of substrings separated by a delimiter, a number k, and the delimiter, and returns the kth substring. For example, `MidDelim("aaa*bb*cccc*d", 3, "*")` returns the string cccc, the third of the substrings separated by the delimiter `"*"`.

23. Write a function `ReplaceChar("flub", 3, "a")` that takes a string, a character `char` and a position and replaces the character in the string at that position with the character. For example `ReplaceChar("flub", 3, "a")` should return `"flab"`.

24. Write a function `ASCIISort` that sorts the characters of a string according to their ASCII values. For example, if the string is `"Z13y56cd24x68A"` then the function should return `"12345668AZcdxy"`. Suggestion: Use the sorting technique of `BubbleSort` (Section 4.1) and a function `SwapChar` based on `ReplaceChar` of the preceding exercise.

25. Write a function `MaxRun` that takes a string `str` and returns the longest substring of strictly increasing consecutive digits. If there are two such runs it should return the first. For example, if the string is `"123abd12456"` the function should return the substring `"123"`. Use the `IsDigit` function.

    Suggestions: Declare a String array `Runs(1 To Len(str))`. For each position i in `str` have `MaxRun` call a program `GetRun(str,i)` that gets the run starting at at i, which could be null.

Chapter 7

Grids

While the Excel spreadsheet is a powerful platform for reading, writing, and crunching data, with a little tweaking it may be also used for interesting visual effects such as dynamic evolution of Markov processes, random walks, and maze solutions. The visual effects take place on what we shall refer to as a *grid*. In the first section of this chapter we develop procedures that allow the user to set up a grid either in the code of the host program or by specifying the dimensions of the grid on the spreadsheet. In later sections we give applications of cell activity on a grid. Additional applications appear throughout the text in one form or another.

7.1   Setting up a Grid

The procedure `MakeGrid` constructs a rectangular grid with user defined specifications. It is passed the top left row and column (T,L), the bottom right row and column (B,R), the column width (CW) and row height (RH) of the cells, and the color index (CX) of the border. These are assembled in the UDT `TGrid`, which must be placed at the head of any module using a grid.

```
Private Type TGrid
    T As Integer: L As Integer: B As Integer: R As Integer
    CW As Single: RH As Single: CX As Integer
End Type
```

The procedure begins by retrieving earlier grid specifications, if any, from Sheet3. If the old specs are the same as the new ones, then the program does nothing. Otherwise the old grid is removed by `ClearGrid`, the new specs are recorded in Sheet3, and a grid with these specs is constructed.

```
Private Function MakeGrid(T As Integer, L As Integer, B As Integer, _
    R As Integer, CW As Single, RH As Single, CX As Integer) As TGrid
    Dim OldGrid As TGrid, NewGrid As TGrid
    NewGrid.T = T: NewGrid.L = L: NewGrid.B = B: NewGrid.R = R
    NewGrid.CW = CW: NewGrid.RH = RH: NewGrid.CX = CX 'specs of new grid
    OldGrid = GetRecordedSpecs()      'specs of old grid found in Sheet3
    If GridsEqual(NewGrid, OldGrid) Then          'if specs the same
```

```
          NewGrid = OldGrid: GoTo lastline              'then do nothing
       End If
       Call ClearGrid(OldGrid, mode)              'otherwise clear old grid,
       Call RecordSpecs(NewGrid)                  'and record specs in Sheet3
                   'cell dimensions of grid:
       Range(Cells(T - 1, L - 1), Cells(B + 1, R + 1)).ColumnWidth = CW
       Range(Cells(T - 1, L - 1), Cells(B + 1, R + 1)).RowHeight = RH
       If T > 1 And L > 1 And CX > -1 Then  'if a colored border is desired
           Call Border(NewGrid)                            'add it
       End If
   lastline:
       MakeGrid = NewGrid                              'return grid
   End Function
```

The underscore symbol in the above code is used to break up lengthy lines. The symbol must be preceded by a space and followed by a carriage return.

The procedure `GridBorder` surrounds the grid with a border, as shown in Figure 7.1, where $CX = 20$. For no color, set $CX = 0$; for no line border set $CX = -1$. In applications all cell activity takes place within the grid proper defined by the parameters T, L, B, R, whether or not the border has color or a line boundary.

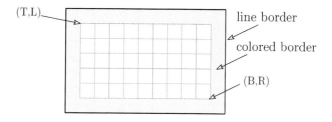

FIGURE 7.1: Grid with colored border.

```
Private Sub Border(G As TGrid)
    If G.T <= 1 Or G.L <= 1 Then Exit Sub
    Range(Cells(G.T - 1, G.L - 1), Cells(G.T - 1, G.R + 1)). _
    Interior.ColorIndex = G.CX
    Range(Cells(G.T - 1, G.R + 1), Cells(G.B + 1, G.R + 1)). _
    Interior.ColorIndex = G.CX
    Range(Cells(G.B + 1, G.L - 1), Cells(G.B + 1, G.R + 1)). _
    Interior.ColorIndex = G.CX
    Range(Cells(G.T - 1, G.L - 1), Cells(G.B + 1, G.L - 1)). _
    Interior.ColorIndex = G.CX
    Range(Cells(G.T - 1, G.L - 1), Cells(G.B + 1, G.R + 1)). _
    BorderAround Weight:=xlThick
End Sub
```

The function `GetRecordedSpecs` retrieves from Sheet3 the specifications of an earlier grid, if any.

```
Private Function GetRecordedSpecs() As TGrid
    Dim G As TGrid
    G.T  = Sheet3.Cells(1, 2).Value: G.L  = Sheet3.Cells(2, 2).Value
    G.B  = Sheet3.Cells(3, 2).Value: G.R  = Sheet3.Cells(4, 2).Value
    G.CW = Sheet3.Cells(5, 2).Value: G.RH = Sheet3.Cells(6, 2).Value
    G.CX = Sheet3.Cells(7, 2).Value
    GetRecordedSpecs = G
End Function
```

The function `GridsEqual` returns True if the grids `G1` and `G2` have the same specifications.

```
Private Function GridsEqual(G1 As TGrid, G2 As TGrid) As Boolean
    GridsEqual = (G1.T = G2.T) And (G1.L = G2.L) And (G1.B = G2.B) And _
                 (G1.R = G2.R) And (G1.CW = G2.CW) And (G1.RH = G2.RH) _
                 And (G1.CX = G2.CX)
End Function
```

The procedure `ClearGrid` is used to remove an old grid if a new one has been specified. The user has the option of removing the border (`mode=1`) or keeping the border (`mode=0`). The user may occasionally have to revise the code of the procedure to accommodate the special needs of a program. The code uses the Selection feature of VBA that allows the user to select a group of cells for a special purpose.

```
Private Sub ClearGrid(G As TGrid, mode As Integer)
    If G.T <= 1 Then Exit Sub
                 'select a block of cells:
    Sheet1.Range(Cells(G.T - mode, G.L - mode), _
                 Cells(G.B + mode, G.R + mode)).Select
    With Selection
        .Borders.LineStyle = xlLineStyleNone
        .Borders(xlDiagonalDown).LineStyle = xlLineStyleNone
        .Borders(xlDiagonalUp).LineStyle = xlLineStyleNone
        .Interior.ColorIndex = xlNone
    End With
    Range(Cells(G.T, G.L), Cells(G.B, G.R)).ClearContents
End Sub
```

The following Boolean function returns False if a cell position is outside the grid. It uses the UDT `TPos` that must be placed at the top of any module using it.

```
Private Type TPos                          'position type
    row As Integer: col As Integer      'row and column of position
End Type

Private Function InGrid(P As TPos, G As TGrid) As Boolean
    InGrid = G.T <= P.row And P.row <= G.B And _
             G.L <= P.col And P.col <= G.R
End Function
```

The following function detects the number of cells in the grid that are not empty. It is useful if memory has to be allocated for an array that depends on the number of entries in the grid.

```
Private Function NumOccupied(Grid As TGrid) As Integer
    Dim i As Integer, j As Integer, m As Integer
    For i = Grid.T To Grid.B                'count no. of occupied cells
        For j = Grid.L To Grid.R
            If Not IsEmpty(Cells(i, j)) Then m = m + 1
        Next j
    Next i
    NumOccupied = m
End Function
```

In applications the host program sets the grid. A second run might be necessary to produce the desired output if the host program is changing to a different grid.

7.2 The Grid as a Torus

The procedure in this section transforms a grid into a *torus*. This is useful in programs with moving cells that try to stray outside the grid. To avert this the procedure causes the cell to appear on the other side of the grid. Figure 7.2 shows a group of three red cells moving toward the right and gradually emerging from the left.

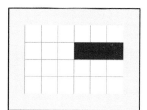 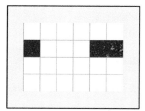 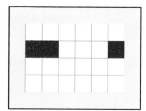

FIGURE 7.2: Red cells moving right on grid.

The procedure takes a cell position (row,col) on the boundary replaces it with a position on the opposite side of the grid.

```
Private Sub Convert2TorusPoint(row As Integer, col As Integer, G As TGrid)
    If row < G.T Then row = G.B              'breached top; go to bottom
    If row > G.B Then row = G.T              'breached bottom; go to top
    If col < G.L Then col = G.R              'breached left; go to right
    If col > G.R Then col = G.L              'breached right; go to left
End Sub
```

7.3 Installing Coordinates in a Grid

It is occasionally convenient to overlay on a grid a traditional x, y coordinate system that supplements the standard row, column spreadsheet coordinates. The procedure `InstallCoord` places coordinates on the outside of a grid created by `MakeGrid`. The origin may be placed anywhere in the grid. For example, if `ORow,OCol` denotes the cell position of the origin, then to put the origin at the center of the grid set

ORow = (Grid.T + Grid.B)/2,   OCol = (Grid.R + Grid.L)/2.

To put the origin at bottom left set

ORow = Grid.B,   OCol = Grid.L.

Figure 7.3 illustrates the two choices. In (a) the asterisk has (x, y) coordinates $(-2, -1)$; in (b) the (x, y) coordinates are $(2, 1)$.

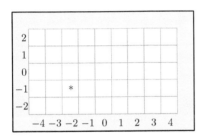
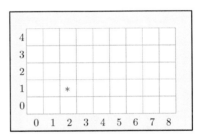

(a) origin at center   (b) origin at lower left

FIGURE 7.3: Grid with coordinates.

The following procedure carries out the installation.

```
Private Sub InstallCoord(Grid As TGrid, ORow As Integer, OCol As Integer)
    Dim i As Integer
    For i = 0 To Grid.R - OCol        'positive coordinates along bottom
        Cells(Grid.B + 1, OCol + i).Value = i
    Next i
    For i = 0 To Grid.L - OCol Step -1    'neg. coordinates along bottom
        Cells(Grid.B + 1, OCol + i).Value = i
    Next i
    For i = 0 To Grid.T - ORow Step -1        'pos.coordinates on left
        Cells(ORow + i, Grid.L - 1).Value = -i
    Next i
    For i = 0 To Grid.B - ORow                'neg. coordinates on left
        Cells(ORow + i, Grid.L - 1).Value = -i
    Next i
End Sub
```

The next procedures convert the row,col position of a cell to x, y coordinates and vice versa.

```
Private Sub RowCol2XY(x As Integer, y As Integer, row As Integer, _
    col As Integer, Grid As TGrid)
    x = Cells(Grid.B + 1, col).Value 'project cell position down, read x
    y = Cells(row, Grid.L - 1).Value 'project cell position left, read y
End Sub

Private Sub XY2RowCol(x As Integer, y As Integer, row As Integer, _
    col As Integer, Grid As TGrid)
    Dim i As Integer, j As Integer
    For i = Grid.T To Grid.B                        'look for y coordinate
        If Cells(i, Grid.L - 1).Value = y Then row = i
    Next i
    For j = Grid.L To Grid.R                         'look for y coordinate
        If Cells(Grid.B + 1, j).Value = x Then col = j
    Next j
End Sub
```

7.4   Sieve of Erastothenes

In this and the following sections we describe programs that require a grid for display. The programs are all launched by a command button.

The sieve of Eratosthenes is an algorithm for finding all prime numbers up to some specified upper limit. The idea is to delete composite numbers (non primes) by first deleting all multiples of 2 except 2, then going to 3, the next number not deleted, and deleting all multiples of 3 except 3, then going to 5, the next number not deleted, and deleting all multiples of 5 except 5, etc. Eventually all that is left are prime numbers. The method is named after Eratosthenes of Cyrene, a Greek mathematician in the third century BC.

The program Sieve sets the height and width of the display grid based on the user-entered upper limit. The procedure Initialize fills the grid with consecutive numbers. The procedure DeleteComposites scans the grid, deleting multiples of primes.

```
Sub Sieve()
    Dim upperlim As Integer, height As Integer, side As Integer
    Dim G As TGrid
    upperlim = Range("B5").Value
    side = upperlim ^ (1 / 2) + 1                    'dimensions of grid side
    G = MakeGrid(4, 4, 4 + side, 4 + side, 5, 13, -1)        'no border
    Call Initialize(G, upperlim)
    Call DeleteComposites(G, upperlim)
End Sub
```

The procedure `Initialize` populates a grid with consecutive integers whose number depends on the upper limit set by the user.

```
Private Sub Initialize(G As TGrid, upperlim As Integer)
    Dim i As Integer, j As Integer, k As Integer: k = 2
    For i = G.T To G.B                          'place numbers in grid
        For j = G.L To G.R
            Cells(i, j).Value = k                   'print integer
            If k = upperlim Then Exit Sub        'quit at upper limit
            k = k + 1
        Next j
    Next i
End Sub
```

The procedure `DeleteComposites` uses the VBA function `Mod` to detect multiples of k other than k itself and then deletes them.

```
Private Sub DeleteComposites(G As TGrid, upperlim As Integer)
    Dim i As Integer, j As Integer, cellval As Integer, k As Integer
    k = 2
    Do While k <= upperlim
        For i = G.T To G.B                          'scan grid for numbers
            For j = G.L To G.R
                If Not IsEmpty(Cells(i, j)) Then
                    cellval = Cells(i, j).Value 'get no. in nonempty cell
                End If
                If k <> cellval And cellval Mod k = 0 Then
                    Cells(i, j).Value = ""          'remove if a multiple of k
                End If
            Next j
        Next i
        k = k + 1                                    'next number to check
    Loop
End Sub
```

7.5 A Changing Rectangle

The program in this section prints a rectangle that shrinks in height about a central point, then shrinks in width, then expands in height, and finally expands in width to achieve its original dimensions. Figure 7.4 shows the process. The user specifies the width and height of the first and largest rectangle; the grid automatically adjusts to accommodate this.

```
Public T As Integer, L As Integer, B As Integer, R As Integer
Public height As Integer, width As Integer
Sub CollapseExpandRectangle()
```

FIGURE 7.4: Collapsing and expanding rectangle.

```
Dim G As TGrid, i As Integer
width = Range("B3").Value                    'read user-entered dimensions
height = Range("B4").Value
G = MakeGrid(5, 5, 5 + height, 5 + width, 2, 11, -1)          'grid specs
T = G.T: L = G.L: B = G.B: R = G.R     'initialize for largest rectangle
Call CollapseWidth                                    'begin the activity
Call CollapseHeight
Call ExpandWidth(G)
Call ExpandHeight(G)
End Sub
```

The four procedures below do the collapsing and expanding. The first keeps reducing R and increasing L until they differ by 1. At the end of the loop R = L, at which time R is incremented and L is decremented so that the rectangle is not solid. The other three procedures work analogously. The procedure DrawRectangle is a variation of GridBorder.

```
Private Sub CollapseWidth()
    Do While R > L
        Call DrawRectangle(T, L, B, R, 3)                'red rectangle
        Call Delay(12)                                   'pause for effect
        Call DrawRectangle(T, L, B, R, 0)                'remove color
        R = R - 1: L = L + 1                             'squeeze rectangle
    Loop
    R = R + 1: L = L - 1                          'reset after last for next
End Sub

Private Sub CollapseHeight()
    Do While T < B
        Call DrawRectangle(T, L, B, R, 3)
        Call Delay(12)
        Call DrawRectangle(T, L, B, R, 0)
        T = T + 1: B = B - 1
    Loop
    T = T - 1: B = B + 1
End Sub

Private Sub ExpandWidth(G As TGrid)
    Do While L >= G.L And R <= G.R
        Call DrawRectangle(T, L, B, R, 3)
        Call Delay(12)
        Call DrawRectangle(T, L, B, R, 0)
```

```
        L = L - 1: R = R + 1
    Loop
    L = L + 1: R = R - 1
End Sub

Private Sub ExpandHeight(G As TGrid)
    Do While T >= G.T And B <= G.B
        Call DrawRectangle(T, L, B, R, 3)
        Call Delay(15)
        If T > G.T And B < G.B Then          'remove color from all but the
            Call DrawRectangle(T, L, B, R, 0)           'last rectangle
        End If
        Call DrawRectangle(T, L, B, R, 0)
        T = T - 1: B = B + 1
    Loop
    T = T + 1: B = B - 1
End Sub

Private Sub DrawRectangle(T As Integer, L As Integer, B As Integer, _
                          R As Integer, C As Integer)
    Range(Cells(T, L), Cells(T, R)).Interior.ColorIndex = C
    Range(Cells(T, R), Cells(B, R)).Interior.ColorIndex = C
    Range(Cells(B, L), Cells(B, R)).Interior.ColorIndex = C
    Range(Cells(T, L), Cells(B, L)).Interior.ColorIndex = C
End Sub
```

The procedure `Delay` uses triple nested For Next loops together with a nonsense calculation to produce a delay. Note that the greater the value of `HowMuch`, the greater the delay. The reader should experiment with various settings, as speed may vary with computer.

```
Sub Delay(HowMuch As Integer)
    Dim i As Integer, j As Integer, k As Integer, x As Double
    For i = 1 To Int(10 * HowMuch)    'triple nested For Loops for delay
        For j = 1 To Int(10 * HowMuch)
            For k = 1 To Int(10 * HowMuch)
                x = 987654321 / 123456789          'arbitrary calculation
            Next k
        Next j
    Next i
End Sub
```

7.6 Table Sum Game

The module `TableSum` allows the user to play the following game: The player specifies a number n. Initialization causes a square of numbers 1 to n^2 to

appear in a grid. (The grid expands or contracts to accommodate the number
of cells.) The player selects a number in the grid from each row and column,
never using a row or column twice. At the end of the game the selected numbers
will always sum to $n(n^2 + 1)/2$. The game is illustrated in Figure 7.5 for the
case $n = 3$.

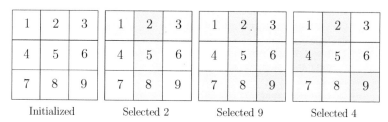

Initialized   Selected 2   Selected 9   Selected 4

FIGURE 7.5: Table sum game.

Here are the command buttons for the program. The first is activated once.
The second is activated each time a number is selected.

```
Sub CommandButton1_Click()
    Call Initialize                     'draw grid and place numbers
End Sub

Sub CommandButton2_Click()
    Call SelectNumber      'press button after number in grid is selected
End Sub
```

The procedure **Initialize** sets up the table in the form of a grid with
bordered cells and places the numbers in the grid. The grid is a square with
side dimension **GridDim**.

```
Sub Initialize()
    Dim G As TGrid, row As Integer, col As Integer
    Dim k As Integer,  GridDim As Integer
    GridDim = Range("B4").Value                    'read grid dimension
    G = MakeGrid(3, 3, 2 + GridDim, 2 + GridDim, 7, 20, -1)
    Range(Cells(G.T, G.L), Cells(G.B, G.R)).Borders.LineStyle _
         = xlContinuous                  'give each cell a border
    row = G.T                            'start numbers here
    col = G.L
    For k = 1 To GridDim ^ 2
        Cells(row, col).Value = k                    'cell gets number
        col = col + 1                     'move right for next cell in row
        If k Mod GridDim = 0 Then          'if at end of grid then
            row = row + 1                      'go to next row
            col = G.L                          'and first column
        End If
    Next k
End Sub
```

After the player selects a number and presses the command button Select , the procedure SelectNumber is called. At this time the grid is changed if necessary, the dimensions depending on user-entered specs. After this, the active cell feature of VBA assigns to the variables row,col, the cell position that the user selected. A double For Next loop assigns color1 to the cells in the row row and the column col, and color2 to the cell in position row,col. A running sum is kept of the numbers on cells with color2. At the end of the game when all cells have colors, the sum equals $n(n^2 + 1)/2$, where n is the grid dimension.

```
Sub SelectNumber()
    Dim i As Integer, j As Integer, row As Integer, col As Integer
    Dim GridDim As Integer, check As Integer, sum As Integer, G As TGrid
    Dim color1 As Integer, color2 As Integer
    G = GetRecordedSpecs()                  'get the grid created in Initialize
    GridDim = Range("B4").Value                   'read the table dimension
    T = 3: L = 3: B = 2 + GridDim: R = 2 + GridDim
    color1 = 27: color2 = 24
    row = ActiveCell.row: col = ActiveCell.Column        'selected by user
    If Not InGrid(row, col, G) Then Exit Sub
    For i = T To B                    'color the row and column of selection
        For j = L To R
            If i = row Then Cells(row, j).Interior.ColorIndex = color1
            If j = col Then Cells(i, col).Interior.ColorIndex = color1
            Cells(row, col).Interior.ColorIndex = color2      'new color
        Next j
    Next i
    For i = T To B                'add the numbers of the selected cells
        For j = L To R
            If Cells(i, j).Interior.ColorIndex = color2 Then
                sum = sum + Cells(i, j).Value
            End If
        Next j
    Next i
    Range("B5").Value = sum                              'print sum
    check = GridDim * (GridDim ^ 2 + 1) / 2       'which should equal this
    Range("B6").Value = check
    If sum = check Then
        MsgBox "game over"
        Range(Cells(G.T, G.L), Cells(G.B, G.R)).Interior.ColorIndex = 0
    End If
End Sub
```

7.7 Finding the Nearest Cells

The program NearestCells is a two-dimensional analog of NearestNumbers. The user enters the letter x in as many grid cells as desired, the letter "o" in a single cell, and a positive integer k. Running the program colors k of the cells marked x that are closest to the cell marked o (see Figure 7.6).

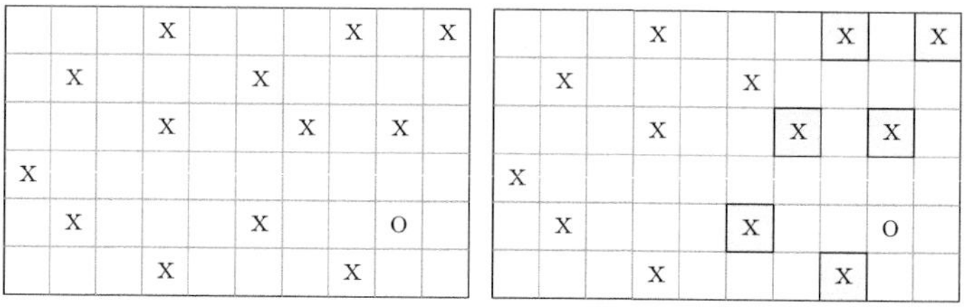

FIGURE 7.6: Six nearest cells.

The main procedure introduces a new UDT TDist, which specifies the position of a cell containing an x and its distance from the cell marked o, the "origin."

```
Private Type TDist
    row As Integer: col As Integer: dist As Integer
End Type
Sub NearestCells()
    Dim k As Integer, n As Integer, D() As TDist, G As TGrid
    Dim orow As Integer, ocol As Integer        'origin row and column
    G = MakeGrid(5, 5, 39, 53, 2.3, 15, 24)
    Call ClearColors(G)
    k = Range("B4").Value              'desired number of nearest cells
    n = NumOccupied(G) - 1                       'number of x's
    Range("B6").Value = n                        'print number of x's
    If n < 1 Or k = 0 Or k > n Then Exit Sub     'flawed parameters
    ReDim D(1 To n)                              'memory for the x data
    Call GetPositions(D, G, orow, ocol)
    If orow = 0 Or D(1).row = 0 Then Exit Sub        'no origin or x's
    Call GetDistances(D, orow, ocol)         'distances from x's to origin
    Call BubbleSort(D)                       sort distances least to greatest
    Call PrintNearest(D, k)                 print k nearest x's to the origin
End Sub
```

The procedure ClearColors removes the colors from the x's from a previous run, if any.

```
Private Sub ClearColors(G As TGrid)
    Dim i As Integer, j As Integer
    For i = G.T To G.B      'scan grid for the locations of o and the x's
        For j = G.L To G.R
```

```
            If Cells(i, j).Value = "x" Then
                Cells(i, j).Interior.ColorIndex = xlNone
            End If
        Next j
    Next i
End Sub
```

The procedure `GetPositions` fills the position part of the `TDist` array with the location of the origin, labeled `orow,ocol`, and the location of the x's.

```
Private Sub GetPositions(D() As TDist, G As TGrid, _
            orow As Integer, ocol As Integer)
    Dim i As Integer, j As Integer, n As Integer
    n = 1
    For i = G.T To G.B      'scan grid for the locations of o and the x's
        For j = G.L To G.R
            If Cells(i, j).Value = "x" Then
                D(n).row = i:  D(n).col = j: n = n + 1        'found an x
            ElseIf Cells(i, j).Value = "o" Then
                orow = i:  ocol = j                          'found the o
            End If
        Next j
    Next i
End Sub
```

The procedure `GetDistances` fills the distance part of the `TDist` array with the distance from an x to the origin o. Distance here is simply the number of horizontal and vertical cells in an elbow path from one cell to another.

```
Private Sub GetDistances(D() As TDist, orow As Integer, ocol As Integer)
    Dim n As Integer
    For n = 1 To UBound(D)
        D(n).dist = Abs(D(n).row - orow) + Abs(D(n).col - ocol)
    Next n
End Sub
```

The procedure `BubbleSort` used in `NearestCells` is a slight variation of the eponymous procedure in Section 5.5. Instead of taking an array of Double items the new version takes an array of `TDist` items. We omit listing the new version.

The procedure `PrintNearest` colors the x's corresponding to the first k entries of the sorted array D.

```
Private Sub PrintNearest(D() As TDist, k)
    Dim i As Integer
    For i = 1 To k                              'print k nearest cells
        Cells(D(i).row, D(i).col).Interior.ColorIsndex = 27
    Next i
End Sub
```

7.8 A Growing Spiral

The program in this section produces two interleaved spirals that grow separately, as illustrated in Figure 7.7. (The numbers in the cells are for later reference and are not printed by the program.) The spirals are then gradually erased.

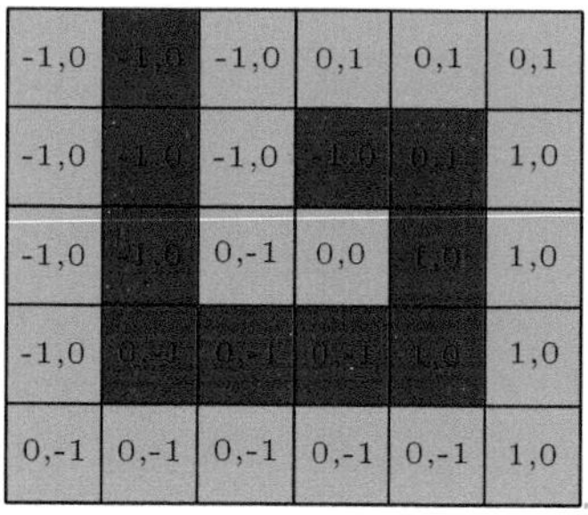

-1,0	-1,0	-1,0	0,1	0,1	0,1
-1,0	-1,0	-1,0	-1,0	0,1	1,0
-1,0	-1,0	0,-1	0,0	1,0	1,0
-1,0	0,-1	0,-1	0,-1	1,0	1,0
0,-1	0,-1	0,-1	0,-1	0,-1	1,0

FIGURE 7.7: Spirals with offsets.

The procedure `Spiral` initially calls `MakeGrid` to establish the boundaries and the column width and row height of the grid containing the spiral. These depend on the user-entered value of the variable `size`. The spiral will then completely fill the grid. Next, the procedure calls `GenerateSpiralPositions`, which prints in columns 1-4 of Sheet2 the positions of the green cells and red cells comprising the spirals. The last two procedures, `CreateSpirals` and `EraseSpirals`, use the information in Sheet2 to produce the interesting visual effect of growing and collapsing interleaved spirals.

```
Private Sub Spiral()
    Dim G As TGrid, size As Integer
    size = Range("B3").Value      'desired size (width, height) of spiral
    G = MakeGrid(5, 5, 10 + 2 * size, 10 + 2 * size, 2, 11, 0)
    Call GenerateSpiralPositions(G)
    Call CreateSpirals
    Call EraseSpirals
End Sub
```

The procedure `GenerateCellPositions` prints the cell positions `row,col` of the green spiral cells in columns 1 and 2 of Sheet2 and positions of the red spiral cells in columns 3 and 4. The green spiral starts at the central cell of the grid, the one labeled $(0, 0)$ in Figure 7.7. Successive cell positions are obtained by adding offsets to the current cell position, `row,col`. To see how this works, consider the pattern exhibited by the row offsets of the green cells, starting with the second cell, the one to the left of the central cell. Notice how pattern

$0, -1, 1, 0$ repeats, but with growing numbers of duplicates:

$$0, \quad -1-1, \quad 000, \quad 1111, \quad 00000, \quad -1-1-1-1-1-1, \quad \text{etc.} \qquad (7.1)$$

These values are stored in the array `offset`:

$$\texttt{offset(0)} = 0, \ \texttt{offset(1)} = -1, \ \texttt{offset(2)} = 0, \ \texttt{offset(3)} = 1,$$

The pattern in (7.1) is produced by repeating `offset((i-1) Mod 4)` exactly i times for $i = 1, 2, \ldots$. The column offsets

$$-1, \quad 00, \quad 111, \quad 0000, \quad -1-1-1-1-1, \quad 000000, \ldots$$

are generated by repeating `offset(i Mod 4)` i times $(i = 1, 2, \ldots)$. Now consider the red cells. The row offsets starting with the second cell of the red spiral, the one marked $(0, 1)$ repeat in the pattern $1, 0, -1, 0$, again with increasing numbers of duplicates:

$$0, 11, \quad 000, \quad -1-1-1-1, \quad 00000, \quad 11111, \quad, \ldots$$

This pattern is produced by using `offset(i+1 Mod 4)` i times $(i = 1, 2, \ldots)$. The column offsets repeat in the pattern $1, 0, -1, 0$, again with increasing numbers of duplicates:

$$1, \quad 00, \quad -1-1-1, \quad 0000, \quad 11111, \quad, \ldots$$

This pattern is produced by using `offset(i+2 Mod 4)` i times $(i = 1, 2, \ldots)$.

Here is the code that generates the positions of the green and red spiral cells in Sheet2.

```
Private Sub GenerateSpiralPositions(G As TGrid)
    Dim offset(0 To 3) As Integer, row As Integer, col As Integer
    Dim centerrow As Integer, centercol As Integer
    Dim i As Integer, j As Integer, k As Integer
    k = 1                                   'first row in sheet 2
    Sheet2.Cells.ClearContents
    offset(0) = 0: offset(1) = -1: offset(2) = 0: offset(3) = 1
    centerrow = (G.T + G.B)/2: centercol = (G.L + G.R)/2    'central cell
    row = centerrow: col = centercol                   'initialize
    Sheet2.Cells(1, 1).Value = row       'position of first green cell is
    Sheet2.Cells(1, 2).Value = col  'placed in columns 1 and 2 of Sheet2
    For i = 1 To G.B - G.T           'generate positions using offsets
        For j = 1 To i                      'repeat each offset i times
            k = k + 1                            'next row in Sheet2
            row = row + offset((i - 1) Mod 4)    'generates 0,-1,0,1
            col = col + offset(i Mod 4)          'generates -1,0,1,0
            If InGrid(row, col, G) Then         'if within grid then
                Sheet2.Cells(k, 1).Value = row      'record position
                Sheet2.Cells(k, 2).Value = col
            End If
        Next j
```

```
         Next i
         k = 1                                              'reset
         row = centerrow - 1: col = centercol        'cell above central cell
         Sheet2.Cells(k, 3).Value = row          'position of first red cell is
         Sheet2.Cells(k, 4).Value = col  'placed in columns 3 and 4 of Sheet2
         For i = 1 To G.B - G.T
             For j = 1 To i                      'repeat each offset i times
                 k = k + 1
                 row = row + offset((i + 1) Mod 4)       'generates 0,1,0,-1
                 col = col + offset((i + 2) Mod 4)       'generates 1,0,-1,0
                 If InGrid(row, col, G) Then             'if within grid then
                     Sheet2.Cells(k, 3).Value = row          'record position
                     Sheet2.Cells(k, 4).Value = col
                 End If
             Next j
         Next i
     End Sub
```

The procedure `CreateSpirals` uses For Next loops to retrieve coordinates of the spiral cells from Sheet2 and to output the colored cells in Sheet1. `EraseSpirals` is similar but the For Next loops run in reverse. We have used the value 5 for the delay factor, which causes a fairly rapid change. To slow things down use a larger value.

```
 Sub CreateSpirals()
     Dim i As Integer, row As Integer, col As Integer, lastrow As Integer
     lastrow = Sheet2.Cells(Rows.Count, 1).End(xlUp).row
     For i = 1 To lastrow            'last nonempty row in Sheet2, column 1
         row = Sheet2.Cells(i, 1).Value              'get cell locations
         col = Sheet2.Cells(i, 2).Value                  'of green spiral
         Cells(row, col).Interior.ColorIndex = 4 'color spiral cell green
         Call Delay(5)
     Next i
     lastrow = Sheet2.Cells(Rows.Count, 3).End(xlUp).row
     For i = 1 To lastrow            'last nonempty row in Sheet2, column 3
         row = Sheet2.Cells(i, 3).Value                  'get cell locations
         col = Sheet2.Cells(i, 4).Value                      'of red spiral
         Cells(row, col).Interior.ColorIndex = 3    'color spiral cell red
         Call Delay(5)
     Next i
 End Sub

 Sub EraseSpirals()
     Dim i As Integer, k As Integer, row As Integer, col As Integer
     lastrow = Sheet2.Cells(Rows.Count, 1).End(xlUp).row
     For i = lastrow To 1 Step -1            'erase green spiral in reverse
         Call Delay(5)
         row = Sheet2.Cells(i, 1).Value     'get locations of green cells
         col = Sheet2.Cells(i, 2).Value
         Cells(row, col).Interior.ColorIndex = 0                'remove color
     Next i
```

```
    lastrow = Sheet2.Cells(Rows.Count, 3).End(xlUp).row
    For i = lastrow To 1 Step -1              'erase red spiral in reverse
        Call Delay(5)
        row = Sheet2.Cells(i, 3).Value
        col = Sheet2.Cells(i, 4).Value
        Cells(row, col).Interior.ColorIndex = 0          'remove color
    Next i
End Sub
```

7.9   Billiard Ball

In this section we construct a program that sends a billiard ball (red cell) diagonally across a pool table (grid). The user specifies the starting point and direction by placing the numbers 1 and 2 in diagonally adjacent cells, thus providing an initial position and a direction for the ball. In Figure 7.8 the initial direction of the ball is southeast (SE); when it hits the edge of the pool table it switches to northeast (NE), etc. Motion continues until either the ball reaches a pocket (corner cell) or its path length reaches the user-specified limit. The figure shows the trail of the cell; in the program only the ball is displayed.

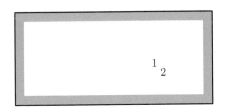 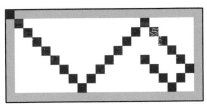

FIGURE 7.8: Billiard ball bank shot; sank in upper left pocket.

The module uses the UDT's TGrid and TPos. The launching program BilliardBall introduces three TPos variables: pos1 and pos2, which are the second to last and last positions of the ball, respectively; and pos3, the new position to be determined from pos1 and pos2. The procedure calls the function GetStart to find pos1 and pos2 for the entered data 1,2, returning False if unsuccessful. If successful the numbers 1,2 are erased and the cell at pos2 is colored red. A Do While loop then moves the ball, stopping when the path length reaches its prescribed maximum (maxlen) or when it is notified by the variable status that the ball has landed in a corner pocket. A Select Case statement calls the appropriate function, DirNE, DirSE, DirNW or DirSW for the next direction. It is these functions that are responsible for finding pos3. The current direction is determined by the function GetDirection, which takes the current values of `pos1,pos2`, which give the direction of the ball, and returns one of the strings `"NE"`, `"SE"`, `"NW"`, or `"SW"`. After this, color is removed from the cell at `pos2`, the variable then `pos3` becomes the new `pos2`, and `pos2` the new `pos1`.

```
Sub BilliardBall()
    Dim G As TGrid, pos1 As TPos, pos2 As TPos, pos3 As TPos
    Dim status As String, pathlen As Integer, maxlen As Integer
    G = MakeGrid(8, 5, 27, 35, 2.5, 11.5, 4)
    Call ResetCorners(G)                        'make corners green again
    maxlen = 500                    'max. no. cells in ball path (arbitrary)
    If Not GetStart(pos1, pos2, G) Then MsgBox "failed": Exit Sub
    dir = GetDirection(pos1, pos2)
    If dir = "" Then MsgBox "failed": Exit Sub
    Cells(pos1.row, pos1.col).Value = ""                  'erase 1 and 2
    Cells(pos2.row, pos2.col).Value = ""
    Cells(pos2.row, pos2.col).Interior.ColorIndex = 3     'initial color
    Do While pathlen < maxlen Or status = "SankBall"
        Select Case GetDirection(pos1, pos2)        'depends on pos1, pos2
            Case Is = "NE": status = DirNE(pos1, pos2, pos3, G)
            Case Is = "SE": status = DirSE(pos1, pos2, pos3, G)
            Case Is = "NW": status = DirNW(pos1, pos2, pos3, G)
            Case Is = "SW": status = DirSW(pos1, pos2, pos3, G)
            Case Is = "": Exit Do
        End Select
        Call Cells(pos2.row, pos2.col).Interior.ColorIndex = 0  'no color
        pos1 = pos2: pos2 = pos3                                 'shift
        Call Cells(pos2.row, pos2.col).Interior.ColorIndex = 3  'new pos.
        Call Delay(12)
        pathlen = pathlen + 1                       'update path length
    Loop
                'remove last red ball on table
    Range(Cells(G.T, G.L), Cells(G.B, G.R)).Interior.ColorIndex = 0
End Sub
```

The procedure `ResetCorners` makes the corners green again in case a ball landed in a pocket on a previous run.

```
Private Sub ResetCorners(G As TGrid)
    Cells(G.T - 1, G.L - 1).Interior.ColorIndex = 4
    Cells(G.T - 1, G.R + 1).Interior.ColorIndex = 4
    Cells(G.B + 1, G.L - 1).Interior.ColorIndex = 4
    Cells(G.B + 1, G.R + 1).Interior.ColorIndex = 4
End Sub
```

Here is the code for `GetStart`. The function scans the grid for the numbers 1,2, recording their positions and returning True if successful.

```
Private Function GetStart(pos1 As TPos, pos2 As TPos,Grid As TGrid) _
        As Boolean
    Dim i As Integer, j As Integer, OK1 As Boolean, OK2 As Boolean
```

```
    For i = Grid.T To Grid.B                    'find the positions of 1 and 2
        For j = Grid.L To Grid.R
            If Cells(i,j).Value = 1 Then
                pos1.row = i: pos1.col = j: OK1 = True
            ElseIf Cells(i,j).Value = 2 Then
                pos2.row = i: pos2.col = j: OK2 = True
            End If
        Next j
    Next i
GetStart = OK1 And OK2
End Function
```

The procedure `GetDirection` takes the locations `pos1,pos2` of 1 and 2 and attempts to find the direction from their relative positions. If unsuccessful the function returns the null string.

```
Private Function GetDirection(pos1 As TPos, pos2 As TPos) As String
Dim dir As String
    If pos2.row = pos1.row - 1 And pos2.col = pos1.col + 1 Then dir = "NE"
    If pos2.row = pos1.row + 1 And pos2.col = pos1.col + 1 Then dir = "SE"
    If pos2.row = pos1.row - 1 And pos2.col = pos1.col - 1 Then dir = "NW"
    If pos2.row = pos1.row + 1 And pos2.col = pos1.col - 1 Then dir = "SW"
GetDirection = dir
End Function
```

The direction functions find the new position `pos3` based on the current direction, as determined by the pair `pos1,pos2`. We illustrate with `DirNe`. There are several cases to consider, depending on the proximity of `pos2` to the grid edge. These are shown in Figure 7.9.

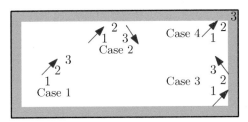

FIGURE 7.9: Billiard ball shot.

```
Private Function DirNE(pos1 As TPos, pos2 As TPos, pos3 As TPos, _
                G As TGrid) As String
    Dim status As String                        'ball in pocket?
    If pos2.col < G.R And pos2.row > G.T Then
        pos3.row = pos2.row - 1: pos3.col = pos2.col + 1 'Case 1: move NE
    ElseIf pos2.col < G.R And pos2.row = G.T Then
        pos3.row = pos2.row + 1: pos3.col = pos2.col + 1 'Case 2: move SE
    ElseIf pos2.col = G.R And pos2.row > G.T Then
        pos3.row = pos2.row - 1: pos3.col = pos2.col - 1 'Case 3: move NW
```

```
    ElseIf pos2.col = G.R And pos2.row = G.T Then
        Cells(pos2.row, pos2.col).Interior.ColorIndex = 0    'Case 4: move
        Cells(pos2.row - 1, pos2.col + 1).Interior.ColorIndex = 3 'to the
            status = "SankBall"                        'upper right corner
    End If
    DirNE = status                                     'return status
End Function
```

The remaining direction functions are handled analogously.

```
Private Function DirSE(pos1 As TPos, pos2 As TPos, pos3 As TPos, _
    G As TGrid) As String
    Dim status As String
    If pos2.col < G.R And pos2.row < G.B Then
        pos3.row = pos2.row + 1: pos3.col = pos2.col + 1         'move SE
    ElseIf pos2.col < G.R And pos2.row = G.B Then
        pos3.row = pos2.row - 1: pos3.col = pos2.col + 1         'move NE
    ElseIf pos2.col = G.R And pos2.row < G.B Then
        pos3.row = pos2.row + 1: pos3.col = pos2.col - 1         'move SW
    ElseIf pos2.col = G.R And pos2.row = G.B Then
        Cells(pos2.row, pos2.col).Interior.ColorIndex = 0
        Cells(pos2.row + 1, pos2.col + 1).Interior.ColorIndex = 3   'move
            status = "SankBall"         'to lower right corner on boundary
    End If
End Function


Private Function DirNW(pos1 As TPos, pos2 As TPos, pos3 As TPos, _
    G As TGrid) As String
    Dim status As String
    If pos2.col > G.L And pos2.row > G.T Then
        pos3.row = pos2.row - 1: pos3.col = pos2.col - 1         'move NW
    ElseIf pos2.col > G.L And pos2.row = G.T Then
        pos3.row = pos2.row + 1: pos3.col = pos2.col - 1         'move SW
    ElseIf pos2.col = G.L And pos2.row > G.T Then
        pos3.row = pos2.row - 1: pos3.col = pos2.col + 1         'move NE
    ElseIf pos2.col = G.L And pos2.row = G.T Then
        Cells(pos2.row, pos2.col).Interior.ColorIndex = 0
        Cells(pos2.row - 1, pos2.col - 1).Interior.ColorIndex = 3   'move
            status = "SankBall"          'to upper left corner on boundary
    End If
End Function


Private Function DirSW(pos1 As TPos, pos2 As TPos, pos3 As TPos, _
    G As TGrid) As String
    Dim status As String
    If pos2.col > G.L And pos2.row < G.B Then
        pos3.row = pos2.row + 1: pos3.col = pos2.col - 1         'move SW
    ElseIf pos2.col > G.L And pos2.row = G.B Then
        pos3.row = pos2.row - 1: pos3.col = pos2.col - 1         'move NW
    ElseIf pos2.col = G.L And pos2.row < G.B Then
        pos3.row = pos2.row + 1: pos3.col = pos2.col + 1         'move SE
    ElseIf pos2.col = G.L And pos2.row = G.B Then
```

```
    Cells(pos2.row, pos2.col).Interior.ColorIndex = 0
    Cells(pos2.row + 1, pos2.col - 1).Interior.ColorIndex = 3    'move
    status = "SankBall"                    'to lower left corner on boundary
  End If
End Function
```

7.10   Exercises

1. Write a program `NumberPainting` that takes a user-entered pattern of color index numbers in a grid and assigns the colors to the cells, as shown in Figure 7.10. Use nested For Next loops to scan the grid for the numbers and to paint the cells.

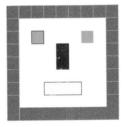

FIGURE 7.10: Number painting.

2. Write a program `ColorChart` that produces the color index chart of Section 1.3.

3. Write a function `PrintDiamond(n)` that prints in a grid a red diamond with an odd number n of rows, as shown in the figure with $n = 7$. The grid dimensions should depend on n.

FIGURE 7.11: PrintDiamond.

4. Write a program `FlashingCells` that colors a grid of 9 cells with alternating cell colors, for example, red, yellow, red, yellow, etc. Switch the colors with a delay to cause a flashing effect in multicolors.

5. Write a program `BounceBallDownStairs` that creates a staircase with a user-entered number n of steps on which the symbol "@" slowly descends the stairs. The grid dimensions should depend on n.

6. Write a program `CollapseExpandSquare` that produces a sequence of collapsing and expanding of concentric squares, as suggested in Figure 7.12 (only one square at a time should appear). The procedure should use a

FIGURE 7.12: Collapsing-expanding square.

For Next loop to draw decreasing concentric squares. Have each iteration of the loop call the procedure `DrawRectangle` to draw a colored square and then, after a delay, erase it. The next iteration should draw a smaller square by adding 1 to the top row and left column and subtracting 1 from the bottom row and right column. The process should be repeated until the square has collapsed into a single cell. Reversing the process should cause the square to expand.

7. Write a program `ModBlinker` that causes a user-entered set of positive integers in a grid to blink according to their remainders modulo a user-entered positive integer m, resulting in interesting patterns. For example, if $m = 2$, then all odd numbers should be set to red for an instant and then all even numbers. The case $m = 3$ is shown in the figure.

1	2		4
5		7	8
	10	11	

	2	3	
5	6		8
9		11	12

1		3	4
	6	7	
9	10		12

FIGURE 7.13: Mod 3 blinker.

8. Write a program `RectangleEnclose` that scans a grid for user entered x's and encloses them in a minimal rectangle, as shown in Figure 7.14

FIGURE 7.14: Minimal rectangle enclosing points.

9. Write a program `Spiral2` like `Spiral` but have the outer and inner spiral start out together.

10. Write a program `ZigZag` that gradually creates and then erases a pattern like the one on the left in Figure 7.15. Model the program on `Spiral`.

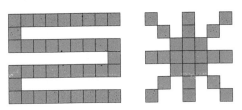

FIGURE 7.15: Spreadsheets for ZigZag and Star.

11. Write a program `ExplodingStar` that gradually expands and contracts a pattern like the one on the right in Figure 7.15 (captured midway through an expansion) starting from a single central cell.

12. (Seven segment display). Write a program `SevenSegment` that prints a user-entered digit 0–9 using seven segments on a grid, as shown in Figure 7.16.

FIGURE 7.16: Seven-segment display.

13. (Langton's ant). The name refers to an algorithm based on a set of simple instructions that causes a cell ("ant") to move on a grid of black and white cells. The cells of the grid are initially white. The ant is allowed to travel

 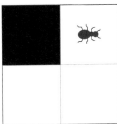 

FIGURE 7.17: Langton's ant.

in the directions north, east, south, or west. The direction is determined by the following rules: At a white cell, the ant turns 90 degrees clockwise, the color of the cell changes to black, and the ant moves forward one cell. At a black cell, the ant turns 90 degrees counterclockwise, the color of the cell changes to white and the ant moves forward one cell. The ant is given an initial position and direction. Write a program `LangtonAnt` that implements the algorithm. The algorithm, invented by Chris Langton in 1986, exhibits interesting patterns depending on the number of moves. The program is an example of a *Turing machine,* that is, a machine with a simple set of rules that can implement computer algorithms. In the case of Langton's ant these simple rules lead to complex *emergent behavior.*

Chapter 8

Recursion

The word "recursion" generally refers to a sequential process that is successively defined in terms of its earlier states. The concept frequently arises in mathematics in the form of *recursive sequences*. For example, the equations

$$a_0 = a_1 = 1; \quad a_n = a_{n-1} + a_{n-2} \ (n > 1)$$

famously define the *Fibonacci sequence*, the first few terms of which are

$$1, \ 1, \ 2, \ 3, \ 5, \ 8, \ 13, \ 21, \ 34, \ldots.$$

Notice that the first two numbers of the sequence are specified but successive numbers are generated from previously calculated terms.

In programming, recursion refers to a procedure calling itself. It is similar to looping in that statements are executed repeatedly. In a loop this happens until a certain condition is satisfied; in recursion a procedure ends when a *base case* or *terminating condition* is met.

Recursion typically requires more memory than looping and can increase the time required to complete a task. On the other hand, recursion can significantly reduce the complexity of a program, providing a relatively simple solution to an otherwise daunting problem. Indeed, it may be the only practical technique for implementing algorithms with many branches that are too complex for a satisfactory looping approach. In this chapter we give several examples of recursive programs.

8.1 The Factorial Function

In Exercise 1 of Section 8.6 the reader was asked to use a For Next loop to construct a function that returns the factorial of a positive integer. In this section we construct a factorial function using recursion. The code implements the formula

$$n! = n \cdot (n-1)!, \quad 0! = 1.$$

We have included in the function a feature that for purposes of illustration prints out the value of the factorial at each step of the recursion.

DOI: 10.1201/9781003351689-8

```
Function Factorial(n As Integer) As Long
    Dim f As Long
    f = 1                                        'base case
    If n > 1 Then
        f = n * FactRecursive(n - 1)             'do the recursion
    End If
    Cells(3 + n, 1).Value = f                    'print out current value
    Factorial = f
End Function
```

To see how the recursion works suppose a program calls `Fact(4)`, where we have abbreviated `Factorial` to `Fact`. During execution, the statement `4*Fact(3)` is encountered. The multiplication is delayed and the function `Fact(3)` is called. In the latter, the statement `3*Fact(2)` is encountered, multiplication is again delayed and `Fact(2)` is called. In this version the program encounters the statement `2*Fact(1)` and calls `Fact(1)`. The latter returns the base value 1, enabling the computation `2*Fact(1)`. The function `Fact(2)` then returns this number to `Fact(3)`, where the calculation `3*Fact(2)` is carried out. `Fact(3)` then returns this number to `Fact(4)`, where the final calculation `4*Fact(3)` occurs. `Fact(4)` then returns this number to the host program. Figure 8.1 illustrates the process.

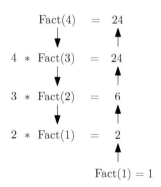

FIGURE 8.1: Factorial recursion.

Excel gives you an overflow in both the loop and recursion versions if n is too large, say $n > 12$. Higher values are possible if the `Long` declarations are replaced by `Double`. Even higher values are possible using strings—see Chapter 27.

All programs requiring iteration, however complex, can be written in terms of loops. However, some of these require emulation of the compiler's method of executing recursion, namely via stack storage of instructions and variables. In these situations, recursion is usually a more reasonable approach.

8.2   Binary Trees

A *binary tree* is a type of hierarchal data structure consisting of *nodes*. Each node N is itself a data structure consisting of a data element and references to at most two additional nodes, a *left node L* and a *right node R*. In this context, N is called the *parent node* and L and R *children nodes*. The tree begins with a single node called the *root node* and ends with childless nodes called *leaf nodes*. The root node is said to be at level 0. Trees are used in computer science to represent linked data and they form the basis for efficient algorithms to insert and retrieve that data. Unlike other data structures such as arrays, which have only one logical way to traverse, trees have many traversal paths, which accounts for their complexity but also their usefulness.

Our goal in this section is to employ a typical recursion algorithm similar to that used in the insertion and retrieval of linked data to construct a program that draws a *perfect* binary tree, that is, one for which all the leaf nodes are at the same level and each internal node has exactly two children. Figure 8.2 depicts a perfect binary tree with levels 0,1,2,3. The node data structures are labelled with capital letters; the left and right references are indicated by edges. The root note is labelled A and the leaf nodes are labelled H to O.

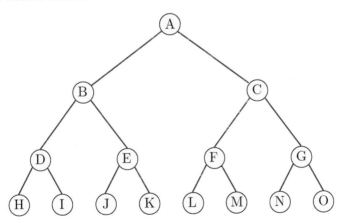

FIGURE 8.2: A perfect binary tree.

The main procedure of the module is `BinaryTree`, which sets various parameters including root position, numbers of levels, and offsets to the successive levels. The latter affect the general shape of the tree.

The procedure `DrawTree`, as you may have deduced, draws the tree. It calls itself to draw the left branches of a node first, then comes out of recursion to draw a right branch. The order in which this is done for the tree in Figure 8.2 in terms of nodes is A,B,D,H,I,E,J,K,C,F,L,M,G,N,O. Figure 8.3 shows the seven steps in the construction of the left branch. This order of node traversal is called *preorder* and is summarized by the description Root, Left, Right. Other important orders are *inorder* (Left, Root, Right) and *postorder*(Left, Right, Root).

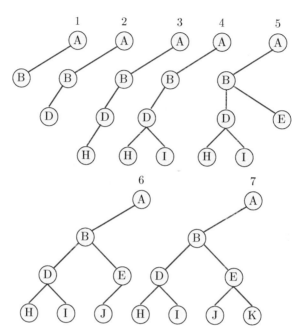

FIGURE 8.3: Steps for main left branch of the tree.

```
Public level As Integer, lastlevel As Integer
Sub CommandButton1_Click()
    lastlevel = Range("B3").Value        '1,2,3,... Determines grid size
    Call BinaryTree
End Sub

Sub BinaryTree()
    Dim Grid As TGrid, rootrow As Integer, rootcol As Integer
    Dim Offset() As Integer
    Dim height As Integer, breadth As Integer, i As Integer
    Dim T As Integer, L As Integer, B As Integer, R As Integer
    height = 2 ^ (lastlevel + 1) + 4         'grid dimensions determined
    breadth = 2 * height + 4                        'by lastlevel
    T = 5: L = 5: B = T + height: R = L + breadth
    Grid = MakeGrid(T, L, B, R, 0.8, 6, 0)
    rootrow = T + 2: rootcol = L + breadth / 2      'middle top of grid
    level = -1
    ReDim Offset(0 To lastlevel)
    For i = 0 To lastlevel                           'offsets from root
        Offset(i) = 2 ^ (lastlevel - i)
    Next i
```

```
      Call DrawTree(rootrow, rootcol, Offset)                        'go!
      MsgBox "clear"
      Call ClearGrid(Grid, 0)
   End Sub
```

The procedure **DrawTree** is the recursive part of the program. It takes advantage of the fact that the tree basically looks the same at any node. The procedure starts by drawing a left line from the root node to the next left node, then calling itself to repeat this until it gets down to the leftmost leaf mode. It then comes out of recursion and backs up to draw the right nodes.

```
Sub DrawTree(row As Integer, col As Integer, Offset() As Integer)
      Dim nextrow As Integer, leftcol As Integer, rightcol As Integer
      Dim leftpart As String, rightpart As String, off As Integer
      idx = 1
      level = level + 1  go to next level
      If level < lastlevel Then
         off = Offset(level)
         nextrow = row + off: rightcol = col + off: leftcol = col - off
         Call DrawLine(row, col, off, "left")    'left line from (row,col)
         Call Delay(20)
         Call DrawTree(nextrow, leftcol, Offset)   'draw tree at next node
         Call DrawLine(row,col,off, "right")     'right line from (row,col)
         Call Delay(20)
         Call DrawTree(nextrow, rightcol, Offset)
      End If
      level = level - 1 'go back to previous level
   End Sub
```

The procedure **Drawline(row,col,off,dir)** draws a diagonal red line from (row,col) to (row,col+off) if dir = "right" and to (row,col-off) if dir = "left". The nodes are marked in blue. The With End With statement specifies the direction of the diagonal line as well as its characteristics.

```
Sub DrawLine(row As Integer, col As Integer, off As Integer, dir As String)
      Dim i As Integer
      Cells(row, col).Borders(xlDiagonalDown).LineStyle = xlNone
      Cells(row, col).Borders(xlDiagonalUp).LineStyle = xlNone
      Cells(row, col).Interior.ColorIndex = 5
      If dir = "right" Then
            'draw a red right line from (row,col) to (row +off ,col+off)
         For i = 1 To off
            With Cells(row + i, col + i).Borders(xlDiagonalDown)
               .LineStyle = xlContinuous
               .Weight = xlThick
               .ColorIndex = 3                        'red edge
            End With
         Next i
         Cells(row + off, col + off).Borders(xlDiagonalDown).LineStyle _
         = xlNone
         Cells(row + off, col + off).Interior.ColorIndex = 5    'blue node
```

```
ElseIf dir = "left" Then
    'draw a red left line from (row,col) to (row +off ,col+off)
    For i = 1 To off
        With Cells(row + i, col - i).Borders(xlDiagonalUp)
            .LineStyle = xlContinuous
            .Weight = xlThick
            .ColorIndex = 3
        End With
    Next i
    Cells(row + off, col - off).Borders(xlDiagonalUp).LineStyle _
    = xlNone
    Cells(row + off, col - off).Interior.ColorIndex = 5
End If
End Sub
```

8.3   The Tower of Hanoi

The Tower of Hanoi is a puzzle that consists of three rods, labeled I, II, and III, and rings of different sizes that can be placed on the rods. Initially, the

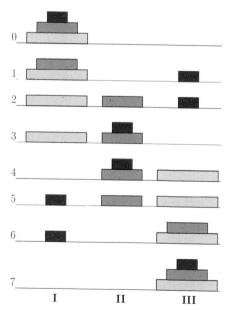

FIGURE 8.4: Tower of Hanoi solution for $n = 3$.

rings are placed on rod I in order of size, forming a conical shape with the larger rings at the bottom.. The objective is to move the rings one at a time to rod III by taking the top ring of a (*source*) rod and placing it on a (*target*) rod. However, there is crucial restriction: no ring may ever be placed above a smaller one, that is, the conical shape of a stack must always be preserved. A third rod (*free*) is available to help with this.

The puzzle is handily solved by recursion: If at any stage it is required to move n rings from a source rod to a target rod, then

(1) move the top $n - 1$ rings to the free rod,

(2) move the nth ring to the target rod,

(3) move the $n - 1$ rings from the free rod to the target rod.

Tasks (1) and (3) are the same as the original task but with one less ring, hence the recursion. Figure 8.4 illustrates the procedure for $n = 3$, with task (1) consisting of steps 1 - 3, task (2) step 4, and task (3) steps 5 - 7. For simplicity we've omitted the rods and depicted the rings as rectangles.

The module `HanoiTower` runs through the steps of the solution on the spreadsheet. The user enters the desired number of rings in cell B5, restricted to 10 or less. (The reader may increase this value, but moving more rings than this increases the time dramatically. See discussion at end of section.) The rings are concretely realized as horizontal cells in a grid. A delay value is entered in B6 to slow down the movements of the rings for a better appreciation of the solution. The user may also step through the solution.

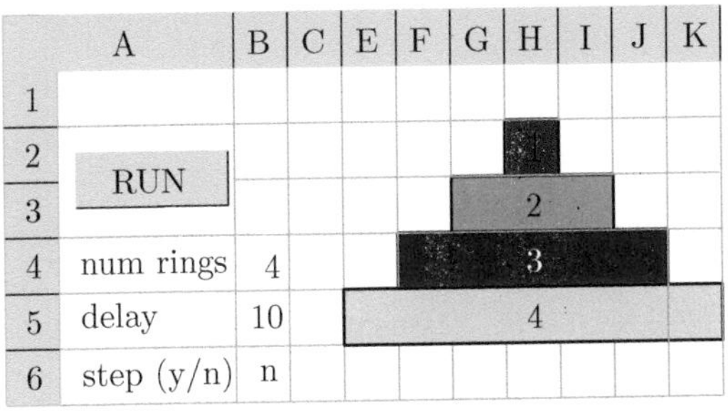

FIGURE 8.5: Moving red ring from source to target.

The main procedure, `HanoiTower`, begins by setting up a grid for the rectangles and reading the user-entered information, including the number of rings and the speed of the moves. The variables `source,free,target` refer to column numbers on the spreadsheet designating the rod positions. They are initially set in that order and correspond to the rods I, II, and III. The procedure then calls `InitTower`, which colors the rings, labels the rods, and places the rings on rod I. Figure 8.5 shows the spreadsheet immediately after `InitTower` was called. Note that the width of disk n counting from the top is $2n - 1$. In particular, the last disk has width `2*NumRings-1`.

Calling the procedure MoveRings starts the recursion. At each stage, the rings are moved from the current source rod to the current target rod using the current free rod as a means to maintain the relative size configurations.

```
Public speed As Integer, NumRings As Integer
Public StepThrough As String, NumMoves As Integer
Public T As Integer, L As Integer, B As Integer, R As Integer
Sub HanoiTower()
    Dim source As Integer, target As Integer, free As Integer
    T = 2: L = 4: B = 12: R = 150
    Range(Cells(T, L), Cells(B, R)).ColumnWidth = 2.2
    Range(Cells(T, L), Cells(B, R)).RowHeight = 12.5
    Call ClearRings()
    NumRings = Range("B4").Value              'desired number of rings
    StepThrough = Range("B6").Value                           'y/n
    speed = Range("B5").Value                 'enter delay value
    If NumRings > 10 Or (NumRings > 5 And StepThrough = "y") Then
        MsgBox ("Sorry. This will take way too long"): Exit Sub
    End If
    NumMoves = 0                              'initial value of count
    source = L + NumRings                 'column number for source rod
    free = source + 2 * NumRings           'column number for free rod
    target = free + 2 * NumRings          'column number for target rod
    Call InitTower(source, target, free)       'put rings on 1st rod
    Call MoveRings(NumRings, source, target, free)           'go!
End Sub


Private Sub ClearRings()
    Range(Cells(T, L), Cells(B, R)).ClearContents
    Range(Cells(T, L), Cells(B, R)).Interior.ColorIndex = xlNone
End Sub


Sub InitTower(source As Integer, target As Integer, free As Integer)
    Dim n As Integer, row As Integer
    Cells(T + NumRings, source).Value = "I"           'label the rods
    Cells(T + NumRings, free).Value = "II"
    Cells(T + NumRings, target).Value = "III"
    For n = 1 To NumRings                     'color the initial rings
        row = T + n - 1    'rows move down
        Range(Cells(row, source - n + 1), Cells(row, source + n - 1)) _
        .Interior.ColorIndex = 2 + n         'cell width (ring size) grows
    Next n
End Sub
```

The procedure MoveRings(n,source,target,free) moves n rings from source to target, using free as the free rod. For $n > 1$ it does this by first calling the version MoveRings(n-1,source,free,target), where free is now the target rod and target the free rod, to move the top $n - 1$ rings from the source rod to the free rod, then moving the leftover nth ring to the target rod, and finally calling MoveRings(n-1,free,target,source), where free is the source rod and source the free rod, to move the $n - 1$ rings from the free rod to the target rod.

```
Sub MoveRings(n As Integer, source As Integer, target As Integer, _
    free As Integer)
    If n = 1 Then
        Call Delay(speed)  'delay, then move a ring from source to target
        Call MoveRing(source, target)
        Call Delay(speed)
    Else
        Call Delay(speed)  'delay then move n-1 rings from source to free
        Call MoveRings(n - 1, source, free, target)
        Call Delay(speed)'delay, then move nth ring from source to target
        Call MoveRing(source, target)
        Call Delay(speed)  'delay, then move n-1 rings from free to target
        Call MoveRings(n - 1, free, target, source)
        Call Delay(speed)                           'delay some more
    End If
End Sub
```

The procedure `MoveRing(source,target)` moves a single ring from current source to current target. For this it uses the function `FirstVacantRow(target)` to find the first vacant row in the target column (row 2, column M in Figure 8.6), and similarly uses `FirstNonVacantRow(source)` to find the first non-vacant row in the source column (row 2, column G in Figure 8.6.) The ring number is used to determine the first and last columns of the ring rectangle (K and O for ring 3), which determine cell width of the rectangle. (Ring k has width $2k - 1$.) The actual copying of the rectangle uses the `Range.Copy` method.

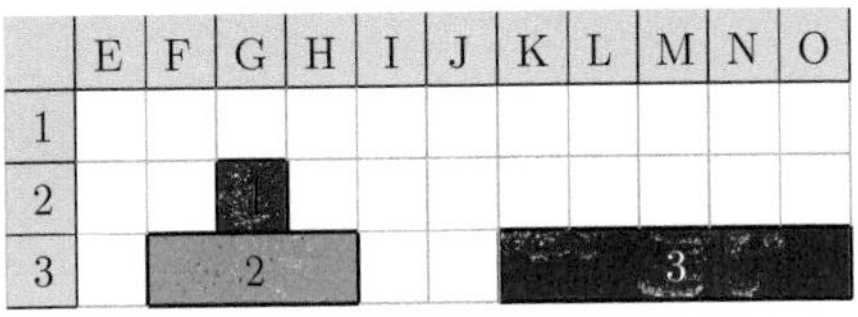

FIGURE 8.6: Moving red ring from source to target.

```
Sub MoveRing(source As Integer, target As Integer)
    Dim i As Integer, j As Integer, k As Integer
    i = FirstVacantRow(target): j = FirstNonVacantRow(source)
    k = Cells(j, source).Value                        'ring number
    Range(Cells(j, source - k + 1), Cells(j, source + k - 1)).Copy _
    Range(Cells(i, target - k + 1), Cells(i, target + k - 1))
    Call Delay(speed)
    Cells(j, source).Value = ""           'remove number and color of ring
    Range(Cells(j, source - j + 1), Cells(j, source + j - 1)). _
    Interior.ColorIndex = 0
    If StepThrough = "y" Then MsgBox ""        'click for next ring move
    NumMoves = NumMoves + 1
```

```
        Range("B7").Value = NumMoves            'keep track of number of moves
    End Sub

    Function FirstVacantRow(col As Integer) As Integer
        Dim row As Integer
        row = T + NumRings - 1                   'start at bottom ring and
        Do While Not IsEmpty(Cells(row, col))    'work up until a vacancy
            row = row - 1
        Loop
        FirstVacantRow = row                                         'vacancy
    End Function

    Function FirstNonVacantRow(col As Integer) As Integer
        Dim row As Integer
        row = T                                  'start at top position and
        Do While IsEmpty(Cells(row, col))        'work down until a non vacancy
            row = row + 1
        Loop
        FirstNonVacantRow = row                                  'non vacancy
    End Function
```

It is of interest to find the number of moves required to carry out the above algorithm. For this let M_n denote the number of moves needed to transfer n rings. From the procedure MoveRings we see that

$$M_n = M_{n-1} + 1 + M_{n-1} = 2M_{n-1} + 1.$$

Since $M_1 = 1$, we have

$$M_2 = 2 \cdot 1 + 1 = 3 = 2^2 - 1, \; M_3 = 2 \cdot 3 + 1 = 7 = 2^3 - 1, \; M_4 = 2 \cdot 7 + 1 = 7 = 2^4 - 1,$$

and in general $M_n = 2^n - 1$. This the least number of steps required to solve the puzzle. For example, 3 rings takes 7 steps, as shown in Figure 8.4. The number of steps escalates quite rapidly with the introduction of additional rings. The reader may check that moving one ring per second in a tower consisting of 31 rings would take almost an average lifetime. Moving 64 disks at a rate of one per second would take over 500 billion years. (Please do not attempt this!)

---

*8.4   A Complex Arithmetic Calculator

A *complex number* is an expression of the form $a + bi$, where a and b are real numbers and i is a symbol with the property $i^2 = 1$. In this section we develop a calculator that uses recursion to evaluate arithmetic expressions with complex numbers such as -.99i/(1+2i) + i(3+4i)^(-5)(6+7i). The program evaluates the expression and returns the result in the standard form $a + bi$. Expressions are evaluated according to standard precedence rules: exponentiation first, multiplication and division next, addition and subtraction last. We begin with a brief review of complex arithmetic.

Complex Arithmetic

Complex numbers $a + bi$ and $c + di$ are added and multiplied in a manner similar to real numbers except that i^2 is replaced by -1 and the terms are collected in the appropriate manner:

$$(a + bi) + (c + di) = (a + c) + (b + d)i$$
$$(a + bi)(c + di) = ac + adi + bci + +bdi^2 = (ac - bd) + (ad + bc)i$$

The complex number $a - bi$ is called the *complex conjugate of* $a + bi$. Note that

$$(a + bi)(a - bi) = a^2 + b^2.$$

For division we need the notion of *absolute value* of a complex number $a + bi$, defined as

$$|a + bi| = \sqrt{a^2 + b^2}.$$

To carry out the division one needs to multiply the numerator and denominator by the complex conjugate of the denominator and collect terms:

$$\frac{a + bi}{c + di} = \frac{(a + bi)(c - di)}{(c + di)(c - di)} = \frac{ac + bd}{c^2 + d^2} + \frac{bc - ad}{c^2 + d^2}i$$

Complex numbers may be represented as points in a two dimensional coordinate system, as shown in Figure 8.7.

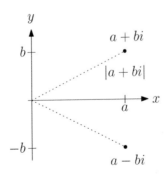

FIGURE 8.7: Graphical representation.

Launching the Program

Here is the command button code that calls the main procedure.

```
Sub CommandButton1_Click()
    Dim expr As String, dplaces As Integer, err As Boolean, z As String
```

```
Range("B3").Value = ""               'erase any previous error message
Range("B4").NumberFormat = "@"              'format as text
expr = Range("B4").Value                    'read expression
z = ComplexCalc(expr, err)                  'do the calculation
If Not err Then
    dplaces = Range("B5").Value    'number of decimal places in answer
    Range("B6").Value = CxRound(z, dplaces)        'rounding
Else: Range("B3").Value = "Error"
End If
End Sub
```

The main procedure `ComplexCalc` is passed a complex arithmetic expression `expr` and an error variable `error`, which is set by `ErrorCheck` and also during the actual calculations. The function `InsertAsterisks(expr)` is used to simplify multiplication code. For example, it transforms the expression i2+3(4+5i)6i into i*2+3*(4+5i)*6i.

```
Function ComplexCalc(expr As String, error As Boolean) As String
    Dim error As Boolean
    expr = RemoveWhiteSpace(expr)                    'compress expression
    error = ErrorCheck(expr)                  'preliminary check for typos
    If Not error Then
        expr = InsertAsterisks(expr)         'simplifies multiplication code
        ComplexCalc = ComplexEval(expr, 1, 0, error) 'evaluate expression
    End If
End Function
```

The Calculation Procedures

The actual calculations are carried out by the following VBA string functions. Each of these must be preceded by the designation WorksheetFunction. (note the period).

- `Complex`. Takes a pair of real numbers a and b and returns the string "a+bi". For example `Complex(3,4)` returns the string "3+4i".

- `ImReal`. Takes a complex number string and returns the real part. For example `ImReal("3+4i")` returns the string "3".

- `Imaginary`. Takes a complex number string and returns the imaginary part. For example `Imaginary("3+4i")` returns the string "4".

- `ImConjugate("3+4i")`. Returns the complex conjugate of a complex number string. For example `ImConjugate("3+4i")` returns the string "3-4i".

- `ImAbs`. Returns the absolute value of a complex number. For example `ImAbs("3+4i")` returns the number 5.

- `ImSum`. Returns the sum of a pair of complex number strings.

- `ImSub`. Returns the difference of a pair of complex number strings.

- `ImProduct`. Returns the product of a pair of complex number strings.

- ImDiv. Divides the first of a pair of complex number strings by the second.

- ImPower. Takes a complex number string "a+bi" and an integer n and returns $(a + bi)^n$. For example, ImPower("1+i",-2) returns "-.5i".

Assigning the Calculations

The function ComplexEval doles out the computational tasks. It consists of a Do While loop that continues as long as there are symbols in expr left to read, that is, as long as idx $\leq$ Len(expr), where idx denotes the position of the current symbol in expr. In each iteration of the loop, a Select-Case statement determines the next calculation. The variable mode, initially set to zero, is responsible for ensuring the correct hierarchy of operations. The function works recursively, calling itself if a subcalculation is required. ExtractString is used to retrieve complex numbers from expr.

```
Function ComplexEval(expr As String, idx As Integer, mode As Integer, _
                err As Boolean) As String
    Dim z As String, w As String, char As String
    Dim LonelyOp As Boolean              'needed in cases like (-2i+3)
    Dim WF As WorksheetFunction          'notation abbreviation
    Set WF = Application.WorksheetFunction  'implement the abbreviation
    Do While idx <= Len(expr) And Not err
        char = Mid(expr, idx, 1)              'character at position idx
        If InStr("0123456789i.", char) > 0 Then char = "#"      'number?
        Select Case char
        Case Is = "#"                         'found a number
            z = ExtractString(expr, idx, ".0123456789i")      'pull it
        Case Is = "+"
            If mode > 0 Then Exit Do          'wait for higher mode ops
            LonelyOp = (idx = 1 Or Midd(expr, idx - 1, 1) = "(")
            idx = idx + 1
            w = ComplexEval(expr, idx, 0, err)
            If err Then Exit Do
            If LonelyOp Then z = "0"
            z = WF.ImSum(z, w)
        Case Is = "-"
            If mode > 0 Then Exit Do
            LonelyOp = (idx = 1 Or Midd(expr, idx - 1, 1) = "(")
            idx = idx + 1
            w = ComplexEval(expr, idx, 1, err)
            If err Then Exit Do
            If LonelyOp Then z = "0"
            z = WF.ImSub(z, w)
        Case Is = "*"
            idx = idx + 1
            w = ComplexEval(expr, idx, 1, err)         'get next factor
            If err Then Exit Do
            z = WF.ImProduct(z, w)
        Case Is = "/"
            idx = idx + 1
```

```
        w = ComplexEval(expr, idx, 1, err)           'get denominator
        err = err Or (WF.ImAbs(w) = 0)
        z = WF.ImDiv(z, w)
    Case Is = "^"                                  'highest op--no wait
        idx = idx + 1                                    'skip "^"
        If Mid(expr, idx, 1) = "(" Then idx = idx + 1    'skip "("
        w = ExtractString(expr, idx, "-0123456789")       'get exp
        If Mid(expr, idx, 1) = ")" Then idx = idx + 1    'skip ")"
        err = err Or WF.ImAbs(z) = 0
        If err Then Exit Do
        z = WF.ImPower(z, CInt(WF.ImReal(w)))
    Case Is = "("
        idx = idx + 1                                    'skip "("
        z = ComplexEval(expr, idx, 0, err)
        idx = idx + 1                                    'skip")"
    Case Is = ")": Exit Do
    Case Else: err = True: Exit Do
    End Select
  Loop
  ComplexEval = z                              'return calculation
End Function
```

The procedure `ExtractString` of Section 6.6 is used to retrieve complex numbers by filtering out non-number symbols.

How Everything Works

It is instructive to follow the main steps in the above procedures to see precisely how a concrete expression is handled. We track the evaluation of `-2i(3+4i)^(-5)` after the function `ComplexCalc` has called `RemoveWhiteSpace` and `InsertAsterisks`. At this stage `expr` has the value `-2i*(3+4i)^(-5)`.

- If `ErrorCheck` (described below) returns no error, then `ComplexCalc` calls `ComplexEval`. In this version the Do While loop is accessed with $idx = 1$, and Select Case then branches to the minus sign. The Boolean variable `LonelyOp` is then set to `True` and version 2 of `ComplexEval` is called.

- In version 2 the Do While loop is entered and Select Case branches to `ExtractString`, which returns `"2i"`. The index `idx` is now pointing to the next symbol, the asterisk.

- In the next iteration of the Do While loop, `Case Is = "*"` is selected. The index `idx` is then advanced to point to the left parenthesis and version 3 of `ComplexEval` is called.

- In version 3 the procedure `ExtractString` returns the complex number string `"3+4i"`. The index `idx` now points to the operator `^`.

- In the next iteration of the Do While loop, `Case Is = "^"` is selected. The index `idx` is then advanced to point to the second left parenthesis and version 4 of `ComplexEval` is called.

- In version 4 the exponent "-5" is retrieved and control returns to version 3.

- Back in version 3 the exponentiation is carried out and the program returns to version 2.

- Back in version 2 the multiplication is carried out and the program returns to version 1, where the final operation of negation is performed. Control then returns to the main procedure `ComplexCalc`.

Figure 8.8 illustrates the entire process.

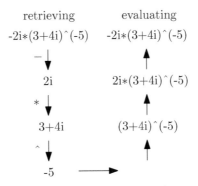

FIGURE 8.8: Recursion process.

Rounding

The function `CxRound` sets the maximum number of decimal places in the final answer by applying the VBA function `Round` to the real and imaginary parts. If `decplaces` is zero no rounding occurs.

```
Function CxRound(zStr As String, decplaces As Integer) As String
    Dim WF As WorksheetFunction, xStr As String, yStr As String
    Dim x As Double, y As Double
    Set WF = Application.WorksheetFunction
    If decplaces <= 0 Then CxRound = zStr: Exit Function
    xStr = WF.ImReal(zStr): yStr = WF.Imaginary(zStr)
    x = CDbl(xStr): y = CDbl(yStr)
    x = Round(x, decplaces): y = Round(y, decplaces)
    CxRound = WF.Complex(x, y)
End Function
```

Checking for Errors

The function `ErrorCheck` returns `True` if an incorrect or out of place symbol is encountered. It is based on the principle that each symbol has a restricted set of eligible immediate predecessors and successors. Recall that the function is called *before* `InsertAsterisks`. Here are the rules:

1. An operator +-/ must be followed by a left parenthesis, a period, a digit, or the letter i.

2. The operator ^ must be followed by a left parenthesis or a positive digit.

3. The operator / must be preceded by a right parenthesis, a digit, a period, or the letter i.

4. A right parenthesis must be followed by a right or left parenthesis, one of the operators +-/^, a digit, a period, the letter i, or nothing.

5. A left parenthesis cannot be followed by one of the operators /^.

6. There must be the same number of left parentheses as right parentheses.

```
Function ErrorCheck(expr As String) As Boolean
    Dim s(-1 To 1) As String, k As Integer, errnum As String
    Dim e(1 To 6) As Boolean, error As Integer
    L = Len(expr)
    For k = 2 To L
        s(0) = Midd(expr, k, 1)                        'current symbol
        s(-1) = Midd(expr, k - 1, 1)                   'its predecessor
        s(1) = Midd(expr, k + 1, 1)                    'its successor
        e(1) = InStr("+-/", s(0)) > 0 And (k = L Or _
            InStr("0123456789i.(", s(1)) = 0)          'list of errors
        e(2) = (s(0) = "^") And InStr("123456789(", s(1)) = 0
        e(3) = (s(0) = "/") And InStr(".0123456789i)", s(-1)) = 0
        e(4) = s(0) = ")" And k <> L And _
            InStr("+-/^(.0123456789i)", s(1)) = 0
        e(5) = s(0) = "(" And InStr("/^", s(1)) > 0
        e(6) = Not IsMatch(expr, "(", ")")
        error = e(1) Or e(2) Or e(3) Or e(4) Or e(5) Or e(6)
        If error Then Exit For
    Next k
    ErrorCheck = error
End Function
```

Inserting Asterisks

The function **InsertAsterisks** uses **InsertString** of Section 6.2 to place an asterisk between suitable characters in a multiplication operation.

```
Function InsertAsterisks(expr As String) As String
    Dim k As Integer, ch As String, nextch As String, insert As Boolean
    k = 1
    Do While k <= Len(expr)
        ch = Mid(expr, k, 1)                           'current character
        nextch = Midd(expr, k + 1, 1, " ")             'next one
        insert = ch = ")" And InStr("0123456789i.(", nextch) > 0 Or _
            InStr("0123456789i.", ch) > 0 And nextch = "(" Or _
          ch = "i" And InStr("0123456789i.", nextch) > 0
        If insert Then expr = InsertString(expr, k, "*")
```

```
        k = k + 1
    Loop
    InsertAsterisks = expr
End Function
```

---

*8.5   A Polynomial Calculator

A (*real*) *polynomial of degree* m is a function of the form

$$P(x) = a_0 + a_1 x + a_2 x^2 + \cdots + a_m x^m,$$

where the a_k are real numbers, called the *coefficients* of P, and $a_m \neq 0$. Polynomials are added, subtracted, and multiplied as illustrated in the following examples:

$$(1 + x + x^2) + (1 + x + x^2 + x^3) = (2 + 2x + 2x^2 + x^3)$$
$$(1 + x + x^2) - (1 + x + x^2 + x^3) = -x^3$$
$$(1 + x + x^2)(1 + x + x^2 + x^3) = 1 + 2x + 3x^2 + 3x^3 + 2x^4 + x^5$$

The general formula for multiplication is

$$(a_0 + a_1 x + a_2 x^2 + \cdots + a_m x^m)(b_0 + b_1 x + b_2 x^2 + \cdots + b_n x^n)$$
$$= c_0 + c_1 x + c_2 x^2 + \cdots + c_{m+n} x^{m+n},$$

where c_k is the sum of all products $a_i b_j$ with $i + j = k$.

In this section we develop a calculator that takes a polynomial expression such as $-x^2 + (3x + 1)(x^2 + x + 1)^3$ and returns the expanded polynomial.

Launching the Program

Here is the command button code that calls the main procedure.

```
Sub CommandButton1_Click()
    Dim expr As String, error As Boolean
    Range("B3").Value = ""                    'erase any previous error message
    Range("B4").NumberFormat = "@"            'format as text
    expr = Range("B4").Value                  'read expression
    expr = PolyCalc(expr, error)              'expand the polynomial
    If Not error Then
        Range("B6").Value = expr              'print expanded polynomial
    Else: Range("B3").Value = "Error"
    End If
End Sub
```

The main procedure `PolyCalc` first carries out a preliminary error check of the input expression. If no error is found, then the terms of the polynomial are encoded for ease of programming and the evaluation procedure `PolyEval` is called. As in the case of `ComplexCalc`, the usual hierarchy of operations is imposed. If no calculation error is found, the result is decoded by the function `DecodePoly` and then returned.

```
Function PolyCalc(expr As String, error As Boolean) As String
    expr = RemoveWhiteSpace(expr)
    error = ErrorCheck(expr)                    'preliminary error check
    If Not error Then
        expr = Convert(expr)                'encode for ease of calculations
        expr = InsertPolyAsterisks(expr)             'for multiplication
        expr = PolyEval(expr, 1, 0, error)           'do the calculations
    End If
    If Not error Then
        expr = DecodePoly(expr)                      'decode the result
        PolyCalc = expr                   'return the expanded polynomial
    End If
End Function
```

Preparing the Expression

To simplify the algebraic operations, terms $a_0, a_1x,\ a_2x^2,\ a_3x^3 \ldots$ are stored as strings $[a_0]$, $[0, a_1]$, $[0, 0, a_2]$, $[0, 0, 0, a_3] \ldots$, etc. Operations are then performed on these encoded terms, ultimately resulting in a string of bracketed coefficients of the output polynomial. For example, the expression $(3 + 2x + 5x^2)(x + 1)$ is encoded as $([3] + [0, 2] + [0, 0, 5]) * ([0, 1] + [1])$. The program carries out the operations producing the intermediate expression $([3, 2, 5]) * ([1, 1])$ and then $[3, 5, 7, 5]$. The last expression is converted into the expanded polynomial $3 + 5x + 7x^2 + 5x^3$.

The function in this subsection converts each term cx^e in the expression into bracketed terms $[0, 0, \ldots, 0, c]$ (e zeros). (Constant terms are processed later.) It uses a Do While loop to find the variable x. Having found one it extracts the coefficient c and the exponent e and creates the bracketed term, which is then substituted for cx^e. The substitution is carried out by the function `RemoveInsertString` of Section 6.2. The following variables are used in the process:

- `expr`: The polynomial expression.

- `coeff`: The coefficient c of x (set to one if none).

- `exp`: The exponent e of x (set to one if none).

- `startpos`: The position in `expr` of the beginning of the coefficient.

- `endpos`: The position in `expr` of the end of the exponent.

- code: The bracketed term $[0, 0, \ldots, 0, c]$.

```
Function Convert(expr As String) As String
    Dim k As Integer, ch As String, coeff As String, exp As String
    Dim code As String, startpos As Integer, endpos As Integer
    k = 1
    Do While k <= Len(expr)              'run through expr to find x's
        ch = Midd(expr, k, 1)                    'current character
        If ch <> "x" Then GoTo continue       'if not an x, move on
        coeff = GetCoeff(expr, k, startpos)     'coefficient of x
        exp = GetExp(expr, k, endpos)           'exponent of x
        code = MakePolyCode(coeff, exp)
        expr = RemoveInsertString(expr, startpos, endpos, code)
continue:
        k = k + 1
    Loop
    Convert = InsertPolyAsterisks(expr)      'for ease of multiplication
End Function
```

The function GetCoeff is passed the polynomial expression and the position of an x and returns its coefficient. It also assigns a value to startpos.

```
Function GetCoeff(expr As String, k As Integer, _
                  startpos As Integer) As String
    Dim coeff As String, j As Integer
    ch = Midd(expr, k - 1, 1)                  'character before x
    If InStr("0123456789.", ch) = 0 Then      'if no coefficient then
        coeff = "1": startpos = k       'set coeff. to 1 and startpos at x
    Else
        j = k - 1                        start with character before x
        Do While j > 0 And InStr("0123456789.", ch) > 0
            coeff = ch & coeff                    'build coeff by
            j = j - 1                             'moving backwards
            ch = Midd(expr, j, 1)
        Loop
        startpos = j + 1          'position of first character in coeff
    End If
    GetCoeff = coeff
End Function
```

The function GetExp is passed the polynomial expression and the position k of an x and returns its exponent. It also assigns a value to endpos.

```
Function GetExp(expr As String, k As Integer, endpos As Integer) As String
    Dim exp As String, j As Integer
    ch = Midd(expr, k + 1, 1)                  'character after x
    If ch <> "^" Then                          'if no exponent then
        exp = "1": endpos = k              set exp to 1 and endpos at x
    Else
        j = k + 2                                      'skip ^
        If Midd(expr, j, 1) = "(" Then j = j + 1      'skip left paren
```

```
      ch = Midd(expr, j, 1)                    'start with 1st char of exp
      Do While j <= Len(expr) And InStr("123456789", ch) > 0
         exp = exp & ch                        'build exponent by
         j = j + 1                             'moving forwards
         ch = Midd(expr, j, 1)
      Loop
      endpos = j - 1                           'at last char of exp
   End If
   GetExp = exp
End Function
```

The function `MakePolyCode` is passed the coefficient and exponent and returns the desired bracketed expression. It uses the function `DupString` of Section 6.1 to fill with zeros.

```
Function MakePolyCode(coeff, exp) As String
   Dim zeros As String
   zeros = DupString("0", CInt(exp), ",")             '= 0,0,...,0
   MakePolyCode = "[" & zeros & "," & coeff & "]"     '= [0,0,...,0,exp]
End Function
```

The function `InsertPolyAsterisks` inserts asterisks between bracketed terms to aid in multiplication of polynomials.

```
Function InsertPolyAsterisks(expr As String) As String
   expr = Replace(expr, ")(", ")*("): expr = Replace(expr, "](", "]*(")
   expr = Replace(expr, ")[", ")*["): expr = Replace(expr, "][", "]*[")
   InsertPolyAsterisks = expr
End Function
```

Assigning the Calculations

The function `PolyEval` is similar to `ComplexEval`. A Do While loop reads the symbols of the expression. In each iteration of the loop, a Select-Case statement determines the next calculation. The variable `mode`, initially set to zero, ensures the correct hierarchy of operations. The procedure uses `ExtractString` of Section 6.6 to extract constants.

```
Function PolyEval(expr As String, idx As Integer, mode As Integer, _
              error As Boolean) As String
   Dim n As Integer, P As String, Q As String, ch As String
   Dim z As String, LonelyOp As Boolean      'for cases like "(-x^3+2)"
   Do While idx <= Len(expr) And Not error
      ch = Midd(expr, idx, 1)                 'character at position idx
      If InStr("0123456789.", ch) > 0 Then ch = "#"           'number
      Select Case ch
         Case Is = "#"              'extract constant; idx advanced
            P = ExtractString(expr, idx, ".1234567890")
            P = "[" & P & "]"       'bracket for later computations
         Case Is = "["                          'extract term
```

```
                P = ExtractString(expr, idx, "[].1234567890,")
        Case Is = "+"
                If mode > 0 Then Exit Do              'wait for higher mode
                LonelyOp = (idx = 1 Or Midd(expr, idx - 1, 1) = "(")
                idx = idx + 1
                Q = PolyEval(expr, idx, 0, error)          'lowest precedence
                If LonelyOp Then P = "[0]"                  'zero polynomial
                P = PolySumDiff(P, Q, 1)                       'add the pols
        Case Is = "-"
                If mode > 0 Then Exit Do                     'wait: mode = 1
                LonelyOp = (idx = 1 Or Midd(expr, idx - 1, 1) = "(")
                idx = idx + 1
                Q = PolyEval(expr, idx, 1, error)
                If LonelyOp Then P = "[0]"                   'zero polynomial
                P = PolySumDiff(P, Q, -1)                     'subtract pols
        Case Is = "*"
                idx = idx + 1
                Q = PolyEval(expr, idx, 1, error) 'otherwise, get product
                P = PolyProd(P, Q)                           'multiply pols
        Case Is = "^"
                idx = idx + 1                                    'skip ^
                If Mid(expr, idx, 1) = "(" Then idx = idx + 1
                n = CInt(ExtractString(expr, idx, "123456789"))
                If Mid(expr, idx, 1) = ")" Then idx = idx + 1
                P = PolyPower(P, n)
        Case Is = "("
                idx = idx + 1                                    'skip "("
                P = PolyEval(expr, idx, 0, error)
                idx = idx + 1                                    'skip")"
        Case Is = ")": Exit Do
        Case Else: error = True: Exit Do
        End Select
    Loop
    PolyEval = P                                      'return calculation
End Function
```

The Calculations

The operations of addition, subtraction, and multiplication of polynomials take a pair of encoded polynomials P,Q, convert them into arrays for convenience, perform the operation, and convert the resulting array into encoded square bracket form.

```
Function PolySumDiff(P As String, Q As String, sign As Integer) As String
    Dim i As Integer, m As Integer, RArr() As Double
    Dim PArr() As String, QArr() As String
    Call CreateArrays(P, Q, PArr(), QArr())           'convert to arrays
    m = UBound(PArr)
    ReDim RArr(0 To m)                                'memory for result
    For i = 0 To m       'do the calculation: + if sign = 1, - if sign = 1
```

```
            RArr(i) = CDbl(PArr(i)) + sign * CDbl(QArr(i))
        Next i
        PolySumDiff = ArrayToPolyCode(RArr)            'convert back to code
    End Function


    Function PolyProd(P As String, Q As String) As String
        Dim i As Integer, m As Integer, RArr() As Double
        Dim PArr() As String, QArr() As String
        Call CreateArrays(P, Q, PArr(), QArr())
        m = UBound(PArr)
        ReDim RArr(0 To m)
        For k = 0 To m
            For i = 0 To UBound(PArr)
                For j = 0 To UBound(QArr)
                    If i + j = k Then _
                    RArr(k) = RArr(k) + CDbl(PArr(i)) * CDbl(QArr(j))
                Next j
            Next i
        Next k
        PolyProd = ArrayToPolyCode(RArr)
    End Function
```

The function `PolyPower` keeps multiplying a polynomial by itself the requisite number of times.

```
    Function PolyPower(P As String, n As Integer) As String
        Dim Q As String, i As Integer
        Q = P
        For i = 1 To n - 1
            Q = PolyProd(P, Q)
        Next i
        PolyPower = Q
    End Function
```

The procedure `CreateArrays` takes a pair `P,Q` of encoded polynomials and returns a pair `PArr,QArr` of corresponding arrays of the same size, which we take as the sum of the degrees. For example, if `P = [0,5]` and `Q = [0,0,7]` then `PArr`$= \{0, 5, 0, 0, 0\}$ and `QArr`$= \{0, 0, 7, 0, 0\}$.

```
    Sub CreateArrays(ByVal P As String, ByVal Q As String, _
                     PArr() As String, QArr() As String)
        Dim i As Integer, j As Integer, k As Integer, m As Integer
        Dim PArr1() As String, QArr1() As String
        P = Midd(P, 2, Len(P)-2)
        Q = Midd(Q, 2, Len(Q)-2)                        'chop brackets
        PArr1 = Split(P, ",")                           'make the arrays
        QArr1 = Split(Q, ",")
        m = UBound(PArr1) + UBound(QArr1)
        ReDim PArr(0 To m)
        ReDim QArr(0 To m)                              'memory for expanded arrays
        For i = 0 To m
```

```
        PArr(i) = "0": QArr(i) = "0"                              'default
        If i <= UBound(PArr1) Then PArr(i) = PArr1(i)            'copy
        If i <= UBound(QArr1) Then QArr(i) = QArr1(i)
    Next i
End Sub
```

The function `ArrayToPolyCode` takes the array obtained from a coded polynomial and converts it into an encoded polynomial. For example, the array $\{0, 0, 2, 0, 0, -5.7\}$ is converted into the string $[0,0,2,0,0,-5.7]$.

```
Function ArrayToPolyCode(RArr() As Double) As String
    Dim i As Integer, R As String
    R = "["
    For i = 0 To UBound(RArr)
        R = R & RArr(i) & ","
    Next i
    ArrayToPolyCode = Mid(R, 1, Len(R) - 1) & "]"
End Function
```

Decoding a Coded Polynomial

The function `DecodePoly` takes an encoded polynomial and converts it to standard form. For example, $[2, 0, 0, 5, 0, -1]$ is converted to $2 + 5x^3 - x^5$.

```
Function DecodePoly(P As String) As String
    Dim PArr() As String, i As Integer, str As String, power As String
    P = Midd(P, 2, Len(P) - 2)                      'strip the brackets
    PArr = Split(P, ",")                        'make an array of coefficients
    str = PArr(0)                                   'constant term
    For i = 1 To UBound(PArr)                        'run through the array
        If PArr(i) = "0" Then GoTo continue         'omit zero terms
        power = "^" & i                              'default
        If PArr(i) = "1" Then PArr(i) = ""          'omit coefficient 1
        If PArr(i) = "-1" Then PArr(i) = "-"         'similar
        If i = 1 Then power = ""                     'omit exponent 1
        str = str & "+" & PArr(i) & "x" & power      'make the term
continue:
    Next i
    str = Replace(str, "++", "+")     'rectify possible ++ in making term
    str = Replace(str, "+-", "-")                    'similar
    If Mid(str, 1, 1) = "0" And Len(str) > 1 Then
        str = Mid(str, 2)                    'remove initial character "0"
    ElseIf Mid(str, 1, 1) = "+" Then
        str = Mid(str, 2)                    'remove initial character "+"
    End If
    DecodePoly = str
End Function
```

Checking for Errors

The function `ErrorCheck` is similar to the eponymous procedure in the module `ComplexEval` in that it checks pairs of characters. The function returns `True` if any of the following rules are violated.

1. Operators +, - must be followed by a left parenthesis, a period, a digit, or the letter x.

2. The operator ^ must be followed by a left parenthesis or a positive digit.

3. The operator ^ must be preceded by a right parenthesis or the letter x.

4. A right parenthesis must be followed by a right or left parenthesis, one of the operators +-^, a digit, a period, the letter x, or nothing.

5. A left parenthesis cannot be followed by ^.

6. There must be the same number of left parentheses as right parentheses.

```
Function ErrorCheck(expr As String) As Boolean
    Dim s(-1 To 1) As String, i As Integer, e(1 To 6) As Boolean
    Dim err As Boolean, L As Integer
    L = Len(expr)
    For i = 2 To L
        s(0) = Midd(expr, i, 1)                         'current symbol
        s(-1) = Midd(expr, i - 1, 1)                    'its predecessor
        s(1) = Midd(expr, i + 1, 1)                     'its successor
        e(1) = InStr("+-^", s(0)) > 0 And (i = L Or _
            InStr("0123456789x.(", s(1)) = 0)
        e(2) = (s(0) = "^") And InStr("123456789(", s(1)) = 0
        e(3) = (s(0) = "^") And InStr(")x", s(-1)) = 0
        e(4) = s(0) = ")" And InStr("+-^(0123456789x)", s(1)) = 0
        e(5) = s(0) = "(" And InStr("^", s(1)) > 0
        e(6) = Not IsMatch(expr, "(", ")")
        err = e(1) Or e(2) Or e(3) Or e(4) Or e(5) Or e(6)
        If err Then Exit For
    Next i
    ErrorCheck = err
End Function
```

8.6 Exercises

1. Write a program `SumRecursion(n)` that uses recursion to calculate the sum of the first n positive integers. Check that this sum is $n(n+1)/2$.

2. Write a recursive function `PowerRecursion(b,p)` that returns b^p, where b is a real number and p is a positive integer.

3. Write a function `OddFactRecursive(n)` that returns $1 \cdot 3 \cdots (2n-1)$. Use this function together with `Factorial` to test the identity

$$1 \cdot 3 \cdots (2n-1) = \frac{(2n)!}{2^n n!}.$$

4. A sequence $\{a_n\}$ is defined recursively by

$$a_{n+1} = 1 + \frac{1}{a_n}, \quad n = 1, 2, \ldots, \quad a_1 = 1.$$

   Implement this as a recursive procedure `Sequence1(n)`. Include in the code a step that prints the current value of a_n. Can you make a conjecture?

5. Write a recursive function `Sequence2(r,n)` that returns a_n for a positive integer n, where $r > 0$ and a_n is defined by the scheme $a_1 = 1$ and $a_n = (a_{n-1} + r/a_{n-1})/2$. Include in the code a step that prints the current value of a_n in a column. Can you make a conjecture?

6. Write a recursive function `Fibonacci(n)` that returns the nth Fibonacci number F_n for a given n. Use the function to calculate the ratio F_{n+1}/F_n for $n = 1, 2, \ldots$. Observe that the ratio get closer and closer to the *golden ratio* $(1 + \sqrt{5})/2$.

7. (Padovan sequence). This is the sequence a_n defined recursively by

$$a_0 = a_1 = a_2 = 1, \quad a_n = a_{n-2} + a_{n-3}.$$

   The first few values of a_n are

$$1, 1, 1, 2, 2, 3, 4, 5, 7, 9, 12, 16, 21, 28, 37, 49, 65, 86, 114, 151, \ldots$$

   Write a recursive function `Padovan` that returns the n member of the sequence. The sequence is named after Richard Padovan who attributed its discovery to Hans van der Laan in 1994.

   The sequence has many interesting algebraic, geometric, and combinatorial properties including, for example, that equilateral triangles whose side lengths follow the Padovan sequence may be positioned to form spiral (Figure 8.9).

8. Write a recursion program `BinomCoeff(n,k)` that returns the *binomial coefficient*

$$\binom{n}{k} = \frac{n!}{k!(n-k)!}, \quad k = 0, 1, \ldots, n.$$

   Use the recursive identity

$$\binom{n}{k} = \binom{n-1}{k} + \binom{n-1}{k-1}, \quad n \geq 1, \ 1 \leq k \leq n-1,$$

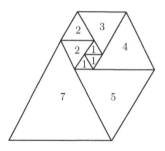

FIGURE 8.9: Padovan triangles.

9. Write a recursion program `RecursionMax` to find the maximum of an array of numbers. Use the principle that the maximum of the list $\{a_1, a_2, \ldots, a_n\}$ is either a_n or the maximum of $\{a_n, a_2, \ldots, a_{n-1}\}$. Print out in a column the steps of the recursion.

10. The *McCarthy 91 function* was defined by John McCarthy in 1970 as a test case for formal verification within computer science. It is defined recursively by

$$M(n) = \begin{cases} M(n) = n - 10 & \text{if } n > 100, \\ M\big(M(n + 11)\big) & \text{if } n \leq 100, \end{cases}$$

where $n = 1, 2, \ldots$ Write a recursive program `M91` that implements the function. Call `M91(1)` and print out in a column the steps of the recursion.

11. The *Catalan numbers* $C_0, C_1, C_2, \ldots$ are defined recursively by

$$C_0 = 1 \quad \text{and} \quad C_{n+1} = \sum_{i=0}^{n-1} C_i C_{n-i-1}, \ n \geq 0.$$

The first few Catalan numbers in the sequence are 1, 1, 2, 5, 14, 42, 132, 429, 1430. The numbers C_n occur in many counting problems, for example,

- the number of ways a convex polygon of $n + 2$ sides can split into triangles by connecting vertices.

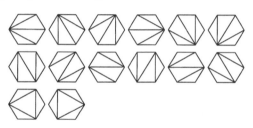

- the number of correct matchings of n pairs of parentheses. For n = 3, the matchings are ((())), ()(()), ()()(), (())(), (()()).

- the number of strings with n X's and n Y's such that no segment starting at the beginning of the string has more Y's than X's. Such a string is called a *Dyck word*. For example, for $n = 3$ the Dyck words are XXXYYY, XYXXYY, XYXYXY, XXYYXY, and XXYXYY.

Write a recursive function `Catalan` that returns C_n for input n.

12. (Quaternion calculator). A *quaternion* is an expression of the form

$$q = a1 + bi + cj + dk,$$

where a, b, c, d are real numbers and $1, i, j, k$ are algebraic symbols with the multiplication properties

$$1^2 = 1, \ i^2 = j^2 = k^2 = ijk = -1,$$
$$ij = k, \ jk = i, \ ki = j$$
$$ji = -k, \ kj = -i, \ ik = -j,$$
$$1i = i1 = i, \ 1j = j1 = j, \ 1k = k1 = k.$$

Quaternions are also written as $a + bi + cj + dk$, without the symbol 1. They were invented by the Irish mathematician William Rowan Hamilton in 1843, who applied them to three dimensional mechanics. They now have uses in computer graphics.

Quaternions may be added, subtracted, and multiplied in a manner similar to complex numbers, namely by carrying out the operations using ordinary arithmetic, reducing the expression using the above properties, and then collecting terms. This yields the formulas

$$q \pm q' = (a \pm a') + (b \pm b')i + (c \pm c')j + (d \pm d')k$$

and

$$qq' = (aa' - bb' - cc' - dd') + (ab' + ba' + cd' - dc')i$$
$$+ (ac' - bd' + ca' + db')j + (ad' + bc' - cb' + da')k$$

It follows from the last formula that 1 is an *identity* with respect to multiplication, that is, $q1 = 1q = q$ for all q. In general, it is *not* true that $qq' = q'q$, that is, multiplication is not *commutative*. If the components a, b, c, d of a quaternion q are not all zero, then q has an *inverse* with respect to multiplication that is, a quaternion q^{-1} such that

$$qq^{-1} = q^{-1}q = 1.$$

The inverse is given by the formula

$$a + bi + cj + dk = \frac{1}{a^2 + b^2 + c^2 + d^2}(a - bi - cj - dk).$$

Division then has two definitions: $q_1 q_2^{-1}$ and $q_2^{-1} q_1$

Write a program `QuaternionCalc` analogous to `ComplexCalc` that evaluates algebraic expressions involving quaternions. For simplicity of coding, represent a quaternion $a + bi + cj + dk$ as $[a, b, c, d]$.

*13. Write a program `ComplexPolyEval` that allows complex numbers as coefficients and constants.

Chapter 9

Charts and Graphs

Excel VBA has a powerful facility that can create a variety of charts and graphs. In this chapter we describe through examples the versions that we shall need in the book.

9.1 Frequency Charts

A (vertical) *frequency chart* is a graphical display of data using vertical bars of different heights to represent the frequency of data placed along a horizontal axis. We illustrate with a program that calculates the frequency of letters in text.

The program `LetterFrequency` below takes text entered in cell B2, finds the number of times each lower case letter of the alphabet appears, and then draws the corresponding frequency chart. For example, running the program with the following verse by Tennessee Williams[1] produces the chart in Figure 9.1.

> *Oh courage! Could you not as well*
> *Select a second place to dwell*
> *Not only in that golden tree*
> *But in the frightened heart of me*

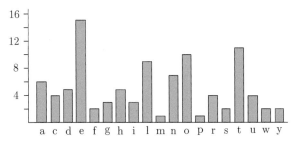

FIGURE 9.1: Frequencies of small letters in verse.

---

[1]The entire poem may be found in his play *The Night of the Iguana*.

DOI: 10.1201/9781003351689-9

The program first clears any chart from a previous run by using the VBA command `ChartObjects.Delete`. The array `LetterArr(1 To 26)` holds the number of times a particular letter is encountered in the text. Thus `LetterArr(1)` holds the "a"-count, `LetterArr(2)` the "b"-count, etc. (we consider only lower case letters). A For Next loop extracts a character `symbol` from `text` and finds its ASCII value `AscVal`. If the symbol is a lower case letter then `LetterArr(AscVal)` is incremented by one. After the frequencies have been tabulated, a second For Next loop prints the letters a to z in column B and the frequencies in column C. The program then invokes the chart facility.

The With End With statement in the code is used to obviate repetition of the prefix Sheet1. The first few parts of the statement tell the program the dimensions of the chart and where to place it on the spreadsheet. A little experimentation may be required in applications. The next part specifies the type of chart, namely `xlColumnClustered`. After this the title is set. The Range statements in the code inform the program where the data is located, in this case column B from row 4 to 29 for the x data that will appear along the horizontal axis, and column C from row 4 to 29 for the y data that will appear along the vertical axis. These statements might seem obscure but the reader may simply use this and other examples as templates.

```
Sub LetterFreq()
    Dim text As String, AscVal As Integer, i As Integer, chrt As Chart
    Dim LetterArr(97 To 122) As Integer
    Range("C4:C29").ClearContents                    'clear old data
    If ChartObjects.Count > 0 Then ChartObjects.Delete       'and chart
    text = Range("B2").Value                          'get the text
    For i = 1 To Len(text)                    'extract letters from text
        AscVal = Asc(Mid(text, i, 1))        'ASCII code for current letter
        If 97 <= AscVal And AscVal <= 122 Then      'if lower case letter
            LetterArr(AscVal) = LetterArr(AscVal) + 1   'update frequency
        End If
    Next i
    For i = 0 To 25              'print letter and frequency of occurrence
        Cells(4 + i,2) = Chr(97 + i): Cells(4 + i,3) = LetterArr(97 + i)
    Next i
    With Sheet1 _
     .ChartObjects.Add(Left:=200, Width:=450, Top:=50, Height:=250)
     .Chart.ChartType = xlColumnClustered
     .Chart.HasTitle = True
     .Chart.ChartTitle.text = "Letter Frequency"
     .Chart.SetSourceData Source:= Range("C" & 4 & ":" & "C" & 29)
     .Chart.SeriesCollection(1).XValues = Range("B" & 4 & ":" & "B" & 29)
     .Chart.Axes(xlCategory).HasTitle = True
     .Chart.Axes(xlCategory).AxisTitle.Text = "Letters"       'x-axis label
     .Chart.Axes(xlValue, xlPrimary).HasTitle = True
     .Chart.Axes(xlValue, xlPrimary).AxisTitle.Characters.Text = _
        "Relative Frequency"                          'y-axis label
    End With
End Sub
```

9.2 Drawing Lines

The program in this section reads the location of numbers $1, 2, \ldots, n$ entered by the user into cells of the spreadsheet and then draws the polygonal line connecting the vertices in the given order. Figure 9.2 illustrates sample input and corresponding output. The grid is given a coordinate system that is heeded by the output graph.

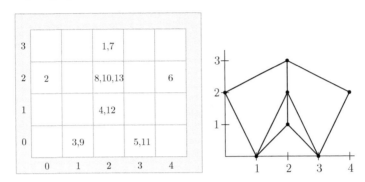

FIGURE 9.2: Input and output for GraphPoly.

The main procedure `GraphPoly` first sets up the grid and provides it with a coordinate system using `MakeGrid` and `InstallCoord` of Chapter 7. The function `GenerateCoord` converts the row-column spreadsheet positions of the entered numbers into grid coordinates, prints these in columns A and B starting in row 7, and returns the row of the last position printed. The chart facility uses this information to graph the polygonal line using the Range statement.

```
Sub GraphPoly()
    Dim Lastrow As Integer, Grid As TGrid
    If ChartObjects.Count > 0 Then ChartObjects.Delete
    Call ClearColumns(7, 1, 2)           'clear cols 1,2 starting at row 7
    Grid = MakeGrid(2, 4, 35, 45, 2.3, 15, 24)
    Call InstallCoord(Grid, Grid.B, Grid.L)           'coordinatize grid
    Lastrow = GenerateCoord(Grid)         'print coord. of entered vertices
    With Sheet1 _
      .ChartObjects.Add(Left:=835, Width:=700, Top:=50, Height:=500)
      .Chart.ChartType = xlXYScatterLines
      .Chart.HasTitle = True
      .Chart.ChartTitle.Text = "Graph Polygon"
      .Chart.SetSourceData Source:=Range("B" & 7 & ":" & "B" & Lastrow)
      .Chart.SeriesCollection(1).XValues = _
              Range("A" & 7 & ":" & "A" & Lastrow)
      .Chart.SeriesCollection(1).MarkerStyle = xlNone 'no markers on graph
    End With
End Sub
```

The function `GenerateCoord` below scans the grid for cells containing positive integers, which are the vertex labels. If `Cell(i,j)` contains a sequence of integers, then the cell is to be visited several times. The numbers in a cell are placed in an array `NumberArray` using the VBA function `Split`. The coordinates of `Cell(i,j)` are computed and printed in columns A and B in the order matching that of the vertex numbers.

```
Private Function GenerateCoord(Grid As TGrid)
    Dim i As Integer, j As Integer, k As Integer, n As Integer
    Dim Lastrow As Integer, NumberSequence As String
    Dim NumberArray() As String
    Lastrow = 7
    For i = Grid.T To Grid.B                'scan grid for number sequence
        For j = Grid.L To Grid.R
            If Not IsEmpty(Cells(i, j)) Then
                NumberSequence = Cells(i, j).Value          'numbers
                NumberArray = Split(NumberSequence, ",")    'make an array
                For k = 0 To UBound(NumberArray)
                    n = NumberArray(k)     'kth vertex number in cell(i,j)
                    Cells(6 + n, 1).Value = j - Grid.L   'coords of vertex
                    Cells(6 + n, 2).Value = Grid.B - i
                    Lastrow = Lastrow + 1
                Next k
            End If
        Next j
    Next i
    GenerateCoord = Lastrow                      'return last data row
End Function
```

The procedure `ClearColumns` clears any data generated by an earlier run. It uses `Cells(Rows.count, 2).End(xlUp).row` to find the row of the last entry in col1 and then clears the block `(row,col1):(lrow,col2)`.

```
Sub ClearColumns(row As Integer, col1 As Integer, col2 As Integer)
    Dim lrow As Long
    If IsEmpty(Cells(row, col1)) Then Exit Sub
    lrow = Cells(Rows.Count, col1).End(xlUp).row
    Range(Cells(row, col1), Cells(lrow, col2)).ClearContents
End Sub
```

9.3   Intersection of Two Lines

In this section we construct a procedure that calculates the point of intersection of two (non parallel) lines given two points on each line. Specifically, the function `LineIntersect(P_1(), Q_1(), P_2(), Q_2(), error)` returns the point of intersection (written as an array) of lines L_1 and L_2 that are specified, respectively, by pairs of points $(P_1(1), P_1(2)), (Q_1(1), Q_1(2))$ and $(P_2(1), P_2(2)),$ $(Q_2(1), Q_2(2))$, as shown in Figure 9.3. We assume that either $P_1(1) \neq Q_1(1)$ or $P_2(1) \neq Q_2(1)$ (not both lines are vertical) and that either $P_1(2) \neq Q_1(2)$ or $P_2(2) \neq Q_2(2)$ (not both lines are horizontal).

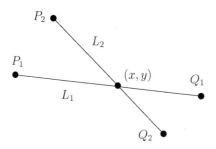

FIGURE 9.3: Intersection of two lines.

The function considers the following cases:

- $P_1(1) = Q_1(1)$ and $P_2(1) \neq Q_2(1)$ (L_1 is vertical). The lines then have the equations

$$L_1 : \; x = P_1(1) \quad \text{and} \quad L_2 : y = P_2(2) + m_2(x - P_2(1)), \text{ where}$$

$$m_2 = \frac{P_2(2) - Q_2(2)}{P_2(1) - Q_2(1)}.$$

Thus the point of intersection is $(P_1(1), P_2(2) + m_2(P_1(1) - P_2(1)))$.

- $P_2(1) = Q_2(1)$ and $P_1(1) \neq Q_1(1)$ (L_2 is vertical). The lines then have the equations

$$L_2 : \; x = P_2(1) \quad \text{and} \quad L_1 : y = P_1(2) + m_1(x - P_1(1)) \text{ where}$$

$$m_1 = \frac{P_1(2) - Q_1(2)}{P_1(1) - Q_1(1)}.$$

The point of intersection is then $(P_2(1), P_1(2) + m_1(P_2(1) - P_1(1)))$.

- $P_2(1) \neq Q_2(1)$ and $P_1(1) \neq Q_1(1)$ (neither line is vertical). The lines then have the equations

$$L_1 : \; y = P_1(2) + m_1(x - P_1(1)) \quad \text{and} \quad L_2 : \; y = P_2(2) + m_2(x - P_2(1)).$$

In this case the point of intersection (x, y) is the solution of the system

$$m_1 x - y = m_1 P_1(1) - P_1(2), \quad m_2 x - y = m_2 P_2(1) - P_2(2.) \quad (9.1)$$

One can solve this system by first subtracting the second equation from the first, producing

$$(m_1 - m_2)x = m_1 P_1(1) - m_2 P_2(1) + P_2(2) - P_1(2)$$

If the lines are not parallel, then $m_1 \neq m_2$ and one may divide by $m_1 - m_2$ to obtain

$$x = \frac{m_1 P_1(1) - m_2 P_2(1) + P_2(2) - P_1(2)}{m_1 - m_2}$$

Then y may be obtained from x using either one of the equations found in (9.1):

$$y = m_1(x - P_1(1)) + P_1(2).$$

Here is the code for the function. It is broken up into the cases outlined above. The intersection point is placed in the array `sol`, which is returned by the function.

```
Function LineIntersect(P_1() As Double, Q_1() As Double, _
        P_2() As Double, Q_2() As Double, error As Boolean) As Double()
    Dim sol(1 To 2) As Double, m_1 As Double, m_2 As Double
    If P_1(1) = Q_1(1) And P_2(1) = Q_2(1) Then
        error = True: Exit Function           'reject lines if both vertical
    End If
            'calculate slopes when possible
    If P_1(1) <> Q_1(1) Then m_1 = (P_1(2) - Q_1(2)) / (P_1(1) - Q_1(1))
    If P_2(1) <> Q_2(1) Then m_2 = (P_2(2) - Q_2(2)) / (P_2(1) - Q_2(1))
    If P_1(1) = Q_1(1) Then                    'L1 is vertical
        sol(1) = P_1(1): sol(2) = P_2(2) + m_2 * (P_1(1) - P_2(1))
    ElseIf P_2(1) = Q_2(1) Then                'L2 is vertical
        sol(1) = P_1(1): sol(1) = P_1(2) + m_1 * (P_2(1) - P_1(1))
    ElseIf m_1 = m_2 Then
        error = True: Exit Function                      'lines parallel
    Else
        sol(1) = (m_1 * P_1(1) - m_2 * P_2(1) + P_2(2) _
            - P_1(2))/(m_1 - m_2)
        sol(2) = m_1 * (sol(1) - P_1(1)) + P_1(2)
    End If
    LineIntersect = sol
End Function
```

The following program graphs the line for user-entered points in the spreadsheet. For this we use the graph type `xlXYScatterLines`.

```
Sub GraphLineIntersect()
    Dim XRange As String, YRange As String
    Dim P_1(1 To 2) As Double, Q_1(1 To 2) As Double  'arrays for points
    Dim P_2(1 To 2) As Double, Q_2(1 To 2) As Double
    Dim Intersection(1 To 2) As Double, error As Boolean
    If ChartObjects.Count > 0 Then ChartObjects.Delete   'clear old chts
    Call ClearColumns(11, 1, 2)    'clear columns 1,2 starting at row 11
            'read in line2 points
    P_1(1) = Cells(6, 1).Value: P_1(2) = Cells(6, 2).Value
    Q_1(1) = Cells(7, 1).Value: Q_1(2) = Cells(7, 2).Value
            'read in line2 points
```

```
P_2(1) = Cells(9, 1).Value: P_2(2) = Cells(9, 2).Value
Q_2(1) = Cells(10, 1).Value: Q_2(2) = Cells(10, 2).Value
sol = LineIntersect(P_1, Q_1, P_2, Q_2, error) 'get intersection pt.
If error Then MsgBox "Check Data": Exit Sub
           'list the Xdata and Ydata for drawing the lines
Cells(12, 1).Value = P_1(1): Cells(12, 2).Value = P_1(2)      'line 1
Cells(13, 1).Value = Q_1(1): Cells(13, 2).Value = Q_1(2)
Cells(14, 1).Value = sol(1): Cells(14, 2).Value = sol(2)
Cells(15, 1).Value = P_2(1): Cells(15, 2).Value = P_2(2)      'line 2
Cells(16, 1).Value = Q_2(1): Cells(16, 2).Value = Q_2(2)
XRange = "A" & 12 & ":" & "A" & 16
YRange = "B" & 12 & ":" & "B" & 16
With Sheet1 _
    .ChartObjects.Add(Left:=280, Width:=460, Top:=15, Height:=400)
    .Chart.SetSourceData Source:=Range(YRange)               'y values
    .Chart.ChartType = xlXYScatterLines
    .Chart.HasTitle = True
    .Chart.ChartTitle.Text = "Line Intersection"
    .Chart.SeriesCollection(1).XValues = Range(XRange)       'x values
End With
End Sub
```

9.4 Projection of a Point onto a Line

The procedure in this section takes three points P, Q, R, written as 1 by 2 arrays, and determines the point S on the line through the first two points that is the perpendicular projection of the third point, as shown in Figure 9.4.

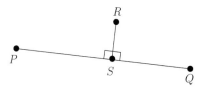

FIGURE 9.4: Perpendicular projection of a point onto a line.

The function uses LineIntersect and the fact that the slopes of perpendicular lines are negative reciprocals of one another.

```
Function Point2Line(P() As Double, Q() As Double, R() As Double, _
                error As Boolean) As Double()
    Dim m As Double, sol() As Double, S(1 To 2) As Double
    If P(1) = Q(1) Then                          'if line is vertical
        ReDim sol(1 To 2)                        'array for projected point
        sol(1) = P(1): sol(2) = R(2)                'projected point
```

```
    ElseIf P(2) = Q(2) Then                        'if line horizontal
        ReDim sol(1 To 2)                       'array for projected point
        sol(1) = R(1): sol(2) = P(2)                'projected point
    Else
        n = (P(2) - Q(2)) / (P(1) - Q(1))            'slope of given line
        m = -1 / n                              'its negative reciprocal
        'get another point on perpendicular line so we have two points
        S(1) = R(1) + 1: S(2) = m + R(2)
        sol = LineIntersect(P, Q, R, S, error)        'return intersection
    End If
    Point2Line = sol
End Function
```

The following program graphs the line for user entered points in the spread-sheet.

```
Sub GraphPoint2Line()
    Dim XRange As String, YRange As String, error As Boolean
    Dim P(1 To 2) As Double, Q(1 To 2) As Double, R(1 To 2) As Double
    Dim sol() As Double
    If ChartObjects.Count > 0 Then ChartObjects.Delete     'clear old chts
    Call ClearColumns(10, 1, 2)     'clear columns 1,2 'starting at row 10
    P(1) = Cells(6, 1).Value: P(2)= Cells(6, 2).Value      'read given line
    Q(1) = Cells(7, 1).Value: Q(2)= Cells(7, 2).Value   'determined by P,Q
    R(1) = Cells(8, 1).Value: R(2)= Cells(8, 2).Value      'read given point
    sol = Point2Line(P, Q, R, error) 'point on line
    If error Then Exit Sub
    Cells(10, 1).Value = P(1):   Cells(10, 2).Value = P(2) 'place data in
    Cells(11, 1).Value = Q(1):   Cells(11, 2).Value = Q(2)     'cells for
    Cells(12, 1).Value = sol(1): Cells(12, 2).Value = sol(2)    'graphing
    Cells(13, 1).Value = R(1):   Cells(13, 2).Value = R(2)
    XRange = "A" & 10 & ":" & "A" & 13
    YRange = "B" & 10 & ":" & "B" & 13
    With Sheet1 _
        .ChartObjects.Add(Left:=280, Width:=400, Top:=15, Height:=400)
        .Chart.SetSourceData Source:=Range(YRange)              'y values
        .Chart.ChartType = xlXYScatterLines             'chart type
        .Chart.HasTitle = True
        .Chart.ChartTitle.Text = ""
        .Chart.SeriesCollection(1).XValues = Range(XRange)        'x values
    End With
End Sub
```

9.5   The Incenter of a Triangle

The *incenter* of a triangle with vertices P, Q, R is intersection of the angle bisectors of the triangle and is also the center of inscribed circle. The incenter may be found as follows: Let the triangle have vertices P, Q, and R and let d_{RQ} denote the distance between the vertices R and Q. Let X be the point along the line from R to P at a distance d_{RQ} from R. The midpoint Y of the line from S to Q is then on the bisector L_R of the angle at R. L_Q may be similarly calculated. `LineIntersect` may then be used find the intersection of L_R and L_Q. Figure 9.5 illustrates the construction

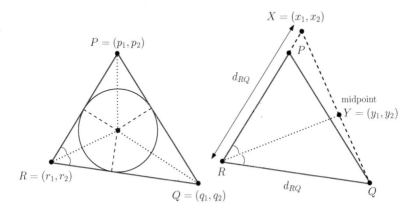

FIGURE 9.5: Incenter of a triangle.

The point $X = (x_1, x_2)$ is found by laying off the distance d_{RQ} in the direction of the line from R to P. Using the Euclidean distance formula we have

$$d_{RQ} = \sqrt{(r_1 - q_1)^2 + (r_2 - q_2)^2}$$

The point X is on the parametric line

$$(1 - t)r_1 + tp_1, (1 - t)r_2 + tp_2 = (t(p_1 - r_1) + r_1, t(p_2 - r_2) + r_2) \qquad (9.2)$$

so for suitable t

$$x_1 = (1 - t)r_1 + tp_1 = t(p_1 - r_1) + r_1, \quad x_2 = (1 - t)r_2 + tp_2 = t(p_2 - r_2) + r_2$$

The parameter t must be chosen so that the distance from X to R is d_{RQ}:

$$\sqrt{t^2[(p_1 - r_1)^2 + (p_2 - r_2)^2]} = d_{RQ}.$$

Solving for t we have

$$t = \frac{d_{RQ}}{\sqrt{(p_1 - r_1)^2 + (p_2 - r_2)^2}}, \qquad (9.3)$$

The desired point Y is the average of the coordinates (x_1, x_2) and (q_1, q_2). We have thus obtained a point on the bisector of the angle at R:

$$Y = \left(\tfrac{1}{2}[t(p_1 - r_1) + r_1 + q_1], \tfrac{1}{2}[t(p_2 - r_2) + r_2 + q_2]\right), \qquad (9.4)$$

where t is given by (9.3). Switching the symbols r and q in this formula gives a point Z (not shown) on the bisector of the angle at Q:

$$Z = \left(\tfrac{1}{2}t[(p_1 - q_1) + r_1 + q_1], \tfrac{1}{2}[t(p_2 - q_2) + r_2 + q_2]\right) \tag{9.5}$$

where now

$$t = \frac{d_{RQ}}{\sqrt{(p_1 - q_1)^2 + (p_2 - q_2)^2}}, \tag{9.6}$$

The function `Incenter` implements this process. It is passed the vertices P, Q, R as arrays and calculates the incenter as the intersection of the angle bisectors from R to Y and Q to Z.

```
Function Incenter(P() As Double, Q() As Double, R() As Double, _
        error As Boolean) As Double()
    Dim d_RQ As Double, t_Y As Double, t_Z As Double
    Dim Y(1 To 2) As Double, Z(1 To 2) As Double
    d_RQ = ((R(1) - Q(1)) ^ 2 + (R(2) - Q(2)) ^ 2) ^ (1 / 2)
    t_Y = (d_RQ / ((P(1) - R(1)) ^ 2 + (P(2) - R(2)) ^ 2)) ^ (1 / 2)
    Y(1) = (t_Y * (P(1) - R(1)) + R(1) + Q(1)) / 2
    Y(2) = (t_Y * (P(2) - R(2)) + R(2) + Q(2)) / 2
    t_Z = (d_RQ / ((P(1) - Q(1)) ^ 2 + (P(2) - Q(2)) ^ 2)) ^ (1 / 2)
    Z(1) = (t_Z * (P(1) - Q(1)) + R(1) + Q(1)) / 2
    Z(2) = (t_Z * (P(2) - Q(2)) + R(2) + Q(2)) / 2
    Incenter = LineIntersect(R(), Y(), Q(), Z(), error)
End Function
```

Here is a program that graphs the line segments from the vertices to the incenter and the perpendicular line segments from the incenter to the sides.

```
Sub GraphIncenter()
    Dim XRange As String, YRange As String
    Dim P() As Double, Q() As Double, R() As Double
    Dim sol() As Double, Grid As TGrid, error As Boolean
    Dim proj_PQ() As Double, proj_QR() As Double, proj_PR() As Double
                'clear old chart and data
    If ChartObjects.Count > 0 Then ChartObjects.Delete
    Call ClearColumns(8, 1, 2)      'clear columns 1,2 starting at row 8
    Grid = MakeGrid(3, 4, 27, 28, 2.3, 15, 24)
    Call InstallCoord(Grid, Grid.B, Grid.L)
    Call GetVertexCoord(Grid, P, Q, R)
    sol = Incenter(P, Q, R, error)
    proj_PQ = Point2Line(P, Q, sol, error)      'proj. from incenter to PQ
    proj_QR = Point2Line(Q, R, sol, error)      'proj. from incenter to QR
    proj_PR = Point2Line(P, R, sol, error)      'proj. from incenter to PR
                'points to be connected by lines
    Cells(8, 1).Value = R(1):       Cells(8, 2).Value = R(2)
    Cells(9, 1).Value = P(1):       Cells(9, 2).Value = P(2)
    Cells(10, 1).Value = Q(1):      Cells(10, 2).Value = Q(2)
    Cells(11, 1).Value = R(1):      Cells(11, 2).Value = R(2)
    Cells(12, 1).Value = sol(1):    Cells(12, 2).Value = sol(2)
```

```
Cells(13, 1).Value = Q(1):        Cells(13, 2).Value = Q(2)
Cells(14, 1).Value = P(1):        Cells(14, 2).Value = P(2)
Cells(15, 1).Value = sol(1):      Cells(15, 2).Value = sol(2)
Cells(16, 1).Value = proj_PQ(1):  Cells(16, 2).Value = proj_PQ(2)
Cells(17, 1).Value = sol(1):      Cells(17, 2).Value = sol(2)
Cells(18, 1).Value = proj_PR(1):  Cells(18, 2).Value = proj_PR(2)
Cells(19, 1).Value = sol(1):      Cells(19, 2).Value = sol(2)
Cells(20, 1).Value = proj_QR(1):  Cells(20, 2).Value = proj_QR(2)
XRange = "A" & 8 & ":" & "A" & 20: YRange = "B" & 8 & ":" & "B" & 20
With Sheet1 _
    .ChartObjects.Add(Left:=580, Width:=460, Top:=15, Height:=400)
    .Chart.SetSourceData Source:=Range(YRange)            'y values
    .Chart.ChartType = xlXYScatterLines
    .Chart.HasTitle = True
    .Chart.ChartTitle.Text = "Incenter"
    .Chart.SeriesCollection(1).XValues = Range(XRange)     'x values
End With
End Sub
```

The procedure `GetVertexCoord` scans the grid for user-entered symbols denoting the vertices, labels the vertices with the letters P, Q and R, and inserts the spreadsheet row, col coordinates into arrays `P, Q, R`, respectively.

```
Private Sub GetVertexCoord(Grid As TGrid, P() As Double, _
            Q() As Double, R() As Double)
    Dim i As Integer, j As Integer, n As Integer
    ReDim P(1 To 2): ReDim Q(1 To 2): ReDim R(1 To 2)
    Dim VX(1 To 3) As Double, VY(1 To 3) As Double       'temp arrays
    n = 1
    For i = Grid.T To Grid.B                    'scan grid for vertices
        For j = Grid.L To Grid.R
            If Not IsEmpty(Cells(i, j)) And n < 4 Then
                If n = 1 Then Cells(i, j).Value = "P"      'label points
                If n = 2 Then Cells(i, j).Value = "Q"
                If n = 3 Then Cells(i, j).Value = "R"
                VX(n) = j - Grid.L: VY(n) = Grid.B - i   'nth X,Y coords.
                n = n + 1
            End If
        Next j
    Next i
    P(1) = VX(1): P(2) = VY(1)
    Q(1) = VX(2): Q(2) = VY(2)
    R(1) = VX(3): R(2) = VY(3)
End Sub
```

9.6   Function Graphs

The graphing facility can draw graphs of functions entered into a cell by the user. We illustrate the method in the following program, which draws a graph of a user-defined function on a specified interval with a specified increment.

The program begins by reading the left and right endpoints of the desired interval and the increment size. This information is used to generate x-values in column C and y-values in column D. The data is then passed to the graphing facility. For example, to graph the function

$$y = 3x^2 + 2/x + 1, \quad 1 \le x \le 2$$

in increments of .01, one enters the values 1, 2 and .01 in cells B3, B4, and B5, respectively, and the expression = 3*x^2 + 2/x + 1 in cell B6 (note the equals sign). The program places the x-values in column 3. The y values are calculated from the x-values by the function `FunctionVal` (Section 4.5) and placed in column 4.

```
Sub GraphFunction()
    Dim LeftPoint As Double, RightPoint As Double, Inc As Double
    Dim NumPoints As Integer, x As Double
    If ChartObjects.Count > 0 Then ChartObjects.Delete  'clear old chart
    Call ClearColumns(8, 3, 4)         'clear cols 3,4 starting from row 8
    LeftPoint = Range("B3").Value
    RightPoint = Range("B4").Value
    Inc = Range("B5").Value: expr = Range("B6").Value
    NumPoints = (RightPoint - LeftPoint) / Inc
    x = LeftPoint
    For j = 1 To NumPoints
        Cells(j + 7, 3).Value = x                 'column of x values
        Cells(j + 7, 4).Value = FunctionEval(expr, "x", x)   'y values
        x = x + Inc
    Next j
    With Sheet1 _
      .ChartObjects.Add(Left:=300, Width:=800, Top:=50, Height:=500)
      .Chart.ChartType = xlLine
      .Chart.HasTitle = True
      .Chart.ChartTitle.Text = "Graph of Function"
      .Chart.SetSourceData Source:= _
            Range("D" & & ":" & "D" & 8 + NumPoints)
      .Chart.SeriesCollection(1).XValues = _
            Range("C" & & ":" & "C" & 8 + NumPoints)
      .Chart.SeriesCollection(1).MarkerStyle = xlNone 'no graph markers
    End With
End Sub
```

The graphing facility generates nice graphs when the increment is small. The reader may wish to enter the following visually interesting wave functions, perhaps experimenting with the parameters, which can dramatically affect the shape of a graph.

- `.05 + (SIN(x))^2 + .04*SIN(50*x)` Use an interval $[0, 5]$ with an increment of .001. The graph is a small, high frequency sine wave riding atop a larger wave.

- `1 + SIN(50*x)*SIN(x)` Use $[0, 5]$ and .001 again. The graph is a high frequency sine wave modulated by a low frequency sine wave.

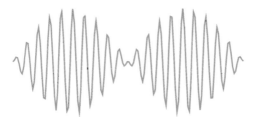

- `1 + SIN(1/x)` Use the interval $[.01, 1]$ and an increment of .001. The graph is a displaced sine wave with decreasing frequency

- `1 + x*SIN(1/x)` Use the interval $[.05, .1]$ and the increment .001. The graph is a displaced sine wave with decreasing frequency and increasing amplitude.

9.7 Fitting a Line to Data

In this section we discuss *scatter plots*, which display points with coordinates (x, y), where x and y are values in data sets. A scatter plot can be of great value, as it may reveal a relationship between the variables. For example, if x is a person's height and y their weight, then a likely correlation exists between x and y, and the nature of that correlation may be inferred from a scatter plot. A scatter plot may also be used to graph parametric equations or to reveal geometric patterns, as in the case of a fractal.

The program in this section makes a scatter plot from either user-entered or randomly generated data and draws the *least squares line*, that is, the line that best fits the data. The terminology reflects how the line is determined, namely by the criterion that the sum of the squares of vertical distances from the line to the data points is minimum. Specifically, given a set of data points

$$(x_1, y_1), (x_2, y_2), \ldots, (x_n, y_n),$$

the least squares line has equation

$$y = mx + b,$$

where the constants m and b, called, respectively, the *slope* and *y-intercept* of the line, are chosen so that the expression

$$\sum_{i=1}^{n} |mx_i + b - y_i|^2,$$

is as small as possible. The absolute value $|mx_i + b - y_i|$ in the sum is the vertical distance from the data point (x_i, y_i) to the line, shown by the double arrow in Figure 9.6. The square of that distance is used so that calculus may be applied to find reasonably simple formulas for m and b:

$$m = \frac{n \sum_{j=1}^{n} x_j y_j - \left(\sum_{j=1}^{n} x_j\right)\left(\sum_{j=1}^{n} y_j\right)}{n \sum_{j=1}^{n} x_j^2 - \left(\sum_{j=1}^{n} x_j\right)^2} \tag{9.7}$$

and

$$b = \frac{n \sum_{j=1}^{n} y_j - m \sum_{j=1}^{n} x_j}{n} \tag{9.8}$$

The least squares line is also called a *trend line*, as it may be used to predict or interpolate values that are not displayed in the plot. However, care must be taken not to read too much into a least squares line. Indeed, such a line exists for *any* data, whether or not there is a trend.

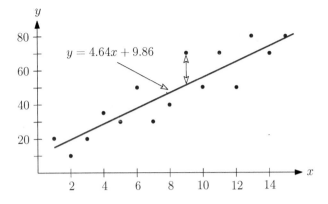

FIGURE 9.6: Data points and least squares line.

The procedure `LeastSquares` generates the least squares line from x-data entered in column B and y-data in column C. The procedure initially calls `DataCount` (Section 4.2) to determine the number of data points in columns B and C. After making a scatter plot from the data, the procedure calls `GetLineParameters`, which calculates the slope and y-intercept of the least squares line, and then calls `GenerateYValues`, to generates the y values of the least squares line corresponding to the original x-values. The line is graphed next to the scatter plot. Figure 9.7 illustrates data input-output in the program less the scatter plot and least squares line, which appear to the right of the data in the actual spreadsheet.

	A	B	C	D
1		X Data	Y Data	Y Modelled
2	Run Program	1.2	.23	.36
3		2.3	.54	.39
4	slope	3.4	.32	.42
5	.03	4.5	.61	.45
6	intercept	5.6	.39	.48
7	.33	6.7	.65	.51
8		7.8	.41	.54

FIGURE 9.7: Spreadsheet for least squares.

```
Public FirstRow As Integer, LastRow As Integer, NData As Integer
Sub LeastSquares()
    Dim slope As Single, intercept As Single
    If ChartObjects.count > 0 Then ChartObjects.Delete      'delete old
    FirstRow = 2
    NData = DataCount(FirstRow, 2)              'get number of data points
```

```
      If NData = 0 Then Exit Sub                      'no data? then exit
      LastRow = FirstRow + NData - 1
                          'draw least squares line:
      With Sheet1 _
          .ChartObjects.Add(Left:=250, Width:=300, Top:=25, Height:=200)
          .Chart.ChartType = xlXYScatter
          .Chart.HasTitle = True
          .Chart.ChartTitle.Text = "Data Points"
          .Chart.SetSourceData Source:=Range("C" & _
          FirstRow & ":" & "C" & LastRow)
          .Chart.SeriesCollection(1).XValues = _
          Range("B" & FirstRow & ":" & "B" & LastRow)
      End With
      Call GetLineParameters(slope, intercept)
      Call GenerateYValues(slope, intercept)  'generate modelled data
      With Sheet1 _
          .ChartObjects.Add(Left:=575, Width:=300, Top:=25, Height:=200)
          .Chart.ChartType = xlLine
          .Chart.HasTitle = True
          .Chart.ChartTitle.Text = "Least Squares Line"
          .Chart.SetSourceData Source:=Range("D" & _
          FirstRow & ":" & "D" & LastRow)
          .Chart.SeriesCollection(1).XValues = _
          Range("B" & FirstRow & ":" & "B" & LastRow)
          .Chart.SeriesCollection(1).MarkerStyle = xlNone      'no markers
      End With
      Range("A5").Value = slope: Range("A7").Value = intercept
  End Sub
```

The slope and intercept of the least squares line are given by the formulas (9.7) and (9.8) translated into code notation as follows:

$$\text{slope} = \frac{\text{Ndata*sumxy} - \text{sumx*sumy}}{\text{Ndata*sumx2} - (\text{sumx})^2}, \quad \text{intercept} = \frac{\text{sumy} - \text{slope*sumx}}{\text{Ndata}},$$

where

- sumx is the sum of the x-coordinates x_i of the data,
- sumy is the sum of the y-coordinates y_i of the data,
- sumxy is the sum of the products $x_i y_i$, and
- sumx2 is the sum of the squares x_i^2.

```
Sub GetLineParameters(slope As Single, intercept As Single)
    Dim sumx As Single, sumy As Single, sumxy As Single
    Dim sumx2 As Single, i As Integer
    For i = FirstRow To LastRow    'find ingredients of slope, intercept
        sumx = sumx + Cells(i, 2).Value
        sumy = sumy + Cells(i, 3).Value
        sumxy = sumxy + Cells(i, 2).Value * Cells(i, 3).Value
```

```
        sumx2 = sumx2 + Cells(i, 2).Value ^ 2
    Next i
    slope = (Ndata*sumxy - sumx*sumy)/(Ndata*sumx2 - sumx^2)
    intercept = (sumy - slope*sumx)/Ndata
End Sub
```

The procedure `GenerateYValues` uses the slope and intercept to calculate the y-values of the least squares line from the x-data in column B.

```
Sub GenerateYValues(slope As Single, intercept As Single)
    Dim i As Integer
    For i = FirstRow To LastRow
        Cells(i, 4).Value = slope * Cells(i, 2).Value + intercept
    Next i
End Sub
```

9.8 Graphing Parametric Curves

The program `GraphFunction` in Section 9.6 may be modified slightly to graph curves defined by a parameter t. Here, the function $y = f(x)$ is replaced by two functions $x = g(t)$ and $y = h(t)$, where $a \le t \le b$, the *parameter interval*. As the parameter t increases from a to b, the point $(g(t), h(t))$ traces out a curve in the plane. Note that by taking $g(t) = t$ one obtains the graph of the function $h(t)$.

The program `GraphCurve` differs from `GraphFunction` in two crucial ways: First, the user must enter two functions rather than one, and second, the chart type must be `xlXYScatterLines` or `xlXYScatter` rather than `xlLine` to adjust for the possibility that two points may have the same x-coordinate. As in `GraphFunction`, the procedure uses `FunctionVal` of Section 4.5.

```
Sub GraphCurve()
    Dim NPoints As Integer, LPoint As Double, RPoint As Double
    Dim Inc As Double, t As Double, Xexpr As String, Yexpr As String
    Call ClearColumns(8, 3, 4)                    'clear old data
    If ChartObjects.Count > 0 Then ChartObjects.Delete    'and chart
    LPoint = Range("B3").Value              'get left end point
    RPoint = Range("B4").Value              'get right end point
    Inc = Range("B5").Value                 'and the increment
    Xexpr = Range("B6").Value: Yexpr = Range("B7").Value  'get functions
    NPoints = (RPoint - LPoint) / Inc
    t = LPoint                              'initial value of parameter
    For j = 1 To NPoints 'evaluate functions Xexpr, Yexpr using t values
        Cells(j + 9, 2).Value = FunctionVal(Xexpr, "t", t)   'x values
        Cells(j + 9, 3).Value = FunctionVal(Yexpr, "t", t)   'y values
        t = t + Inc
    Next j
```

```
    With Sheet1 _
        .ChartObjects.Add(Left:=300, Width:=500, Top:=50, Height:=400)
        .Chart.ChartType = xlXYScatterLines
        .Chart.HasTitle = True
        .Chart.ChartTitle.Text = "Parametric Curve"
        .Chart.SetSourceData Source:=Range _
        ("C" & 10 & ":" & "C" & 10 + NPoints)
        .Chart.SeriesCollection(1).XValues = _
        Range("B" & 10 & ":" & "B" & 10 + NPoints)
        .Chart.SeriesCollection(1).MarkerStyle = xlNone
    End With
End Sub
```

The reader may wish to generate the following interesting curves, varying the parameters and noting the effect. Enter the expressions in A8 and A9 with an increment .01. Let the interval for the first curve be $[0, 7\pi]$ and the interval for the last three $[0, 4\pi]$.

- t*(Cos(t), t*Sin(t)   The graph is a spiral emanating from the origin.

- Sin(2*t), Sin(3*t)   The graph is called a *Lisajous pattern.*

- 7*Cos(t) - Cos(7*t), 7*Sin(t) - Sin(7*t)   The graph is called an *epicycloid* and is the locus of a point on the rim of a wheel rolling on the outside of another wheel.

- `7*Cos(t) + Cos(7*t), 7*Sin(t) - Sin(7*t)` The graph is called a *hypocycloid* and is the locus of a point on the rim of a wheel rolling on the inside of another wheel.

9.9 A Fractal Game

The game in the title was first proposed by Michael Barnsley in his book *Fractals Everywhere*. It is played on a set of vertices of a polygon and consists of the following steps:

(1) Select a point (x, y) at random.
(2) Select one of the vertices (x_0, y_0) at random.
(3) Plot a new point at a specified fraction f of the distance along the line from (x, y) to (x_0, y_0). Call the new point (x, y).
(4) Repeat steps (2) and (3) a specified number of times.

The algorithm sometimes results in what is called a *fractal*, that is, a geometric pattern that is self-similar at any magnification. For a triangle, the choice $f = .5$ produces the so called *Sierpinski triangle*, the first few steps in the evolution of which are shown in the figure. The Sierpinski triangle may be viewed as the result of repeatedly removing inner triangles of half the dimensions from the blue ones, ultimately leaving only a "skeleton".

The program `ChaosTriangleGame` carries out the above steps (1)–(4) for a triangle with vertices $(0,0)$, $(2,0)$, and $(1,2)$. The initial point is $(x,y) = (0,0)$. At each stage the function `Rnd` (see Chapter 10) is used to randomly select a vertex with equal probabilities. A new point (x,y) is then chosen according to rule (3).

```
Sub ChaosTriangleGame()
    Dim x0 As Double, y0 As Double, x As Double, y As Double
    Dim NumIt As Integer, j As Integer, f As Double
    If ChartObjects.Count > 0 Then ChartObjects.Delete      'clear charts
    Call ClearColContent(8, 3, 4)                           'and old data
    f = 0.5                                          'distance fraction
    NumIt = 10000                          'repeat algorithm this many times
    For j = 1 To NumIterations             'start initialization (x,y) = (0,0)
        Randomize
        Select Case Rnd               'pick vertex at random with equal prob.
            Case Is < 0.33:  x0 = 0: y0 = 0
            Case Is < 0.67:  x0 = 2: y0 = 0
            Case Else:       x0 = 1: y0 = 2
        End Select
        'print new (x,y) along line from old (x,y) to (x0,y0)
        x = f * x0 + (1 - f) * x: y = f * y0 + (1 - f) * y
        Cells(j + 7, 3).Value = x: Cells(j + 7, 4).Value = y
    Next j
    With Sheet1 _
        .ChartObjects.Add(Left:=300, Width:=800, Top:=50, Height:=500)
        .Chart.SetSourceData Source:=Range("D" & 8 & ":" & "D" & NumIt)
        .Chart.ChartType = xlXYScatter
        .Chart.HasTitle = True
        .Chart.ChartTitle.Text = "Sierpinski Triangle"
        .Chart.SeriesCollection(1).XValues = _
        Range("C" & 8 & ":" & "C" & NumIt)
        .Chart.SeriesCollection(1).MarkerStyle = xlMarkerStyleDot
    End With
End Sub
```

9.10   Overlaying Graphs

The VBA chart facility allows one to construct multiple charts with a common x-axis. We illustrate this with an example that describes the interaction between two species.

Consider an ecosystem supporting a population of predators and a population of prey, say foxes and rabbits. In such a system one would expect the numbers of the two species to cycle in a dependent fashion: as prey increases, the number of predators increases, resulting in a decrease of food per predator, causing a decrease of predators, resulting in an increase in prey, etc. This complex interaction has been famously modelled by the Lotka-Voltera equations, a system of differential equations describing the rate of change of the two populations. Such a model is beyond the scope of this book, but there is a discrete-time approach that uses recursively defined sequences. The specific model we consider is based on the coupled recursion formulas

$$x_{n+1} = ax_n(1 - x_n) - bx_ny_n$$
$$y_{n+1} = -cy_n + dx_ny_n,$$

where

- x_n is the density of the prey population (with density $= 1$ characterizing the maximum sustainable prey population),

- y_n is the density of the predator population (relative to prey population),

- a is the rate of growth of prey in the absence of predators,

- b is the rate at which predators consume prey,

- c is the rate of death of predators in the absence of prey, and

- d is the rate at which the predator population increases as a result of consuming prey.

The module `PredatorPrey` graphs predator-prey interactions for user-entered parameters. An example is given in Figure 9.8.

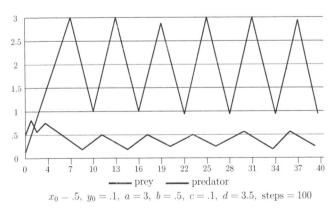

$x_0 - .5,\ y_0 = .1,\ a = 3,\ b = .5,\ c = .1,\ d = 3.5,\ \text{steps} = 100$

FIGURE 9.8: Predator–prey cycles.

```
Sub PredatorPrey()
    Dim chrt As Chart                        'declare a Chart object
    Dim a As Double, b As Double, c As Double, d As Double
    Dim x As Double, y As Double, oldx As Double, oldy As Double
```

```
Dim NumSteps As Integer, LastRow As Integer, i As Integer
If ChartObjects.Count > 0 Then ChartObjects.Delete
Set chrt = Sheet1.Shapes.AddChart.Chart              'add chart
x = Range("B3").Value: y = Range("B4").Value       'initial values
a = Range("B5").Value: b = Range("B6").Value      'parameter values
c = Range("B7").Value: d = Range("B8").Value              'ditto
NumSteps = Range("B9").Value          'number of steps in the cycles
LastRow = NumSteps + 11
For i = 1 To NumSteps              'generate the population densities
    Cells(11 + i, 1).Value = i
    Cells(11 + i, 2).Value = x: oldx = x                  'prey
    Cells(11 + i, 3).Value = y: oldy = y             'predator
    x = a * oldx * (1 - oldx) - b * oldx * oldy    'new prey number
    y = -c * oldy + d * oldx * oldy          new predator number
Next
With chrt
    .ChartArea.Width = 600
    .ChartArea.Left = 300
    .ChartArea.Top = 100
    .ChartArea.Height = 400
    .ChartType = xlLine
    .SeriesCollection.NewSeries                       'prey series
    .SeriesCollection(1).Name = "prey"
    .SeriesCollection(1).XValues = _
    Range("A" & 12 & ":" & "A" & LastRow)
    .SeriesCollection(1).Values = _
    Range("B" & 12 & ":" & "B" & LastRow)
    .SeriesCollection(1).MarkerStyle = xlNone
    .SeriesCollection.NewSeries                     'predator series
    .SeriesCollection(2).Name = "predator"
    .SeriesCollection(2).Values = _
    Range("C" & 12 & ":" & "C" & LastRow)
    .SeriesCollection(2).MarkerStyle = x1None
End With
End Sub
```

The For Next loop generates the data using the predator prey equations with i representing time steps. The remainder of the code is devoted to creating the superimposed charts. The code illustrates an alternate method for specifying chart characteristics, namely, by declaring a Chart object `chrt`.

9.11 Bezier Curves

A *Bezier curve of order n* is a parametric curve defined by a set of $n + 1$ points

$$(x_0, y_0), (x_1, y_1), \ldots, (x_{n-1}, y_{n-1}), (x_n, y_n).$$

The first point of the curve is the *start point* and the last point the *end point*. The intermediate points $(x_1, y_1), \ldots, (x_{n-1}, y_{n-1})$ control the shape of the curve and as such are called *control points*. We shall refer to the $n + 1$ points as *(Bezier) data points*. Bézier curves are frequently used in computer graphics to model smooth curves. Figure 9.9 shows an example for the case $n = 3$. The start and end points of the curve are shown in black. The control points match the color of the curve. There are two for each curve. The small boldface numbers indicate the order of the points, which is important. The figure illustrates the basic principle of Bezier curves: the control points act as a "magnet," pulling the curve towards them.

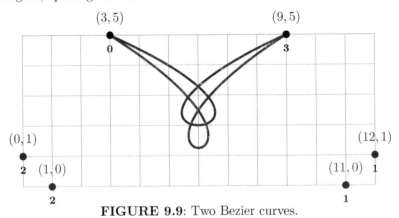

FIGURE 9.9: Two Bezier curves.

To see how a Bezier curve is generated, consider the simplest case $n = 2$ with data points (x_0, y_0), (x_1, y_1), and (x_2, y_2). A Bezier curve for this case is depicted in red in Figure 9.10. A point (x, y) on the curve is obtained as

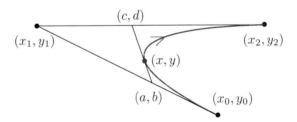

FIGURE 9.10: Typical point on a second order Bezier curve.

follows: Choose a point (a, b) on the line segment from (x_0, y_0) to (x_1, y_1). Such a point may be described parametrically as

$$a = (1 - t)x_0 + tx_1, \quad b = (1 - t)y_0 + ty_1, \tag{9.9}$$

where t is a number between 0 and 1. For the same value of t define the point (c, d) by

$$c = (1 - t)x_1 + tx_2, \quad d = (1 - t)y_1 + ty_2, \tag{9.10}$$

which lies on the line segment from (x_1, y_1) to (x_2, y_2) at the same proportional distance. Now set

$$x = (1 - t)a + tc, \quad y = (1 - t)b + td, \tag{9.11}$$

a point on the line segment from (a, b) to (c, d), again at the same proportional distance. Varying t from 0 to 1 produces the Bezier curve (in the direction of the arrow.) To find the representation of (x, y) in terms of the data points, substitute the values (9.9) and (9.10) into (9.11):

$$x = (1 - t)\big[(1 - t)x_0 + tx_1\big] + t\big[(1 - t)x_1 + tx_2\big]$$
$$y = (1 - t)\big[(1 - t)y_0 + ty_1\big] + t\big[(1 - t)y_1 + ty_2\big]$$

Expanding and collecting terms we arrive at the parametric equation for a Bezier curve of order two:

$$\begin{aligned} x(t) &= (1 - t)^2 x_0 + 2t(1 - t)x_1 + t^2 x_2 \\ y(t) &= (1 - t)^2 y_0 + 2t(1 - t)y_1 + t^2 y_2 \end{aligned} \qquad 0 \le t \le 1. \tag{9.12}$$

Figure 9.11 shows how moving the point (x_1, y_1) changes the shape of the curve.

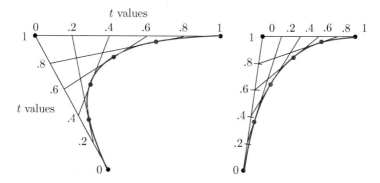

FIGURE 9.11: Quadratic Bezier curve shifting shape.

The equations in (9.12) for a second order Bezier curve may be written in terms of binomial coefficients (see Section 19.6) as

$$x(t) = \sum_{i=0}^{2} \binom{n}{i}(1 - t)^{n-i} t^i x_i, \quad y(t) = \sum_{i=0}^{2} \binom{n}{i}(1 - t)^{n-i} t^i y_i, \quad 0 \le t \le 1.$$

This suggests the following definition for parametric equations for a general Bezier curve:

$$x(t) = \sum_{i=0}^{n} \binom{n}{i}(1 - t)^{n-i} t^i x_i, \quad y(t) = \sum_{i=0}^{n} \binom{n}{i}(1 - t)^{n-i} t^i y_i, \quad 0 \le t \le 1.$$

The start and end points are gotten by setting t equal to 0 and 1, respectively. For $n = 3$ the equations simplify to

$$x(t) = (1-t)^3 x_0 + 3t(1-t)^2 x_1 + 3t^2(1-t)x_2 + t^3 x_3$$
$$y(t) = (1-t)^3 y_0 + 3t(1-t)^2 y_1 + 3t^2(1-t)y_2 + t^3 y_3.$$
$$0 \le t \le 1.$$

The module `Bezier` draws a Bezier curve using data points entered as integers $0, 1, 2, \ldots, n$ into the cells of a coordinatized grid. Figure 9.12 indicates grid input for the blue Bezier curve in Figure 9.9. Here, the user has entered

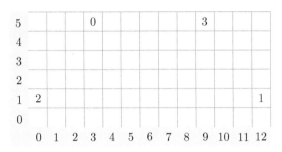

FIGURE 9.12: Spreadsheet input for Bezier example.

the integers $0, 1, 2, 3$ into cells of the grid, as shown. Running the program prints the coordinates of the integers in columns A and B and then uses these to generate the coordinates of the Bezier curve. The coordinates are obtained by letting the parameter t take the values $0, .01, .02, \ldots, 1$, resulting in 101 points. The curve is generated from the coordinates.

The main procedure declares as `Public` variables the number of data points and the beginning and end rows of the Bezier curve coordinates generated by the module. After clearing previous data and setting up the grid, the procedure calls `GetDataPoints` to print out the coordinates of the Bezier data points entered by the user into the grid. The VBA chart facility produces the curve.

```
Sub Bezier()
    Dim Grid As TGrid
    If ChartObjects.count > 0 Then ChartObjects.Delete
    Call ClearColumns(8, 1, 2)
    Grid = MakeGrid(2, 4, 25, 35, 2.3, 15, 24)
    Call InstallCoord(Grid, Grid.B, Grid.L)
    NPoints = GetDataPoints(Grid) 'read pts. from grid, return number
    FirstRow = 9 + NPoints                'curve data starts here
    LastRow = FirstRow + 100                    'and ends here
    Call MakeCurve
End Sub
```

The function `GetDataPoints` scans the grid for the user-entered integers, prints their coordinates in columns A and B, and returns the number of these points.

```
Private Function GetDataPoints(Grid As TGrid) As Integer
    Dim i As Integer, j As Integer, n As Integer, count As Integer
    For i = Grid.T To Grid.B                        'scan grid for vertices
        For j = Grid.L To Grid.R
            If Not IsEmpty(Cells(i, j)) Then
                n = Cells(i, j).Value                   'found a data point
                Cells(8 + n, 1).Value = j - Grid.L    'print coordinates
                Cells(8 + n, 2).Value = Grid.B - i
                count = count + 1                     'update point count
            End If
        Next j
    Next i
    GetDataPoints = count                           'return number of points
End Function
```

The procedure `MakeCurve` generates points of the Bezier curve and prints them in columns A and B under the data points. The parameter t defining the x and y coordinates of the curve takes values from 0 to 1 in steps of .01. For each value of t the procedure reads the control points and from these computes the terms of the sums defining the curve. The procedure uses a worksheet function to compute the binomial coefficients. `MakeChart` graphs the curve using the generated points.

```
Sub MakeCurve()
    Dim i As Integer, n As Integer, x As Double, y As Double
    Dim sumx As Double, sumy As Double, t As Single
    Dim WF As WorksheetFunction                     'notation abbreviation
    Set WF = Application.WorksheetFunction            'implement abbreviation
    n = NPoints - 1                                 'degree of the Bezier curve
    For t = 0 To 1 Step 0.01                        'increment in steps of .01
        sumx = 0: sumy = 0                              'initialize each time
        For i = 0 To n          'read data points and calculate Bezier sums
            x = Cells(8 + i, 1).Value: y = Cells(8 + i, 2).Value
            sumx = sumx + WF.Combin(n, i) * (1 - t) ^ (n - i) * t ^ i * x
            sumy = sumy + WF.Combin(n, i) * (1 - t) ^ (n - i) * t ^ i * y
        Next i
        Cells(FirstRow + 100 * t, 1).Value = sumx       'print coordinates
        Cells(FirstRow + 100 * t, 2).Value = sumy
    Next t
    With Sheet1 _
        .ChartObjects.Add(Left:=700, Width:=650, Top:=25, Height:=500)
        .Chart.ChartType = xlXYScatterLines
        .Chart.HasTitle = True
        .Chart.ChartTitle.Text = "Bezier Curve"
        .Chart.SetSourceData Source:=Range _
        ("B" & FirstRow & ":" & "B" & LastRow)
        .Chart.SeriesCollection(1).XValues =
        Range("A" & FirstRow & ":" & "A" & LastRow)
        .Chart.SeriesCollection(1).MarkerStyle = xlNone
    End With
End Sub
```

9.12 Exercises

1. (Projectile motion). This exercise is for readers with a trigonometry background. It may be shown that a projectile launched at an angle of θ radians from the horizontal from a point $(0, y_0)$ and with an initial velocity of v_0 meters per second has a path given by the equations (in meters)

$$x = (v_0 \cos \theta)t, \quad y = y_0 + (v_0 \sin \theta)t - \tfrac{1}{2}gt^2,$$

where $g = 9.8$ meters per second squared, the acceleration due to gravity. (These are derived from Newton's second law of motion under the ideal assumption that there is no drag from air friction.) From these equations

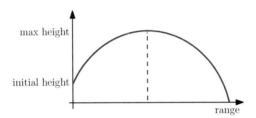

one can show that the maximum height $y_{\max}$ of the projectile and the time t_R it takes the projectile to hit the ground are given by the formulas

$$y_{\max} = y_0 + \frac{(v_0 \sin \theta)^2}{2g}, \quad t_R = \frac{v_0 \sin \theta + \sqrt{(v_0 \cos \theta)^2 + 2gy_0}}{2g}$$

The range R (horizontal distance traveled) is then obtained by substituting t_R into the equation for x:

$$R = (v_0 \cos \theta)t_R$$

Write a program ProjectileMotion that calculates $y_{\max}$ and R for user-entered values of θ (in radians). Use the VBA functions Sin, Cos. Graph the motion of the projectile using the VBA graphing facility.

2. (Projectile motion with air resistance). The model in the preceding exercise neglects the effect of air resistance. It may be shown that if the friction caused by air resistance is assumed to be proportional to the velocity (valid for low velocities) then the path of the projectile is given by the equations

$$x = \frac{v_0 \cos \theta}{a} \left(1 - e^{-at}\right), \quad y = y_0 - \frac{gt}{a} + \frac{1}{a} \left(v_0 \sin \theta + \frac{g}{a}\right) \left(1 - e^{-at}\right),$$

where e is the base of the natural logarithm and a is a positive constant proportional to the air friction. Write a program `ProjectileMotion2` that graphs the motion of the projectile for various values of a using the VBA graphing facility. Use the VBA function `Exp`. Notice the change in the graph as a is increased. Compare the graphs with those produced by `ProjectileMotion`.

3. Write a program `GraphRegularPoly` that reads a user-entered positive integer n and graphs a regular polygon with n vertices. Use either vertices with coordinates of the form

$$x = \cos(k\theta), \quad y = \sin(k\theta)$$

or, for a slightly different effect,

$$x = \cos(k\theta) - \sin(k\theta), \quad y = \cos(k\theta) + \sin(k\theta)$$

where $\theta = 2\pi/n, \quad k = 1, 2, \ldots, n$.

4. Write a program `GraphPolyRotate` that incorporates into `GraphPoly` the ability to rotate the figure through a specified angle θ and about a specified axis point (x_0, y_0) on a grid with coordinates. Use the transformation

$$x = x_0 + (x - x_0)\cos\theta + (y - y_0)\sin\theta, \quad y = y_0 - (x - x_0)\sin\theta + (y - y_0)\cos\theta,$$

which rotates a point (x, y).

5. Write a program `LeastSquaresFunction` that incorporates into the procedure `LeastSquares` the option of generating y-data from a user defined function for x-data of the form $1, 2, \ldots$ For the function use the VBA function `Evaluate`.

6. Write a program `ChaosSquareGame` based on the same idea as the `ChaosTriangleGame` but with the triangle replaced by a square with vertices $(0, 0)$, $(1, 0)$ $(1, 1)$, and $(0, 1)$. In this case fractals appear only if certain restrictions regarding the next choice of vertex are imposed. Three such restrictions are (1) avoid last point chosen; (2) avoid opposite diagonal point; (3) avoid next point in counterclockwise direction. Allow the user to enter any one of the restrictions as well as the fraction f. Try $f = .5$ and $f = .6$ for interesting fractals.

7. Write a variation `GraphPolyMid` of `GraphPoly` in Section 9.2 that allows the user to specify midpoints to be calculated by the program. For example, the figure on the right below was generated by entering the indicated vertex numbers in the grid cells and in addition entering in a column the following sequences: 1,5,2; 1,8,2; 2,6,3; 1,7,3. Here, 5 refers to the midpoint of the line segment from the grid cell labeled 1 to the grid cell labelled 2, etc. The order of the drawing is 1 to 2, which draws the segment from x, y coordinates $(1, 0)$ to $(2, 3)$; then 2 to 3, which draws

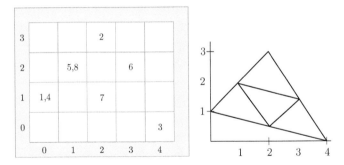

the segment from $(2,3)$ to $(4,0)$; then 3 to 4, which draws the segment from $(4,0)$ to $(0,1)$; then 4 to 5, which draws the segment from $(0,1)$ to the midpoint of the first line segment, etc.

Use the program to verify that the three medians of a triangle intersect in a single point (called the *centroid* of the rectangle), where a *median* is the line segment joining a vertex with the midpoint of the opposite side. Verify also that the line connecting the midpoints of two sides is parallel to the third side (Figure 9.13).

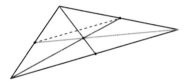

FIGURE 9.13: Medians meet at a point.

8. (Orthocenter). An *altitude* of a triangle is a line from a vertex to the opposite side. The *orthocenter* of a triangle is the intersection of its altitudes (Figure 9.14). Use `Point2Line` to construct a procedure `GraphOrthocenter` that graphs the triangle and the lines from the orthocenter to the vertices.

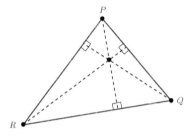

FIGURE 9.14: Orthocenter of a triangle.

Chapter 10

Random Numbers

In this chapter we give some applications of the Excel VBA random number generator Rnd. Random number generators have many uses, not the least of which is Monte Carlo simulations, which are used to approximate theoretical outcomes of various processes. It should be remarked that the numbers generated by Rnd and similar functions in other languages are not truly random, as they rely on deterministic algorithms and *seed numbers*, the latter to initiate the algorithm. Thus knowing both the algorithm and the seed one can theoretically predict the output. Nevertheless, these *pseudorandom* numbers, as they are called, are sufficiently reliable to find important uses in such diverse areas as computer simulations, cryptography, and electronic games.

10.1 The VBA Function Rnd

The function Rnd returns a random decimal x with $0 \le x < 1$. This might seem too restrictive to be very useful, but by suitable arithmetic manipulation the function can be used to generate random decimals or integers in arbitrary ranges. In this section we give two general functions that extend the capability of Rnd and give some simple applications. More complex applications are given in the remaining sections.

Random Decimals

Since $0 \le 8*$ Rnd< 8, the code -3 + 8*Rnd returns a random decimal x with $-3 \le x < 5$. The function RndDec elaborates on this idea. It takes as arguments integer bounds lower and upper returns a random decimal x with lower $\le x <$ upper.[1]

```
Function RndDec(lower As Double, upper As Double) As Double
    Dim x As Double
    Randomize: x = Rnd                          'generate the number
    RndDec = lower + x *(upper - lower)
End Function
```

---

[1] One can also use the VBA function WorksheetFunction.RandBetween for this.

In the above code we have used the statement `Randomize` before invoking `Rnd`. This creates a new seed for `Rnd` to prevent successive random numbers from following a pattern.

Random Integers

To generate a random integer one can use `Rnd` together with `Int`, which returns the integer part of a number. For example, to generate random integers n with $50 \leq n \leq 60$ use the code `50 + Int(11*Rnd)`, since $0 \leq 11*$ `Rnd` < 11. More generally, the following function returns a random integer between `lower` and `upper` inclusive.

```
Function RndInt(lower As Integer, upper As Integer) As Integer
    Randomize: RndInt = lower + Int(Rnd * (upper - lower + 1))
End Function
```

Random Permutations

The function `RndPerm(n)` takes a positive integer n and returns a permutation (rearrangement) of the integers $1, 2, \ldots, n$ in the form of an array.

```
Function RndPerm(n As Integer) As Integer()
Dim perm() As Integer, i As Integer, last, repeat As Boolean
ReDim perm(1 To n)
i = 1: last = 1
perm(1) = RndInt(1, n)                    'first member of permutation
Do While last < n
    repeat = False
    x = RndInt(1, n)                               'get random integer
    For i = 1 To last                           'check if already used
        If x = perm(i) Then repeat = True: Exit For       'dup, no good
    Next
    If Not repeat Then last = last + 1: perm(last) = x          'good
Loop
RndPerm = perm
End Function
```

Random Positions

Here's an application that will find uses in later programs that generate random cell movements. The function takes a cell position `Pos` and returns the cell position obtained by adding one of the four random offsets $(\pm 1, 0)$, $(0, \pm 1)$, each with probability $1/4$, to `Pos.row` and `Pos.col`.

```
Private Function NextRndPos(Pos As TPos) As TPos
    Dim i As Integer, j As Integer, NewPos As TPos
    Select Case RndInt(1, 4)          'get a random integer between 1 and 4
        Case Is = 1: j = 1        'E
        Case Is = 2: j = -1       'W
```

```
        Case Is = 3: i = -1       'N
        Case Is = 4: i = 1        'S
    End Select
    NewPos.row = i + Pos.row                           'new row
    NewPos.col = j + Pos.col                           'new column
    NextRndPos = NewPos
End Function
```

To allow diagonal motion replace `RndInt(1, 4)` by `RndInt(1,8)` and add the following statements to Select Case:

```
    Case Is = 5:  i = 1: j = 1       'SE
    Case Is = 6:  i = 1: j = -1:     'SW
    Case Is = 7:  i = -1: j = 1:     'NE
    Case Is = 8:  i = -1: j = -1:    'NW
```

Probability Distributions

Rolling a die produces a number from 1 to 6. If the die is fair, then each number has the same (theoretical) probability of appearing, namely 1/6. We can summarize this with a *probability array* `P(1 To 6)`, where each entry is 1/6. If, instead, the die is loaded in such a way that, say, the odd numbers are twice as likely to appear as the even numbers, then the probability array describing the experiment has entries $P(k) = 2/9$ if $k = 1, 3, 5$ and $P(k) = 1/9$ if $k = 2, 4, 6$.

The function `RndOut`, short for "random outcome," uses `Rnd` to simulate an abstract version of the die tossing experiment. The function takes as input a probability array `P` whose only requirements are that the entries be positive and add up to 1, and returns an integer in accordance with these probabilities. Thus if `Rnd` produces number x that lies in the interval from 0 to $P(1)$, the function returns the integer 1; if x lies in the interval from $P(1)$ to $P(2)$, the function returns the integer 2; etc. (Figure 10.1). Since the interval lengths are proportional to the probabilities, the function simulates the toss of a many-sided, possibly loaded, die.

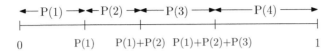

$$0 \qquad P(1) \qquad P(1)+P(2) \quad P(1)+P(2)+P(3) \qquad 1$$

FIGURE 10.1: Interval lengths as probabilities.

Here is the code. A For Next loop calculates the *cumulative probabilities*, that is, numbers obtained by successively adding the members of `P`.

```
Function RndOut(P() As Double) As Integer
    Dim k As Integer, CumProb As Double, x As Double
    Randomize: x = Rnd                    'random number between 0 and 1
```

```
For k = 1 To UBound(P)              'find the interval in which x lies
    If CumProb <= x And x < CumProb + P(k) Then Exit For
    CumProb = CumProb + P(k)            'beginning of next interval
    Next k
    RndOut = k                            'x lies in kth interval
End Function
```

10.2   A Nonsense Generator

Here's an application of `RndInt` that generates gibberish sentences. The program assumes that a sentence consists of seven components: *Subject, Auxiliary Verb, Adverb, Verb, Adjective Phrase, Noun,* and *Adverbial* or *Prepositional Phrase.* The components are chosen from lists entered by the user in spreadsheet columns. The program randomly selects an item from each column and then concatenates these to form a grammatically correct, but usually meaningless, sentence. Column A contains a list of possible subjects, Column B a list of possible auxiliary verbs, Column C a list of adverbs, etc. The reader may supplement these if desired.

A: I, He, She, They, You, No one, Everyone

B: can, may, might, should, will

C: almost certainly, deliberately, frequently, fully, never, sometimes partially, reluctantly, strategically, with few exceptions

D: alleviate, contemplate, disambiguate, expedite, facilitate, attenuate, remediate, intermodulate, intercorrelate

E: diverse, dynamically adjusted, emerging, correlated, multifaceted, developmentally expeditious, interactively positioned

F: frameworks, functionalities, goals, interfaces, synergies, paradigms, trajectories, integrated procedures

G: consistent with high impact protocols, employing suitable metrics, guided by non parametric considerations, via the application of technological polemics, within a suitable zone of limited acuity, within a balanced paradigm, within the framework of spatial and temporal dynamics.

The program `NonsenseGenerator` generates a random row number for each column and selects the sentence component in that row. These are concatenated to form what one might generously call a "sentence." For example, referring to the above lists, if the random integers 5,3,1,2,2,5,3 are generated for columns A to G, respectively, then the following useful advice appears:

> *You might almost certainly contemplate dynamically*
> *adjusted synergies guided by non parametric considerations.*

The program assumes that each list starts in row 6. For each of the columns 1-7 the function `RndInt` generates a random row number between 6 and the last entry in the column. The sentence component in that row is then attached to the current sentence. The complete sentence is printed in cell C3.

```
Sub NonsenseGenerator()
    Dim j As Integer, sentence As String, n As Integer, lRow As Integer
    For j = 1 To 7                          'run through the columns
        lRow = Cells(Rows.count, j).End(xlUp).row    'last nonempty row
        n = RndInt(6, lRow)                 'get random row in column j
        sentence = sentence & " " & Cells(n, j).Value  'attach component
    Next j
    Range("C3").Value = sentence            'return composite sentence
End Sub
```

10.3   Area under a Curve

One can approximate the area under the graph of a positive function on an interval $[a, b]$ by enclosing the graph in a box with lower side on the interval, repeatedly selecting a point in the box at random, and counting the number of points that fall below the curve. If the box has area A and the relative frequency of points under the curve is x, then one can expect that xA approximates the area under the curve. [2] Figure 10.2 illustrates this for the curve $y = x^2$, $a = .25$ and $b = 1$.

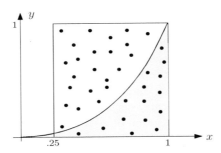

FIGURE 10.2: Random points in the rectangle.

The function `AreaUnderCurve` takes as arguments the function expression, the left and right interval endpoints, the top of the rectangle, and the number of random points. The function `RndDec` randomly selects an x-coordinate and a y-coordinate. The function value $z = F(x)$ is calculated by the VBA function

---

[2]This method of approximation is an example of *Monte Carlo simulation,* used in later chapters as well.

Evaluate. If $y < z$ the random point (x, y) is counted by `ptsBelow` as a point below the graph. The variable `npts` counts the total points.

```
Function AreaUnderCurve(expr As String, left As Double, top As Double _
                       right As Double, npts As Long) As Double
    Dim newexpr As String, boxarea As Double, ptsBelow As Long, j As Long
    Dim x As Double, y As Double, z As Double, relativeFreq As Double
    For j = 1 To npts
        x = RndDec(left, right)      'select value at random in interval
        y = RndDec(0, top)                               'ditto
        newexpr = Replace(expr, "x", x)     'assign x-value to "x"
        z = Evaluate(newexpr)                 'evaluate result
        If y < z Then ptsBelow = ptsBelow + 1      'points below graph
    Next j
    boxarea = top * (right - left)
    relFreq = ptsBelow / npts   'relative frequency of points under graph
    AreaUnderCurve = boxarea * relFreq
End Function
```

The precise area under the curve in the example is .328125 (derived by calculus methods). Running the program using `npts=` 1000, say, yields fair approximations.

The reader may wish to generalize the program by allowing a lower curve other than the x-axis. For example, one could use $y = \sin x$ for the upper curve and $y = 2x/\pi$ for the lower curve, where $0 \le x \le \pi/2$.

10.4   A Simple Random Walk

A random walk is a mathematical model of a path consisting of a sequence of random steps. For example, stock price changes or molecular motion have been modelled by random walks. Random walks have applications in a variety of disciplines including physics, computer science, economics, finance, biology, and sociology. Random walks may occur in any dimension. In this section we use `NextRndPos` to direct the motion of a single red cell in a spreadsheet grid. At each step the program chooses a random direction for the cell, colors the new cell and removes the color and borders from the old cell. Figure 10.3 shows the trail left by a walk.

FIGURE 10.3: Single cell random walk.

The following functions facilitate the movement of a cell. The function CellColor, not needed in this section but used elsewhere in the chapter, returns the color index of a cell. The function ColorCell colors a cell with a given color index. The remaining procedures add and remove cell borders. All of the procedures use the UDT TPos, defined in § 7.1.

```
Private Function CellColor(Pos As TPos) As Integer
    CellColor = Cells(Pos.row, Pos.col).Interior.ColorIndex
End Function
    Private Sub ColorCell(Pos As TPos, color As Integer)
    Cells(Pos.row, Pos.col).Interior.ColorIndex = color
End Sub
Private Sub AddCellBorder(Pos As TPos)
    Cells(Pos.row, Pos.col).BorderAround ColorIndex:=1
End Sub
Private Sub RemoveCellBorder(Pos As TPos)
    Cells(Pos.row, Pos.col).Borders.LineStyle = xlLineStyleNone
End Sub
```

The main procedure RandomWalk sets the initial position of the cell in the center of the grid. A For Next loop generates the random walk using the procedures NextRandPos, which returns a new position, and MoveCell, which erases the old cell and colors the new cell. The cell is moved only if the new cell is in the grid.

```
Sub RandomWalk()
    Dim Steps As Integer, Speed As Integer, k As Integer
    Dim Grid As TGrid, Pos As TPos
    Steps = 100: Speed = 15                        'motion parameters
    L = 5: T = 5: R = 30: B = 40                   'boundaries of motion
    Grid = MakeGrid(T, L, B, R, 2.5, 13, 24)
    Pos.row = (L + R) / 2: Pos.col = (T + B) / 2   'initial pos. of cell
    For k = 1 To 200                               'do the walk
        Call MoveOneStep(Pos, Grid)
        Call Delay(Speed)
    Next k
    MsgBox "clear"
    Call ClearGrid(Grid, 0)
End Sub

Private Sub MoveOneStep(Pos As TPos, Grid As TGrid)
    Dim NewPos As TPos
    NewPos = NextRndPos(Pos)
    If InGrid(NewPos.row, NewPos.col, Grid) Then
        Call ColorCell(Pos, xlNone)                     'remove color
        Call RemoveCellBorder(Pos)            'and border from old red cell
        Pos = NewPos
        Cells(Pos.row, Pos.col).Value = "*"             'track entire path
        Call ColorCell(Pos, 3)
        Call AddCellBorder(Pos)
    End If
End Sub
```

10.5   Spreading Disease

In this section we construct a program that simulates the spread of disease in a population. The program causes red and green cells to execute independent random walks. A red cell (diseased individual) coming into contact with a green cell (healthy individual) transfers its color to the latter, simulating the spread of an infection.

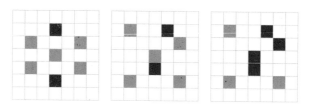

FIGURE 10.4: Transfer of disease.

Figure 10.4 depicts a typical scenario. In the grid on the left, the socially distanced population consists of 5 healthy individuals and 2 infected people. In the middle grid, random motion has started, producing contact between infected and noninfected individuals. The third grid reveals the unfortunate consequences.

The program assumes that the user has entered letters g for green and r for red in various cells of the grid on which the activity takes place. The main procedure SpreadOfDisease calls the function NumOccupied of Section 7.1 to find the number of individuals (red and green cells) and then calls Initialize to populate the array Pos with their positions. A For Next loop repeatedly calls the procedure NewPositions to change the positions of the cells. The latter procedure also checks if a red and green cell are contiguous and if so turns the green cell red to simulate the transfer of an infection.

```
Public NumCells As Integer, NumGreen As Integer, NumSteps
Sub SpreadOfDisease()
    Dim Pos() As TPos, Grid As TGrid, i As Integer
    Grid = MakeGrid(6, 6, 40, 40, 2.3, 11.5, 24)
    NumRed = 0: NumGreen = 0                          'initialize
    NumSteps = Range("B3").Value        'desired number of days of spread
    NumCells = NumOccupied(Grid)               'number of individuals
    If NumCells = 0 Then Exit Sub
    ReDim Pos(1 To NumCells)             'array for cell position and color
    Call Initialize(Pos, Grid)               'get cell characteristics
    For i = 1 To NumSteps
        Call NewPositions(Pos, Grid)  'get a set of new random positions
        If NumGreen = 0 Then Exit For             'everyone infected
    Next i
    MsgBox "clear"
```

```
        Call ClearGrid(Grid, 0)
End Sub
```

The procedure `Initialize` scans the grid for cells with color letters r, g and records the position and color of the cell. It also colors the cell and adds a border using `ColorCell` and `AddCellBorder`.

```
Private Sub Initialize(Pos() As TPos, Grid As TGrid)
    Dim i As Integer, j As Integer, m As Integer: m = 1
    For i = Grid.T To Grid.B                     'scan grid for "g", "r"
        For j = Grid.L To Grid.R
            If IsEmpty(Cells(i, j)) Then GoTo continue
            Pos(m).row = i: Pos(m).col = j              'record position
            If Cells(i, j).Value = "g" Then
                Call ColorCell(Pos(m), 4)
                NumGreen = NumGreen + 1      'update number of green cells
            ElseIf Cells(i, j).Value = "r" Then
                Call ColorCell(Pos(m), 3)
                NumRed = NumRed + 1          'update number of red cells
            End If
            m = m + 1
continue:
        Next j
    Next i
End Sub
```

The procedure `NewPositions` finds new random positions for the red and green cells using the procedure `NextRndPos`. The procedure runs through the position array of the cells with a For Next loop, getting a random position for a cell and checking if the new position is occupied. If not then the new cell gets the color of the old cell and is colored red if it has a red neighbor.

```
Private Sub NewPositions(Pos() As TPos, Grid As TGrid)
    Dim m As Integer, NewPos As TPos, color As Integer
    For m = 1 To NumCells
        NewPos = NextRndPos(Pos(m))
        If Not InGrid(Pos(m), Grid) Then GoTo continue
        color = CellColor(Pos(m))
        If CellColor(NewPos) = xlNone Then
            Call ColorCell(Pos(m), xlNone)
            Call ColorCell(NewPos, color)
            Pos(m) = NewPos                          'update position
            If NeighborRed(Pos(m)) Then              'infect neighbor
                Call ColorCell(Pos(m), 3)
                NumGreen = NumGreen - 1
            End If
        End If
continue:
    Next m
End Sub
```

The function `NeighborRed` returns True if a green cell at position `Pos` has a red neighbor.

```
Private Function NeighborRed(Pos As TPos) As Boolean
    Dim m As Integer, n As Integer, IsRed As Boolean, Nbor As TPos
    If CellColor(Pos) = 4 Then
        For m = -1 To 1        'use offsets to check for red cell neighbors
            For n = -1 To 1
                Nbor.row = Pos.row + m: Nbor.col = Pos.col + n
                If CellColor(Nbor) = 3 Then IsRed = True
            Next n
        Next m
    End If
    NeighborRed = IsRed
End Function
```

10.6   Particle Swarm Optimization

In this section we use multiple random walks to find the minimum value of a function $f(x, y)$, called the *objective function*.[3] The idea is to calculate $f(x, y)$ for many randomly changing pairs (x, y) (*particles*), adjusting direction probabilities at each step so that the particles gradually cluster around those giving the smallest values of f. Our version is a simplification of the method developed by Eberhart and Kennedy in 1995, which was inspired by the social behavior of an organized population such as a flock birds or a school of fish perhaps attempting to minimize their distance to food or safety. Since then it has become an important tool in computational optimization. The goal of this section is to simulate the process using VBA.

The main procedure of the module is `ParticleSwarmOpt`. The program assumes that the user has entered values for the following parameters:

- `unit`: The incremental change in the variables x and y.

- `attraction`: A probability that influences degree of attraction to the best position of a particle so far. This could be interpreted as, say, a bird's predilection to follow a presumed leader while still retaining some independent impulses allowing it to perhaps follow a better leader or to become a new leader

- `NParticles`: The desired number of particles.

- `NSteps`: The desired number of random movements.

---

[3]The choice of minimum here is arbitrary. Maximum could be used as well. Or one could use the fact that the maximum of f occurs at the minimum of $-f$.

- LowerSearch,UpperSearch: The search range. (Lower and upper bounds for particle coordinates (x, y)).

- expr: The function $f(x, y)$ to be minimized.

The procedure first reads these values and then assigns memory for the array Particles of particle positions of type TPart. The x and y values of the mth particle are Particles(m).x and Particles(m).y, respectively. Next the procedure calls InitSwarm, which assigns random values in the search range to the particle positions. The main part of PSwarmOpt is a For Next loop that executes the motion of the particles, each of these pictured as a bold, red "x" on the grid. At each iteration of the loop, two procedures are called: GetBestPos and MoveSwarm. The former obtains the position BestPos at which the objective function has a minimum value over all current particle positions. This information is imparted to the procedure MoveSwarm, which randomly changes the position of each particle, favoring the direction that most decreases the gap between the particle and the best position so far. After completion of the loop, PSwarmOpt prints the best position and the function value at that position. The latter is calculated using FunctionVal2, a two variable version of FunctionVal (§ 4.5).

```
Private Type TPart                      'particle type for coordinates
            x As Double: y As Double
Public NParticles, expr As String, attraction As Single
Public LowerSearch As Double, UpperSearch As Double, unit As Double
Sub PSwarmOpt()
    Dim G As TGrid, Particles() As TPart, BestPart As TPart
    Dim k As Integer, NSteps As Integer, Pos As TPos
    Dim gridheght As Integer, gridwidth As Integer
    G = MakeGrid(12, 4, 50, 60, 1, 9, 24)
    Range(Cells(G.T, G.L), Cells(G.B, G.R)).Font.ColorIndex = 3
    Range(Cells(G.T, G.L), Cells(G.B, G.R)).Font.FontStyle = "Bold"
    unit = Range("B1").Value              'decimal unit: 1, .1, .01, etc.
    attraction = Range("B2").Value     'degree of attraction to BestPart
    NParticles = Range("B3").Value          'desired number of particles
    NSteps = Range("B4").Value              'desired number of movements
    LowerSearch = Range("B5").Value                      'search range
    UpperSearch = Range("B6").Value
    expr = Range("B7").Value
    ReDim Particles(1 To NParticles)
    Call InitSwarm(Particles, BestPart, G)    'generate random particles
    For k = 1 To NSteps
        BestPart = GetBest(Particles)              'best position so far
        Call MoveSwarm(BestPart, Particles, G)
    Next k
    Pos = ConvertCoordinates(BestPart, G)
    Call ColorCell(Pos, 4)                    'color best position green
    Range("A9").Value = BestPart.x     'print final best coord. and value
    Range("B9").Value = BestPart.y
    Range("B10").Value = FunctionVal2(expr, BestPart.x, BestPart.y)
```

```
        MsgBox ("clear")
        Call ClearGrid(G, 0)
    End Sub
```

The function `InitSwarm` uses `RndDec` to generate random values between `LowerSearch` and `UpperSearch` for the positions of particles and prints these as red x's in the grid. It also initializes the best position coordinates `Best.x`, `Best.y` to the values in cells A9, B9, respectively.

```
    Private Sub InitSwarm(Part() As TPart, Best As TPart, G As TGrid)
        Dim m As Integer, Pos As TPos
        For m = 1 To NParticles
            Part(m).x = RndDec(LowerSearch, UpperSearch)
            Part(m).y = RndDec(LowerSearch, UpperSearch)
            Pos = ConvertCoordinates(Part(m), G)
            Cells(Pos.row,Pos.col).Value = "x"              'print particle
        Next m
        Best.x = Range("A9").Value                          'initial values
        Best.y = Range("B9").Value
    End Sub
```

The function `ConvertCoordinates` converts the x, y coordinates of a particle to row, column spreadsheet coordinates. The conversion is needed because the former might not be integers or may have values that place them outside the grid. The function subtracts `LowerSearch` from a given position, thereby shifting to nonnegative values, then divides by `unit` to obtain integers, and finally reduces the result modulo the grid dimensions to put the coordinates inside the grid.

```
    Private Function ConvertCoordinates(p As TPart, G As TGrid) as TPos
        Dim Pos As TPos
        Pos.row = G.T + (p.y - LowerSearch) / unit Mod (G.B - G.T)
        Pos.col = G.L + (p.x - LowerSearch) / unit Mod (G.R - G.L)
        ConvertCoordinates = Pos
    End Function
```

The procedure `GetBest` is passed the array `Particles` of current x, y positions of the particles and calculates the position `Best` that produces the minimum value of the function over all the particle positions. A For Next loop runs through the particle positions calculating the value of the function at each position and the value of the function at the best position so far. If the former is less than the latter, the best position is updated.

```
    Private Function GetBest(Particles() As TPart) As TPart
        Dim m As Integer, Best As TPart, BestValue As Double
        Dim ThisValue As Double
        For m = 1 To NParticles
            ThisValue = FunctionEval(expr, Particles(m).x, Particles(m).y)
            BestValue = FunctionEval(expr, Best.x, Best.y)
            If ThisValue < BestValue Then    'if found a smaller value, then
```

```
            BestValue = ThisValue 'update best value and best pos. so far
            Best.x = Particles(m).x:  Best.y = Particles(m).y
         End If
      Next m
      GetBest = Best
   End Function
```

The function `FunctionEval(expr,x,y)` replaces the characters x,y in `expr` with the Double values x, y, respectively. It then evaluates the resulting expression using the VBA function `Evaluate`.

```
Function FunctionEval(ByVal expr As String, x As Double, y As Double) _
         As Double
      expr = Replace(expr, "x", x): expr = Replace(expr, "y", y)
      FunctionEval = Evaluate(expr)      'VBA function evaluates expression
   End Function
```

The procedure `MoveSwarm` moves each particle to a new position determined by `Best`.

```
Private Sub MoveSwarm(BestPart As TPart, Part() As TPart, G As TGrid)
      Dim m As Integer, Pos As TPos
      For m = 1 To NParticles
         Pos = ConvertCoordinates(Part(m), G)        'from particle to cell
         Cells(Pos.row, Pos.col).Value = " " 'Call ColorCell(Pos, xlNone)
         Part(m) = MoveParticle(BestPart, Part(m))
         Pos = ConvertCoordinates(Part(m), G)
         Cells(Pos.row, Pos.col).Value = "x"         'Call ColorCell(Pos, 3)
      Next m
   End Sub
```

The function `MoveParticle` is the workhorse of the module. It takes as arguments the current position of a particle and the best position so far, and determines which one-step direction the particle should take from the former to put it closest to the latter. It then assigns a user-specified probability (`attraction`) to that direction. A relatively large probability causes the particles to move more rapidly toward a good position. However, if `attraction` is too large, a particle might not explore its immediate neighborhood well enough, possibly causing better positions to be overlooked.

The procedure first gets the x and y offsets from the current position `Curr` to the best position `Best`. It then determines the direction, denoted by an integer `idx` between 1 and 4, that most decreases distance between the current and best positions. Probabilities are assigned that favor the direction designated by `idx`.

```
Private Function MoveParticle(BestPart As TPart, CurrPart As TPart) As TPart
      Dim DirProbs(1 To 4) As Double, x As Double, y As Double
      Dim NewPart As TPart, xOff As Double, yOff As Double
      Dim idx As Integer, i As Integer
      xOff = BestPart.x - CurrPart.x              'offsets from current to best
```

```
    yOff = BestPart.y - CurrPart.y
    'get direction that most decreases distance between current and best
    If Abs(xOff) > Abs(yOff) And xOff > 0 Then idx = 1          'move right
    If Abs(xOff) > Abs(yOff) And xOff <= 0 Then idx = 2         'move left
    If Abs(xOff) <= Abs(yOff) And yOff <= 0 Then idx = 3          'move up
    If Abs(xOff) <= Abs(yOff) And yOff > 0 Then idx = 4         'move down
    For i = 1 To 4                              'assign probabilities
        DirProbs(i) = (1 - attraction) / 3
    Next i
    DirProbs(idx) = attraction      'higher probability in this direction
    Select Case RndOut(DirProbs)                  'random index 1 to 4
      Case Is = 1: x = unit                          'move right
      Case Is = 2: x = -unit                         'move left
      Case Is = 3: y = -unit                           'move up
      Case Is = 4: y = unit                          'move down
    End Select
    NewPart.x = CurrPart.x + x: NewPart.y = CurrPart.y + y    'add offset
    NewPart.x = Truncate(LowerSearch, NewPart.x, UpperSearch)
    NewPart.y = Truncate(LowerSearch, NewPart.y, UpperSearch)
    MoveParticle = NewPart
End Function
```

The procedure `Truncate` is used to ensure that the particle stays within the search area, resulting in its display staying in the grid.

```
Function Truncate(x As Double, y As Double, z As Double) As Double
    Dim u As Double
    u = y
    If y < x Then u = x: If y > z Then u = z
    Truncate = u
End Function
```

Figure 10.5 depicts a typical scenario. The red particle is given a higher probability to move right toward the green particle (best position so far) rather than up, since the gap is greater in the former direction.

$$BestPos.x - Pos.x = 12, BestPos.y - Pos.y = 3$$

FIGURE 10.5: Probability highest in direction to right.

The reader may wish to test the program using the function

$$(x - 1)^2 + (y - 2)^2 + (x - 3)^2 + (y - 4)^2$$

with, say, 30 particles and an attraction value of .35. The actual minimum occurs at $(x, y) = (2, 3)$; the program yields very close agreement.

10.7 A Randomly Sprouting Plant

The program in this section causes an initial "seed" to grow a stem and then sprout in random directions, producing a "plant." The positions of the plant cells are stored in an array so that the growth may be reversed, causing the plant to wither. At each stage of growth, one of the cells of the existing plant is randomly selected, as is a direction of sprouting from that cell: right, left, up or down. The probabilities of left and right growth are chosen to be equal but less than the probability of upward growth, so that the plant grows upward faster than it spreads out. The plant is prohibited from folding over on itself, ensuring that its component cells have enough sunlight to prosper. Figure 10.6 shows initial and later stages of plant growth.

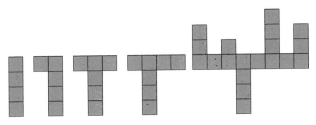

FIGURE 10.6: Sprouting plant.

The module declares as Public the desired number `NCells` of plant cells, the current number `CurrNCells` of cells at any given stage of growth, and the rate of growth `rate`. Their values are entered into the spreadsheet by the user. After allocating the array `Pos` for the cell positions, the main procedure `SproutWither` calls the procedures that cause the plant to sprout and then wither.

```
Public NCells As Integer, CurrNCells As Integer, rate As Integer
Sub SproutWither()
    Dim Pos() As TPos, Grid As TGrid
    Grid = MakeGrid(5, 5, 40, 40, 2.3, 12, 24)
    rate = Range("B5").Value                    'rate of growth
    NCells = Range("B4").Value         'desired number of plant cells
    ReDim Pos(1 To NCells)                   'positions of cells
    Call GrowStem(Pos, Grid, 4)
    Call Sprout(Pos, 4)                     'grow green plant
    Delay (2 * rate)
    Call Wither(Pos, 9)                     'turn plant brown
    Delay (2 * rate)
    Call Wither(Pos, 0)                  'plant disintegrates
End Sub
```

The plant stem is located at the bottom center of the grid. The procedure `GrowStem` stores the positions of the stem cells in the first four entries of the array `Pos` and colors the stem.

```
Private Sub GrowStem(Pos() As TPos, Grid As TGrid, color As Integer)
    Dim i As Integer
    For i = 1 To 4
        Pos(i).row = Grid.B - i + 1:  Pos(i).col = (Grid.L + Grid.R) / 2
        ColorCell(Pos(i), color)
        Delay (rate + 12)          'stem grows more slowly than the leaves
    Next i
    CurrNCells = 4                 'number of cells in plant at this stage
End Sub
```

The procedure **Sprout** consists of a Do While loop that repeatedly calls the function **GrowNewCell** to generate the cells of the plant. The function returns **False** if growth of a particular cell is impossible due to lack of sunlight, as determined by the procedure **CellOk**. The procedure **Wither** changes the color in the reverse order of growth by running backward through the array **Pos**.

```
Private Sub Sprout(Pos() As TPos, color As Integer)
    Dim i As Integer: i = 1
    Do While i <= NCells - 4          'grow a total of NCells with stem
        If GrowNewCell(Pos, color) Then        'new cell ok?
            i = i + 1                          'yes, advance
            Call Delay(rate)
        End If
    Loop
End Sub

Private Sub Wither(Pos() As TPos, color As Integer)
    Dim i As Integer
    For i = CurrNCells To 1 Step -1
        ColorCell(Pos(i), color)
        Call Delay(rate)
    Next i
End Sub
```

The function **GrowNewCell** invokes the function **RndInt** to generate a random index **PlantIdx** for the, array **Pos**, which holds the plant cell positions. This produces a random spot on the body of current plant for a potential new cell. The index is always at least 4 to avoid the stem. The procedure **NextRandPosAlt** generates a new plant cell position with the probabilities .2 (left), .2 (right), and .58 (up). The larger probability of .58 ensures that growth is generally upward. The probabilities may be altered by the user for other effects, perhaps accounting for lack of sunlight on one side of the plant. The procedure **CellOk** checks for self-intersections, rejecting any potential plant cell position that encroaches on the existing plant. This ensures sufficient exposure to sunlight for healthy growth.

```
Private Function GrowNewCell(Pos() As TPos, color As Integer) As Boolean
    Dim i As Integer, j As Integer, OK As Boolean, PlantPos As TPos
    Dim Idx As Integer, NewPos As TPos, Pr(1 To 4) As Double
    Idx = RndInt(4, CurrNCells)                    'array index
```

```
PlantPos.row = Pos(Idx).row                    'position on plant
PlantPos.col = Pos(Idx).col
    'assign probabilities for right, left, up, down: growth mainly up
Pr(1) = .2: Pr(2) = .2: Pr(3) = .59: Pr(4) = .01
NewPos = NextRandPosAlt(Pr, PlantPos)
OK = CellOk(PlantPos, NewPos)              'check if potential position OK
If OK Then                                     'add new cell to plant
    CurrNCells = CurrNCells + 1     'successively attached another cell
    Pos(CurrNCells) = NewPos                    'get its position
    ColorCell(NewPos, color)
End If
GrowNewCell = OK
End Function
```

The function `NextRandPosAlt` takes the place of `NextRandPos`, since different probabilities are needed for the growth directions. The procedure is given an array of probabilities and selects the outcome accordingly using `RndOut`.

```
Private Function NextRandPosAlt(Pr() As Double, Pos As TPos) As TPos
    Dim NewPos As TPos, i As Integer, j As Integer
    Select Case RndOut(Pr)                     'get random offset
        Case Is = 1: j = 1:    'right
        Case Is = 2: j = -1:   'left
        Case Is = 3: i = -1:   'up
        Case Is = 4: i = 1:    'down
    End Select
    NewPos.row = Pos.row + i
    NewPos.col = Pos.col + j                    'add the offset
    NextRandPosAlt = NewPos
End Function
```

The procedure `CellOk` returns `False` when an attempt is made to place a new cell in positions like those marked nc in the first two cell diagrams in Figure 10.7. It does this by checking the neighbors of the potential new cell nc

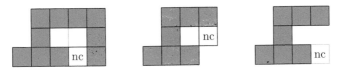

FIGURE 10.7: Cell nc (new cell) rejected in first two figures.

to see if they are part of the plant. Not all neighbors are checked; only those that surround the front and sides of the potential new cell, which we shall call the cell's *aura* (pictured in light yellow). If any of these are part of the plant, that is if they are colored green, then the cell is rejected.

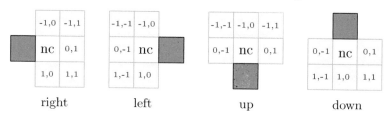

right   left   up   down

FIGURE 10.8: Offsets producing aura checked by `CellOK()`.

In Figure 10.8, the relevant neighbors (the yellow aura) of a potential new cell are described in terms of their offsets from the cell. For example, if a new cell is to be attached on the right, then all row offsets are checked except those with column offset -1, since that offset refers to the column containing the original cell. If the cells in the aura have no color or if the color is that of the border (crawling up the border is acceptable), then `CellOK` returns `True`.

```
Private Function CellOk(Pos As TPos, NewPos As TPos) As Boolean
    Dim i As Integer, j As Integer, C As Integer, InAura As Boolean
    Dim OK As Boolean, ro As Integer, co As Integer, Neighbor As TPos
    If CellColor(NewPos) <> xlNone Then      'new cell already has color
        OK = False: GoTo lastline                        'so reject
    End If
    ro = NewPos.row - Pos.row               'offsets of NewPos from Pos
    co = NewPos.col - Pos.col
    For i = -1 To 1          'check surrounding aura of potential new cell
        For j = -1 To 1
            InAura = (co > 0 And (j <> -1)) Or (co < 0 And (j <> 1)) Or _
                     (ro < 0 And (i <> 1))  Or (ro > 0 And (i <> -1))
            If InAura Then
                Neighbor.row = NewPos.row + i            'pos of aura cell
                Neighbor.col = NewPos.col + j
                C = CellColor(Neighbor)              'get its cell color
                OK = (C = xlNone) Or (C = 24)   'no color or border color
                If Not OK Then GoTo lastline
            End If
        Next j
    Next i
    lastline:
    CellOk = OK
End Function
```

10.8   Random Walk of a Rotating Pattern

The program in this section takes a user-entered pattern of x's in a grid, colors the pattern, and then successively reflects the pattern in diagonal, vertical, and horizontal axes passing through an origin, causing a spinning effect in a counterclockwise motion. In Figure 10.9 the dotted lines passing through the origin divide the grid into 8 regions that abut on the lines. The pattern in the figure labeled with the letters a,b,c,d starts in what we have designated as region 1, is reflected in a diagonal axis into region 2, then is reflected in a vertical axis into region 3, etc., until finally the pattern is in region 8. At this stage the origin randomly moves to another position and the cycle repeats.

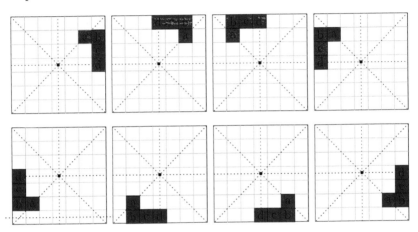

FIGURE 10.9: One rotation.

Figure 10.10 shows the rotation of the cell labelled 1 as it travels through the regions. The row, column transformations of the points 1–8 from one region

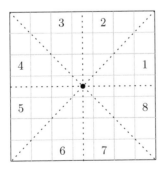

FIGURE 10.10: Journey of initial cell 1 in region 1 through regions 2–8.

to the next are given by the following rules, where (r_k, c_k) are the spreadsheet coordinates of cell labeled k and (r, c) are the spreadsheet coordinates of the origin, marked by a black dot in the figure.

$$(r_2, c_2) = (r_1 + c_1 - c, r + c - r_1) \qquad \text{from region 1 to region 2.}$$
$$(r_3, c_3) = (r_1, 2c - c_1) \qquad \text{from region 2 to region 3.}$$
$$(r_4, c_4) = (r - c + c_3, c - r + r_3) \qquad \text{from region 3 to region 4.}$$
$$(r_5, c_5) = (2r - r_4, c_4) \qquad \text{from region 4 to region 5.}$$
$$(r_6, c_6) = (r + c - c_5, r + c - r_5) \qquad \text{from region 5 to region 6.}$$
$$(r_7, c_7) = (r_7, 2c - c_6) \qquad \text{from region 6 to region 7.}$$
$$(r_8, c_8) = (r - c + c_7, c - r + r_7) \qquad \text{from region 7 to region 8.}$$
$$(r_1, c_1) = (2r - r_7, c_8) \qquad \text{from region 8 to region 1.}$$

After setting up the grid, the main procedure `RotateAndMove` uses the function `GetPattern` to populate the first part of the array `pattern` with the initial pattern entered by the user. The remaining parts are filled by `GetReflections`. The procedure `PrintReflections` successively prints the patterns in the eight regions of the coordinate system in a counterclockwise direction.

```
Public SpinRate As Integer, revolutions As Integer
Public orow As Integer, ocol As Integer
Sub RotateAndMove()
    Dim G As TGrid, patterns() As TPos, size As Integer
    Dim InitOrigin As TPos
    G = MakeGrid(6, 6, 34, 34, 3, 13, 24)
    revolutions = WorksheetFunction.Max(Range("B3").Value, 1)
    SpinRate = WorksheetFunction.Max(Range("B4").Value, 1)
    InitOrigin.row = (G.T + G.B) / 2:   InitOrigin.col = (G.L + G.R) / 2
    orow = InitOrigin.row: ocol = InitOrigin.col        'initial origin
    Call ColorCell(InitOrigin, 33)
    size = NumOccupied(G)                    'number of cells in pattern
    If size = 0 Then Exit Sub
    ReDim patterns(1 To 8, 1 To size)          '8 regions for pattern
    Call GetPattern(patterns, G)     'initialize region 1 of pattern array
    Call PrintReflections(patterns, G)             'do the reflections
    Call ClearGrid(G, 0)                         'clear current origin
    Call ColorCell(InitOrigin, 33)              'restore origin origin
End Sub
```

The procedure `GetPattern` fills the first of the eight parts of the array `patterns`. The procedure `GetReflections` constructs a reflected pattern from the previous one, starting with the initial pattern entered by the user.

```
Private Sub GetPattern(patterns() As TPos, G As TGrid)
    Dim i As Integer, j As Integer, m As Integer
    m = 1
    For i = G.T To G.B                      'scan grid for original pattern
        For j = G.L To G.R
            If Not IsEmpty(Cells(i, j)) Then
                'get position mth cell in initial pattern:
```

```
                patterns(1, m).row = i: patterns(1, m).col = j
                m = m + 1
            End If
        Next j
    Next i
End Sub

Private Sub GetReflections(patterns() As TPos)
    Dim region As Integer, ToCell As TPos, FromCell As TPos
    For region = 2 To 8
        For m = 1 To UBound(patterns, 2)
            FromCell = patterns(region - 1, m)
            Call ReflectCell(FromCell, ToCell, region - 1)
            patterns(region, m) = ToCell
        Next m
    Next region
End Sub
```

The procedure **ReflectCell** reflects cell position **FromCell** through the one of the eight axes into cell position **ToCell** of the next region.

```
Private Sub ReflectCell(FromCell As TPos, ToCell As TPos, r As Integer)
    Select Case r
        Case Is = 1: ToCell.row = orow + ocol - FromCell.col      '1 to 2
                     ToCell.col = ocol + orow - FromCell.row
        Case Is = 2: ToCell.row = FromCell.row                    '2 to 3
                     ToCell.col = 2 * ocol - FromCell.col
        Case Is = 3: ToCell.row = orow - ocol + FromCell.col      '3 to 4
                     ToCell.col = ocol - orow + FromCell.row
        Case Is = 4: ToCell.row = 2 * orow - FromCell.row         '4 to 5
                     ToCell.col = FromCell.col
        Case Is = 5: ToCell.row = orow + ocol - FromCell.col      '5 to 6
                     ToCell.col = ocol + orow - FromCell.row
        Case Is = 6: ToCell.row = FromCell.row                    '6 to 7
                     ToCell.col = 2 * ocol - FromCell.col
        Case Is = 7: ToCell.row = orow - ocol + FromCell.col      '7 to 8
                     ToCell.col = ocol - orow + FromCell.row
        Case Is = 8: ToCell.row = 2 * orow - FromCell.row         '8 to 1
                     ToCel.col = FromCell.col
    End Select
End Sub
```

The following two procedures print the reflections in a counterclockwise direction.

```
Private Sub PrintReflections(patterns() As TPos, G As TGrid)
    Dim region As Integer, k As Integer, Origin As TPos
    Origin.row = orow: Origin.col = ocol
    For k = 0 To 8 * revolutions
        Call Delay(SpinRate)
        region = k Mod 8 + 1                         'repeat every 8
```

```
        Call PrintReflection(patterns, region, 3, G)        'print pattern
        Call Delay(SpinRate)
        Call PrintReflection(patterns(), region, xlNone, G)    'erase it
        If k > 0 And region = 1 Then        'next revolution so translate
            Call ColorCell(Origin, xlNone)
            Call TranslateReflections(patterns, G)
            Origin.row = orow: Origin.col = ocol
            Call ColorCell(Origin, 33)
        End If
    Next k
End Sub


Private Sub PrintReflection(patterns() As TPos, region As Integer, _
                    color As Integer, G As TGrid)
    Dim m As Integer
    For m = 1 To UBound(patterns, 2)
        If InGrid(patterns(region, m), G) Then
            Call ColorCell(patterns(region, m), color)
        End If
        Call Delay(SpinRate)
    Next m
End Sub
```

The final procedure randomly translates the entire system, origin and pattern.

```
Private Sub TranslateReflections(patterns() As TPos, G As TGrid)
    Dim region As Integer, m As Integer, roff As Integer, coff As Integer
    Dim Origin As TPos, newOrigin As TPos
    Origin.row = orow: Origin.col = ocol
    newOrigin = NextRndPos(Origin)
    If Not InGrid(newOrigin, G) Then Exit Sub
    roff = newOrigin.row - orow: coff = newOrigin.col - ocol
    orow = newOrigin.row:  ocol = newOrigin.col
    For region = 1 To 8                              'offset patterns
        For m = 1 To UBound(patterns, 2)
            patterns(region, m).row = patterns(region, m).row + roff
            patterns(region, m).col = patterns(region, m).col + coff
        Next m
    Next region
End Sub
```

10.9   Random Tessellations

A *tessellation of the plane* is a paving by geometric shapes called *tiles*, juxtaposed so that there are no gaps or overlaps. For example, Figure 10.11

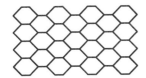 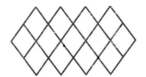 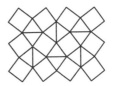

FIGURE 10.11: Tessellation figures

shows tessellations by hexagons, parallelograms, and mixed figures. The reader may wish to go online to discover many interesting tessellations, in particular those of historical significance found in paintings and in floor and wall coverings.

The module `RandomTess` in this section generates tessellations on a spreadsheet grid. The program fits different-colored congruent tiles together, the tiles having a shape determined by a *code table*. The numbers in the table indicate the shape of the tile. The number 1 in a spreadsheet cell indicates that the cell should be part of the tile. Cells in positions labelled 2 should be attached at the bottom. Cells in positions labelled 3 should be attached at the right Figure 10.12 indicates the scheme. The the program converts the code table into a tile table, as shown. The user has the option of either entering the code

CodeTable (5 × 7)   TileTable (10 × 14)   Tile

```
1 1 1 1 2 1 1        1 1 1 1     1 1
1 1 1 2 2 2 1        1 1 1           1
3 3 1 1 1 1 1            1 1 1 1 1 1 1
1 3 1 1 1 1 1        1   1 1 1 1 1     1
1 1 1 1 1 1 1        1 1 1 1 1 1 1
                                1
                        1 1 1
```

FIGURE 10.12: Tile code and tile.

table manually or having the program generate it randomly.

Because of the finite nature of a spreadsheet grid, the module generates tessellations on a torus. Figure 10.13 shows a tessellation with six tiles on a grid. The dashes refer to parts that have been placed in torus fashion on opposite sides of the grid.

The main procedure `RandomTess` begins by reading various user-entered parameters. The code table is placed into the array `CodeTab` and then converted by `MakeTile` to an array `TileTab`, which is used to print the tile. The conversion scheme is indicated in Figure 10.12, where blanks in the figure on the right represent zeros, not indicated for clarity. Ones in `TileMat` represent cells that are colored by `PrintTile` with the current color, while zeros represent cells that are not colored. The procedure `PrintTessellation` fits the tiles together to form the tessellation.

```
Public NumCodeRows As Integer, NumCodeCols As Integer
Public NumTileRows As Integer, NumTileCols As Integer
```

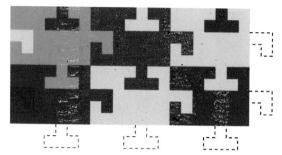

FIGURE 10.13: Tessellation on torus.

```
Public CodeRow As Integer, CodeCol As Integer
Public T As Integer, L As Integer, B As Integer, R As Integer
Sub Tessellations()
    Dim CodeTab() As Integer, TileTab() As Integer
    Rows("9:100"). RowHeight = 10
    Columns("C:CZ").ColumnWidth = 1.3
    CodeRow = 3: CodeCol = 4              'top left position of code table
    NumCodeRows = Range("B4")             'number of rows of code table
    NumCodeCols = Range("B5")             'number of columns of code table
    NumTileRows = Range("B6")             'number of rows of tile figures
    NumTileCols = Range("B7")             'number of columns of tile figures
    T = CodeRow + NumCodeRows + 1    ' dimensions of tessellation display
    L = CodeCol
    B = T + NumTileRows * NumCodeRows - 1
    R = L + NumTileCols * NumCodeCols - 1
    If Range("B3").Value = "y" Then      'if "y" then
        CodeTab = MakeRandomCodeTab()  'make a random code table for tile
    Else
        CodeTab = GetCodeTab()      'otherwise read user-entered code table
    End If
    TileTab = MakeTileTab(CodeTab)      'convert code table to tile table
    Call DisplayCodeTab(CodeTab)             'for random table
    'Call DisplayTileTab(TileTab)            'for testing
    Range("A1:CZ200").Interior.ColorIndex = 0   'clear old tessellation
    Call PrintTessellation(TileTab)
End Sub
```

The procedure `MakeTile` uses the array `CodeTable` to construct an array `table` of 1's and 0's (the latter placed at initialization) that reflects the shape of the tile.

```
Function MakeTileTab(CodeTab() As Integer) As Integer()
    Dim i As Integer, j As Integer, table() As Integer
    ReDim table(1 To 2 * NumCodeRows, 1 To 2 * NumCodeCols)
    For i = 1 To NumCodeRows
        For j = 1 To NumCodeCols
            Select Case CodeTab(i,  j)
                Case Is = 1: table(i, j) = 1                'keep as is
```

```
                  Case Is = 2: table(i + NumCodeRows,j) = 1    'put on bottom
                  Case Is = 3: table(i, j + NumCodeCols) = 1   'put at right
               End Select
           Next j
       Next i
       MakeTileTab = table
   End Function
```

The procedure **PrintTessellation** prints the tiles in different colors, calling **PrintTile** for each tile. The rows and columns are adjusted so that the tiles are printed as on a torus.

```
Sub PrintTessellation(TileTab() As Integer)
    Dim i As Integer, j As Integer, k As Integer
    k = RndInt(3,56)                            'color of first tile
    For i = 0 To NumTessRows - 1
        For j = 0 To NumTessCols - 1
            k = k + 1
            k = k Mod 56                              'next color
            If k = 0 Or k = 2 Then k = 3             'avoid white
            Call PrintTile(TileTab, T + i * NumCodeRows, _
                L + j * NumCodeCols, k)     'jump by dimensions of tile
        Next j
    Next i
    End Sub
```

```
Sub PrintTile(table() As Integer, row As Integer, col As Integer, color)
    Dim i As Integer, j As Integer, nrow As Integer, ncol As Integer
    Dim TileRow As Integer, TileCol As Integer
    nrow = UBound(table, 1): ncol = UBound(table, 2)
    For i = 1 To nrow        '(row,col) = position of upper left of tile
        For j = 1 To ncol
            If table(i, j) = 1 Then
                TileRow = i + row - 1              '(row,col) = position of
                TileCol = j + col - 1             'upper left of tile
                If TileRow > B Then TileRow = T + TileRow - B - 1
                If TileCol > R Then TileCol = L + TileCol - R - 1
                Cells(TileRow, TileCol).Interior.ColorIndex = color
            End If
        Next j
    Next i
End Sub
```

The construction of the random code table consists of two short random walks through a table initialized with 1's, these turning into 2's along one walk and into 3's along the other.

```
Function MakeRandomCodeTab() As Integer()
    Dim i As Integer, j As Integer, table() As Integer, count As Integer
    Dim row As Integer, col As Integer 'current position of random walks
    Dim newrow As Integer, newcol As Integer          'new position
```

```
Dim numTwos As Integer, numThrees As Integer      'count 2's and 3's
ReDim table(1 To NumCodeRows, 1 To NumCodeCols)    'for random code
For i = 1 To NumCodeRows                        'initialize with 1's
    For j = 1 To NumCodeCols
        table(i, j) = 1
    Next j
Next i
'''''''''''' the following code places random 2's in table ''''''''
k = RndInt(2, NumCodeCols - 1)                'get a random column number
table(1, k) = 2      'initialize with two adjacent cells in rows 1,2
table(2, k) = 2                  'other choices produce other effects
numTwos = 2              'keep track of number of twos placed in table
row = 2: col = k                      'position of beginning of walk
Do While numTwos <= 6 And count < 50   'change for different effects
    newrow = row: newcol = col
    Select Case RndInt(1, 4)       'get a random offset between 1 and 4
            Case Is = 1: newcol = col + 1
            Case Is = 2: newcol = col - 1
            Case Is = 3: newrow = row - 1
            Case Is = 4: newrow = row + 1
    End Select
    If 1 <= newrow And newrow <= NumCodeRows And 1 <= newcol _
    And newcol <= NumCodeCols Then   'if new position is in table then
        row = newrow: col = newcol                    'update position
        table(row, col) = 2                        'enter a 2 in table
        numTwos = numTwos + 1                          'update count
    End If
    count = count + 1
Loop
'''''''''''' the following code places random 3's in table ''''''''
k = RndInt(NumCodeRows - 1, 2)
table(k, 1) = 3: table(k, 2) = 3
numThrees = 2: row = k: col = 2
Do While numThrees <= 6 And count < 100
    newrow = row: newcol = col
    Select Case RndInt(1, 4)
            Case Is = 1: newcol = col + 1
            Case Is = 2: newcol = col - 1
            Case Is = 3: newrow = row - 1
            Case Is = 4: newrow = row + 1
    End Select
    If 1 <= newrow And newrow <= NumCodeRows And 1 <= newcol
    And newcol <= NumCodeCols Then
        row = newrow: col = newcol
        If table(row, col) <> 2 Then            'if not already a 2 then
            table(row, col) = 3                     'enter 3 in table
            numThrees = numThrees + 1
        End If
    End If
    count = count + 1
Loop
```

```
    MakeRandomCodeTab = table
End Function
```

Here is the procedure that reads the code table from the spreadsheet if one was entered by the user.

```
Function GetCodeTab() As Integer()
    Dim i As Integer, j As Integer, table() As Integer
    Dim m As Integer, n As Integer
    Do Until IsEmpty(Cells(CodeRow + m, CodeCol))    'get number of rows
        m = m + 1                                    'of table
    Loop
    Do Until IsEmpty(Cells(CodeRow, CodeCol+n))    'get number of columns
        n = n + 1                                  'of table
    Loop
    ReDim table(1 To m, 1 To n)                     'table array
    For i = 1 To m
        For j = 1 To n
            table(i, j) = Cells(CodeRow + i - 1, CodeCol + j - 1).Value
        Next j
    Next i
    GetCodeTab = table                              'return table array
End Function
```

The following procedure displays the random code table if that option was selected by the user.

```
Sub DisplayCodeTab(CodeTab() As Integer)
    Dim i As Integer, j As Integer
    For i = 1 To NumCodeRows
        For j = 1 To NumCodeCols
        Cells(CodeRow - 1 + i, CodeCol - 1 + j) = CodeTab(i, j)
        Next j
     Next i
End Sub
```

10.10   Maxwell's Entropy Demon

In this section we implement a thought experiment devised by the British physicist Clerk Maxwell, which purports to refute the second law of thermodynamics. The law asserts that an isolated system consisting of two bodies of different temperatures, say an ice cube and glass of hot water, will eventually, through the motion of the water molecules, approach an equilibrium state in which both bodies have the same temperature. Thus the ice cube will eventually melt producing a glass of water with uniform temperature, but the process can never reverse. This is an instance of the second law of thermodynamics: in a closed system entropy can never decrease.

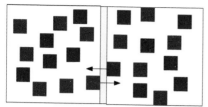
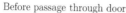
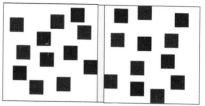

Before passage through door

After passage through door

FIGURE 10.14: The demon at work.

Maxwell's thought experiment to decrease entropy involved a vessel partitioned into right and left sections, each containing randomly moving molecules (Figure 10.14). The partition separating the sections contains a door operated by a demon. If a faster-than-average (pictured as red) molecule in the left section moves towards the door the demon opens the door to allow the molecule to enter the right section. Similarly, if a slower-than-average (blue) molecule in the right section moves towards the door the demon opens the door to allow the molecule to enter the left section. In this way the vessel's right section becomes hot and the left section cold: the metaphorical ice cube is reassembled. This apparent negation of the second law of thermodynamics has been refuted by the observation that the demon's efforts to distinguish swift from slow molecules violate the requirement that the system be isolated.

The following program starts with a partitioned grid containing a random mixture of red (fast) and blue (slow) molecules. For simplicity, instead of an actual demon operating the door (demons are notoriously difficult to code), the program allows any red molecule that touches the left side of the partition to pass to the right and any blue molecule that touches the right side of the partition to pass to the left. To facilitate the coding we shall use the UDT `TMole` to describe the grid position, color, and the side of a door that a molecule currently finds itself on.

The procedure `MaxwellsDemon` launches the experiment. After setting up the grid and the partition column `pCol`, it reads the user entered values `numSteps` and `density`. The latter determines how many molecules in the container: the larger the setting the less crowded the container becomes. It should be set to at least 10 to avoid overcrowding.

```
Private Type TMole     'molecule type: position, color, side of container
    Pos As TPos: color As Integer: side As String
End Type
Public pCol As Integer                      'center column for partition
Sub MaxwellsDemon()
    Dim Mole() As TMole, G As TGrid, i As Integer, numSteps As Integer
    Dim T As Integer, L As Integer, B As Integer, R As Integer
    Dim density As Integer
    T = 5: L = 4: B = 25: R = 32: pCol = (L + R)/2 'grid & center column
    G = MakeGrid(T, L, B, R, 2.3, 12, 24)
```

```
density = Range("B3").Value      'the higher the value the less packed
If density < 10 Then density = 10              'not too packed
numSteps = Range("B4").Value
Mole = Initialize(G, density)       'populate container with molecules
For i = 1 To numSteps                   'run the experiment
    Call NewPositions(Mole)
Next i
MsgBox "clear"
Range(Cells(G.T, G.L), Cells(G.B, G.R)).Interior.ColorIndex = xlNone
End Sub
```

The function `Initialize` randomly fills the grid with red and blue cells
using `RndInt`. While the integers produced have values from 1 to `density`, only
the integers 1 and 2 are used. Thus the larger the value of `density` the less
frequently a molecule (cell) is created, hence the less dense the system.

```
Private Function Initialize(G As TGrid, density As Integer) As TMole()
    Dim i As Integer, j As Integer, m As Integer, k As Integer
    Dim Mole() As TMole, temp() As TMole
               'install partition column
    Range(Cells(G.T, pCol), Cells(G.B, pCol)).Interior.ColorIndex = 24
    ReDim temp(1 To (G.B - G.T) * (G.R - G.L))       'for molecule specs.
    m = 1
    For i = G.T To G.B
        For j = G.L To G.R
            If j = pCol Then GoTo continue
            k = RndInt(1, density)       'random number for coloring cell
            If k = 1 Then
                temp(m).color = 8                       'blue
            ElseIf k = 2 Then
                temp(m).color = 3                       'red
            Else: GoTo continue
            End If
            If j < pCol - 1 Then
                temp(m).side = "L"              'molecule on left side
            Else: temp(m).side = "R"            'molecule on right side
            End If
            temp(m).Pos.row = i: temp(m).Pos.col = j   'location of cell
            Call ColorCell(temp(m).Pos, temp(m).color)
            m = m + 1
    continue:
        Next j
    Next i
    ReDim Mole(1 To m - 1) 'now have m-1 molecules
    For i = 1 To m - 1
        Mole(i) = temp(i)        'transfer to smaller array for efficiency
    Next i
    Initialize = Mole                       'return molecule spec array
End Function
```

The next procedure is heart of the module. It moves the molecules randomly and lets the molecules through the partition, red from left to right, blue from right to left.

```
Private Sub NewPositions(Mole() As TMole)
    Dim m As Integer, NewPos As TPos, OldColor As Integer
    For m = 1 To UBound(Mole)              'move all the molecules one step
        NewPos = NextRandPos(Mole(m).Pos)
            'deal with blue molecules on right side of door
        If Mole(m).Pos.col = pCol + 1 _
        And Mole(m).side = "R" And Mole(m).color = 8 Then
            Call ColorCell(Mole(m).Pos, xlNone)   'remove color from blue
            Mole(m).Pos.col = Mole(m).Pos.col - 1      'now inside door
            Call ColorCell(Mole(m).Pos, Mole(m).color)   'color cell and
            Call Delay(15)              'delay it as it passes through door
            Call ColorCell(Mole(m).Pos, 24)       'restore partition color
            Mole(m).Pos.col = Mole(m).Pos.col - 2  'on left side of door
            Call ColorCell(Mole(m).Pos, Mole(m).color)        'color it
            Mole(m).side = "L"
            'deal similarly with red molecules on left side of door
        ElseIf Mole(m).Pos.col = pCol - 1 _
        And Mole(m).side = "L" And Mole(m).color = 3 Then
            Call ColorCell(Mole(m).Pos, xlNone)
            Mole(m).Pos.col = Mole(m).Pos.col + 1
            Call ColorCell(Mole(m).Pos, Mole(m).color)
            Call Delay(15)
            Call ColorCell(Mole(m).Pos, 24)
            Mole(m).Pos.col = Mole(m).Pos.col + 2
            Call ColorCell(Mole(m).Pos, Mole(m).color)
            Mole(m).side = "R"
            'deal with cells not at door
        ElseIf CellColor(NewPos) = xlNone Then
            Call ColorCell(Mole(m).Pos, xlNone)
            Call ColorCell(NewPos, Mole(m).color)
            Mole(m).Pos = NewPos      'update position of mth particle
        End If
    Next m
End Sub
```

10.11   Multi-Random Walk

The module in this section extends the random walk of Section 10.4 by replacing the single cell with several many-celled "organisms." The user specifies the length, position, and color of each organism as well as the number of steps and the speed of the motion. Running the program causes single-celled organisms to emerge from their "birth cells," grow to their specified lengths, and then move randomly.

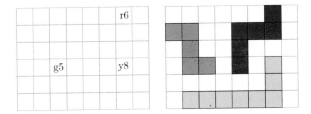

FIGURE 10.15: Multi-cell random walks.

Figure 10.15 depicts an abbreviated version of the grid on which the motion takes place. In the figure on the left, the user has entered into three cells data in the form of a letter specifying the color of the organism and a number specifying the length. Thus the label g5 in a cell designates that a green organism of length 5 should emerge from that cell. For later reference we refer to such a combination of letter and number as a *color-length code*. The figure on the right depicts fully grown organisms executing random walks. The program includes code to prevent the organisms from colliding.

The module uses the new data type

```
Private Type TOrg
    Body(1 To 100) As TPos: fullLen As Integer
    currLen As Integer: color As Integer
End Type
```

Here, `Body` contains the positions of the cells of the organism with `Body(1)` the position of the head. The variable `currLen` is the current length of the organism as it is growing, and `fullLen` its length at adulthood. The latter will always be no bigger then `maxLen`, which is arbitrarily set at 100.

The main procedure is `RandomWalks`. The user-entered values of `Speed` and `Steps` set the rate of motion and the number of steps the organisms will take. The procedure declares an array `Org` with entries of type `TOrg` for the organisms characteristics. Thus `Org(m).color` is the color of the mth organism, `Org(m).currLen` its current size, and `Org(m).fullLen` its final size. The kth cell of organism m, starting from the head, is located at `Org(m).Body(k).row`, `Org(m).Body(k).col`. After "birth," the organism grows from a single cell to its specified length. During growth the value `Org(m).currLen` is incremented until the organism reaches the value `Org(m).fullLen`. The `Public` variable `NumOrg` contains the number of color-length codes that were entered by the user.

```
Public NumOrg As Integer, maxLen As Integer
Sub RandomWalks()
    Dim Grid As TGrid, Org() As TOrg
    Dim k As Integer, m As Integer, steps As Integer, rate As Integer
    Dim gridheight As Integer, gridwidth As Integer
    gridheight = Range("B6").Value: gridwidth = Range("B7").Value
    Grid = MakeGrid(5, 5, 5 + gridheight, 5 + gridwidth, 2.3, 15, 24)
```

```
      rate = Range("B4").Value: steps = Range("B5").Value
      maxLen = 100            'max. no. of cells in an organism (arbitrary)
      NumOrg = NumOccupied(Grid)
      If NumOrg = 0 Then Exit Sub
      ReDim Org(1 To NumOrg)
      Call Initialize(Org, Grid)
      For k = 1 To steps                                 'do the walk
          For m = 1 To NumOrg                     'run through the organisms
              Call MoveOneStep(Org, m)
          Next m
          Call Delay(rate)
      Next k
      MsgBox "clear"
      Call ClearGrid(Grid, 0)
  End Sub
```

The function `NumOccupied` from Chapter 7 counts the number of cells containing color-length codes. The procedure `Initialize` scans the grid looking for these. Each time it finds one it records the color in `Org(m).color` and the desired length in `Org(m).fullLen`. The current length `Org(m).currLen` is set to 1, which is the length at birth.

```
  Private Sub Initialize(Org() As TOrg, Grid As TGrid)
      Dim i As Integer, j As Integer, m As Integer, cl As String
      m = 1                                              'organism 1
      For i = Grid.T To Grid.B       'scan grid for color length (cl) codes
          For j = Grid.L To Grid.R
              If m <= NumOrg And Not IsEmpty(Cells(i, j)) Then
                  Org(m).Body(1).row = i Org(m).Body(1).col = j  'head pos.
                  cl = Cells(i, j).Value         'color-length specification
                  Org(m).color = LetterToColor(Mid(cl, 1, 1))
                  Org(m).fullLen = Int(Mid(cl, 2, Len(cl) - 1))
                  If Org(m).fullLen > MaxLen Then Org(m).fullLen = MaxLen
                  Org(m).currLen = 1
                  Cells(i, j).Interior.ColorIndex = Org(m).color
                  m = m + 1                                'next organism
              End If
          Next j
      Next i
  End Sub
```

The procedure `LetterToColor` converts the letter part of the color-length code to a `ColorIndex`. These may be supplemented by the reader.

```
  Function LetterToColor(letter As String)
      Dim color As Integer
      Select Case letter
          Case Is = "r": color = 3     'red
          Case Is = "g": color = 4     'green
          Case Is = "p": color = 7     'pink purple
          Case Is = "b": color = 32    'blue
```

```
            Case Is = "o": color = 46    'orange
            Case Is = "y": color = 6     'yellow
            Case Is = "s": color = 44    'sunset
            Case Is = "a": color = 12    'army
            Case Is = "q": color = 28    'aqua
            Case Else: color = 1         'black
        End Select
        LetterToColor = color
    End Function
```

The procedure `MoveOneStep` moves all the organisms one step. For each m the procedure `NextRandPos` of Section 10.4 finds a position for the new head of organism m. This position is checked by the function `CellOk`, described in Section 10.7, to make sure that the organism doesn't collide with itself or other organisms. If `CellOk` approves of the new position, then the organism is allowed to move. Because the restrictions imposed by `CellOk` may make forward motion impossible, after 10 attempts the tail and head are switched and motion is reversed. It could still be the case that neither forward nor backward motion is possible, resulting in a frustrated frozen organism, as illustrated in Figure 10.16.

FIGURE 10.16: Organism can go neither forward nor backward.

```
Private Sub MoveOneStep(Org() As TOrg, m As Integer)
    Dim Attempt As Integer, TailPos As TPos, NewHeadPos As TPos
    Dim i As Integer, j As Integer
    For Attempt = 1 To 10
        NewHeadPos = NextRndPos(Org(m).Body(1))
        If CellOk(Org(m).Body(1), NewHeadPos) Then
            Call MoveOrg(m, NewHeadPos, Org)
            Exit For                        'organism successfully moved
        End If
        If Attempt = 10 Then        'if can't move forward, then set for
            Call ReverseCells(Org(m).Body,Org(m).currLen) 'reverse motion
        End If
    Next Attempt
End Sub
```

The procedure `MoveOrg` carries out the movement of an individual organism `Org(m)`. This involves three steps: First, if the organism is not fully grown, as indicated by the inequality `Org(m).currLen < Org(m).fullLen`, then the current number of cells is incremented. Otherwise, the tail of the organism, which is at position `Org(m).Body(currLen)`, is deleted. Second, the entries in the array `Org(m).Body` are shifted down to allow the position `NewHeadPos` of the new head to be assigned to `Org(m).Body(1)`. Third, the new head is given the appropriate color.

```
Private Sub MoveOrg(m As Integer, NewHeadPos As TPos, Org() As TOrg)
    If Org(m).currLen < Org(m).fullLen Then       'if organism incomplete
        Org(m).currLen = Org(m).currLen + 1        'then increase length
    Else                                          'otherwise delete tail
        Call ColorCell(Org(m).Body(Org(m).fullLen), xlNone)
    End If
    Call ShiftDown(Org(m).Body)                   'make room for the new head
    Org(m).Body(1) = NewHeadPos                         'update
    Call ColorCell(NewHeadPos, Org(m).color)
End Sub
```

Here are the array procedures called by `MoveOneStep`. The first shifts down the position entries of the *m*th organism to make room for a new head. The second reverses the entries of the position array.

```
Private Sub ShiftDown(Pos() As TPos)
    Dim k As Integer
    For k = UBound(Pos) To 2 Step -1       'move entries down for new head
        Pos(k) = Pos(k - 1)
    Next k
End Sub
```

```
Private Sub ReverseCells(Pos() As TPos, length As Integer)
    Dim k As Integer, TempPos() As TPos
    ReDim TempPos(1 To length)
    For k = 1 To length                        'current number of cells
        TempPos(k) = Pos(k)
    Next k
    For k = 1 To length
        Pos(k) = TempPos(length - k + 1)       'reverse order of entries
    Next k
End Sub
```

10.12   A Shedding Organism

The program in this section causes a single multi-celled, randomly moving red-celled "organism" to shed its tail cells until there are only two cells left, whereupon the creature gradually reassembles itself. The organism grows from a single cell and moves randomly until at some predetermined stage the shedding takes place. It then continues to move randomly while shedding.

At the same time the tail cells are also moving randomly. These "offspring" eventually reattach to the "mother organism" in the reverse order in which they were shed. Figure 10.17 illustrates the various phases of the organism.

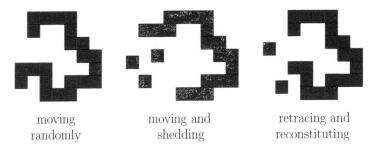

moving   moving and   retracing and
randomly   shedding   reconstituting

FIGURE 10.17: Evolution of the organism.

The main procedure `RandomWalkShed` sets up the grid for the organism and then reads the user-entered values of `NumCells` (the number of cells in the organism), `rate` (the rate at which the organism moves, e.g. 5 for very fast, 20 for very slow), and `NumSteps` (the number of steps the mother organism takes in her journey). Next, the grid is cleared of any previous cells and Sheet2 is cleared of any data related to the reconstitution. Memory is then allocated for the mother and eventual offspring. The procedure `Initialize` causes the birth of the mother. Shedding begins with `RunShed` sometime thereafter.

```
Public T As Integer, L As Integer, B As Integer, R As Integer
Public LogRow As Integer, rate As Integer, NumCells As Integer
Public NumOffSpring As Integer, CurrLength  As Integer, NumSteps
Sub RandomWalkShed()
    Dim MotherPos() As TPos, Grid As TGrid, Head As TPos
    Dim OffSpringPos() As TPos
    gridheight = Range("B6").Value    'allow user to set grid dimensions
    gridwidth = Range("B7").Value
    T = 5: L = 5:  B = T + gridheight: R = L + gridwidth    'boundaries
    Grid = MakeGrid(T, L, B, R, 2.3, 15, 24)
    NumCells = Range("B3").Value: NumSteps = Range("B5").Value
    rate = Range("B4").Value
    Sheets("Sheet2").Cells.Clear                        'clear old record
    NumOffSpring = 0: CurrLength = 1: LogRow = 1             'initialize
    ReDim MotherPos(1 To NumCells)        'memory for original organism
    ReDim OffSpringPos(1 To NumCells)          'memory for shed cells
    Call GrowMother(MotherPos)
    Call RunShed(MotherPos, OffSpringPos)            'start movement
    Call Delay(6)                   'wait a bit before reconstituting
    Call Rewind                          'reconstitute mama
End Sub
```

The procedure `GrowMother` successively adds cell positions to the mother array, colors the cells, and records the events for later playback. The procedure uses `ShiftDown` and `ReverseCells` of the preceding section with `maxLen` in the code replaced by `NumCells`.

```
Private Sub GrowMother(MPos() As TPos)
    Dim attempt As Integer, NewHead As TPos, Head As TPos
    Head.row = (T + B) / 2: Head.col = (L + R) / 2  'mom embryo position
    Call ColorCell(Head, 3)
    Call Record(Head, 3)      'record location and index color 3 in sheet2
    MPos(1) = Head                        'initial position of mom
    For m = 2 To NumCells
        For attempt = 1 To 10         'try 10 times to grow another cell
            NewHead = NextRndPos(MPos(1))
            If CellOk(MPos(1), NewHead) Then
                Call ColorCell(NewHead, 3)          'new head gets color
                Call Record(NewHead, 3)              'record event
                CurrLength = CurrLength + 1     'mother grew by one cell
                Call ShiftDown(MPos)         'make room for the new head
                MPos(1) = NewHead                     'update position
                Call Delay(rate)
                Exit For
            End If
            If attempt = 10 Then 'if can't move forward then setup array
                Call ReverseCells(MPos, CurrLength)          'for reverse
            End If
        Next attempt
    Next m
End Sub
```

The procedure `RunShed` starts the shedding when the mother has moved one half of the prescribed number of steps. Shedding takes place every other step per the statement `k Mod 2 = 0`. Remove the statement for faster shedding, or replace it with, say `k Mod 5 = 0` for slow shedding. (Or add code that lets the user make the decision.)

```
Private Sub RunShed(MPos() As TPos, OPos() As TPos)
    Dim k As Integer
    For k = 1 To NumSteps
        Call MoveOrgs(MPos, OPos)                  'move current organisms
        If k > NumSteps / 2 And k Mod 2 = 0 And CurrLength > 2 Then
            Call ShedCell(MPos, OPos)
        End If
        Call Delay(rate)
    Next k
End Sub
```

```
Private Sub ShedCell(MPos() As TPos, OPos() As TPos)
    Dim Tail As TPos
    Tail = MPos(CurrLength)
    Call ColorCell(Tail, xlNone)                       'remove tail
    Call Record(Tail, xlNone)                          'record event
    If CurrLength > 1 Then CurrLength = CurrLength - 1  'chop off a cell
```

```
        NumOffSpring = NumOffSpring + 1        'now have one more offspring
        OPos(NumOffSpring) = Tail              'new offspring gets old tail cell
    End Sub
```

The procedure `MoveOrgs` moves the entire set of organisms, mother and offspring, one step at a time.

```
Private Sub MoveOrgs(MPos() As TPos, OPos() As TPos)
    Dim m As Integer
    Call MoveMotherOneStep(MPos)
    If NumOffSpring > 0 Then
        For m = 1 To NumOffSpring
            Call MoveOffSpringOneStep(m, OPos)
        Next m
    End If
End Sub
```

The procedure `MoveMotherOneStep` uses `NextRndPos` to get the next head position of the mother. If the position is ok, then she is moved.

```
Private Sub MoveMotherOneStep(MPos() As TPos)
    Dim attempt As Integer, NewHead As TPos, Tail As TPos
    For attempt = 1 To 10                  'try 10 times to move mother
        NewHead = NextRndPos(MPos(1))
        If CellOk(MPos(1), NewHead) Then
            Call ColorCell(NewHead, 3)              'new head gets color
            Call Record(NewHead, 3)                    'record event
            Tail = MPos(CurrLength)
            Call ColorCell(Tail, xlNone)            'remove tail cell
            Call Record(Tail, xlNone)                  'record event
            Call ShiftDown(MPos)             'make room for the new head
            MPos(1) = NewHead
            Exit For
        End If
        If attempt = 10 Then     'if can't move forward, then setup array
            Call ReverseCells(MPos, CurrLength)       'for reverse motion
        End If
    Next attempt
End Sub
```

The procedure `MoveOffspringOneStep` moves the *m*th offspring one step in a manner similar to the mother's movement.

```
Private Sub MoveOffspringOneStep(m A Integer, Pos() As TPos)
    Dim i As Integer, j As Integer, attempt As Integer, NewPos As TPos
    For attempt = 1 To 10                  'try 10 times to get an ok cell
        NewPos = NextRndPos(Pos(m))
        If CellOk(Pos(m), NewPos) Then
            Call ColorCell(Pos(m), xlNone)   'remove color from old cell
            Call Record(Pos(m), xlNone)            'and record event
            Pos(m) = NewPos                        'update position
```

```
            Call ColorCell(Pos(m), 3)              'color new cell
            Call Record(Pos(m), 3)                 'and record event
            Exit For
        End If
    Next attempt
End Sub
```

The procedure `Record` keeps a log in Sheet2 of the color events of the program. The procedure `Rewind` simply reverses the order of events.

```
Private Sub Record(Pos As TPos, color As Integer)
    Sheet2.Cells(LogRow, 1).Value = Pos.row
    Sheet2.Cells(LogRow, 2).Value = Pos.col
    Sheet2.Cells(LogRow, 3).Value = color
    LogRow = LogRow + 1
End Sub

Sub Rewind()
    Dim i As Integer, row  As Integer, col As Integer, color As Integer
    For i = LogRow To 1 Step -1        'start at the bottom and work up
        row = Sheet2.Cells(i, 1).Value    'get position, color of event
        col = Sheet2.Cells(i, 2).Value
        color = Sheet2.Cells(i, 3).Value
        If color = 3 Then                        'reverse color event
            Cells(row, col).Interior.ColorIndex = 0
        ElseIf color = xlNone Then
            Cells(row, col).Interior.ColorIndex = 3
        Call Delay(5)
        End If
    Next i
End Sub
```

10.13   A Splitting Organism

The program in this section causes a single randomly moving "organism" to split into two randomly moving organisms, each of these splitting in two, etc. The program then "rewinds" the motion, reconstituting the organism in reverse order. The original organism is given a spiral shape of 16, 32, 64, 128, 256 or 512 cells, the choice made by the user.

The main procedure `RandomWalkSplit` begins by allocating memory for the organisms. The procedure `Initialize` creates the original organism in the shape of a spiral. The procedure `RunSplits` directs the movement and the splitting of the organisms, recording each event in Sheet2. The entire aggregate motion is played backwards by `Rewind` using the color event history in Sheet2.

```
Public LogRow As Integer, maxLen As Integer, rate As Integer
Public power As Integer
Sub RandomWalkSplit()
    Dim Grid As TGrid, Org() As TOrg
    Dim gridheight As Integer, gridwidth As Integer
    Sheets("Sheet2").Cells.Clear
    gridheight = Range("B6").Value: gridwidth = Range("B7").Value
    Grid = MakeGrid(5, 5, 5 + gridheight, 5 + gridwidth, 2.3, 15, 24)
    power = Range("B4").Value: rate = Range("B5").Value
    If power > 9 Or power < 4 Then MsgBox "check power"
    maxLen = 2 ^ power                          'size of initial organism
    ReDim Org(1 To maxLen)
    LogRow = 1
    Call InitMother(Org, Grid)
    Call RunSplits(Org)
    Call Rewind
End Sub
```

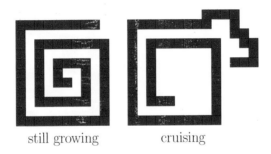

still growing   cruising

FIGURE 10.18: Evolution of the organism.

In contrast to `GrowMother` in Section 10.12, the procedure `InitMother` creates the initial organism in the shape of a spiral. It uses the offset method described in the module `GrowSpiral` of Section 4.2. The spiral begins with a single cell in the center of the grid and expands outward until it reaches its full length. Figure 10.18 shows the organism at mid growth and then starting to randomly unwind.

Sheet2 is used to store the position and color of a cell immediately before the color is changed, allowing the original state of the cell to be recovered during the rewinding process. The information is recorded by the procedure `Record(Pos,color,advance)`, described later. The variable `advance` is set to 1 if the row number in Sheet2 that contains the information for an organism is to be incremented. Otherwise it is set to 0.

```
Private Sub InitMother(Org() As TOrg, Grid As TGrid)
    Dim i As Integer, j As Integer, CellCount As Integer, Head As TPos
    Dim offset(0 To 3) As Integer
    offset(0) = 1: offset(1) = 0: offset(2) = -1: offset(3) = 0
    Head.row = (Grid.T + Grid.B) / 2                      'initial cell
```

```
    Head.col = (Grid.L + Grid.R) / 2
    For i = 1 To Grid.B - Grid.T                    'generate positions
        For j = 1 To i                           'repeat each offset i times
            Call Record(Head, 0, 1)      'record original 0 color; advance
            Call ColorCell(Head, 3)
            Call Record(Head, 3, 1)                    'record new color and
            Org(1).Body(maxLen - CellCount) = Head      'advance position
            CellCount = CellCount + 1
            If CellCount = maxLen Then Exit Sub
            Head.row = Head.row + offset(i Mod 4)      'generates 0,-1,0,1
            Head.col = Head.col + offset((i + 1) Mod 4)          '-1,0,1,0
            Call Delay(8)
        Next j
    Next i
End Sub
```

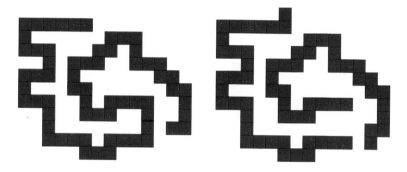

FIGURE 10.19: Splitting of the organism.

The procedure `RunSplits` below uses a Select Case statement to alternately move and split the organisms. For the first 99 steps, only the original organism moves. At step 100, the organism is split into two organisms. For the next 99 steps the two organisms execute separate random motions. The process of splitting and moving continues until the value of `NumSteps` is reached. The numbers selected for the latter were arrived at by experimentation to ensure pleasing visual effects. The reader may wish to fiddle with these. Figure 10.19 illustrates the state of affairs one step after the 64-celled organism on the left divides (at the dark blue line segment).

```
Private Sub RunSplits(Org() As TOrg)
    Dim k As Integer, NumSteps As Integer
    Select Case power   'select number of steps according to size of org.
        Case Is = 4: NumSteps = 399
        Case Is = 5: NumSteps = 424
        Case Is = 6: NumSteps = 434
        Case Is = 7: NumSteps = 499
        Case Is = 8: NumSteps = 509
        Case Is = 9: NumSteps = 440
```

```
End Select
For k = 1 To NumSteps
Select Case k
      Case Is < 100:  Call MoveOrgs(Org, 1)        'move original org.
      Case Is = 100:  Call SplitOrgs(Org, 1)        'split into 2
      Case Is < 200:  Call MoveOrgs(Org, 2)         'move these
      Case Is = 200:  Call SplitOrgs(Org, 2)        'split into 4
      Case Is < 300:  Call MoveOrgs(Org, 4)         'move these
      Case Is = 300:  Call SplitOrgs(Org, 4)        'split into 8
      Case Is < 400:  Call MoveOrgs(Org, 8)         'move these
      Case Is = 400:  Call SplitOrgs(Org, 8)        'split into 16
      Case Is < 425:  Call MoveOrgs(Org, 16)        'move these
      Case Is = 425:  Call SplitOrgs(Org, 16)       'split into 32
      Case Is < 435:  Call MoveOrgs(Org, 32)        'move these
      Case Is = 435:  Call SplitOrgs(Org, 32)       'split into 64
      Case Is < 440:  Call MoveOrgs(Org, 64)
End Select
'Call Delay(rate)
Next k
End Sub
```

The procedure `SplitOrgs` is passed the array `Org` and the current number `NumOrgs` of organisms. It then proceeds to divide the arrays into twice as many, each with half the length. Figure 10.20 indicates how this works for the row arrays of two 32-celled organisms. (The position entries `Org(i).Body(j)` are abbreviated $O(i,j)$.) Nested For Next loops carry out the splitting, copying

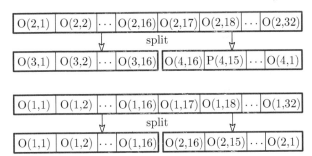

FIGURE 10.20: Splitting two row arrays into four.

the array `Org(2).Body` into the arrays `Org(3).Body` and `Org(4).Body` in such a way so that `Org(3).Body` contains bottom half of the old `Org(2).Body` and `Org(3).Body` contains the top half of the old `Org(2).Body` in reverse (causing the new organisms to initially repel each other).

```
Private Sub SplitOrgs(Org() As TOrg, NumOrgs As Integer)
    Dim m As Integer, k As Integer, OrgSize, Head As TPos
    OrgSize = maxLen / NumOrgs
    For m = NumOrgs To 1 Step -1              'm goes to 2m,2m-1
        For k = 1 To OrgSize / 2 'bottom half in reverse, then 'top half
            Org(2 * m).Body(k) = Org(m).Body(OrgSize - k + 1)
```

```
            If m > 1 Then Org(2 * m - 1).Body(k) = Org(m).Body(k)
            End If
        Next k
    Next m
    LogRow = LogRow + 1
End Sub
```

The procedure `MoveOrgs` uses a For Next loop to move the current group of organisms precisely one step. This is done by the function `MoveOneStep` using the procedures `NextRandPos`, `CellOk`, `ShiftDown`, and `ReverseCells` of the module `RandomWalks`.

```
Private Sub MoveOrgs(Org() As TOrg, NumOrgs As Integer)
    Dim m As Integer
    For m = 1 To NumOrgs
        Call MoveOneStep(Org(m).Body, NumOrgs)          'move organism m
    Next m
    LogRow = LogRow + 1       'new log row after all organisms have moved
End Sub
```

```
Private Sub MoveOneStep(Pos() As TPos, NumOrgs As Integer)
    Dim attempt As Integer, OrgSize As Integer, i As Integer
    Dim j As Integer  NewHead As TPos, OldHead As TPos, OldTail As TPos
    OldHead = Pos(1): OrgSize = maxLen / NumOrgs
    For attempt = 1 To 10
        Call RndOff(i, j, 1)
        NewHead = NextRndPos(OldHead)
        If CellOk(OldHead, NewHead) Then
            Call Record(OldHead, 3, 0)          'record old position, color
            Call Record(NewHead, 0, 0)
            Call ColorCell(NewHead, 3)
            OldTail = Pos(OrgSize)
            Call ColorCell(OldTail, xlNone)                 'remove tail
            Call Record(OldTail, 3, 0)
            Call ShiftDown(Pos)              'make room for the new head
            Pos(1) = NewHead
            Exit For
        End If
        If attempt = 10 Then              'if can't move forward, then set
            Call ReverseCells(Pos, OrgSize)    'arrays for reverse motion
        End If
    Next attempt
End Sub
```

The procedure `Record` notes in row `LogRow` of Sheet2 the color change of each cell using the data triple `row`, `column`, `oldcolor`. The variable `LogRow` is incremented after all of the organisms have moved one step. Thus (except for the initialization phase) each row of Sheet2 contains, in groups of three, the old color information for all organisms during a single step.

```
Private Sub Record(Pos As TPos, oldcolor As Integer, advance As Integer)
```

```
      Dim j As Integer: j = 1
      Do Until IsEmpty(Sheet2.Cells(LogRow, j))      'find an empty spot
         j = j + 1                                    'in LogRow
      Loop
      Sheet2.Cells(LogRow, j).Value = Pos.row  'first empty spot in LogRow
      Sheet2.Cells(LogRow, j + 1).Value = Pos.col
      Sheet2.Cells(LogRow, j + 2).Value = oldcolor
      If advance = 1 Then LogRow = LogRow + 1          'next row in Sheet2
   End Sub
```

The procedure **Rewind** reverses the steps of the organisms by restoring the old color of the cells. It uses a For Next loop to run through the rows in Sheet2 in reverse.

```
   Sub Rewind()
      Dim i As Integer, j As Integer, row As Integer, col As Integer
      Dim oldcolor As Integer
      For i = LogRow - 1 To 1 Step -1  'run through Sheet2 rows in reverse
         j = 1
         Do Until IsEmpty(Sheet2.Cells(i, j))
            row = Sheet2.Cells(i, j).Value
            col = Sheet2.Cells(i, j + 1).Value
            oldcolor = Sheet2.Cells(i, j + 2).Value
            Cells(row, col).Interior.ColorIndex = oldcolor
            j = j + 3
         Loop
         Call Delay(Speed)
      Next i
   End Sub
```

10.14   Random Mazes

The module **Maze** in this section generates a random maze and then solves it, that is, finds a path from an arbitrarily chosen starting point S to an arbitrarily chosen finish point F. Figure 10.21, depicts the spreadsheet for the module after pressing the $\boxed{\text{Create Maze}}$ button. Pressing the command button $\boxed{\text{Solve Maze}}$ creates three paths from S to F, one using the tree method, another using the right hand rule, and the third using a random walk. Figure 10.22 shows the tree method path. The maze size (virtually unlimited) is set by the command button $\boxed{\text{Set Dimensions}}$.

Generating a Maze

There are several algorithms that may be used to generate the walls of a maze. We shall use the following randomized version of *Prim's algorithm*,

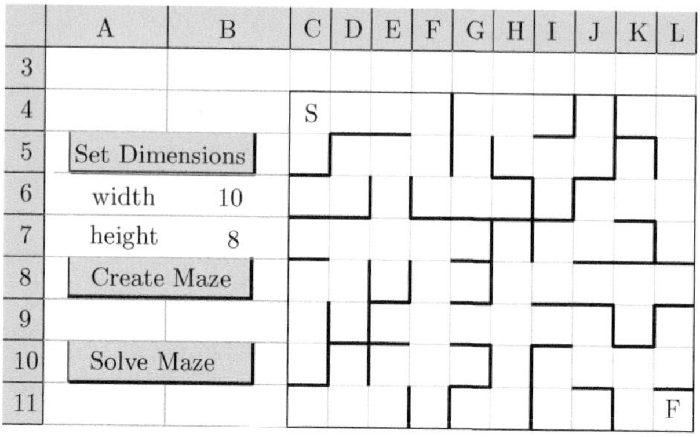

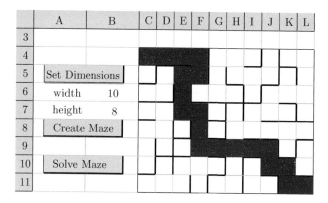

FIGURE 10.21: Creation of maze.

FIGURE 10.22: Tree method solution.

which starts with a grid of cells, an empty list of walls, and an empty list of visited cells and proceeds as follows:

1. Choose a cell and mark it as visited.
2. Add the walls of the cell to the wall list.
3. While there are still unvisited cells:

    1. Choose a random wall from the list. If only one of the cells separated by the wall is visited, then:

        1. Mark the unvisited cell as visited.
        2. Add the neighboring walls of the cell to the wall list.

    2. Remove the wall from the list.

Figure 10.23 illustrates the process for a small grid consisting of 16 cells. We have numbered the cells for ease of reference; visited cells are designated by red numbers. Initially, cell 11 was chosen, marked as visited, and then its walls

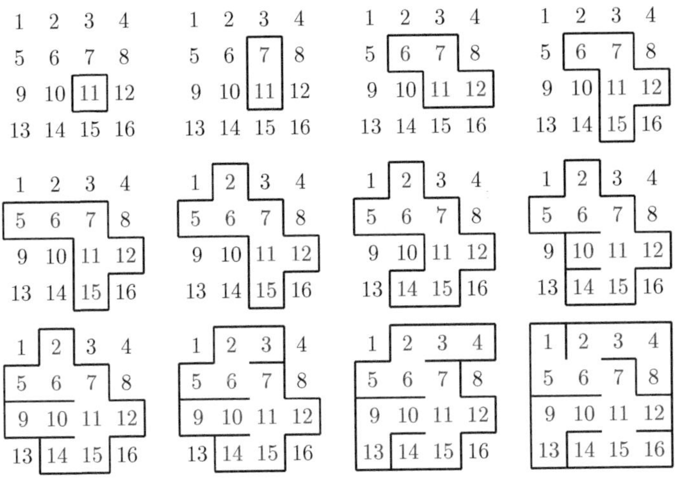

FIGURE 10.23: Prim's algorithm.

added. Next, the wall between 11 and 7 was randomly chosen, cell 7 marked as visited, its walls added and the wall between 7 and 11 removed. Next, the wall between 6 and 7 was randomly chosen, cell 6 marked as visited, its walls added and the wall between 6 and 7 removed. The process continues until all cells have been visited.

To implement the algorithm in VBA we shall designate a wall internally by the pair of adjacent cells that the wall separates; externally on the spreadsheet a wall is an edge between the cells. The data type `TWall` is used for this:

```
Private Type TWall
    CellA As TPos: CellB As TPos
End Type
```

We shall also need the type `TEdge`, which refers to the sides (top, left, bottom, right) of a cell.

```
Private Type TEdge
    T As Boolean: L As Boolean: B As Boolean: R As Boolean
End Type
```

The main procedure `Maze` sets up a grid from user information (width and height) and initializes some parameters pertaining to the grid. It then calls `MazeGenerator` to generate a random maze and later calls `MazeSolver` to solve the maze, that is, find a path from user-designated start to finish.

```
Public NumWalls As Integer, NumVisited As Integer
Public NumRows As Integer, NumCols As Integer, NumCells As Integer
Sub Maze()
```

```
Dim gridwidth As Integer, gridheight As Integer, G As TGrid
gridwidth = Range("B6").Value: gridheight = Range("B7").Value
G = MakeGrid(4, 4, 3 + gridheight, 3 + gridwidth, 2.5, 12, 24)
NumCols = G.R - G.L + 1
NumRows = G.B - G.T + 1
NumCells = NumRows * NumCols
Call MazeGenerator(G)                        'generate a maze
MsgBox "solve maze"
Call MazeSolver(G)                     'solve it by 3 methods
MsgBox "clear"
Call ClearGrid(G, 0)
End Sub
```

The procedure, `MazeGenerator`, first reads the desired width and height of the grid from cells B6 and B7. The grid is constructed using the procedure `MakeGrid` from Chapter 7. Next, memory is allocated for the Boolean array `Visited`, which keeps track of visited cells, and an array `Wall`, which holds the current set of walls. The procedure then calls `AddInitialWalls`, which selects a random cell, adds its walls to the list and then marks the cell as visited. A Do While loop continues the process, repeatedly calling `AddNewWalls` until all cells have been visited.

```
Private Sub MazeGenerator(G As TGrid)
    Dim Wall() As TWall, Visited() As Boolean
    ReDim Wall(1 To 4 * NumCells)             'memory for cell walls
    ReDim Visited(G.T To G.B, G.L To G.R)   'memory for visited cells
    NumWalls = 0: NumVisited = 0                      'initialize
    Call AddInitialWalls(Wall, Visited, G)                  'ditto
    Do While NumVisited < NumCells
        Call AddNewWalls(Wall, Visited, G)
    Loop
End Sub
```

The collection of walls is an array `Wall` of type `TWall`. The kth wall is physically an edge on the spreadsheet between the cells `Wall(k).CellA` and `Wall(k).CellB`. The procedure `AddInitialWalls` chooses a random cell, adds walls (cell pairs /edges) to the cell, and removes a random wall (cell pair /edge) from those just added.

```
Private Sub AddInitialWalls(Wall() As TWall, Visited() As Boolean, _
        Grid As TGrid)
    Dim GridCell As TPos, k As Integer
    GridCell.row = RndInt(Grid.T, Grid.B)            'pick a random cell
    GridCell.col = RndInt(Grid.L, Grid.R)
    Call AddWallsToCell(GridCell, Wall, Grid)        'add walls to cell
    Visited(GridCell.row, GridCell.col) = True        'mark as visited
    NumVisited = 1                                 'initial cell visited
    k = RndInt(1, NumWalls)       'pick a random wall from those just added
    Call RemoveWall(Wall, k)                          'and remove it
End Sub
```

The procedure `AddNewWalls` chooses a random wall `Wall(k)` and adds walls to whichever of the cells `Wall(k).CellA` or `Wall(k).CellB` has not yet been visited, marking that cell as visited. It then removes the wall, as both cells have now been visited.

```
Private Sub AddNewWalls(Wall() As TWall, Visited() As Boolean, G As TGrid)
    Dim k As Integer
    k = RndInt(1,NumWalls)                        'pick a random wall
    If Not Visited(Wall(k).CellA.row, Wall(k).CellA.col) Then 'add walls
        Call AddWallsToCell(Wall(k).CellA, Wall, G)   'to unvisited cells
        Visited(Wall(k).CellA.row, Wall(k).CellA.col) = True
        NumVisited = NumVisited + 1
        Call RemoveWall(Wall, k)
    ElseIf Not Visited(Wall(k).CellB.row, Wall(k).CellB.col) Then
        Call AddWallsToCell(Wall(k).CellB, Wall, G)
        Visited(Wall(k).CellB.row, Wall(k).CellB.col) = True
        NumVisited = NumVisited + 1
        Call RemoveWall(Wall, k)
    End If
End Sub
```

The procedure `AddWallsToCells` adds to the array and to the spreadsheet new walls between a cell `CellA` and each neighbor `CellB`.

```
Private Sub AddWallsToCell(CellA As TPos, Wall() As TWall, G As TGrid)
    Dim CellB As TPos, i As Integer, j As Integer
    For i = -1 To 1
        For j = -1 To 1
            CellB.row = CellA.row + i: CellB.col = CellA.col + j
            If i ^ 2 + j ^ 2 <> 1 Then GoTo continue   'i or j must be 0
            If Not InGrid(CellB, G) Then GoTo continue
            NumWalls = NumWalls + 1            'increase the number of walls
            Wall(NumWalls).CellA = CellA      'add wall CellA/CellB to list
            Wall(NumWalls).CellB = CellB
            Call AddEdge(CellA, i, j)                  'add physical edge
continue:
        Next j
    Next i
End Sub
```

The procedure `AddEdge` inserts an edge in the spreadsheet between `CellA` and its neighbor `CellB`, defined by offsets `(i,j)` from `CellA`. Several cases must be considered based on the relative positions of the cells (Figure 10.24)).

```
Private Sub AddEdge(CellA As TPos, i As Integer, j As Integer)
    Dim edge As Integer
    If i = 1 Then
        edge = xlEdgeBottom                    'CellB below CellA
    ElseIf i = -1 Then
        edge = xlEdgeTop                       'CellB above CellA
    ElseIf j = 1 Then
```

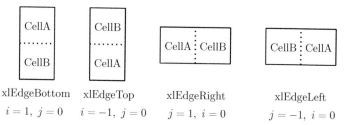

FIGURE 10.24: Adding an edge.

```
    edge = xlEdgeRight              'CellB to right of CellA
ElseIf j = -1 Then
    edge = xlEdgeLeft               'CellB to left of CellA
End If
With Cells(CellA.row, CellA.col).Borders(edge)
  .LineStyle = xlContinuous
  .Weight = xlThick
End With
End Sub
```

The following procedure removes the kth wall from the wall list and the spreadsheet.

```
Private Sub RemoveWall(Wall() As TWall, k As Integer)
    Dim i As Integer
    Call RemoveEdge(Wall(k).CellA, Wall(k).CellB)
    For i = k To NumWalls - 1     'shift wall list down to remove wall k
        Wall(i) = Wall(i + 1)
    Next i
    NumWalls = NumWalls - 1
End Sub
```

```
Private Sub RemoveEdge(CellA As TPos, CellB As TPos)
    Dim edge As Integer
    If CellA.col < CellB.col Then
        edge = xlEdgeRight
    ElseIf CellA.col > CellB.col Then
        edge = xlEdgeLeft
    ElseIf CellA.row < CellB.row Then
        edge = xlEdgeBottom
    ElseIf CellA.row > CellB.row Then
        edge = xlEdgeTop
    End If
    Cells(CellA.row, CellA.col).Borders(edge).LineStyle = xlNone
End Sub
```

Solving a Maze

The remaining procedures in the section are devoted to solving mazes generated by MazeGenerator. Specifically, they find a path from a designated start cell marked S to a designated finish cell marked F. The main procedure here is `MazeSolver`. The procedure calls `StartFinish` to find the start and finish cells. It then calls procedures to solve the maze by the *tree method*, procedures to solve the maze by the *right hand rule*, and finally a procedure to solve the maze by random motion. The first method is the fastest, the last the slowest.

```
Private Sub MazeSolver(G As TGrid)
    Dim spos As TPos, fpos As TPos, Tree() As Integer, Edges() As TEdge
    Range(Cells(G.T, G.L), Cells(G.B, G.R)).Interior.ColorIndex = xlNone
    Call StartFinish(spos, fpos, G)          'get start-finish positions
    Tree = CreateTree(spos, G)
    Call CreateTreePath(fpos, Tree, G)
    Edges = CreateEdgeList(G)
    Call CreateRightRulePath(Edges, G)
    Call CreateRandomPath(Edges, G)
End Sub
```

The procedure `StartFinish` searches the grid for the letters S and F. Failing to find them, it uses the upper left and lower right corners, respectively.

```
Private Sub StartFinish(spos As TPos, fpos As TPos, Grid As TGrid)
    Dim i As Integer, j As Integer
    spos.row = Grid.T: spos.col = Grid.L                     'default
    fpos.row = Grid.B: fpos.col = Grid.R
    For i = Grid.T To Grid.B
        For j = Grid.L To Grid.R
            If Cells(i, j).Value = "S" Then spos.row = i: spos.col = j
            If Cells(i, j).Value = "F" Then fpos.row = i: fpos.col = j
        Next j
    Next i
    Cells(spos.row, spos.col) = "S": Cells(fpos.row, fpos.col) = "F"
End Sub
```

The Tree Method

This algorithm begins by placing the number 1 in the start position (marked S). The number 2 is then placed in every neighboring cell accessible (not blocked by a wall) from the cell marked 1. The number 3 is then placed in every neighboring cell accessible from each 2, etc. In this way all the cells receive numbers forming a *tree*. To find a path from S to F one simply backtracks from F to S by first finding the number n assigned to the cell containing F and then following the neighbors with the numbers $n-1, n-2, \cdots, 2, 1$. Figure 10.25 illustrates the process.

The function `CreateTree` creates a table of numbers in accordance with the tree algorithm. The table `Tree` precisely reflects the grid. The function begins by placing the number $k = 1$ in the start position `spos`. A Do While loop places the number $k+1$ in the accessible neighbors of each cell labeled k using the procedure `Nabrs`.

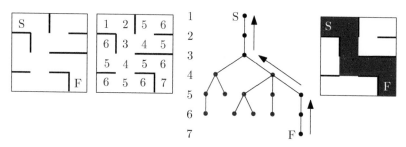

FIGURE 10.25: Maze tree.

```
Private Function CreateTree(spos As TPos, G As TGrid) As Integer()
    Dim Tree() As Integer, k As Integer, m As Integer, n As Integer
    ReDim Tree(G.T To G.B, G.L To G.R)
    Tree(spos.row, spos.col) = 1                'put 1 in start position
    k = 1
    Do While k < NumCells             'surround number k by number k+1
        For m = G.T To G.B     'check all locations (m,n) for the number k
            For n = G.L To G.R
                If Tree(m, n) = k Then Call Nabrs(m, n, k, Tree, G)
            Next n
        Next m
        k = k + 1
    Loop
    CreateTree = Tree
End Function

Private Sub Nabrs(m As Integer, n As Integer, k As Integer, _
            Tree() As Integer, G As TGrid)
    Dim i As Integer, j As Integer, Pos As TPos
    For i = -1 To 1          'check all offsets from cell position (m,n)
        For j = -1 To 1                          'for missing walls
            Pos.row = m + i: Pos.col = n + j
            If i ^ 2 + j ^ 2 = 1 And InGrid(Pos, G) Then
                If WallMissing(m, n, i, j) And _
                Tree(Pos.row, Pos.col) = 0 Then   'label unlabeled nbr.
                Tree(Pos.row, Pos.col) = k + 1
            End If
        Next j
    Next i
End Sub
```

The function `WallMissing` is passed a location (m, n) and an offset (i, j) and returns `True` if no edge separates the cells at (m, n) and $(m + i, n + j)$.

```
Private Function WallMissing(m As Integer, n As Integer, i As Integer, _
            j As Integer) As Boolean
    If i = 1 Then
        WallMissing = (Cells(m + i, n).Borders(xlEdgeTop).LineStyle = _
```

```
                                        xlLineStyleNone)
    ElseIf i = -1 Then
        WallMissing = (Cells(m + i, n).Borders(xlEdgeBottom).LineStyle = _
                                        xlLineStyleNone)
    ElseIf j = 1 Then
        WallMissing = (Cells(m, n + j).Borders(xlEdgeLeft).LineStyle = _
                                        xlLineStyleNone)
    ElseIf j = -1 Then
        WallMissing = (Cells(m, n + j).Borders(xlEdgeRight).LineStyle = _
                                        xlLineStyleNone)
    End If
End Function
```

The procedure `TreePath` is passed the final position `fpos` and the tree. It reads the number k at `fpos` and then backtracks up the tree, each time seeking an accessible cell with number $k - 1$ until it has found the cell marked 1.

```
Private Sub CreateTreePath(fpos As TPos, Tree() As Integer, G As TGrid)
    Dim m As Integer, n As Integer, i As Integer, j As Integer
    Dim k As Integer, TreePath() As TPos, Pos As TPos
    m = fpos.row: n = fpos.col: k = Tree(m, n)
    ReDim TreePath(1 To k)
    TreePath(k).row = m
    TreePath(k).col = n
    Do While Tree(m, n) > 1
        For i = -1 To 1
            For j = -1 To 1
                Pos.row = m + i: Pos.col = n + j
                If i ^ 2 + j ^ 2 = 1 And InGrid(Pos, G) Then
                    If WallMissing(m, n, i, j) And _
                        Tree(m + i, n + j) = k - 1 Then
                        m = m + i: n = n + j: k = k - 1
                        TreePath(k).row = m: TreePath(k).col = n
                        GoTo line1
                    End If
                End If
            Next j
        Next i
line1:
    Loop
    For k = 1 To UBound(TreePath)              'print out the path in red
        Call ColorCell(TreePath(k), 3)
        Call Delay(10)
    Next k
End Sub
```

The Right-Hand Rule Method

The right-hand rule asserts that if you start at any part of a maze with hedges connected to a boundary hedge and always keep your right hand in

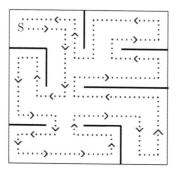

FIGURE 10.26: Right-hand rule path.

contact with a hedge you will reach every other part of the maze. Figure 10.26 shows such a path starting and ending at S.

The first step in the VBA implementation of the rule is to create an array that lists the edges around each cell of the maze. The following function generates a two dimensional array `Edges` of type `TEdge` that marks `Edges(m,n).B` True if `Cell(m,n)` has a bottom edge and proceeding similarly with the other edges.

```
Private Function CreateEdgeList(Grid As TGrid) As TEdge()
    Dim m As Integer, n As Integer, Edges() As TEdge
    ReDim Edges(Grid.T To Grid.B, Grid.L To Grid.R)
    For m = Grid.T To Grid.B
        For n = Grid.L To Grid.R
            If Cells(m, n).Borders(xlEdgeBottom).LineStyle = _
                xlContinuous Then Edges(m, n).B = True
            If Cells(m, n).Borders(xlEdgeTop).LineStyle = _
                xlContinuous Then Edges(m, n).T = True
            If Cells(m, n).Borders(xlEdgeLeft).LineStyle = _
                xlContinuous Then Edges(m, n).L = True
            If Cells(m, n).Borders(xlEdgeRight).LineStyle = _
                xlContinuous Then Edges(m, n).R = True
        Next n
    Next m
    CreateEdgeList = Edges
End Function
```

The right-hand rule requires a direction at each juncture. We shall use the abbreviations N,E,S,W for north (up), east (right), south (down), and west (left). Thus a right turn means one of the following changes of direction: N to E, E to S, S To W, or W to N. The following function returns an initial direction by finding a missing edge in the start cell.

```
Private Function StartDirection(spos As TPos, Edges() As TEdge, _
                    G As TGrid) As String
    Dim dir As String
```

```
  If Not Edges(spos.row, spos.col).T And spos.row > G.T Then
     dir = "N"                                'no top edge so go north
  ElseIf Not Edges(spos.row, spos.col).R And spos.col < G.R Then
     dir = "E"                                'no right edge so go east
  ElseIf Not Edges(spos.row, spos.col).L And spos.col > G.L Then
     dir = "W"                                'no left edge so go west
  ElseIf Not Edges(spos.row, spos.col).B And spos.row < G.B Then
     dir = "S"                                'no bottom edge so go south
  End If
  StartDirection = dir
End Function
```

The right-hand rule is implemented by the procedure `CreateRightRulePath`, which first specifies an initial direction from the start cell and then modifies the direction at each juncture according to the right-hand rule.

```
Private Sub CreateRightRulePath(Edges() As TEdge, Grid As TGrid)
   Dim spos As TPos, fpos As TPos, done As Booleanr, dir As String
   Dim m As Integer, n As Integer, cm As Integer, cn As Integer
   Call StartFinish(spos, fpos, Grid)
   dir = StartDirection(spos, Edges, Grid)    'get an initial direction
   m = spos.row: n = spos.col: cm = m: cn = n        'old-new positions
   Cells(m, n).Interior.ColorIndex = 28
   Do Until done
      If dir = "N" Then                      'if current direction is north
         If Not Edges(m, n).R And n < Grid.R Then
            cn = n + 1: cm = m: dir = "E"       'turn east if possible
         ElseIf Not Edges(m, n).T And m > Grid.T Then
            cm = m - 1: cn = n: dir = "N":   'otherwise continue north
         ElseIf Not Edges(m, n).L And n > Grid.L Then
            cn = n - 1: cm = m: dir = "W":         'otherwise go west
         ElseIf Not Edges(m, n).B And m < Grid.B Then
            cm = m + 1: cn = n: dir = "S":      'last resort: reverse
         End If
      ElseIf dir = "E" Then                  'if current direction is east
         If Not Edges(m, n).B And m < Grid.B Then
            cm = m + 1: cn = n: dir = "S":      'turn south if possible
         ElseIf Not Edges(m, n).R And n < Grid.R Then
            cn = n + 1: cm = m: dir = "E":    'otherwise continue east
         ElseIf Not Edges(m, n).T And m > Grid.T Then
            cm = m - 1: cn = n: dir = "N":        'otherwise go north
         ElseIf Not Edges(m, n).L And n > Grid.L Then
            cn = n - 1: cm = m: dir = "W":      'last resort: reverse
         End If
      ElseIf dir = "S" Then                  'if current direction is south
         If Not Edges(m, n).L And n > Grid.L Then
            cn = n - 1: cm = m: dir = "W":      'turn west if possible
         ElseIf Not Edges(m, n).B And m < Grid.B Then
            cm = m + 1: cn = n: dir = "S":    'otherwise continue south
         ElseIf Not Edges(m, n).R And n < Grid.R Then
            cn = n + 1: cm = m: dir = "E":         'otherwise go east
```

```
            ElseIf Not Edges(m, n).T And m > Grid.T Then
                cm = m - 1: cn = n: dir = "N":         'last resort: reverse
            End If
        ElseIf dir = "W" Then                'if current direction is south
            If Not Edges(m, n).T And m > Grid.T Then
                cm = m - 1: cn = n: dir = "N":     'turn north if possible
            ElseIf Not Edges(m, n).L And n > Grid.L Then
                cn = n - 1: cm = m: dir = "W":     'otherwise continue west
            ElseIf Not Edges(m, n).B And m < Grid.B Then
                cm = m + 1: cn = n: dir = "S":       'otherwise go south
            ElseIf Not Edges(m, n).R And n < Grid.R Then
                cn = n + 1: cm = m: dir = "E":     'last resort: reverse
            End If
        End If
        Cells(m, n).Interior.ColorIndex = 0          'uncolor last cell
        Cells(cm, cn).Interior.ColorIndex = 28       'color new cell
        m = cm: n = cn
        If m = fpos.row And n = fpos.col Then
            done = True                              'found finish cell
        End If
        Call Delay(15)
    Loop
    Cells(m, n).Interior.ColorIndex = 0
End Sub
```

Random Walk Method

This method is the slowest and may not always be successful, depending on the complexity of the maze and the available time. The procedure `RandomPath` is essentially a random walk except that the moving cell is not allowed to penetrate an edge. A Do While loop moves the cell using the `Edges` list to prevent a breach. A cutoff on looping is provided so as not to overtax the patience of the user.

```
Private Sub CreateRandomPath(Edges() As TEdge, G As TGrid)
    Dim spos As TPos, fpos As TPos, Pos As TPos
    Dim NewPos As TPos, ctr As Integer
    Call StartFinish(spos, fpos, G)
    Call CellColor(spos, 5)
    Pos = spos
    Do While Not (Pos.row = fpos.row And Pos.col = fpos.col) And _
                ctr < 10000
        NewPos = NextMazePos(Pos, Edges, G)
        If NewPos.row <> 0 Then
            Call ColorCell(Pos, xlNone)
            Call ColorCell(NewPos, 5)
            Pos = NewPos
        End If
        ctr = ctr + 1
    Loop
```

```
End Sub

Private Function NextMazePos(Pos As TPos, Edges() As TEdge, G As TGrid) _
            As TPos
    Dim NewPos As TPos, Temp As TPos, i As Integer, j As Integer
    Temp = NextRndPos(Pos, 1)
    If InGrid(Temp, G) And Not Blocked(Pos, Temp, Edges) Then
        NewPos = Temp
    End If
    NextMazePos = NewPos
End Function

Private Function Blocked(pos As TPos, newpos As TPos, Edges() As TEdge) _
            As Boolean
    Blocked = pos.row > newpos.row And Edges(pos.row, pos.col).T Or _
            pos.row < newpos.row And Edges(pos.row, pos.col).B Or _
            pos.col < newpos.col And Edges(pos.row, pos.col).R Or _
            pos.col > newpos.col And Edges(pos.row, pos.col).L
End Function
```

10.15 Exercises

1. Write a program that generates an array of, say, 1000 random Double numbers and calculates the average, median and mode of the numbers. (See Exercises 7, 8, 9 of Section 5.8.)

2. Write a program Digit3Avg(numIterations) that generates random digits $0 - 9$ and returns the percentage of times the digit $= 3$. Is this what you would expect? (Try numIterations$= 10,000$).

3. Write a function RandomVowel(mode) that returns a randomly chosen uppercase vowel if mode $=$ "upper" or a randomly chosen lowercase vowel if mode $=$ "lower". Write the analogous functions RandomConsonant and RandomLetter

4. Write a program RandomVowelAvg(numiterations) that generates random letters and returns the percentage of times the letter is a vowel.

5. Write a program ThreeLetters that prints in a column a list of three-letter lower case words with the middle letter a vowel and the first and last letters consonants, all randomly chosen. How often do the words make sense?

6. Write a program FourLetters that prints in a column a list of four-letter lower case words with first and third letters consonants the second and

fourth letters vowels, all randomly chosen. How often do the words make sense?

7. For this exercise the reader should refer back to Exercise 9 of Section 6.9. Write a program `RandomPalindromes` that generates a list of random words (nonsensical or otherwise) of user-entered length that are palindromes.

8. Write a function `RandOddInt` that returns a random odd number strictly between given lower and upper limits.

9. Write a procedure `RandomRectangle` that takes a fixed cell Cell(i, j) in the top left corner of a grid, generates two random integer offsets `r,c`, and draws the rectangle as shown in Figure 10.27. Make a loop with delay that draws a bunch of these for an interesting visual effect,

$$(i,j) \qquad \qquad (i, j+c)$$

$$(i+r,j) \qquad \qquad (i+r, j+c)$$

FIGURE 10.27: Random rectangle.

10. Write a program `GraphRandomFunction` that incorporates into the procedure `GraphFunction` the option of attaching a random component to the function using `RndDec` of 10.1 to generate random values. For example, entering `r*SIN(x)` should produce a sine wave with randomly varying amplitude determined by `r` = `RndDec`.

11. Write a program `GraphRandomCurve` that is the analog of the program `GraphRandomFunction` in the previous exercise.

12. Write a program `ConnectRandomNumbers` that generates random numbers in a grid up to a user-specified value and then connects the numbers in order by elbow paths.

13. A common method for pricing a stock option assumes that the stock price follows a *geometric binomial law*. This means that each day the stock goes up by a factor u with probability p or down by a factor d with probability $1 - p$, where $0 < d < 1 < u$. For example, if the price of the stock on Sunday night is S_0, and if the stock goes up on Monday, down Tuesday and Wednesday, and up again on Thursday, then on Friday morning the stock will be worth $u^2 d^2 S_0$. Write a program `Stockpath` that generates a random sequence of u's and d's and then graphs the stock price path. Allow the user to enter the factors u, d, the probability p, the initial stock price, and the number of days over which the price is observed. The figure illustrates a stock price path over 9 days (with value $u^5 d^4 S_0$ at the end of the 9th day.)

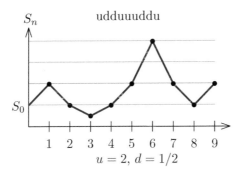

14. Write a program `Clusters` that does the following: The user enters into cells of a grid letters referring to colors, as in the procedure `LetterToColor`. Running the program should color the cells and cause the cells to execute random motion as in the program `SpreadDisease` in Section 10.5. Motion of a cell should cease as soon as it finds itself attached to a cell of the same color, thus forming clusters of cells of the same color. Figure 10.28 shows an example. The program is another example of *emergent behavior*,

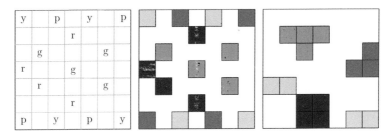

FIGURE 10.28: Three stages of clusters.

defined as many objects reacting together to produce group behavior that the individual objects do not possess on their own. (Birds flocking and fish schooling are more traditional examples.)

15. (Twelve coin puzzle). Suppose you have twelve coins numbered 1–12. Eleven of the coins have equal weights but the remaining coin, a counterfeit, weighs either more or less than the others. You don't know which coin is counterfeit or whether it is lighter or heavier. To determine this you have an equal balance scale that you may use to weigh the coins against each other, but you are allowed only three weighings.

(a) Construct an algorithm that finds the counterfeit coin and determines whether it is lighter or heavier than the others. An example of such a weighing is given in Figure 10.29, which reveals that the counterfeit coin is among coins 7–12 but nothing more.

FIGURE 10.29: Weighing 3 against 3.

(b) Implement the puzzle on VBA in a program `TwelveCoins`. *Suggestions*: Let the program randomly select the counterfeit coin and whether it's lighter or heavier than the others. Let column A represent the left part

	A	B
2	weigh	
4		
5	7	10
6	8	11
7	9	12
8		

	A	B
2	weigh	
4		10
5		11
6	7	12
7	8	
8	9	

FIGURE 10.30: Spreadsheet for Figure 10.29.

of the scale, column B the right. The user should enter coin numbers in these columns. Pressing [weigh] should cause the program to tilt the scale by shifting the numbers. Figure 10.30 illustrates input and output. User should enter the solution in row 3, e.g., "3 heavy" or "5 light." The program should acknowledge whether the answer is correct or not.

16. Write a program `RandomWalkOnPattern` that produces a random walk on a user-entered pattern of contiguous x's in a grid.

17. Write a program `RandomWalksWithBarriers`, based on `RandomWalks`, in which the organisms attempt to escape from enclosures or navigate simple mazes (Figure 10.31). The user should enter these as x's in cells; the program should then color the blocking cells with the grid boundary color.

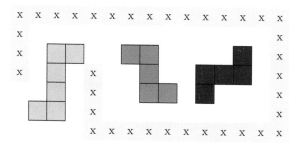

FIGURE 10.31: Random walks with barriers. Yellow escapes.

18. Write a program `RandomWalksWithAbsorbingBarriers`, a variation of `RandomWalksWithBarriers`, in which the organisms, after hitting a barrier, are absorbed into the barrier, never to emerge again. The program should terminate after so many steps or when there are no more organisms, whichever comes first.

19. Write a program `RandomWalksReflected`, based on `RandomWalks`, that divides a grid into four quadrants, runs random walks of several organisms and reflects the motion in the axes. (See Figure 10.32.)

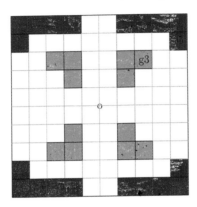

FIGURE 10.32: Random reflected walks.

Part II

Linear Analysis

Chapter 11

Linear Equations

In this chapter we describe a technique for solving systems of linear equations and show how the method may be implemented in VBA Excel. The principle ideas of the chapter are applied in later chapters on linear programming, matrix algebra, and determinants.

11.1 Matrix Arrays

To formulate and solve the problems in this and the next three chapters, we shall need the notion of a matrix. In this chapter we use matrices as convenient devices to solve systems of linear equations. In the next chapter we show how matrices may be combined to form an algebraic system similar to that of numbers.

An $m \times n$ *dimensional matrix* is a rectangular array of mn numbers denoted by the progressively more detailed notation

$$A = \left[a_{ij}\right]_{m \times n} = \begin{bmatrix} a_{11} & a_{12} & \cdots & a_{1j} & \cdots & a_{1n} \\ a_{21} & a_{22} & \cdots & a_{2j} & \cdots & a_{2n} \\ \vdots & \vdots & \ddots & \vdots & \ddots & \vdots \\ a_{i1} & a_{i2} & \cdots & a_{ij} & \cdots & a_{in} \\ \vdots & \vdots & \ddots & \vdots & \ddots & \vdots \\ a_{m1} & a_{m2} & \cdots & a_{mj} & \cdots & a_{mn} \end{bmatrix}. \tag{11.1}$$

The number a_{ij} in row i and column j is called the (i, j) *entry* of A. For example, the $(2, 3)$ entry of the 2×4 matrix $\left[\begin{smallmatrix} 1 & 2 & 3 & 4 \\ 5 & 6 & 7 & 8 \end{smallmatrix}\right]$ is the number 7. Two matrices A and B are said to be *equal* if they have the same dimensions and equal entries: $a_{ij} = b_{ij}$ for all indices i and j. A $1 \times n$ matrix is called a *row matrix* and an $m \times 1$ matrix a *column matrix*. The terms *row vector* and *column vector* are also used.

In the programs that follow, matrices are stored in *matrix arrays*, that is, two dimensional arrays whose indices start at $(1, 1)$. For example, the declaration `Dim Mat(1 To 5,1 To 6) As Double` provides storage for a 5×6 matrix array `Mat` (with entries automatically initialized to 0). Note that a row matrix is not the same as a one dimensional array. For example, the declarations `Dim`

`A(1 To 5) As Double, B(1 To 1,1 To 5) As Double` produces two arrays, each of which has 5 entries. However, these entries are accessed differently: the jth entry of `A` is `A(j)`, while the jth entry of `B` is `B(1,j)`.

The following two procedures will be used frequently in this and subsequent chapters. The first, `MatrixIn`, reads a spreadsheet matrix into a matrix array. The function is passed the $(1,1)$ position `row,col` of the spreadsheet matrix and returns the array. It uses two Do Until loops to determine the dimensions of the matrix. Both loops use the VBA function `IsEmpty`, which returns `True` when an empty cell is encountered.

```
Function MatrixIn(row As Integer, col As Integer) As Double()
    Dim i As Integer, j As Integer, Mat() As Double
    Dim m As Integer, n As Integer
    Do Until IsEmpty(Cells(row + m, col))      'get no. of rows of matrix
        m = m + 1
    Loop
    Do Until IsEmpty(Cells(row, col + n))    'get no. of columns of matrix
        n = n + 1
    Loop
    ReDim Mat(1 To m, 1 To n)                              'matrix array
    For i = 1 To m
        For j = 1 To n
            Mat(i, j) = Cells(row + i - 1, col + j - 1).Value
        Next j
    Next i
    MatrixIn = Mat                                      'return matrix
End Function
```

The companion procedure `MatrixOut(Mat,row,col)` prints the matrix array `Mat` with the $(1,1)$ entry at spreadsheet position `row,col`. It uses the VBA function `Ubound` discussed in Section 5.2.

```
Sub MatrixOut(Mat() As Double, row As Integer, col As Integer)
    Dim i As Integer, j As Integer
    For i = 1 To Ubound(Mat, 1)                            'print matrix
        For j = 1 To Ubound(Mat, 2)
            Cells(row + i - 1, col + j - 1).Value = Mat(i, j)
        Next j
    Next i
End Sub
```

11.2 Systems of Linear Equations

A *system of linear equations* is a collection of m equations in n unknowns of the form

$$a_{11}x_1 + a_{12}x_2 + \cdots + a_{1n}x_n = b_1$$
$$a_{21}x_1 + a_{22}x_2 + \cdots + a_{2n}x_n = b_2$$
$$\vdots$$
$$a_{m1}x_1 + a_{m2}x_2 + \cdots + a_{mn}x_n = b_m. \tag{11.2}$$

The symbols a_{ij} and b_i represent constants and the symbols x_j variables. A *solution* of the system is an assignment of values of the variables $x_1, x_2, \ldots, x_n$ that satisfies the equations. As we shall see, a system may have no solutions, a unique solution, or infinitely many solutions.

If the variables $x_1, x_2, \ldots, x_n$ are understood, then the system (11.2) may be unambiguously "encoded" by its so-called *augmented matrix*:[1]

$$\begin{bmatrix} a_{11} & a_{12} & \cdots & a_{1n} & b_1 \\ a_{21} & a_{22} & \cdots & a_{2n} & b_2 \\ \vdots & \vdots & \ddots & \vdots & \\ a_{m1} & a_{m2} & \cdots & a_{mn} & b_n \end{bmatrix}$$

Clearly, all the essential information of the system is represented by the matrix. Thus augmented matrices provide a shorthand way to represent systems of linear equations.

The following examples show how systems of linear equations arise in applications.

Example 11.1. (Mixtures). A one-liter jar is filled with 5 chemicals labeled 1 to 5. The proportions are as follows: The amount of chemical 1 equals the amount of chemicals 2 and 5 together (in liters), and one-half the amount of 3, 4 and 5 together. The amount of chemical 4 equals the amount of 2 and 3 together, and is one-half the amount of 3 and 5 together. Find the amount of each chemical.

To describe these conditions mathematically, let x_j denote the amount of chemical j in the jar. The above conditions translate into the equations

$$x_1 = x_2 + x_5 = \tfrac{1}{2}(x_3 + x_4 + x_5), \quad x_4 = x_2 + x_3 = \tfrac{1}{2}(x_3 + x_5).$$

Rearranging produces the system

$$
\begin{aligned}
x_1 - x_2 \qquad\qquad\quad - x_5 &= 0 \\
2x_1 \qquad - x_3 - x_4 - x_5 &= 0 \\
x_2 + x_3 - x_4 \qquad\quad &= 0 \\
x_3 - 2x_4 + x_5 &= 0 \\
x_1 + x_2 + x_3 + x_4 + x_5 &= 1
\end{aligned}
\tag{11.3}
$$

---

[1] The only purpose of the vertical bar in the matrix is to emphasize that the matrix arose from a system of equations. The bar is omitted if no confusion can arise.

with augmented matrix

$$\left[\begin{array}{ccccc|c} 1 & -1 & 0 & 0 & -1 & 0 \\ 2 & 0 & -1 & -1 & -1 & 0 \\ 0 & 1 & 1 & -1 & 0 & 0 \\ 0 & 0 & 1 & -2 & 1 & 0 \\ 1 & 1 & 1 & 1 & 1 & 1 \end{array}\right].$$

Using methods developed later in the chapter one can show that

$$x_1 = .316, \quad x_2 = .053, \quad x_3 = .158, \quad x_4 = .211, \quad x_5 = .263,$$

rounded to three decimal places.   ◇

Example 11.2. (Stock Portfolios). A portfolio with total value $60,500 contains a variety of stocks, each of which has current value $10, $50, $100, $200, or $500. There are half as many $500 stocks as the total number of the other stocks, one-fourteenth as many $200 stocks as the total number of the other stocks, one-ninth as many $100 stocks as the total number of the other stocks, and one-fifth as many $50 stocks as the total number of the other stocks.

To determine the number of each type of stock, let x_1 denote the number of $10 stocks, x_2 the number of $50 stocks, x_3 the number of $100 stocks, x_4 the number of $200 stocks, and x_5 the number of $500 stocks. The given information then translates into the system

$$\begin{aligned}
x_1 + x_2 + x_3 + x_4 - 2x_5 &= 0 \\
x_1 + x_2 + x_3 - 14x_4 + x_5 &= 0 \\
x_1 + x_2 - 9x_3 + x_4 + x_5 &= 0 \\
x_1 - 5x_2 + x_3 + x_4 + x_5 &= 0 \\
10x_1 + 50x_2 + 100x_3 + 200x_4 + 500x_5 &= 60500
\end{aligned} \tag{11.4}$$

which has augmented matrix

$$\left[\begin{array}{ccccc|c} 1 & 1 & 1 & 1 & -2 & 0 \\ 1 & 1 & 1 & -14 & 1 & 0 \\ 1 & 1 & -9 & 1 & 1 & 0 \\ 1 & -5 & 1 & 1 & 1 & 0 \\ 10 & 50 & 100 & 200 & 500 & 60500 \end{array}\right].$$

Applying methods developed later gives the solution

$$x_1 = 100, \quad x_2 = 50, \quad x_3 = 30, \quad x_4 = 20, \quad x_5 = 100. \qquad ◇$$

Example 11.3. (Manufacturing). A factory produces five models, labeled F1 to F5, of the popular Framosham fish filter. Each filter passes through five production stages labeled P1 to P5.

Manufacturing Data

	F1	F2	F3	F4	F5	Total Hours
P1	.3	.9	.5	.3	.6	78
P2	.1	.2	.8	.2	.2	47
P3	.4	.6	.9	.3	.6	85
P4	.1	.5	.7	.8	.4	84
P5	.5	.3	.9	.4	.3	69

The column under Fj gives the number of hours required at each production stage to produce one filter of type Fj. The last column gives the total number of hours available every week for each production stage Pi. We wish to determine the number x_j of filters of type Fj that may be produced if all of the available hours are utilized.

Here are the equations that must be satisfied:

$$.3x_1 + .9x_2 + .5x_3 + .3x_4 + .6x_5 = 78$$
$$.1x_1 + .2x_2 + .8x_3 + .2x_4 + .2x_5 = 47$$
$$.4x_1 + .6x_2 + .9x_3 + .3x_4 + .6x_5 = 85 \qquad (11.5)$$
$$.1x_1 + .5x_2 + .7x_3 + .8x_4 + .4x_5 = 84$$
$$.5x_1 + .3x_2 + .9x_3 + .4x_4 + .3x_5 = 69$$

The augmented matrix is

$$\begin{bmatrix} .3 & .9 & .5 & .3 & .6 & 78 \\ .1 & .2 & .8 & .2 & .2 & 47 \\ .4 & .6 & .9 & .3 & .6 & 85 \\ .1 & .5 & .7 & .8 & .4 & 84 \\ .5 & .3 & .9 & .4 & .3 & 69 \end{bmatrix}.$$

Using the methods developed later produces the solution

$$x_1 = 10, \quad x_2 = 20, \quad x_3 = 30, \quad x_4 = 40, \quad x_5 = 50. \qquad \diamond$$

11.3   The Gauss Jordan Method

In this section we describe an algorithm that will allow us to solve a system of linear equations or else determine that the system has no solution. The basic idea is to transform the system into a one that has precisely the same solutions as the original system but is trivial to solve.

The transformation to a simpler system is accomplished by a sequence of *equation operations* of the following form:

<u>EqOp 1</u> Interchange two equations.

EqOp 2 Multiply an equation by a nonzero number.

EqOp 3 Add a multiple of one equation to another.

Here's an example that illustrates these operations using some convenient shorthand notation:

$$\begin{array}{l} 2x_1 + 4x_2 = 6 \\ x_1 + x_2 = 2 \end{array} \xrightarrow{\frac{1}{2}E_1} \begin{array}{l} x_1 + 2x_2 = 3 \\ x_1 + x_2 = 2 \end{array} \xrightarrow{E_1 \leftrightarrow E_2} \begin{array}{l} x_1 + x_2 = 2 \\ x_1 + 2x_2 = 3 \end{array}$$

$$\xrightarrow{(-1)E_1 + E_2} \begin{array}{l} x_1 + x_2 = 2 \\ x_2 = 1 \end{array} \xrightarrow{(-1)E_2 + E_1} \begin{array}{l} x_1 = 1 \\ x_2 = 1 \end{array}$$

While the notation is largely self-explanatory, it should be emphasized that in last operation, for example, it is equation *one* that is to be altered, not equation two.

The important feature of equation operations is they do not change the solutions of the system. For example, the reader may check that the solution to each of the above systems is the trivial solution of the last one, namely $x_1 = 1, x_2 = 1$. Equally important is the fact that operations may be reversed. In our example, the reversed operations take the form

$$\begin{array}{l} x_1 = 1 \\ x_2 = 1 \end{array} \xrightarrow{1E_2 + E_1} \begin{array}{l} x_1 + x_2 = 2 \\ x_2 = 1 \end{array} \xrightarrow{1E_1 + E_2} \begin{array}{l} x_1 + x_2 = 2 \\ x_1 + 2x_2 = 3 \end{array}$$

$$\xrightarrow{E_1 \leftrightarrow E_2} \begin{array}{l} x_1 + 2x_2 = 3 \\ x_1 + x_2 = 2 \end{array} \xrightarrow{2E_1} \begin{array}{l} 2x_1 + 4x_2 = 6 \\ x_1 + x_2 = 2 \end{array}$$

It follows from these observations that if one system has been transformed into another by a sequence of equation operations, then the two systems have precisely the same solutions, that is, they are *equivalent systems*. This principle is the basis for the Gauss-Jordan method of solving systems of linear equations discussed in the remainder of the section.

To streamline the algorithm we convert the equation operations on a system of equations to *row operations* on the augmented matrix of the system. The operations then take the following form:

RowOp 1 Interchange two rows.

RowOp 2 Multiply a row by a nonzero number.

RowOp 3 Add a multiple of one row to another row.

The first sequence of equations operations above may then be expressed as

$$\begin{bmatrix} 2 & 4 & 6 \\ 1 & 1 & 2 \end{bmatrix} \xrightarrow{\frac{1}{2}R_1} \begin{bmatrix} 1 & 2 & 3 \\ 1 & 1 & 2 \end{bmatrix} \xrightarrow{R_1 \leftrightarrow R_2} \begin{bmatrix} 1 & 1 & 2 \\ 1 & 2 & 3 \end{bmatrix}$$

$$\xrightarrow[\quad]{(-1)R_1 + R_2} \begin{bmatrix} 1 & 1 & 2 \\ 0 & 1 & 1 \end{bmatrix} \xrightarrow[\quad]{(-1)R_2 + R_1} \begin{bmatrix} 1 & 0 & 1 \\ 0 & 1 & 1 \end{bmatrix}$$

As with equation operations, the notation in the last row operation, for example, is meant to indicate that it is row *one* that is to be altered, not row two.

The *Gauss-Jordan method* consists of a sequence of row operations applied to the augmented matrix of a system of equations that transforms the matrix into one with the following properties:

- All zero rows (rows with only zeros) are below all nonzero rows (rows with at least one zero).

- The first nonzero entry (called the *leading entry*) in a nonzero row is 1.

- The leading entry in one row is to the left of all leading entries below it.

- All entries above and below a leading entry are zero.

A matrix with these properties is said to be in *reduced row echelon form*. For example, the first matrix below is in reduced row echelon form but the second is not. For emphasis we have enclosed the leading entries in the first matrix by rectangles.

$$\begin{bmatrix} \boxed{1} & 3 & 0 & 0 & 0 \\ 0 & 0 & \boxed{1} & 0 & 0 \\ 0 & 0 & 0 & \boxed{1} & 2 \\ 0 & 0 & 0 & 0 & 0 \end{bmatrix} \qquad \begin{bmatrix} 1 & 3 & 5 & 1 \\ 0 & 0 & 7 & 2 \\ 0 & 0 & 0 & 0 \\ 0 & 0 & 0 & 3 \end{bmatrix}.$$

Any matrix may be transformed into reduced row echelon form by a sequence of row operations. In particular this can be done for the augmented matrix of a system of linear equations. Since the system of linear equations corresponding to a matrix in reduced row echelon form is essentially trivial, we now have a way of finding the solutions of any system (or determining that the system has no solutions).

The procedure used to transform a matrix A into row reduced echelon form is described in the following steps. (We exclude the trivial case where every entry of A is zero.)

(a) Find the leftmost column that has at least one nonzero entry. This is called a *pivot column*. The *pivot position* is at the top of the column in what is called the *pivot row*.

(b) Choose a nonzero entry in the pivot column.

(c) Use a type 1 row operation on the matrix to move the entry to the pivot position.

(d) Use a type 2 row operation to put a 1 in the pivot position.

(e) Use type 3 row operations to put zeros in all but the pivot position of the pivot column.

(f) Find the leftmost non-zero column in the matrix consisting of the rows below the last pivot row and apply steps (b)–(e). Continue the process until there are no more rows left to modify.

We shall call the process of placing a one in the pivot position of a column and zeros elsewhere *clearing the column*. Here some examples of how to apply the algorithm to the augmented matrix of a system of linear equations.

Example 11.4. To solve the system

$$2x_1 + 2x_2 + 3x_3 = 9$$
$$4x_1 + 5x_2 + 6x_3 = 12$$
$$7x_1 + 8x_2 + 9x_3 = 15$$

we use row operations to find the reduced row echelon form of the augmented matrix. We have enclosed the entry used to clear the column in a rectangle.

$$\begin{bmatrix} 2 & 2 & 3 & 9 \\ 4 & 5 & 6 & 12 \\ 7 & 8 & 9 & 15 \end{bmatrix} \xrightarrow{\frac{1}{2}R_2} \begin{bmatrix} \boxed{1} & 1 & \frac{3}{2} & \frac{9}{2} \\ 4 & 5 & 6 & 12 \\ 7 & 8 & 9 & 15 \end{bmatrix} \begin{array}{c} (-4)R_1 + R_2 \\ \xrightarrow{\hspace{1.5cm}} \\ (-7)R_1 + R_3 \end{array}$$

$$\begin{bmatrix} 1 & 1 & \frac{3}{2} & \frac{9}{2} \\ 0 & \boxed{1} & 0 & -6 \\ 0 & 1 & -\frac{3}{2} & -\frac{33}{2} \end{bmatrix} \begin{array}{c} (-1)R_2 + R_1 \\ \xrightarrow{\hspace{1.5cm}} \\ (-1)R_2 + R_3 \end{array} \begin{bmatrix} 1 & 0 & \frac{3}{2} & \frac{21}{2} \\ 0 & 1 & 0 & -6 \\ 0 & 0 & -\frac{3}{2} & -\frac{21}{2} \end{bmatrix}$$

$$\xrightarrow{-\frac{2}{3}R_3} \begin{bmatrix} 1 & 0 & \frac{3}{2} & \frac{21}{2} \\ 0 & 1 & 0 & -6 \\ 0 & 0 & \boxed{1} & 7 \end{bmatrix} \xrightarrow{-\frac{3}{2}R_3 + R_1} \begin{bmatrix} 1 & 0 & 0 & 0 \\ 0 & 1 & 0 & -6 \\ 0 & 0 & 1 & 7 \end{bmatrix}$$

The last matrix is the augmented matrix of the trivial system

$$\begin{aligned} x_1 \quad &= 0 \\ x_2 \quad &= -6 \\ x_3 &= 7 \end{aligned}$$

which therefore gives the solution of the original system.   ◇

Example 11.5. Consider the following modification of the system in Example 11.4, obtained by changing the coefficient of x_1 in the first equation from 2 to 1:

$$x_1 + 2x_2 + 3x_3 = 9$$
$$4x_1 + 5x_2 + 6x_3 = 12$$
$$7x_1 + 8x_2 + 9x_3 = 15$$

The row operations are now

$$\begin{bmatrix} 1 & 2 & 3 & 9 \\ 4 & 5 & 6 & 12 \\ 7 & 8 & 9 & 15 \end{bmatrix} \begin{array}{c} (-4)R_1 + R_2 \\ \xrightarrow{\hspace{1.5cm}} \\ (-7)R_1 + R_3 \end{array} \begin{bmatrix} 1 & 2 & 3 & 9 \\ 0 & -3 & -6 & -24 \\ 0 & -6 & -12 & -48 \end{bmatrix} \xrightarrow{-\frac{1}{3}R_2}$$

$$\begin{bmatrix} 1 & 2 & 3 & 9 \\ 0 & 1 & 2 & 8 \\ 0 & -6 & -12 & -48 \end{bmatrix} \xrightarrow[\;\; 6R_2 + R_3 \;\;]{(-2)R_2 + R_1} \begin{bmatrix} 1 & 0 & -1 & -7 \\ 0 & 1 & 2 & 8 \\ 0 & 0 & 0 & 0 \end{bmatrix}$$

The last matrix is the augmented matrix of the system

$$x_1 \quad - \quad x_3 = -7$$
$$x_2 + 2x_3 = 8$$

which has the solutions $x_1 = -7 + x_3$, $x_2 = 8 - 2x_3$, x_3 arbitrary. Thus the system has infinitely many solutions, one for each value of the parameter x_3. ◊

Example 11.6. Consider the system

$$x_1 + 2x_2 + 3x_3 = 9$$
$$4x_1 + 5x_2 + 6x_3 = 12$$
$$7x_1 + 8x_2 + 9x_3 = 14$$

obtained from the system in Example 11.5 by changing the number 15 in the last equation to 14. This changes the number -48 in the above calculations to -49, leading to the echelon form

$$\begin{bmatrix} 1 & 0 & -1 & -7 \\ 0 & 1 & 2 & 8 \\ 0 & 0 & 0 & 1 \end{bmatrix}$$

The last row corresponds to the equation $0 \cdot x_1 + 0 \cdot x_2 + 0 \cdot x_3 = 1$, which clearly has no solution. It follows that the original system has no solutions. ◊

Example 11.7. Consider the system

$$x_1 + \quad 2x_2 + \quad 3x_3 + \quad 4x_4 = 5$$
$$6x_1 + \quad 7x_2 + \quad 8x_3 + \quad 9x_4 = 10$$
$$11x_1 + 12x_2 + 13x_3 + 14x_4 = 15$$
$$16x_1 + 17x_2 + 18x_3 + 19x_4 = 20$$

Clearing column 1 and then column 2 in the augmented matrix we have

$$\left[\begin{array}{cccc|c} 1 & 2 & 3 & 4 & 5 \\ 6 & 7 & 8 & 9 & 10 \\ 11 & 12 & 13 & 14 & 15 \\ 16 & 17 & 18 & 19 & 20 \end{array}\right] \longrightarrow \left[\begin{array}{cccc|c} 1 & 2 & 3 & 4 & 5 \\ 0 & -5 & -10 & -15 & -20 \\ 0 & -10 & -20 & -30 & -40 \\ 0 & -15 & -30 & -45 & -60 \end{array}\right]$$

$$\longrightarrow \left[\begin{array}{cccc|c} 1 & 0 & -1 & -2 & -3 \\ 0 & 1 & 2 & 3 & 4 \\ 0 & 0 & 0 & 0 & 0 \\ 0 & 0 & 0 & 0 & 0 \end{array}\right].$$

The last matrix corresponds to the equivalent system

$$x_1 \quad - \quad x_3 - 2x_4 = -3$$
$$x_2 + 2x_3 + 3x_4 = 4$$

which has solutions

$$x_1 = x_3 + 2x_4 - 3, \quad x_2 = -2x_3 - 3x_4 + 4, \quad x_3, x_4 \text{ arbitrary.}$$

Thus the set of solutions is described by two parameters, x_3 and x_4. $\Diamond$

11.4   Row Operations with VBA

In this section we develop a program that reads a user entered matrix and a column of row operations symbols and outputs the corresponding transformed matrices. This allows the user to step through the row echelon algorithm, letting Excel do the arithmetic. Figure 11.1 depicts a typical input-output

	A	B	C	D
4	operations	matrices		
5	R1<->R2	7	8	9
6		1	2	3
7		4	5	6
9	-7R1+R2	1	2	3
10		7	8	9
11		4	5	6
13	(-1/6)R2	1	2	3
14		0	-6	-12
15		4	5	6

	A	B	C	D
17	-4R1+R3	1	2	3
18		0	1	2
19		4	5	6
21	3R2+R3	1	2	3
22		0	1	2
23		0	-3	-6
25	answer	1	2	3
26		0	1	2
27		0	0	0

FIGURE 11.1: Spreadsheet for `RowOperations()`.

spreadsheet. The original matrix was entered in the block B5:D7 and the first operation entered in row 5. Running the program once results in the printout of the matrix starting in row 9. The operation in row 9 was then entered; running the program again produces the matrix starting in row 13. The process may be continued as long as desired. (In the next section the process is further automated so that the row echelon form appears automatically.)

The main procedure `RowOperations` declares the following variables:

- op: the operation string.

- `optype`: the type of operation (RowOps 1,2,3).

- `oprow`: the row number of the current operation.

- `scalar`: the scalar multiplier of a row in an operation.

- `rowa,rowb`: the rows in an operation.

- `idx`: the position of the current symbol in `op`.

- `Mat`: the current matrix.

- `nrows,ncols`: dimensions of `Mat`.

After reading the spreadsheet matrix, the procedure finds the row of the last operation entered, which is also the row of the last matrix in the matrix column. The procedure `ExtractOpData` finds the information needed to perform the operation, which is ultimately carried out by `RowOpCalc`.

```
Sub RowOperations()
    Dim Mat() As Double, nrows As Integer, ncols As Integer
    Dim scalar As Double, rowa As Integer, rowb As Integer
    Dim optype As Integer, oprow As Integer, op As String
    If IsEmpty(Cells(5, 2)) Then Exit Sub          'no matrix entered
    oprow = Cells(Rows.Count, 1).End(xlUp).row     'last non blank row
    If oprow < 5 Then Exit Sub                     'no row op entered
    Mat = MatrixIn(oprow, 2)                       'read the matrix in oprow
    op = Cells(oprow, 1).Value                     'read operation
    Call ExtractOpData(op, optype, scalar, rowa, rowb)
    Call RowOpCalc(optype, scalar, rowa, rowb, Mat())     'do the row op
    nrows = UBound(Mat, 1): ncols = UBound(Mat, 2)
    Call MatrixOut(Mat(), oprow + nrows + 1, 2)           'print matrix
End Sub
```

The procedure `ExtractOpData` uses a Do While loop to run through the symbols in the operation `op`. As shown above, the symbols are of the form

$$\text{R1<->R2 (RowOp 1), sR1 (RowOp 2), and sR1+R2 (RowOp 3),}$$

where s denotes a scalar. After a symbol is read, a Select Case instruction determines what action should be taken. If the symbol is R then the procedure assumes a row number follows and the number is extracted by `GetRowNum`. If a row number has not previously been found then the number is assigned to the variable `rowa`; otherwise it is assigned to `rowb`. If the symbol is < then `optype` is assigned the value 1 (row interchange); if the symbol is + then `optype` is given the value 3 (scalar times row added to another row). The default value of `optype` is 2 (scalar times row). Finally, if none of these symbols appear, then the procedure assumes that the symbol is the beginning of a scalar, and `GetScalar` retrieves the scalar.

```
Sub ExtractOpData(op As String, optype As Integer, scalar As Double, _
   rowa As Integer, rowb As Integer)
   Dim k As Integer, idx As Integer
   op = RemoveWhiteSpace(op)                      'remove extra space
   idx = 1                              'position of first symbol in op
   optype = 2                                    'default (scalar mult)
   Do While idx <= Len(op)                  'loop through symbols in op
      Select Case Mid(op, idx, 1)    'read the character at position idx
         Case Is = "R"
            idx = idx + 1                              'skip to number
            k = GetRowNum(op, idx)
            If rowa = 0 Then
               rowa = k                        'first row number found
            Else
               rowb = k                       'second row number found
            End If
         Case Is = "<": optype = 1: idx = idx + 3      'skip past "<->"
         Case Is = "+": optype = 3: idx = idx + 1       'skip past "+"
         Case Else: scalar = GetScalar(op, idx)         'idx now at "R"
      End Select
   Loop
End Sub
```

The function `GetScalar` builds a string `ScalarString` starting at the current value of `idx`, evaluates the string using the VBA function `Evaluate`, which returns a Variant, and converts it to Double.

```
Function GetScalar(op As String, idx As Integer) As Double
   Dim ScalarString As String
   Do While idx < Len(op) And Mid(op, idx, 1) <> "R"
      ScalarString = ScalarString & Mid(op, idx, 1)
      idx = idx + 1
   Loop                                            'idx now at R
   GetScalar = CDbl(Evaluate(ScalarString))
End Function
```

The procedure `GetRowNum` assumes the current symbol is the beginning of a row number, which is extracted as a string by a Do While loop and converted to an integer.

```
Function GetRowNum(op As String, idx As Integer) As Integer
   Dim RowNumString As String, char As String
   Do While idx <= Len(op)
      char = Mid(op, idx, 1)
      If Asc(char) < 49 Or Asc(char) > 57 Then: Exit Do
      RowNumString = RowNumString & char
      idx = idx + 1
   Loop                                 'idx now one past row number
   GetRowNum = CInt(RowNumString)
End Function
```

The actual row operations are carried out by `RowOpCalc`. The procedure is passed the operation type `optype` and returns the scalar multiplier (if relevant) and the row numbers stated in the operation. The procedure uses a Select Case statement to branch to the desired operation.

```
Sub RowOpCalc(optype As Integer, scalar As Double, rowa As Integer, _
              rowb As Integer, Mat() As Double)
    Dim nRows As Integer, nCols As Integer, temp As Double, j As Integer
    Dim decplaces As Integer
    decplaces = 10        'arbitrary but needed for floating point quirk
    nRows = UBound(Mat(), 1): nCols = UBound(Mat(), 2)
    Select Case optype
        Case Is = 1                              'switch rowa and rowb
            For j = 1 To nCols
                temp = Mat(rowa, j)                  'store rowa entry
                Mat(rowa, j) = Mat(rowb, j)   'copy rowb entry into rowa
                Mat(rowb, j) = temp        'copy old rowa entry into rowb
            Next j
        Case Is = 2                              'multiply rowa by scalar
            For j = 1 To nCols
                Mat(rowa, j) = scalar * Mat(rowa, j)
            Next j
        Case Is = 3                              'add scalar*rowa to rowb
            For j = 1 To nCols
                Mat(rowb, j) = Round(Mat(rowb, j) _
                + scalar * Mat(rowa, j), decplaces)
            Next j
    End Select
End Sub
```

11.5   Row Echelon Form with VBA

The function `RowEchelon` described in this section takes a matrix as input and returns its row reduced echelon form. It may therefore be used to solve systems of linear equations, as explained in Section 11.3.

The function uses Do While loops to search for pivot columns and rows in the input matrix. The outer Do While loop starts by looking for the first row that has a nonzero entry in column 1. If such a row exists, it is moved into position `leadrow`, which is initially set to 1. The procedure `ClearMatCol` is then called to make the corner entry 1 and place zeros above and below that entry. After incrementing `leadrow`, the loop then goes on to the next column and the process is repeated. The looping continues until the last row or column is reached. The function `RowEchelon` also uses the variable `switches` to keep track of the number of row switches used in the procedure, and calculates the product `prod` of the matrix entries whose reciprocals are the row multipliers used in type 2 operations of the pivoting process. These variables will be used in the chapter on determinants and are not currently relevant.

```
Function RowEchelon(A() As Double, prod As Double, switches As Integer) _
        As Double()
    Dim row As Integer, col As Integer, nrows As Integer
    Dim ncols As Integer, leadrow As Integer, B() As Double
    nrows = UBound(A, 1): ncols = UBound(A, 2)        'dimensions of A()
    prod = 1: switches = 0     'variables for future use--ignore for now
    toprow = 1: row = 1: col = 1                        'initialize
    B = CopyMat(A)                                 'copy to preserve A
    Do While toprow <= nrows And col <= ncols       'find pivot columns
        row = toprow
      'search below current top row for a row with a entry <> 0
        Do While row < nrows And B(row, col) = 0
            row = row + 1
        Loop
        If B(row, col) <> 0 Then                 'if such a row was found
            prod = B(row, col) * prod          'update prod(for future use)
            If row <> toprow Then
                switches = switches + 1   'update switches(for future use)
                'move row up to toprow position:
                Call RowOpCalc(1, 0, row, toprow, B)
            End If
                'put 1 in pivot position, 0's elsewhere:
            Call ClearMatCol(B, toprow, col)
            toprow = toprow + 1                    'next leading entry row
        End If
        col = col + 1                  'next col to search for next pivot
    Loop
    RowEchelon = B
End Function

Function CopyMat(A() As Double) As Double()    'return a copy of matrix A
    Dim B() As Double, nrows As Integer, ncols As Integer
    Dim i As Integer, j As Integer
    nrows = UBound(A, 1): ncols = UBound(A, 2)
    ReDim B(1 To nrows, 1 To ncols)
    For i = 1 To nrows
        For j = 1 To ncols
            B(i, j) = A(i, j)                          'copy A to B
        Next j
    Next i
    CopyMat = B                          'returned copied matrix
End Function
```

The procedure `ClearMatCol` first calls `RowOpCalc` to perform the type 2 operation of dividing the pivot row by the pivot entry, thus placing a 1 in the pivot position. A For Next loop calls `RowOpCalc` to perform type 3 operations, placing zeros above and below the ones in the pivot positions.

```
Sub ClearMatCol(Mat() As Double, prow As Integer, pcol As Integer)
    Dim nrows As Integer, ncols As Integer, pe As Double
    Dim i As Integer, x As Double
    nrows = UBound(Mat(), 1)
    ncols = UBound(Mat(), 2)                         'dimensions
    pe = Mat(prow, pcol)                             'pivot entry
    Call RowOpCalc(2, 1/pe, prow, prow, Mat)    'divide prow by pivot
    For i = 1 To nrows    'for each i, replace row i by row i + x*prow
        x = -Mat(i, pcol)
        If Not i = prow Then
            Call RowOpCalc(3, x, prow, i, Mat())
        End If
    Next i
End Sub
```

Here is a command button procedure for running the program. The procedure assumes that entry $(1, 1)$ of the input matrix is in cell C3. The output matrix appears one column to the right of the input matrix. An example is depicted in Figure 11.2. The input matrix is in rows 3–5 and columns C–G. The reduced row echelon form of the matrix appears in columns I–M.

```
Sub CommandButton1_Click()
    Dim A() As Double, prod As Double, switches As Integer
    A = MatrixIn(3, 3)                          'get the matrix
    A = RowEchelon(A, prod, switches)       'make the echelon form
    Call MatrixOut(A, 3, 4 + UBound(A, 2))      'one col separation
End Sub
```

	C	D	E	F	G	H	I	J	K	L	M
3	1	2	3	4	5		1	0	-1	-2	-3
4	6	7	8	9	10		0	1	2	3	4
5	11	12	13	14	15		0	0	0	0	0

FIGURE 11.2: Spreadsheet for `RowEchelon()`.

The function `RowEchelon` is able to find the row echelon form of very large matrices, limited essentially only by spreadsheet capacity and memory.

11.6   Exercises

1. Find the solutions, if any, of the following systems by stepping through the row echelon algorithm using the program `RowOperations`.

(a)

$$2x_1 + x_2 + 2x_3 + 2x_4 + 2x_5 = 0$$
$$x_1 + x_2 + 2x_3 + 2x_4 \qquad = 1$$
$$x_1 + 2x_2 + 2x_3 + x_4 + 2x_5 = 0$$
$$x_3 + x_4 + x_5 = 1$$
$$x_1 + x_2 + x_3 + x_4 + 2x_5 = 1$$

(b)

$$3x_1 + 3x_2 + 3x_3 + 2x_4 + x_5 = 2$$
$$x_2 + x_3 + 2x_4 + 3x_5 = 3$$
$$2x_1 + x_3 + x_4 = 2$$
$$2x_1 + 2x_3 + 3x_4 = 0$$
$$2x_1 + 3x_2 + 2x_3 + x_4 + 3x_5 = 1$$

(c)

$$-x_2 + 2x_3 - x_4 + 3x_5 = 0$$
$$2x_1 + 3x_2 + 2x_3 + 4x_4 - x_5 = 1$$
$$2x_1 + 2x_2 + 4x_3 - x_5 = 3$$
$$4x_1 + x_2 + 2x_3 + 5x_4 - 2x_5 = 4$$
$$x_1 + x_5 = -1$$

(d)

$$7x_1 - 7x_2 + 36x_3 + 48x_4 + 49x_5 = 47$$
$$-17x_1 + 33x_2 + 24x_3 + 33x_4 + 4x_5 = 22$$
$$21x_1 + 43x_2 + 48x_3 - 20x_4 + 40x_5 = 3$$
$$24x_1 + 14x_2 - 8x_3 + 38x_4 + 42x_5 = -1$$
$$38x_1 + 44x_2 - 8x_3 + 6x_4 + 34x_5 = -18$$

2. Use the program RowEchelon to verify the solutions to the systems in Examples 11.1, 11.2, and 11.3.

3. (Traffic flow). Consider the system of one-way streets in Figure 11.3. The arrows represent traffic flow and the numbers are vehicles per time period obtained by traffic sensors. The subscripted x's are unknown. The basic principle of traffic flow is that the traffic going into an intersection (pictured as black dots in the figure) must equal the traffic coming out (over some fixed time period). Thus for the intersection at the top left, one must have $500 + 200 = x_1 + x_3$. Find the augmented matrix of the system of linear equations that must be satisfied by the variables x_i. Proceed at the traffic intersections from top to bottom and left to right. Use RowEchelon to show that the system has infinitely many solutions. Find the one for which $x_{17} = 325$

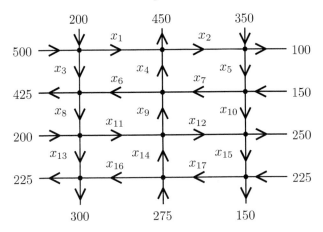

FIGURE 11.3: Traffic flow pattern.

4. (Nutritional supplements). A certain supplement containing three ingredients labeled 1,2, and 3 consists of a total of 300 grams. The supplement must contain as much of ingredient 1 as the sum of ingredients 2 and 3, and half as much of ingredient 2 as the combined amounts of ingredients 1 and 3. Set x_j = number of grams of supplement j. Write the conditions as a system of linear equations and find the solution using RowEchelon.

5. Write a function IsEqualMat(A,B) that determines if given matrices A,B are equal.

6. The *trace* of a square matrix is the sum of the entries in the upper left to lower right diagonal. Write a function TraceMat(A) that returns the trace of A.

7. The *transpose* A^T of an $m \times n$ matrix $A = [a_{ij}]_{m \times n}$ is defined as the $n \times m$ matrix $[b_{ij}]_{n \times m}$, where $b_{ij} = a_{ji}$. Thus A^T is obtained from A by changing columns to rows and rows to columns. For example,

$$\begin{bmatrix} 1 & 2 & 3 \\ 4 & 5 & 6 \end{bmatrix}^T = \begin{bmatrix} 1 & 4 \\ 2 & 5 \\ 3 & 6 \end{bmatrix}.$$

Note that the transpose of a row matrix is a column matrix and vice versa and that $(A^T)^T = A$. Write a function Transpose that returns the transpose of a matrix array A.

8. The $n \times n$ *identity matrix* I_n is defined as

$$I_n = \begin{bmatrix} 1 & 0 & 0 & \cdots & 0 \\ 0 & 1 & 0 & \cdots & 0 \\ \vdots & \vdots & \vdots & \ddots & \vdots \\ 0 & 0 & 0 & \cdots & 1 \end{bmatrix}_{n \times n}$$

that is, ones along the main diagonal and zeros elsewhere. We shall see the importance of this matrix in Chapter 13. Write a program `IdentityMat` that takes a positive integer n and returns I_n

9. A matrix with integer entries is called an *integer matrix*. These may be generated by applying row operations to the identity matrix. Write a program `RndIntMat` that successively generates random integer matrices starting with the identity matrix and proceeding for as many steps as desired. After each step the program should print out the result. At

	A	B	C	D	E	F
4	steps	20		4	18	-1
5	size	3		-43	-176	-6
6	lower	-4		9	35	3
7	upper	4				

FIGURE 11.4: Spreadsheet for `RndIntMat`.

the next step the program should read the previous result from the spreadsheet and start again from there. Figure 11.4 shows the input and output, including the size of the matrices and the upper and lower bounds for the random multiplier in operation type 3. The spreadsheet shows the result of running 20 steps starting from the identity matrix. Running the program again with this matrix as input would result in much larger entries. Use the program `RowOperations` and `RndInt` to generate the random row operations.

10. Write a program that generates random integer matrices and finds their row echelon from. The user should enter the dimensions of the matrix as well as the upper and lower ranges of the random entries. Use the function `RndInt`.

*11. Write a program `ComplexRowOperations` that is the complex number analog of `RowOperations`. (Use the module `ComplexEval` of Section 8.4.)

*12. Write a program `ComplexRowEchelon` that is the complex number analog of `RowEchelon`. (See the module `ComplexEval` of Section 8.4.)

Chapter 12

Linear Programming

A *linear programming* (LP) *problem* asks to find the maximum or minimum value of a linear function in several variables subject to certain restrictions in the form of linear inequalities. In this chapter we describe some of the many contexts in which LP problems arise, develop the simplex method for solving LP problems, and implement the method in VBA.

12.1 Linear Inequalities and Feasible Regions

A *linear inequality* in the variables $x_1, \ldots, x_n$ is an inequality of the form

$$a_1 x_1 + a_2 x_2 + \cdots + a_n x_n \leq b \quad \text{or} \quad a_1 x_1 + a_2 x_2 + \cdots + a_n x_n \geq b,$$

where $a_1, \ldots a_n, b$ are real numbers and not all of the coefficients a_i are zero. The *solution set of a linear inequality* is the collection of all sequences $(x_1, \ldots, x_n)$ satisfying the inequality. For $n = 2$ the solution set of a linear inequality may be graphed as a *half-plane*. For example, the solution set of $2x_1 + 3x_2 \geq 6$ consists

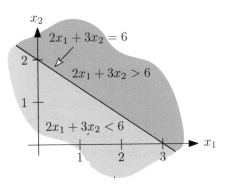

FIGURE 12.1: The half-planes of the line $2x_1 + 3x_2 = 6$.

of the points on or above the line $2x_1 + 3x_2 = 6$ (Figure 12.1). This may be seen by noting that if a point on the line is moved vertically upward, thereby increasing x_2, the value of $2x_1 + 3x_2$ increases. Alternatively, one may take a

DOI: 10.1201/9781003351689-12

convenient *test point* either above or below the line and evaluate the expression $2x_1 + 3x_2$ at that point to see whether it is positive or negative. For example, the point $(x_1, x_2) = (0, 0)$ lies below the line and satisfies $2x_1 + 3x_2 < 6$. Therefore, *all* points below the line satisfy $2x_1 + 3x_2 < 6$ and so all points above the line must satisfy the reverse inequality.

A *system of linear inequalities* in the variables $x_1, \ldots, x_n$ is a finite collection of linear inequalities in these variables. The *solution set of the system* is the collection of all values of $(x_1, \ldots, x_n)$ that simultaneously satisfy the inequalities. This set is called the *feasible region* of the system.

As seen from the above discussion, a feasible region in two dimensions is the intersection of half-planes. For example, the green region in Figure 12.2 represents the solution set of the system

$$2x_1 + 3x_2 \leq 6, \quad 3x_1 + 2x_2 \leq 6, \quad x_1 + x_2 \geq 1, \quad x_1 \geq 0, \quad x_2 \geq 0.$$

The yellow region represents the solution set of the system

$$2x_1 + 3x_2 \geq 6, \quad 3x_1 + 2x_2 \geq 6, \quad 4x_1 - x_2 \geq 0, \quad x_2 \geq 0.$$

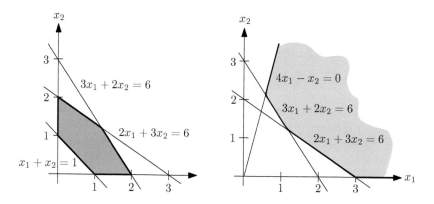

FIGURE 12.2: Feasible regions.

The vertices of these regions are called *corner points* and play a fundamental role in linear programming. Since these points represent solutions of pairs of linear equations, we shall eventually need an efficient algorithm for solving such systems. This is developed in Section 12.4. But first we give a precise description of the general LP problem and some examples.

12.2 Formulation of the General LP Problem

LP problems fall into three categories: the *standard max LP problem*, the *standard min LP problem*, and *nonstandard LP problems*. Fortunately, as we shall see, the method that solves the first type may be used to solve the others. In the following three subsections we give the general formulation as well as examples of each type of LP problem.

The Standard LP Max Problem

This problem is formally stated as follows:

$$\text{Maximize } z = c_1 x_1 + c_2 x_2 + \cdots + c_n x_n \text{ subject to}$$
$$a_{11} x_1 + a_{12} x_2 + \cdots + a_{1n} x_n \leq b_1$$
$$a_{21} x_1 + a_{22} x_2 + \cdots + a_{2n} x_n \leq b_2$$
$$\vdots \tag{12.1}$$
$$a_{m1} x_1 + a_{m2} x_2 + \cdots + a_{mn} x_n \leq b_m$$
$$x_j \geq 0, \ j = 1, \ldots n.$$

The constants b_j are required to be nonnegative. The linear inequalities in (12.1) describe the feasible region of the problem. The linear function z is called the *objective function*. Solutions of the problem are points $(x_1, x_2, \ldots, x_n)$ in the feasible region at which the maximum value of z occurs.

Example 12.1. (Manufacturing). A company makes n products, each of which requires raw materials labeled $1, 2, \ldots, m$. Suppose each item of product j requires a_{ij} units of raw material i in the manufacturing process, as indicated in Figure 12.3.

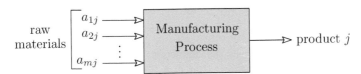

FIGURE 12.3: The manufacture of one item of product j.

This implies that for each i, $a_{i1} x_1$ units of raw material i are required to produce x_1 items of product 1, $a_{i2} x_2$ units of raw material i are required to produce x_2 items of product 2, etc. If there are only b_i units of raw material i on hand, then the inequality

$$a_{i1} x_1 + a_{i2} x_2 + \cdots + a_{in} x_n \leq b_i$$

must be satisfied. Now suppose that the company makes a profit of c_j dollars on each item of product j. It follows that the company's total profit is

$$z = c_1 x_1 + c_2 x_2 + \cdots + c_n x_n.$$

Maximizing total profit while respecting the raw material limitations is therefore an LP problem of the type (12.1). ◊

The Standard LP Min Problem

The problem has the following general form:

$$\text{Minimize} \quad w = c_1 y_1 + c_2 y_2 + \cdots + c_n y_m \text{ subject to}$$
$$a_{11} y_1 + a_{12} y_2 + \cdots + a_{1n} y_m \geq b_1$$
$$a_{21} y_1 + a_{22} y_2 + \cdots + a_{2n} y_m \geq b_2$$
$$\vdots \tag{12.2}$$
$$a_{n1} y_1 + a_{n2} y_2 + \cdots + a_{nm} y_m \geq b_n$$
$$y_j \geq 0, \ j = 1, \ldots m,$$

where $b_j \geq 0$. The linear function w to be minimized is called the *objective function*, as in the LP maximum case. A *solution* is a point (or points) in the feasible region at which the minimum value of w occurs.

Example 12.2. (Packaging) Consider a company that produces m different food supplements, each containing nutrition elements labelled $1, 2 \ldots, n$. Suppose each item of supplement j contains a_{ij} units of nutrition element i, so y_j such items contain $a_{ij} y_j$ such units. (Figure 12.4.) It is desired to combine

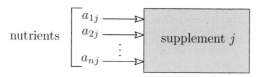

FIGURE 12.4: Nutrition elements in one item of supplement j.

the supplement into a package that contains y_j items of food supplement j in such a way that, for each i, the package contains at least b_i units of nutrient i. Thus one requires of the package that

$$a_{i1} y_1 + a_{i2} y_2 + \cdots + a_{im} y_m \geq b_i, \quad i = 1, 2, \ldots, n.$$

If the cost of producing one item of supplement j is c_j dollars, then the quantity

$$w = c_1 y_1 + c_2 y_2 + \cdots + c_n y_n$$

is the cost of producing the package of supplements. The desire to minimize w leads to problem (12.2). $\Diamond$

Nonstandard LP Problems

These arise in maximization or minimization problems with mixed constraints. For example, if in Example 12.2 it is deemed that the package should contain no more than b_1 units of nutrient element 1, then the first inequality in (12.2) must be reversed, leading to a nonstandard problem. Here is another example:

Example 12.3. (Distribution of products). Suppose there are production facilities in various locations that manufacture a product sold at certain outlets. Each facility can produce only so many items weekly and each outlet needs to satisfy customer demand by selling so many items weekly. There is a cost to ship the items from facility to outlet. The problem is to determine the number of items that should go to each outlet to minimize transportation costs while respecting supply and demand constraints. Figure 12.5 illustrates the case for

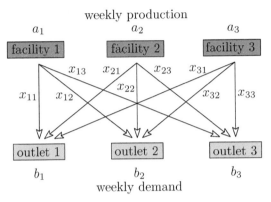

FIGURE 12.5: Transportation problem.

three production facilities and three outlets. The quantity a_i is the weekly production of facility i, b_j is the weekly demand at outlet j, and x_{ij} is the number of items that need to shipped from facility i to outlet j. If the cost of shipping one item from facility i to outlet j is denoted by c_{ij}, we can then state the problem of minimizing cost as the following LP problem:

Minimize
$$w = c_{11}x_{11} + c_{12}x_{12} + c_{13}x_{13} + c_{21}x_{21} + c_{22}x_{22} + c_{23}x_{23}$$
$$+ c_{31}x_{31} + c_{32}x_{32} + c_{33}x_{33}$$

subject to
$$x_{11} + x_{12} + x_{13} \leq a_1$$
$$x_{21} + x_{22} + x_{23} \leq a_2$$
$$x_{31} + x_{32} + x_{33} \leq a_3$$
$$x_{11} + x_{21} + x_{31} \geq b_1$$
$$x_{12} + x_{22} + x_{32} \geq b_2$$
$$x_{13} + x_{23} + x_{33} \geq b_3$$
$$x_{11} \geq 0, \quad x_{12} \geq 0, \quad \ldots, \quad x_{33} \geq 0.$$ ◊

12.3   Graphic Solution for Two Dimensions

A two-dimensional LP problem with only a few constraints may be solved graphically. For an example consider the problem

$$\text{Maximize} \quad z = 3x_1 + 7x_2 \quad \text{subject to}$$
$$5x_1 + 3x_2 \leq 50$$
$$3x_1 + 3x_2 \leq 29$$
$$x_1 + 4x_2 \leq 32$$
$$x_1 \geq 0 \; x_2 \geq 0.$$

The feasible region is shown in green in Figure 12.6. The region is bounded by the lines whose equations are obtained by changing the above inequalities to equalities:

$$5x_1 + 3x_2 = 50, \quad 2x_1 + 3x_2 = 29, \quad x_1 + 4x_2 = 32, \quad x_1 = 0, \quad x_2 = 0.$$

The vertices of the region, called *corner points*, are obtained by solving pairs of linear equations. For example, the point $(4, 7)$ is the solution of the system

$$x_1 + 4x_2 = 32, \quad 2x_1 + 3x_2 = 29.$$

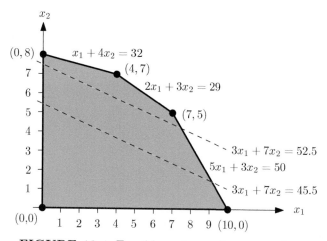

FIGURE 12.6: Feasible region and corner points.

The parallel dotted lines in the figure are the graphs of $3x_1 + 7x_2 = z$, where $z = 45.5$ for the lower line and $z = 52.5$ for the upper line. As the dotted line moves up, the value of x_2 increases, hence so does the value of z. Thus the largest possible value of z (satisfying the given constraints) is attained at the instant the entire line emerges from the region, namely (by inspection) at the point $(4, 7)$. At this point, the objective function z has the value $3 \cdot 4 + 7 \cdot 7 = 61$, which is necessarily the maximum value of z on the region. A similar analysis holds for the minimum.

The preceding example illustrates the following basic principle of linear programming in two variables:

The maximum and minimum value of a linear function in x_1 and x_2 on a bounded feasible region occurs at corner points of the region.

This suggests the following geometric method for solving a two-dimensional LP problem on a bounded region:

- Graph the feasible region.

- Find the corner points by solving suitable pairs of linear equations.

- Test the value of the objective function at each corner point.

- The largest value obtained in the preceding step is the maximum and the smallest is the minimum.

Note that in this algorithm it is possible to have infinitely many points at which the maximum occurs. For example, the maximum value of the function $z = 2x_1 + 3x_2$ in the above LP problem occurs at each point of the line segment from $(4, 7)$ to $(7, 5)$.

The geometric method just described is not practical if there are many constraints and not even possible for more than two variables. To solve LP problems with arbitrarily many variables and constraints we need a general algorithm. The remainder of the chapter is devoted to this end.

12.4 The Simplex Method

We illustrate the simplex method with the example of the previous section:

$$\text{Maximize} \quad z = 3x_1 + 7x_2 \quad \text{subject to}$$
$$5x_1 + 3x_2 \leq 50$$
$$2x_1 + 3x_2 \leq 29$$
$$x_1 + 4x_2 \leq 32$$
$$x_1 \geq 0 \quad x_2 \geq 0.$$

The first step is to convert the system of inequalities into a system of *equalities*. This is accomplished by adding nonnegative *slack variables* s_1, s_2, and s_3 to the left sides of the first three inequalities above, resulting in the system of linear equations

$$
\begin{aligned}
5x_1 + 3x_2 + s_1 &= 50 \\
2x_1 + 3x_2 + s_2 &= 29 \\
x_1 + 4x_2 + s_3 &= 32 \\
-3x_1 - 7x_2 + z &= 0
\end{aligned}
\tag{12.3}
$$

We have included in the system the objective equation $z = 3x_1 + 7x_2$, written in a manner consistent with that of the first three equations. We now have a system of four linear equations in the nonnegative unknowns x_1, x_2, s_1, s_2, s_3 and z. The system generally has infinitely many solutions; we seek one in the feasible region that produces the largest value of z.

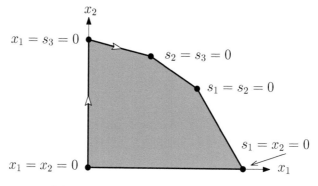

There are several important things to notice. First, the feasible region is described by the inequalities

$$
x_1 \geq 0, \quad x_2 \geq 0, \quad s_1 \geq 0, \quad s_2 \geq 0, \quad s_3 \geq 0.
$$

Second, corner points are obtained by setting certain pairs of these variables equal to zero. For example, the corner point that is the intersection of the lines $5x_1 + 3x_2 = 50$ and $2x_1 + 3x_2+ = 29$ may described by the pair of equations $s_1 = 0$ and $s_2 = 0$. However, not every pair of such equations gives a corner point. For example, the pair $s_2 = 0$ and $x_2 = 0$ leads to the equation $2x_1 = 29$, which produces a point outside the feasible region. Third, setting suitable pairs of the variables x_1, x_2, s_1, s_2, s_3 equal to zero in the first three equations of (12.3) leads to systems that can be solved uniquely for the remaining variables, producing a value of z. The fundamental idea behind the simplex method is to start at the known corner point $x_1 = 0$, $x_2 = 0$ (at which z has the value zero) and move to corner points that increase z the fastest until further movement no longer produces an increase. Here are the details:

Since the maximum value of z occurs at a corner point and since these are obtained by setting pairs of the variables x_1, x_2, s_1, s_2, and s_3 equal to zero, at each stage of the process it must be determined which pair of these variables should be set equal to zero. These are called the *inactive variables*; the remaining variables are called *active variables*. We start by taking x_1 and

x_2 to be inactive, producing the corner point $x_1 = 0$, $x_2 = 0$. The variables s_1, s_2, and s_3 are then active. All of this may be conveniently summarized in what is called the *initial tableau* of the problem, a beefed-up version of the augmented matrix of system (12.3):

	x_1	x_2	s_1	s_2	s_3	z	
s_1	5	3	1	0	0	0	50
s_2	2	3	0	1	0	0	29
s_3	1	4	0	0	1	0	32
	-3	-7	0	0	0	1	0

Notice that the system variables are written along the top of the tableau and the active variables in the left-most column (marked in red). With $x_1 = x_2 = 0$ one can read the remaining solutions $s_1 = 50$, $s_2 = 29$, $s_3 = 32$, and $z = 0$ directly from the tableau. These values form what is called a *basic solution* associated with the tableau. Note that the active variables are distinguished by the fact that their columns contain a single one and zeros elsewhere. (The z column also has this property.)

Having initially chosen x_1 and x_2 as the inactive variables, we must now decide if this choice gives us the maximum value of z or if we should move to another corner point, that is, make another pair of variables inactive, to obtain a larger value of z. For this we look at the equation $z = 3x_1 + 7x_2$ and notice that an increase of 1 unit in x_1 results in an increase of 3 units in z, and an increase of one unit in x_2 results in an increase of 7 units in z. That these increases are possible is reflected in the tableau by the fact that the entries in the last row in the columns under x_1 and x_2 are negative. Since an increase of one unit in x_2 results in the largest increase of z, for efficiency we should move in the positive x_2 direction. This will keep x_1 inactive (its value is still 0), but will change x_2 to an active variable, requiring one of the variables s_1, s_2, or s_3 to become inactive. Because x_2 is to become active, we must construct a system equivalent to (12.3) whose augmented matrix has its x_2 column consisting of a single one and zeros elsewhere, the hallmark of an active variable. This can be accomplished by row operations on the tableau, pivoting on a suitable entry chosen from the first three positions of x_2 column and clearing that column. To choose the pivot entry in that column we form the ratios of the first three entries of the last column of the tableau with the corresponding entries of the x_2 column:

$$\frac{50}{3}, \quad \frac{29}{3}, \quad \text{and} \quad \frac{32}{4}. \tag{12.4}$$

The last ratio is the smallest, and since this corresponds to the third entry in the x_2 column, we choose that entry as the pivot. (The rationale for this choice is explained later.) Performing row operations on the augmented matrix of (12.3) we have

$$
\begin{bmatrix}
5 & 3 & 1 & 0 & 0 & 0 & 50 \\
2 & 3 & 0 & 1 & 0 & 0 & 29 \\
1 & \boxed{4} & 0 & 0 & 1 & 0 & 32 \\
-3 & -7 & 0 & 0 & 0 & 1 & 0
\end{bmatrix}
\xrightarrow{\frac{1}{4}R_3}
\begin{bmatrix}
5 & 3 & 1 & 0 & 0 & 0 & 50 \\
2 & 3 & 0 & 1 & 0 & 0 & 29 \\
\frac{1}{4} & 1 & 0 & 0 & \frac{1}{4} & 0 & 8 \\
-3 & -7 & 0 & 0 & 0 & 1 & 0
\end{bmatrix}
$$

$$
\begin{array}{c}
-3R_3 + R_1 \\
-3R_3 + R_2 \\
7R_3 + R_4 \\
\xrightarrow{\hspace{2cm}}
\end{array}
\begin{bmatrix}
\frac{17}{4} & 0 & 1 & 0 & -\frac{3}{4} & 0 & 26 \\
\frac{5}{4} & 0 & 0 & 1 & -\frac{3}{4} & 0 & 5 \\
\frac{1}{4} & 1 & 0 & 0 & \frac{1}{4} & 0 & 8 \\
-\frac{5}{4} & 0 & 0 & 0 & \frac{7}{4} & 1 & 56
\end{bmatrix}
\tag{12.5}
$$

From (12.5) we obtain the new tableau

	x_1	x_2	s_1	s_2	s_3	z	
s_1	$\frac{17}{4}$	0	1	0	$-\frac{3}{4}$	0	26
s_2	$\frac{5}{4}$	0	0	1	$-\frac{3}{4}$	0	5
x_2	$\frac{1}{4}$	1	0	0	$\frac{1}{4}$	0	8
	$-\frac{5}{4}$	0	0	0	$\frac{7}{4}$	1	56

The variables x_1 and s_3 are now the inactive variables and s_1, s_2, x_2 are active. The change in the status of x_2 and s_3 is described by saying that x_2 is the *entering variable* and s_3 the *departing variable*. The basic solution to the system, which may be read directly from the tableau, is

$$x_1 = 0, \ s_3 = 0, \ s_1 = 26, \ s_2 = 5, \ x_2 = 8, \ z = 56.$$

Since all variables are nonnegative, the corresponding point lies in the feasible region and is the corner point $x_1 = 0$, $s_3 = 0$, that is, $x_1 = 0$, $x_2 = 8$. At this point $z = 56$. We remark that it is precisely because we needed all the variables to be nonnegative that we chose the entry in the x_2 column corresponding to the *smallest* ratio as the pivot entry. Indeed, the entries 26 and 5 in the last column of (12.5) result from the operations $(-3)8 + 50$ and $(-3)8 + 29$, respectively. Dividing these by 3 we obtain the expressions $-8 + 50/3$ and $-8 + 29/3$, which are positive precisely because the ratio $32/4 = 8$ in (12.4) is less than the ratios $50/3$ and $29/3$. To illuminate further, suppose instead that we had pivoted on the entry corresponding the ratio $50/3$. We would then have

$$
\begin{bmatrix}
5 & \boxed{3} & 1 & 0 & 0 & 0 & 50 \\
2 & 3 & 0 & 1 & 0 & 0 & 29 \\
1 & 4 & 0 & 0 & 1 & 0 & 32 \\
-3 & -7 & 0 & 0 & 0 & 1 & 0
\end{bmatrix}
\xrightarrow{\frac{1}{3}R_1}
\begin{bmatrix}
\frac{5}{3} & \boxed{1} & \frac{1}{3} & 0 & 0 & 0 & \frac{50}{3} \\
2 & 3 & 0 & 1 & 0 & 0 & 29 \\
1 & 4 & 0 & 0 & 1 & 0 & 32 \\
-3 & -7 & 0 & 0 & 0 & 1 & 0
\end{bmatrix}
$$

$$
\begin{array}{c}
-3R_1 + R_2 \\
-4R_1 + R_3 \\
7R_1 + R_4 \\
\xrightarrow{\hspace{2cm}}
\end{array}
\begin{bmatrix}
\frac{5}{3} & 1 & \frac{1}{3} & 0 & 0 & 0 & \frac{50}{3} \\
-3 & 0 & -1 & 1 & 0 & 0 & -21 \\
-\frac{17}{3} & 0 & -\frac{4}{3} & 0 & 1 & 0 & -\frac{104}{3} \\
\frac{26}{3} & 0 & \frac{7}{3} & 0 & 0 & 1 & \frac{350}{3}
\end{bmatrix}
$$

which has basic solution

$$x_1 = 0, \ x_2 = \frac{50}{3}, \ s_1 = 0, \ s_2 = -21, \ s_3 = -\frac{104}{3}, \ z = \frac{350}{3}.$$

Since there are negative values for s-variables, the point obtained lies outside the feasible region and so the resulting increased value of z cannot be used.

We have now moved to another corner point and in doing so have increased the value z. Can we increase z even more? The answer is yes: there is a negative entry of $-5/4$ in the last row under the x_1 column in the second tableau, implying that increasing the current value $x_1 = 0$ will result in an increase in z. Increasing x_1 will make it active, hence row operations must be applied to change the x_1 column of the second tableau into one containing a 1 and 0's. The ratios of the first three entries of the last column in the tableau with the corresponding members in the x_1 column are, respectively,

$$\frac{26}{17/4} = \frac{104}{17}, \quad \frac{5}{9/4} = \frac{4}{45}, \quad \text{and} \quad \frac{8}{1/4} = 32.$$

As the second ratio is the smallest, we choose the second entry of the x_1 column as the pivot entry:

$$\begin{bmatrix} \frac{17}{4} & 0 & 1 & 0 & -\frac{3}{4} & 0 & 26 \\ \boxed{\frac{5}{4}} & 0 & 0 & 1 & -\frac{3}{4} & 0 & 5 \\ \frac{1}{4} & 1 & 0 & 0 & \frac{1}{4} & 0 & 8 \\ -\frac{5}{4} & 0 & 0 & 0 & \frac{7}{4} & 1 & 56 \end{bmatrix} \xrightarrow{\frac{4}{5}R_2} \begin{bmatrix} \frac{17}{4} & 0 & 1 & 0 & -\frac{3}{4} & 0 & 26 \\ \boxed{1} & 0 & 0 & \frac{4}{5} & -\frac{3}{5} & 0 & 4 \\ \frac{1}{4} & 1 & 0 & 0 & \frac{1}{4} & 0 & 8 \\ -\frac{5}{4} & 0 & 0 & 0 & \frac{7}{4} & 1 & 56 \end{bmatrix}$$

$$\begin{matrix} -\frac{17}{4}R_2 + R_1 \\ -\frac{1}{4}R_2 + R_3 \\ \frac{5}{4}R_2 + R_4 \\ \xrightarrow{\hspace{2cm}} \end{matrix} \begin{bmatrix} 0 & 0 & 1 & -\frac{17}{5} & \frac{9}{5} & 0 & 9 \\ 1 & 0 & 0 & \frac{4}{5} & -\frac{3}{5} & 0 & 4 \\ 0 & 1 & 0 & -5 & \frac{2}{5} & 0 & 7 \\ 0 & 0 & 0 & 1 & 1 & 1 & 61 \end{bmatrix}$$

This gives the tableau

	x_1	x_2	s_1	s_2	s_3	z	
s_1	0	0	1	$-\frac{17}{5}$	$\frac{9}{5}$	0	9
x_1	1	0	0	$\frac{4}{5}$	$-\frac{3}{5}$	0	4
x_2	0	1	0	-5	$\frac{2}{5}$	0	7
	0	0	0	1	1	1	61

which has basic solution

$$x_1 = 4, \quad x_2 = 7, \quad s_1 = 9, \quad s_2 = s_3 = 0, \quad z = 61.$$

Since there are no negative entries in the last row, z cannot be further increased. Thus the maximum value 61 of z occurs when $x_1 = 4$ and $x_2 = 7$, in agreement with results of the preceding section.

12.5   Standard Max Problem with VBA

In this section we develop a module that generates the tableaus and finds the solution of the standard LP maximum problem. The program can also find the solution of the standard LP minimum problem, explained later in Section 12.6. We illustrate some of the procedures in the module with the example in Section 12.4:

$$\text{Maximize} \quad z = 3x_1 + 7x_2 \ \text{subject to}$$
$$5x_1 + 3x_2 \le 50$$
$$2x_1 + 3x_2 \le 29$$
$$x_1 + 4x_2 \le 32$$
$$x_1 \ge 0 \ x_2 \ge 0.$$

The main procedure `Simplex` assumes that the augmented matrix of the system is entered with the $(1, 1)$ entry in cell D6 and the coefficients of the objective function are entered in the row below. This provides coded input for the program. The input for the above example is displayed in Figure 12.7. The

	D	E	F
6	5	3	50
7	2	3	29
8	1	4	32
9	3	7	

FIGURE 12.7: Spreadsheet for LP max example.

user also enters the text "max" or "min" in cell B6, informing the program whether a maximum or a minimum is to be calculated. The following are the main variables and arrays of the module:

- `mode`: set equal to "max" or "min".
- `TabRow`: first row of current tableau, initially set to 6.
- `TabCol`: first column of current tableau. Always equals 4.
- `NRows`: number of rows of tableau.
- `NCols`: number of columns of input and, later, of tableaus.
- `NumX`: number of x-variables.
- `NumS`: number of slack variables.
- `NumV`: total number of variables ($=$ `NumX` + `NumS`).

- `InMat`: array for input.
- `Tableau`: array for tableaus.
- `VLabs`: array for labels of x-variables and slack variables.
- `Active`: array to note which variables are active.
- `VStat`: array to note the status of each variable.

The procedure `Simplex` first sets the initial values of `TabRow` and `TabCol`. It then reads the entry in cell B6 into the variable `mode`. If `mode=min`, the procedure calls `ConvertToMax`, discussed in Section 12.6. Next, the procedure calls `MatrixIn`, which returns the spreadsheet entries in `InMat`. The numbers `NumX` and `NumS` are calculated from the dimensions `NRows,NCols` of the matrix: `NumX=NCols-1` and `NumS=NRows-1`. For the above example, `NRows=4`, `NCols=3`, `NumX=2`, and `NumS=3`. These parameters provide the necessary information for the dynamic allocation of the arrays `InMat,Tableau,VLabs,Active`, and `VStat`, which are all (automatically) initialized to zero. The procedure `FormInitialTab` constructs the first tableau and `PrintTab` prints it. The remaining code in `Simplex` consists of a Do While loop to produce and print successive tableaus until there are no more negative entries in the last row of the tableau, indicated by the function `GetNextTab` returning zero. The Do While loop also guards against infinite loops by stopping at 40 iterations (this number is arbitrary). The solution to the problem is printed by the procedure `PrintMaxSol`.

```
Public NumS As Integer, NumX As Integer, NumV As Integer
Public TabRow As Integer, TabCol As Integer
Public NRows As Integer, NCols As Integer
Sub Simplex()
    Dim VLabs() As String, Tableau() As Double, Mat() As Double
    Dim Active() As Integer, VStat() As Integer, n As Integer
    Dim mode As String
    TabRow = 6: TabCol = 4              '(1,1) position of initial tableau
    Mat() = MatrixIn(TabRow, TabCol)             'skeleton tableau
    mode = Range("B6").Value
    If mode = "min" Then Mat = Transpose(Mat)
    nrows = UBound(Mat, 1): ncols = UBound(Mat, 2)     'dimensions of Mat
    NumX = ncols - 1                      'number of x variables
    NumS = nrows - 1                      number of slack variables
    NumV = NumX + NumS                    total number of variables
    ncols = NumV + 2              'update to number of columns in tableau
    ReDim Tableau(1 To nrows, 1 To ncols)        'initialized to zero
    ReDim VLabs(1 To NumV + 1)            'memory for variable labels
    ReDim Active(1 To NumS)               'and for active variables
    ReDim VStat(1 To NumV)                'and for variable status
    Call MakeInitialTab(Tableau, Mat)      'construct first tableau
    Call GetLabels(VLabs)                  'labels x1,...xn, s1,...,sm
    Call InitActiveVars(Active)                 'initially at n+1,...,m
    Call InitStatus(VStat) 'set x vars = 0(inactive), s vars = 1(active)
    Call PrintTab(Tableau, Active, VLabs)              'initial tableau
    Do While GetNextTab(Tableau, Active, VStat) > 0 And n < 40
```

```
      Call PrintTab(Tableau, Active, VLabs)            'next tableau
      n = n + 1
   Loop
   Call PrintMaxSol(Tableau, VStat, VLabs)
   If mode = "min" Then
      Call PrintMinSol
   End If
End Sub
```

The procedure `MakeInitialTab` reads the spreadsheet data into the array `Tableau`, adds suitable ones and zeros, and configures the last row, which holds the negatives of the coefficients of the objective function.

```
Sub MakeInitialTab(T() As Double, Mat() As Double)
   Dim i As Integer, j As Integer
   For i = 1 To NRows                           'rows of Tableau
      For j = 1 To NumX                         'first NumX cols of InMat
         T(i, j) = Mat(i, j)                    'read x-coefficients
         If i = NRows Then T(i, j) = -T(i, j)   'make negative
      Next j
      T(i, NumV + 2) = Mat(i, NumX + 1)         'get the constraints
      For j = NumX + 1 To NumV + 1
         If i + NumX = j Then T(i, j) = 1       'add the 1's
      Next j
   Next i
End Sub
```

At this stage, the `Tableau` array for our example looks like

$$
\begin{bmatrix}
5 & 3 & 1 & 0 & 0 & 0 & 50 \\
2 & 3 & 0 & 1 & 0 & 0 & 29 \\
1 & 4 & 0 & 0 & 1 & 0 & 32 \\
-3 & -7 & 0 & 0 & 0 & 0 & 1
\end{bmatrix}
$$

Before printing the first tableau the procedure `GetLabels` is called, which places in the array `VLabs` the labels x1, ..., xn, s1, ..., sm, and z, where $n =$ `NumX` and $m =$ `NumS`. These are used for the top and left sides of the tableau. The "subscripts" of the labels are appended to x and s by the concatenation operator &.

```
Sub GetLabels(Labels() As String)
   Dim i As Integer
   For i = 1 To NumX
      Labels(i) = "x" & i                       'form xi labels
   Next i
   For i = 1 To NumS
      Labels(i + NumX) = "s" & i                'form si labels
   Next i
   Labs(NumV + 1) = "z"
End Sub
```

The procedure `InitActiveVars` in the next listing initializes the array `Active`. At any stage, the array tells the program which variables are active. Specifically, `Active(i)` is the position of the ith active variable in the list $x_1, \ldots, x_n, s_1, \ldots, s_m$. Thus, initially `Active(1)= n+1`, `Active(2)=n+2`, ... In our example, the list of variables is x_1, x_2, s_1, s_2, s_3, hence the array `Active` has initial value {3,4,5}, reflecting the fact that the initial active variables are s_1, s_2, s_3.

```
Sub InitActiveVars(Active() As Integer)
    Dim i As Integer
    For i = 1 To NumS
        Active(i) = NumX + i 'position of si in list x1,...,xn,s1,...,sm
    Next i
End Sub
```

The procedure `InitStatus` initializes the array `VStat`. At any stage, the array tells the program the status of the ith variable in list $x_1, \ldots, x_n, s_1, \ldots, s_m$. Thus `VStat(i)= 0` means that the ith variable in the list is inactive; while `VStat(i)= 1` means that it is active. For our example, calling `InitStatus` produces `VStat = {0,0,1,1,1}`.

```
Sub InitStatus(ByRef VStat() As Integer)
    Dim i As Integer
    For i = 1 To NumX
        VStat(i) = 0
    Next i
    For i = NumX + 1 To NumV
        VStat(i) = 1
    Next i
End Sub
```

The procedure `PrintTab` prints the tableaus, taking into account which variables are active. For our example it produces

	x_1	x_2	s_1	s_2	s_3	z	
s_1	5	3	1	0	0	0	50
s_2	2	3	0	1	0	0	29
s_3	1	4	0	0	1	0	32
	-3	-7	0	0	0	1	0

```
Sub PrintTab(T() As Double, Active() As Integer, ColLabs() As String)
    Dim i As Integer, j As Integer
    TabRow = TabRow + NRows + 2        'pos. of new tableau in spreadsheet
    Call PrintColumnLabels(ColLabs)
    Call PrintRowLabels(Active, ColLabs)
    Call MatrixOut(T, TabRow, TabCol)
End Sub
```

```
Sub PrintRowLabels(Active() As Integer, VarLabs() As String)
    Dim i As Integer
    For i = 1 To NumS
        Cells(TabRow + i - 1, TabCol - 1).Value = VarLabs(Active(i))
    Next i
End Sub

Sub PrintColumnLabels(VLabs() As String)
    Dim i As Integer
    For i = 1 To NumX + NumS + 1
        Cells(TabRow - 1, TabCol - 1 + i).Value = VLabs(i)
    Next i
End Sub
```

The function `GetNextTab` constructs a new tableau from the last tableau. The key component is the procedure `ClearMatCol`, introduced in Section 11.5. The function first determines the pivot column and the pivot row and then calls `Pivot` to clear the column. The arrays `Active` and `VStat` are then updated.

```
Function GetNextTab(T() As Double, Active() As Integer, _
                    VStat() As Integer) As Integer
    Dim PivotRow As Integer, PivotCol As Integer, PivotEntry As Double
    Dim current_act As Integer, new_act As Integer, Ratio As Double
    PivotCol = GetPivotCol(T)
    If PivotCol = 0 Then
        GoTo lastline                       'no more negative entries
    End If
    PivotRow = GetPivotRow(T, PivotCol)
    Call ClearMatCol(T, PivotRow, PivotCol)
    current_act = Active(PivotRow)   'number of current active variable
    new_act = PivotCol               'number of new active variable
    Active(PivotRow) = new_act                        'updates
    VStat(new_active) = 1: VStat(current_act) = 0
lastline:
    GetNextTab = PivotCol
End Function
```

The function `GetPivotCol` used in the preceding function finds the most negative entry in the last row of the tableau if one exists, and returns its column number. Otherwise the function returns zero.

```
Function GetPivotCol(T() As Double) As Integer
    Dim j As Integer, PivotCol As Integer, MostNeg As Double
    MostNeg = Tableau(NRows, 1)                  'provisional value
    PivotCol = 1
    For j = 2 To NCols
        If T(NRows, j) < MostNeg Then
            MostNeg = T(NRows, j)                        'update
            PivotCol = j
        End If
    Next j
```

```
      If MostNeg >= 0 Then PivotCol = 0    'no more negatives in last row
      GetPivotCol = PivotCol
End Function
```

The function `GetPivotRow` forms ratios of the last column with the pivot column and returns the row number corresponding to the smallest ratio. It consists of two For Next loops. The first loop finds *some* ratio, while the second finds the smallest.

```
Function GetPivotRow(T() As Double, PivotCol) As Integer
    Dim i As Integer, PivotRow As Integer, Ratio As Double
    For i = 1 To NRows - 1                       'get any ratio to start
        If T(i, PivotCol) > 0 Then
            Ratio = T(i, NCols) / T(i, PivotCol): PivotRow = i
            Exit For
        End If
    Next i
    For i = 1 To NRows - 1                       'now get smallest ratio
        If T(i, PivotCol) > 0 Then
            If T(i, NCols) / T(i, PivotCol) < Ratio Then
                Ratio = T(i, NCols) / T(i, PivotCol): PivotRow = i
            End If
        End If
    Next i
    GetPivotRow = PivotRow
End Function
```

The procedure `PrintMaxSol` prints out the solution to the LP max problem, printing the labels `x1`, ..., `xn`, `s1`, ..., `sm`, `z` first. Solutions will be placed under these. The columns are then scanned for inactive and active variables. If `vstat(j)` = 0, that is, if the jth variable in the list $x_1, \ldots, x_n, s_1, \ldots, s_m$ is inactive, then a zero is placed under the jth label. Otherwise, the jth variable is active and the position i of the 1 in the jth column is determined. The rightmost value in the tableau `T(i,NCols)` is then printed under the jth label. The last value printed is z.

```
Sub PrintMaxSol(T() As Double, VStats() As Integer, ColLabs() As String)
    Dim i As Integer, j As Integer, SolRow As Integer
    SolRow = TabRow + NRows + 2  'position of solution in spreadsheet
    Call PrintColumnLabels(ColLabs)
    For j = 1 To NumS + NumX
        If VStats(j) = 0 Then               'jth var list member is inactive
            Cells(SolRow, TabCol - 1 + j).Value = 0    'inactive solution
        Else
            For i = 1 To NRows     'find nonzero entry in active column j
                If Not T(i, j) = 0 Then
                    Cells(SolRow, TabCol - 1 + j).Value = _
                        T(i, NCols) / T(i, j)          'active solution
                    Exit For
                End If
```

```
            Next i
        End If
    Next j
    Cells(SolRow, TabCol - 2 + NCols).Value = Tabl(NRows, NCols)  '= z
    Cells(SolRow, TabCol - 1).Value = "solutions:"
End Sub
```

Here is the final printout for our example:

	x_1	x_2	s_1	s_2	s_3	z	
s_1	0	0	1	$-\frac{17}{5}$	$\frac{9}{5}$	0	9
x_1	1	0	0	$\frac{4}{5}$	$-\frac{3}{5}$	0	4
x_2	0	1	0	-5	$\frac{2}{5}$	0	7
	0	0	0	1	1	1	61
solution:	4	7	9	0	0	61	

12.6　Duality and the Standard Min Problem

The standard LP minimization problem is solved using the notion of *duality*. To see how this works, consider the problem

$$\text{Minimize} \quad w = 50y_1 + 29y_2 + 32y_3 \quad \text{subject to}$$
$$5y_1 + 2y_2 + y_3 \geq 3$$
$$3y_1 + 3y_2 + 4y_3 \geq 7$$
$$y_1 \geq 0, \ y_2 \geq 0, \ y_3 \geq 0.$$

We "encode" the problem by the matrix

$$A = \begin{bmatrix} 5 & 2 & 1 & 3 \\ 3 & 3 & 4 & 7 \\ 50 & 29 & 32 & 0 \end{bmatrix}$$

and then take its transpose (Exercise 7 of Section 11.6):

$$A^T = \begin{bmatrix} 5 & 3 & 50 \\ 2 & 3 & 29 \\ 1 & 4 & 32 \\ 3 & 7 & 0 \end{bmatrix},$$

The resulting matrix is then "decoded" to form the max problem

$$\text{Maximize} \quad z = 3x_1 + 7x_2 \quad \text{subject to}$$
$$5x_1 + 3x_2 \le 50$$
$$2x_1 + 3x_2 \le 29$$
$$x_1 + 4x_2 \le 32$$
$$x_1 \ge 0 \ x_2 \ge 0.$$

This is the problem of Section 12.5. The last tableau of the solution for that problem together with a new row of labels is

	x_1	x_2	s_1	s_2	s_3	z	
s_1	0	0	1	$-\frac{17}{5}$	$\frac{9}{5}$	0	9
x_1	1	0	0	$\frac{4}{5}$	$-\frac{3}{5}$	0	4
x_2	0	1	0	-5	$\frac{2}{5}$	0	7
	0	0	0	1	1	1	61
			y_1	y_2	y_3		$w_{\min}$

We now invoke the *LP duality principle*:

> *The minimum value is the maximum value of the dual problem and occurs at the entries $y_1, y_2, \ldots$ in the last row under the slack variables $s_1, s_2, \ldots$.*

Thus in our example the minimum value is 61 and it occurs at the point where $y_1 = 0$, $y_2 = 1$, $y_3 = 1$ as indicated in the tableau.

To use Simplex for minimization problems the user enters the problem exactly as in the LP max case (see Figure 12.7). Setting mode to min causes the program to replace the entered matrix by its transpose.

The final procedure PrintMinSol places subscripted y's and a w under the last row of the tableau.

```
Sub PrintMinSol()
    Dim i As Integer, j As Integer
    TabRow = TabRow + NRows            'position of solution in spreadsheet
    For j = 0 To NumS - 1
        Cells(TabRow, TabCol + NumX + j).Value = "y" & j + 1
    Next j
    Cells(TabRow, TabCol + NCols - 1).Value = "w"
End Sub
```

12.7 Solving Nonstandard LP Problems

In this section we consider LP max and LP min problems with mixed constraints. Since minimizing an objective function is equivalent to maximizing the negative of the function, it suffices to consider the maximum case. We illustrate with the following example, whose feasible region is shown in Figure 12.8.

$$\text{Maximize} \quad z = 3x_1 + 2x_2 \quad \text{subject to}$$
$$3x_1 + 2x_2 \geq 6$$
$$2x_1 + 3x_2 \geq 6$$
$$-4x_1 + x_2 \leq 1$$
$$x_1 + x_2 \leq 4$$
$$x_1 \leq 2$$
$$x_1 \geq 0, \quad x_2 \geq 0.$$

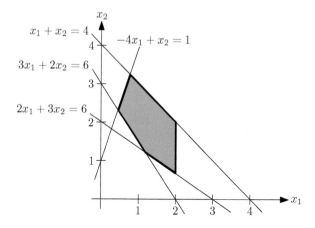

FIGURE 12.8: Feasible region for example.

The first step is to append nonnegative variables to the left sides of the constraint inequalities to produce equalities. This requires adding nonnegative slack variables s_3, s_4, and s_5 to the last three constraints and *subtracting* nonnegative *surplus variables* s_1 and s_2 from the first two constraints. The result is the system

$$
\begin{aligned}
3x_1 + 2x_2 - s_1 &= 6 \\
2x_1 + 3x_2 - s_2 &= 6 \\
-4x_1 + x_2 + s_3 &= 1 \\
x_1 + x_2 + s_4 &= 4 \\
x_1 + + s_5 &= 2 \\
-3x_1 - 2x_2 + z &= 0,
\end{aligned}
\tag{12.6}
$$

which has infinitely many solutions. We seek one in the feasible region that gives the largest value of z. Proceeding as before, we form the tableau

	x_1	x_2	s_1	s_2	s_3	s_4	s_5	z	
s_1	3	2	-1	0	0	0	0	0	6
s_2	2	3	0	-1	0	1	0	0	6
s_3	-4	1	0	0	1	0	0	0	1
s_4	1	1	0	0	0	1	0	0	4
x_1	1	0	0	0	0	0	1	0	2
	-3	-2	0	0	0	0	0	1	0

The basic solution is

$$
x_1 = x_2 = 0, \quad s_1 = s_2 = -6, \quad s_3 = 1, \quad s_4 = 4, \quad s_5 = 2.
$$

However, the point $x_1 = x_2 = 0$ is outside the feasible region, not to mention the fact that s_1 and s_2 are negative. We need to find an equivalent system that leads to a point inside the feasible region. This requires finding a pivot entry. The pivot column is found as follows: Choose any negative active variable, say s_1, in the above tableau. Since the sole nonzero entry in its column is in row 1, go to a column other than the last one that has a positive entry in row 1, say column 1. This is the pivot column. The pivot row is then determined as before by considering ratios: the smallest ratios with a *positive* denominator are $6/3$ and $2/1$. Since these are equal we may choose either row 1 or row 5 as the pivot row. We choose the latter, giving us the entry in column 1 and row 5 as the pivot. Performing row operations using the program `RowOps` of Section 11.4 results in the tableau

	x_1	x_2	s_1	s_2	s_3	s_4	s_5	z	
s_1	0	2	-1	0	0	0	-3	0	0
s_2	0	3	0	-1	0	0	-2	0	2
s_3	0	1	0	0	1	0	4	0	9
s_4	0	1	0	0	0	1	-1	0	2
x_1	1	0	0	0	0	0	1	0	2
	0	-2	0	0	0	0	3	1	6

The basic solution is

$$
x_2 = s_1 = s_5 = 0, \quad x_1 = 2, \quad s_2 = -2, \quad s_3 = 9, \quad s_4 = 2.
$$

Since $s_2 < 0$, we apply the rule again. The sole nonzero entry under s_2 in the preceding tableau is in row 2, so we choose column 2, the only column with a positive entry in row 2 (excluding the last). Thus the pivot entry is in row 2 and column 2. Applying row operations results in the tableau

	x_1	x_2	s_1	s_2	s_3	s_4	s_5	z	
s_1	0	0	-1	$\frac{2}{3}$	0	0	$-\frac{5}{3}$	0	$-\frac{4}{3}$
x_2	0	1	0	$-\frac{1}{3}$	0	0	$-\frac{2}{3}$	0	$\frac{2}{3}$
s_3	0	0	0	$\frac{1}{3}$	1	0	$\frac{14}{3}$	0	$\frac{25}{3}$
s_4	0	0	0	$\frac{1}{3}$	0	1	$-\frac{1}{3}$	0	$\frac{4}{3}$
x_1	1	0	0	0	0	0	1	0	2
	0	0	0	$-\frac{2}{3}$	0	0	$\frac{5}{3}$	1	$\frac{22}{3}$

The basic solution is

$$x_1 = 2, \quad x_2 = 2/3, \quad s_2 = s_5 = 0, \quad s_1 = 4/3, \quad s_3 = 25/3, \quad s_4 = 4/3.$$

All solutions are nonnegative so we are in the feasible region and we can now solve the LP max problem by applying the program `Simplex`. This yields the final tableau

	x_1	x_2	s_1	s_2	s_3	s_4	s_5	z	
s_1	0	0	1	0	0	2	-1	0	4
x_2	0	1	0	0	0	1	-1	0	2
s_3	0	0	0	0	1	-1	5	0	7
s_2	0	0	0	1	0	3	-1	0	4
x_1	1	0	0	0	0	0	1	0	2
	0	0	0	0	0	2	1	1	10

From this we see that the maximum value of z is 10 and that the maximum occurs at the point $x_1 = x_2 = 2$.

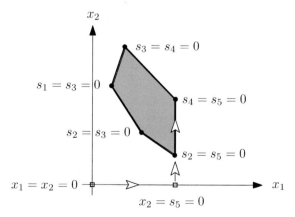

12.8   Exercises

1. Describe the region in the figure in terms of linear inequalities.

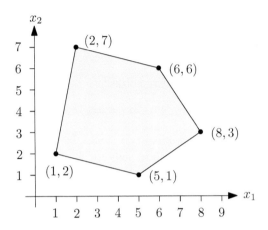

2. Use the geometric method to find the maximum and minimum of the objective functions (a) $z = 3x - 7y$ and (b) $z = 6x + 7y$ subject to constraints

$$x + 3y \geq 11, \ 4x + y \geq 11, \ x + 8y \leq 57, \ 5x - y \leq 39.$$

3. Use the geometric method to find the maximum and minimum of the objective functions (a) $z = x - y$ and (b) $z = 5x + 3y$ subject to constraints

$$x + 2y \geq 7, \ 2x - y \geq -1, \ 2x - 3y \geq -11, \ x + 2y \leq 19, \ 3x - y \leq 22, \ x - 3y \leq 2.$$

4. Solve the following standard LP max problems by hand and check your answers with `Simplex`.

(a) Maximize $z = 5x_1 + 6x_2$ subject to

$$2x_1 + x_2 \leq 2, \ 2x_1 + 3x_2 \leq 4, \ x_1 + x_2 \leq 2.$$

(b) Maximize $z = 5x_1 + 2x_2$ subject to

$$5x_1 + x_2 \leq 5, \ x_1 + x_2 \leq 5, \ x_1 + x_2 \leq 2.$$

(c) Maximize $z = 6x_1 + 7x_2$ subject to

$$5x_1 + 2x_2 \leq 7, \ 2x_1 + 4x_2 \leq 4 \ x_1 + 4x_2 \leq 4.$$

5. Solve the following standard LP min problems by hand and check your answers with `Simplex`.

   (a) Minimize $w = 2x_1 + 3x_2$ subject to

   $$4x_1 + 3x_2 \geq 5, \ x_1 + 4x_2 \geq 2, \ x_1 + 7x_2 \geq 7.$$

   (b) Minimize $w = 3x_1 + 5x_2$ subject to

   $$x_1 + 2x_2 \geq 4, \ 3x_1 + x_2 \geq 6, \ 4x_1 + 7x_2 \geq 4.$$

   (c) Minimize $w = 7x_1 + 5x_2$ subject to

   $$5x_1 + 5x_2 \geq 3, \ 2x_1 + 5x_2 \geq 6, \ 3x_1 + x_2 \geq 3.$$

6. An ice cream store produces two flavors, chocolate and vanilla. Suppose that each quart of chocolate sells for \$7.00 and each quart of vanilla sells for \$6.00. Suppose also that a quart of chocolate requires 5 eggs, 4 cups of cream, and 3 cups of sugar, while a quart of vanilla requires 4 eggs, 3 cups of cream, and 4 cups of sugar. If the store has on hand 50 eggs, 29 cups of cream, and 32 cups of sugar, how many quarts x_1 of chocolate and x_2 of vanilla should be produced from this supply to maximize revenue from the sale.

7. A company produces m different models of portable smoke shifters. A shifter goes through a sequence of n manufacturing steps, each requiring a certain number of hours. Suppose shifter model j requires a_{ij} hours to complete step i and there are only b_i hours available to carry out the steps. Suppose also that the company makes a profit of c_j dollars on each item of model j. Set up the LP problem whose solution produces the maximum total profit given the stated constraints.

8. A poultry farmer wishes to buy no less than 100 pounds of grain feed that contains barley, corn, rye, and wheat. There are 9 feed stores in his county, each selling premixed grain. The amount per pound of each grain in the mixtures is given in the table. The cost per pound of each mixture is given in the last row.

	1	2	3	4	5	6	7	8	9
barley	.5	.1	.5	.3	.4	.1	.1	.6	.7
corn	.3	.4	.2	.4	.3	.3	.5	.2	.1
rye	.1	.2	.2	.1	.2	.4	.2	.2	.1
wheat	.1	.3	.1	.2	.1	.2	.3	0	.1
cost	.074	.068	.072	.076	.060	.057	.044	.041	.040

The farmer requires that his purchase contain at least 40 pounds of barley and 30 pounds of corn. Find the number x_j of pounds purchased from store j so that the requirements are met and the cost is minimum.

Chapter 13

Matrix Algebra

In this chapter we show how matrices may be added, subtracted, multiplied, and divided, resulting in a rich algebraic system. VBA Excel procedures are developed throughout to automate calculations. Several applications are given, including input-output models and polynomial curve fitting.

13.1   Matrix Scalar Multiplication

The *scalar multiple* of a matrix A by a number t, denoted by the symbol tA, is the matrix obtained by multiplying each entry of A by t. For example,

$$(-2) \begin{bmatrix} 1 & 2 & 3 & 4 \\ 5 & 6 & 7 & 8 \end{bmatrix} = \begin{bmatrix} -2 & -4 & -6 & -8 \\ -10 & -12 & -14 & -16 \end{bmatrix}$$

Performing the operation in reverse provides a way of *factoring* numbers from a matrix. The following procedure returns the result of multiplying a matrix array A by a scalar t.

```
Function ScalarMult(A() As Double, t As Double) As Double()
    Dim i As Integer, j As Integer, B() As Double
    Dim Arows As Integer, Acols As Integer
    Arows = UBound(A, 1): Acols = UBound(A, 2)
    ReDim B(1 To Arows, 1 To Acols)              'memory for return array
    For i = 1 To Arows
        For j = 1 To Acols
            B(i, j) = t * A(i, j)                'multiply each entry of by t
        Next j
    Next i
    ScalarMult = B                               'return scalar multiple
End Function
```

DOI: 10.1201/9781003351689-13

13.2   Matrix Addition and Substraction

The *sum* of two $m \times n$ matrices

$$A = \begin{bmatrix} a_{11} & a_{12} & \cdots & a_{1n} \\ a_{21} & a_{22} & \cdots & a_{2n} \\ \vdots & \vdots & \ddots & \vdots \\ a_{m1} & a_{m2} & \cdots & a_{mn} \end{bmatrix}, \quad B = \begin{bmatrix} b_{11} & b_{12} & \cdots & b_{1n} \\ b_{21} & b_{22} & \cdots & b_{2n} \\ \vdots & \vdots & \ddots & \vdots \\ b_{m1} & b_{m2} & \cdots & b_{mn} \end{bmatrix}$$

is defined as

$$A + B = \begin{bmatrix} a_{11} + b_{11} & a_{12} + b_{12} & \cdots & a_{1n} + b_{1n} \\ a_{21} + b_{21} & a_{22} + b_{22} & \cdots & a_{2n} + b_{1n} \\ \vdots & \vdots & \ddots & \vdots \\ a_{m1} + b_{m1} & a_{m2} + b_{m2} & \cdots & a_{mn} + b_{mn} \end{bmatrix}.$$

Thus to find the sum of two matrices with the same dimensions simply add the entries. For example,

$$\begin{bmatrix} 1 & 2 & 3 & 4 \\ 5 & 6 & 7 & 8 \end{bmatrix} + \begin{bmatrix} 1 & 1 & 1 & 1 \\ 1 & 1 & 1 & 1 \end{bmatrix} = \begin{bmatrix} 2 & 3 & 4 & 5 \\ 6 & 7 & 8 & 9 \end{bmatrix}.$$

Similarly, the *difference* of two m by n matrices as above is defined as

$$A - B = \begin{bmatrix} a_{11} - b_{11} & a_{12} - b_{12} & \cdots & a_{1n} - b_{1n} \\ a_{21} - b_{21} & a_{22} - b_{22} & \cdots & a_{2n} - b_{1n} \\ \vdots & \vdots & \ddots & \vdots \\ a_{m1} - b_{m1} & a_{m2} - b_{m2} & \cdots & a_{mn} - b_{mn} \end{bmatrix}.$$

The function `AddSubMat(A,B, sign)` returns the sum or difference of matrix arrays `A` and `B` depending on whether the integer `sign` is 1 or -1.

```
Function AddSubMat(A() As Double, B() As Double, sign As Integer, _
          err As Boolean) As Double()
    Dim i As Integer, j As Integer, C() As Double, Arows As Integer
    Dim Acols As Integer Brows As Integer, Bcols As Integer
    Arows = UBound(A, 1): Acols = UBound(A, 2)
    Brows = UBound(B, 1): Bcols = UBound(B, 2)
    err = Arows <> Brows Or Acols <> Bcols
    If err Then Exit Function              'quit if dimensions not ok
    ReDim C(1 To Arows, 1 To Acols)             'memory for return array
    For i = 1 To Arows
        For j = 1 To Acols
            C(i,j) = A(i,j) + sign*B(i,j) 'add:sign=1, subtract:sign=-1
        Next j
    Next i
    AddSubMat = C
End Function
```

13.3 Matrix Multiplication

The operations of scalar multiplication, addition, and subtraction of matrices are fairly straightforward and may seem quite natural to the reader. Perhaps less so is the operation of matrix multiplication. We begin with the following definition.

The *dot product* c of n-tuples $(a_1, a_2, \cdots, a_n)$ and $(b_1, b_2, \cdots, b_n)$ is defined as the sum of the products of corresponding components:

$$c := a_1 b_1 + a_2 b_2 + \cdots + a_n b_n = \sum_{k=1}^{n} a_k b_k.$$

Here we have used *summation notation* as a short way of writing the sum. A *dot* is usually placed between the n-tuples to indicate a dot product. For example,

$$(1, 2, 3, 4, 5) \cdot (1, 1/2, 1/3, 1/4, 1/5) = 1 + 1 + 1 + 1 + 1 = 5.$$

The *product AB* of an $m \times n$ matrix A and an $n \times p$ matrix B is the $m \times p$ matrix C whose (i, j)th entry is the dot product of row i of A and row j of B. We indicate this by the following scheme:

$$\begin{bmatrix} a_{11} & \cdots & a_{1k} & \cdots & a_{1n} \\ \vdots & & \vdots & & \vdots \\ \boxed{a_{i1} \cdots a_{ik} \cdots a_{in}} \\ \vdots & & \vdots & & \vdots \\ a_{m1} & \cdots & a_{mk} & \cdots & a_{mn} \end{bmatrix} \begin{bmatrix} b_{11} & \cdots & \boxed{b_{1j}} & \cdots & b_{1p} \\ \vdots & & \vdots & & \vdots \\ b_{k1} & \cdots & \boxed{b_{kj}} & \cdots & b_{kp} \\ \vdots & & \vdots & & \vdots \\ b_{n1} & \cdots & \boxed{b_{nj}} & \cdots & b_{np} \end{bmatrix} = \begin{bmatrix} c_{11} & \cdots & c_{1j} & \cdots & c_{1p} \\ \vdots & & \vdots & & \vdots \\ c_{i1} & \cdots & \boxed{c_{ij}} & \cdots & c_{ip} \\ \vdots & & \vdots & & \vdots \\ c_{m1} & \cdots & c_{mj} & \cdots & c_{mp} \end{bmatrix}.$$

In summation notation,

$$c_{ij} = \sum_{k=1}^{n} a_{ik} b_{kj}.$$

Note that the product is defined only when the "inner dimensions" of the matrices (in this case n) are the same.

Here's an example of a matrix multiplied by its transpose:

$$\begin{bmatrix} 1 & 2 & 3 & 4 \\ 4 & 3 & 2 & 1 \end{bmatrix} \begin{bmatrix} 1 & 4 \\ 2 & 3 \\ 3 & 2 \\ 4 & 1 \end{bmatrix} = \begin{bmatrix} 30 & 20 \\ 20 & 30 \end{bmatrix} = 10 \begin{bmatrix} 3 & 2 \\ 2 & 3 \end{bmatrix};$$

and the other way around

$$\begin{bmatrix} 1 & 4 \\ 2 & 3 \\ 3 & 2 \\ 4 & 1 \end{bmatrix} \begin{bmatrix} 1 & 2 & 3 & 4 \\ 4 & 3 & 2 & 1 \end{bmatrix} = \begin{bmatrix} 17 & 14 & 11 & 8 \\ 14 & 13 & 12 & 11 \\ 11 & 12 & 13 & 14 \\ 8 & 11 & 14 & 17 \end{bmatrix}.$$

The example shows that, unlike multiplication of numbers, AB and BA need not be the same even if the dimensions are set up correctly for the multiplications.

The function MultMat(A,B,error) below returns the product of the matrix arrays A and B in that order. The procedure sets error to True if the matrices do not have the proper dimensions required for multiplication. If one of the matrices has dimensions 1×1 we invoke scalar multiplication.

```
Function MultMat(A() As Double, B() As Double, errer As Boolean) As Double()
    Dim i As Integer, j As Integer, k As Integer, C() As Double
    Dim Arows As Integer, Acols As Integer
    Dim Brows As Integer, Bcols As Integer
    Arows = UBound(A, 1): Acols = UBound(A, 2)
    Brows = UBound(B, 1): Bcols = UBound(B, 2)
    ReDim C(1 To Arows, 1 To Bcols)              'memory for return array
    If Arows = 1 And Acols = 1 Then       'if A is 1 x 1 use scalar mult.
        C = ScalarMult(B, A(1, 1))
    ElseIf Brows = 1 And Bcols = 1 Then    if B is 1 x 1 use scalar mult.
        C = ScalarMult(A, B(1, 1))
    ElseIf Acols <> Brows Then
        error = True: Exit Function                     'dimensions not ok
    Else
        For i = 1 To Arows
            For j = 1 To Bcols
                For k = 1 To Acols
                    C(i, j) = C(i, j) + A(i, k) * B(k, j)
                Next k
            Next j
        Next i
    End If
    MultMat = C
End Function
```

A particularly important example of matrix multiplication is illustrated by the calculation

$$
\begin{bmatrix} 1 & 0 & 0 \\ 0 & 1 & 0 \\ 0 & 0 & 1 \end{bmatrix}
\begin{bmatrix} 1 & 2 & 3 \\ 4 & 5 & 6 \\ 7 & 8 & 9 \end{bmatrix} =
\begin{bmatrix} 1 & 2 & 3 \\ 4 & 5 & 6 \\ 7 & 8 & 9 \end{bmatrix} =
\begin{bmatrix} 1 & 2 & 3 \\ 4 & 5 & 6 \\ 7 & 8 & 9 \end{bmatrix}
\begin{bmatrix} 1 & 0 & 0 \\ 0 & 1 & 0 \\ 0 & 0 & 1 \end{bmatrix}
$$

The matrix with ones and zeros in this calculation is called the 3×3 *identity matrix*. (See Exercise 8.) The key fact here is that multiplication by the identity matrix leaves the original matrix unchanged. More generally, the $n \times n$ *identity matrix* I_n is defined as

$$
I_n = \begin{bmatrix} 1 & 0 & 0 & \cdots & 0 \\ 0 & 1 & 0 & \cdots & 0 \\ \vdots & \vdots & \vdots & \ddots & \vdots \\ 0 & 0 & 0 & \cdots & 1 \end{bmatrix}_{n \times n}
$$

and has the properties

$$AI_n = A \text{ and } I_n B = B$$

for any $m \times n$ matrix A and $n \times p$ matrix B. Thus the identity matrix acts like the number 1 in ordinary multiplication.

13.4   Inverse of a Matrix

The notion of *matrix inversion* is a process akin to forming the fraction $1/a$ of a nonzero number a. Here is the precise definition: The *inverse* of an $n \times n$ matrix A is another $n \times n$ matrix X with the property $AX = I_n$, where I_n is the $n \times n$ identity matrix. An inverse X need not exist, but if it does then it is unique and the equation $XA = I_n$ also holds. Notice that the inverse is defined for *square* matrices only. Hereafter, we drop the subscript n from the notation I_n.

To see how the inverse is constructed, consider the case $n = 3$ and the explicit form of a typical equation $AX = I$:

$$\begin{bmatrix} 1 & 1 & 1 \\ 1 & 3 & 2 \\ 1 & 2 & 1 \end{bmatrix} \begin{bmatrix} x_{11} & x_{12} & x_{13} \\ x_{21} & x_{22} & x_{23} \\ x_{31} & x_{32} & x_{33} \end{bmatrix} = \begin{bmatrix} 1 & 0 & 0 \\ 0 & 1 & 0 \\ 0 & 0 & 1 \end{bmatrix}.$$

Here, the x_{ij} are the entries of the unknown inverse X. We may write this matrix equation as three separate equations:

$$\begin{bmatrix} 1 & 1 & 1 \\ 1 & 3 & 2 \\ 1 & 2 & 1 \end{bmatrix} \begin{bmatrix} x_{11} \\ x_{21} \\ x_{31} \end{bmatrix} = \begin{bmatrix} 1 \\ 0 \\ 0 \end{bmatrix}, \quad \begin{bmatrix} 1 & 1 & 1 \\ 1 & 3 & 2 \\ 1 & 2 & 1 \end{bmatrix} \begin{bmatrix} x_{12} \\ x_{22} \\ x_{32} \end{bmatrix} = \begin{bmatrix} 0 \\ 1 \\ 0 \end{bmatrix}, \quad \begin{bmatrix} 1 & 1 & 1 \\ 1 & 3 & 2 \\ 1 & 2 & 1 \end{bmatrix} \begin{bmatrix} x_{13} \\ x_{23} \\ x_{33} \end{bmatrix} = \begin{bmatrix} 0 \\ 0 \\ 1 \end{bmatrix}.$$

If we perform the indicated matrix multiplications we obtain three systems of equations, the first in the unknowns x_{11}, x_{21}, x_{31}, the second in the unknowns x_{12}, x_{22}, x_{32}, and the third in the unknowns x_{13}, x_{23}, x_{33}. The systems have the following augmented matrices:

$$\left[\begin{array}{ccc|c} 1 & 1 & 1 & 1 \\ 1 & 3 & 2 & 0 \\ 1 & 2 & 1 & 0 \end{array}\right], \quad \left[\begin{array}{ccc|c} 1 & 1 & 1 & 0 \\ 1 & 3 & 2 & 1 \\ 1 & 2 & 1 & 0 \end{array}\right], \text{ and } \left[\begin{array}{ccc|c} 1 & 1 & 1 & 0 \\ 1 & 3 & 2 & 0 \\ 1 & 2 & 1 & 1 \end{array}\right].$$

Notice that these differ only in the right column. We may therefore solve the above three systems simultaneously by applying the row echelon algorithm to the combined augmented matrix

$$[\,A \mid I\,] = \left[\begin{array}{ccc|ccc} 1 & 1 & 1 & 1 & 0 & 0 \\ 1 & 3 & 2 & 0 & 1 & 0 \\ 1 & 2 & 1 & 0 & 0 & 1 \end{array}\right].$$

This yields

$$
\left[\begin{array}{ccc|ccc}
1 & 0 & 0 & 1 & -1 & 1 \\
0 & 1 & 0 & -1 & 0 & 1 \\
0 & 0 & 1 & 1 & 1 & -2
\end{array}\right],
\tag{13.1}
$$

which resolves into the following three systems, written in matrix form, equivalent to the original ones:

$$
\begin{bmatrix} 1 & 0 & 0 \\ 0 & 1 & 0 \\ 0 & 0 & 1 \end{bmatrix}
\begin{bmatrix} x_{11} \\ x_{21} \\ x_{31} \end{bmatrix}
=
\begin{bmatrix} 1 \\ -1 \\ 1 \end{bmatrix},
\quad
\begin{bmatrix} 1 & 0 & 0 \\ 0 & 1 & 0 \\ 0 & 0 & 1 \end{bmatrix}
\begin{bmatrix} x_{12} \\ x_{22} \\ x_{32} \end{bmatrix}
=
\begin{bmatrix} -1 \\ 0 \\ 1 \end{bmatrix}
$$

and

$$
\begin{bmatrix} 1 & 0 & 0 \\ 0 & 1 & 0 \\ 0 & 0 & 1 \end{bmatrix}
\begin{bmatrix} x_{13} \\ x_{23} \\ x_{33} \end{bmatrix}
=
\begin{bmatrix} 1 \\ 1 \\ -2 \end{bmatrix}.
$$

Using the identity matrix property in these equations we see that the columns of x's are precisely the columns of the right side of the augmented matrix (13.1).

The method of the preceding example may be used to find the inverse of any square matrix A (if it exists): Simply apply the row echelon algorithm to the matrix $[A \mid I]$ to obtain $[I \mid A^{-1}]$. The right side is then the inverse of A. Here we have used the customary notation A^{-1} for the inverse.

What if a matrix has no inverse? The method just described reveals that fact as well. For example, the matrix on the right below is the reduced row echelon form of the matrix $[A \mid I]$ on the left:

$$
[A \mid I] =
\left[\begin{array}{ccc|ccc}
1 & 2 & 3 & 1 & 0 & 0 \\
4 & 5 & 6 & 0 & 1 & 0 \\
7 & 8 & 9 & 0 & 0 & 1
\end{array}\right]
\longrightarrow
\left[\begin{array}{ccc|ccc}
1 & 0 & -1 & 0 & -\frac{8}{3} & \frac{5}{3} \\
0 & 1 & 2 & 0 & \frac{7}{3} & -\frac{4}{3} \\
0 & 0 & 0 & 1 & -2 & 1
\end{array}\right].
$$

Since the identity matrix does not appear in the left half of the matrix, A has no inverse.

The inverse matrix operation may be used to solve systems of equations that have the same number of unknowns as equations. This involves writing the system in (11.2) as a matrix equation $AX = B$, where

$$
A =
\begin{bmatrix}
a_{11} & a_{12} & \cdots & a_{1n} \\
a_{21} & a_{22} & \cdots & a_{2n} \\
\vdots & \vdots & \ddots & \vdots \\
a_{m1} & a_{m2} & \cdots & a_{mn}
\end{bmatrix},
\quad
X =
\begin{bmatrix} x_1 \\ x_2 \\ \vdots \\ x_n \end{bmatrix},
\quad \text{and} \quad
B =
\begin{bmatrix} b_1 \\ b_2 \\ \vdots \\ b_m \end{bmatrix}.
$$

The matrix A is called the *coefficient matrix* of the system. If $m = n$ and A^{-1} exists, then, as in ordinary algebra, one can multiply the equation $AX = B$ on both sides by A^{-1}, resulting in $A^{-1}AX = A^{-1}B$. Since $A^{-1}A = IX = X$, the solution of the system is given by the matrix equation $X = A^{-1}B$. This observation is sometimes useful in matrix calculations. (See for, example, Section 13.6.)

In the remainder of the section we construct a program that inverts a matrix. The function `InvertMat(A,error)` below returns A^{-1} if it exists and sets `error` to `True` otherwise. The function first acquires memory for the combined augmented matrix `AugMat`, which has double the number of columns of the input matrix A. Next, the function places A in the left half of `AugMat` and the identity matrix I in the right half, so that `AugMat` is of the form $[A \mid I]$. The procedure `RowEchelon` of Section 11.5 then converts `AugMat` to reduced row echelon form. If the left half of the resulting matrix is the identity matrix, a condition indicated by `AugMat(nrows,ncols)` = 1, then the right half, which is the desired inverse, is returned by the function. Otherwise, `error` is set to `True`. The variables `PivotProd` and `Switches` are included for future use and may be ignored for now.

```
Function InvertMat(Mat() As Double, error As Boolean) As Double()
    Dim i As Integer, j As Integer, nrows As Integer, ncols As Integer
    Dim MatInv() As Double, AugMat() As Double, ok As Boolean
    Dim PivotProd As Double, switches As Integer          'for future use
    nrows = UBound(Mat(), 1):  ncols = UBound(Mat(), 2)
    If nrows <> ncols Then error = True: Exit Function    'Mat not square
    ReDim AugMat(1 To nrows, 1 To 2 * ncols)
    ReDim MatInv(1 To nrows, 1 To ncols)
    For i = 1 To nrows       'put Mat in left half of AugMat, I in right
        For j = 1 To 2 * ncols
            If j <= ncols Then AugMat(i, j) = Mat(i, j)
            If j > ncols And j - ncols = i Then AugMat(i, j) = 1
        Next j
    Next i
    AugMat = RowEchelon(AugMat, PivotProd, switches)
    error = AugMat(nrows, ncols) <> 1        'last entry of left half = 0?
    If Not error Then
        For i = 1 To nrows                   'put right half of AugMat in Mat
            For j = 1 To ncols
                MatInv(i, j) = AugMat(i, ncols + j)
            Next j
        Next i
    End If
    InvertMat = MatInv                                    'return the inverse
End Function
```

The function may easily be implemented with a command button procedure that retrieves a matrix from the spreadsheet and prints the inverse.

13.5 Matrix Powers

If A is an $n \times n$ matrix and p is an integer ≥ 2, we define $A^p = AA \cdots A$ (p factors). We also define $A^0 = I$ and $A^1 = A$. If $p < 0$, then A^p is defined as B^{-p}, where $B = A^{-1}$.

The following function `PowerMat` takes a square matrix A and an integer p and returns A^p.

```
Function PowerMat(A() As Double, p As Integer, error As Boolean) As Double()
    Dim B() As Double, C() As Double, n As Integer
    If UBound(A, 1) <> UBound(A, 2) Then Exit Function
    n = UBound(A, 1)
    If p = 0 Then PowerMat = IdentityMat(n): Exit Function    'return I
    If p = 1 Then PowerMat = A: Exit Function                 'return A
    ReDim B(1 To n, 1 To n): ReDim C(1 To n, 1 To n)
    B = CopyMat(A)                              'make a copy of A
    If p < 0 Then                            'if the exponent is < 0
        B = InvertMat(B, error)                     'invert first
        p = -p
    End If
    If error Then Exit Function
    C = CopyMat(B)
    For n = 1 To p - 1                     'keep multiplying C by B
        C = MultMat(B, C, error)
    Next n
    PowerMat = C
End Function
```

13.6 Input-Output Models

An *input-output model* is a mathematical description of the interdependence among the sectors of an economy, such as manufacturing, transportation, or agriculture. The model was developed by Wassily Leontief (1906-1999) and earned him the Nobel Prize in economics in 1973. It is based on an *input-output* (IO) *matrix* whose columns correspond to the commodity requirements of the sectors.

To illustrate the idea we consider first an economy with just three sectors, say agriculture, natural gas, and electricity. Assume (not necessarily realistically) that these sectors have the following production needs, where a unit may be taken to be one dollar:

- To produce 1 unit of agriculture requires .2 units of agriculture, .3 units of energy, and .2 units of manufacturing.

- To produce 1 unit of energy requires 0 units of agriculture, .1 units of energy, and .2 units of manufacturing.

- To produce 1 unit of manufacturing requires 0 units of agriculture, .3 units of energy, and .1 units of manufacturing.

The data is conveniently summarized in the following IO matrix:

$$
\begin{array}{c}
\text{output} \\
\text{input}\begin{array}{c} \text{agriculture} \\ \text{energy} \\ \text{transportation} \end{array}
\begin{array}{ccc} \text{agriculture} & \text{energy} & \text{transportation} \\
\begin{bmatrix} .2 & 0 & 0 \\ .3 & .1 & .3 \\ .2 & .2 & .1 \end{bmatrix} \end{array}
\end{array}.
$$

In general the interdependence among n sectors of an economy is summarized by the IO matrix

$$
\begin{array}{c}
\text{output} \\
\text{input}\begin{array}{c} \text{Sector 1} \\ \text{Sector 2} \\ \vdots \\ \text{Sector } n \end{array}
\begin{array}{cccc} \text{Sector 1} & \text{Sector 2} & \cdots & \text{Sector } n \\
\begin{bmatrix} a_{11} & a_{12} & \cdots & a_{1n} \\ a_{21} & a_{22} & \cdots & a_{2n} \\ \vdots & \vdots & \ddots & \vdots \\ a_{n1} & a_{n2} & \cdots & a_{nn} \end{bmatrix} \end{array}
\end{array}.
$$

Here, a_{ij} is the number of units from sector i needed to produce one unit from sector j.

The input-output matrix represents the *internal demand* of the economy. There may also be additional demand from outside the system. This is represented by the so-called *external demand vector*

$$
D = \begin{bmatrix} d_1 \\ \vdots \\ d_n \end{bmatrix},
$$

where, d_i is the demand in units for the commodity from sector i (over some fixed period of time). For the economy to meet both external and internal demands, a certain level of production is required. This is represented by the so-called *production vector*

$$
X = \begin{bmatrix} x_1 \\ \vdots \\ x_n \end{bmatrix},
$$

where x_j is the total number of units produced by sector j (over the same time period). The problem then is to determine the values x_j that allow the sectors to satisfy both internal and external demands.

The solution of the problem rests on the observation that since a_{ij} is the number of units from sector i required to produce one unit from sector j, the quantity $a_{ij}x_j$ is the number of units from sector i required to produce x_j units from sector j. Letting j vary we see that for each i the sum

$$a_{i1}x_1 + a_{i2}x_2 + \cdots + a_{in}x_n$$

represents the total number of units from sector i required to produce x_1 units from sector 1, x_2 units from sector 2, etc. Adding to this the external demand d_i for commodity i we find that the total production x_i from sector i must satisfy

$$x_i = a_{i1}x_1 + a_{i2}x_2 + \cdots + a_{in}x_n + d_i, \quad i = 1, \ldots, n.$$

These equations may be assembled into a matrix equation $X = AX + D$, where

$$X = \begin{bmatrix} x_1 \\ x_2 \\ \vdots \\ x_n \end{bmatrix}, \quad A = \begin{bmatrix} a_{11} & a_{12} & \cdots & a_{1n} \\ a_{21} & a_{22} & \cdots & a_{2n} \\ \vdots & \vdots & \ddots & \vdots \\ a_{n1} & a_{n2} & \cdots & a_{nn} \end{bmatrix}, \quad \text{and} \quad D = \begin{bmatrix} d_1 \\ d_2 \\ \vdots \\ d_n \end{bmatrix}.$$

By standard matrix algebra, the equation may be written as $(I - A)X = D$. Multiplying by $(I - A)^{-1}$ produces the solution

$$X = (I - A)^{-1}D.$$

This represents the production levels of the sectors needed to satisfy both the internal and external demands.

The columns of $(I - A)^{-1}$ have an interesting economic interpretation: Suppose the demand D changes by an amount ΔD, so the new demand is $D + \Delta D$. The production vector then changes from X to $X + \Delta X$. Thus we have the equations

$$(I - A)^{-1}D = X \quad \text{and} \quad (I - A)^{-1}(D + \Delta D) = \Delta X.$$

Subtracting we get

$$(I - A)^{-1}\Delta D = \Delta X.$$

Now suppose total demand for items from sector one increases by 1 but total demand for items from the remaining sectors stays the same. Thus ΔD has a 1 in the first position and 0's elsewhere. The reader may check that ΔX is then the first column of $(I - A)^{-1}$. In general, the jth column of $(I - A)^{-1}$ is the (vector) amount that production must change in order to accommodate an increase of 1 unit of demand from the jth sector.

The solution X of an IO model is easily obtained using the procedure InvertMat. The module IOModel below assumes that the IO matrix A is entered into the spreadsheet with its $(1,1)$ entry in cell D3 and that the demand matrix D is entered in column 2 starting at B3. The program then prints

	B	D	E	F	H	I
3	100	.10	.50	0	.20	1905
4	400	.20	.30	.40	.01	1673
5	300	0	.20	.10	.04	878
6	600	.30	.20	.50	.50	3890

FIGURE 13.1: Spreadsheet for `IOModel`.

the production vector X one column to the right of A. Figure 13.1 shows an example. Columns B and D–H were entered by the user and column I contains the production vector calculated by the program (rounded).

The module begins by allocating memory for various matrix arrays. The function `AddSubMat` forms the matrix array `Diff=I-A`, and `InvertMat` inverts it to form `DiffInv`. The procedure `MultMat` then produces the desired production vector `DiffInv D`.

```
Sub IOModel()
    Dim x() As Double, D() As Double, Diff() As Double
    Dim DiffInv() As Double, A() As Double, Id() As Double
    Dim nrows As Integer, ncols As Integer, err As Boolean
    A = MatrixIn(3, 4)                        'get IO matrix A
    D = MatrixIn(3, 2)                        'get demand vector D
    nrows = UBound(A, 1): ncols = UBound(A, 2)    'IO matrix dimensions
    If nrows <> ncols Then MsgBox "Matrix not square!": Exit Sub
    Id = IdentityMat(nrows)                'make the identity matrix I
    Diff = AddSubMat(Id, A, -1, err)                'make I-A
    DiffInv = InvertMat(Diff, err)
    If Not err Then
        x = MultMat(DiffInv, D, err)        'find X = (I-A)^(-1)D
        Call MatrixOut(x, 3, 5 + ncols)        'print X
    Else
        MsgBox "Solution not found."
    End If
End Sub
```

The IO model discussed above is frequently referred to as an *open model*, since some of the output goes to the external demand D. If all of the output in the production process is used internally, characterized by the fact that each column of the IO matrix sums to 1, then the model is said to be *closed*. The production matrix X in a closed model satisfies $AX = X$ or $(I - A)X = O$, where O is the matrix all of whose entries are zero. The latter system typically has infinitely many solutions.

13.7   Polynomial of Best Fit

We saw in Section 9.7 that it is possible to find a line that best fits, in the least squares sense, data points $(x_1, y_1), \ldots, (x_n, y_n)$. Apart from its simplicity, there is nothing particularly special about using a line. Indeed, another type of curve may be a better predictor of a possible trend suggested by the data. In this section we consider the problem of fitting data to polynomials and develop a program that graphs a polynomial of arbitrary degree that best fit the data in the least squares sense.

We begin by considering the problem of finding a cubic polynomial

$$P(x) = c_0 + c_1 x + c_2 x^2 + c_3 x^3$$

that best fits the data

$$(x_1, y_1), \ldots, (x_n, y_n),$$

where the x_i are distinct numbers. Specifically, we seek the values of the four coefficients c_0, c_1, c_2, c_3 with the property that the sum of the squares

$$\sum_{i=1}^{n} \left[P(x_i) - y_i \right]^2 = \sum_{i=1}^{n} \left[c_0 + c_1 x_i + c_2 x_i^2 + c_3 x_i^3 - y_i \right]^2 \tag{13.2}$$

is as small as possible. It may be shown using calculus that the coefficients c_i are the solutions of the linear system

$$\sum_{i=1}^{n} \left[c_0 + c_1 x_i + c_2 x_i^2 + c_3 x_i^3 - y_i \right] = 0$$

$$\sum_{i=1}^{n} x_i \left[c_0 + c_1 x_i + c_2 x_i^2 + c_3 x_i^3 - y_i \right] = 0$$

$$\sum_{i=1}^{n} x_i^2 \left[c_0 + c_1 x_i + c_2 x_i^2 + c_3 x_i^3 - y_i \right] = 0$$

$$\sum_{i=1}^{n} x_i^3 \left[c_0 + c_1 x_i + c_2 x_i^2 + c_3 x_i^3 - y_i \right] = 0.$$

Note the reversal of notation: the c's are the unknowns and the x's and y's (the given data) are the constants. We can simplify the description of the system by introducing the sums

$$S_k = \sum_{i=1}^{n} x_i^k \quad \text{and} \quad T_k = \sum_{i=1}^{n} x_i^k y_i. \tag{13.3}$$

The system may then be written

$$c_0 S_0 + c_1 S_1 + c_2 S_2 + c_3 S_3 = T_0$$
$$c_0 S_1 + c_1 S_2 + c_2 S_3 + c_3 S_4 = T_1$$
$$c_0 S_2 + c_1 S_3 + c_2 S_4 + c_3 S_5 = T_2$$
$$c_0 S_3 + c_1 S_4 + c_2 S_5 + c_3 S_6 = T_3$$

Matrix methods may now be used to solve the system for the c's in terms of the S's and T's.

It is just as easy to formulate the solution for general polynomials of degree m. In this case one has the analogous system

$$c_0 S_0 + c_1 S_1 + \cdots + c_m S_m = T_0$$
$$c_0 S_1 + c_1 S_2 + \cdots + c_m S_{m+1} = T_1$$
$$\vdots$$
$$c_0 S_m + c_1 S_{m+1} + \cdots + c_m S_{2m} = T_m$$

$$(13.4)$$

If we set

$$S = \begin{bmatrix} S_0 & S_1 & \cdots & S_m \\ S_1 & S_2 & \cdots & S_{m+1} \\ \vdots & \vdots & \cdots & \vdots \\ S_m & S_{m+1} & \cdots & S_{2m} \end{bmatrix}, \quad C = \begin{bmatrix} c_0 \\ c_1 \\ \vdots \\ c_m \end{bmatrix} \quad \text{and } T = \begin{bmatrix} T_0 \\ T_1 \\ \vdots \\ T_m \end{bmatrix}$$

then system (13.4) may be written $SC = T$. The polynomial of best fit therefore has coefficients given by the matrix $C = S^{-1}T$.

It is interesting to note that if the degree m of the desired polynomial is chosen to be one less than the number of data points n, resulting in a system of n equations in n unknowns, then one obtains an *exact* fit, that is, $P(x_i) = y_i$ for each i. This is the same thing as saying that the error term

$$\sum_{i=1}^{n} \left[P(x_i) - y_i \right]^2 = \sum_{i=1}^{n} \left[c_0 + c_1 x_i + c_2 x_i^2 \cdots + c_{n-1} x_i^{n-1} - y_i \right]^2,$$

is zero. This means that each of the terms in the sum is zero. It follows that the c_i's are then solutions of the system

$$\begin{bmatrix} 1 & x_1 & x_1^2 & \cdots & x_1^{n-1} \\ 1 & x_2 & x_2^2 & \cdots & x_2^{n-1} \\ & & \vdots & & \\ 1 & x_n & x_n^2 & \cdots & x_n^{n-1} \end{bmatrix} \begin{bmatrix} c_0 \\ c_1 \\ \vdots \\ c_{n-1} \end{bmatrix} = \begin{bmatrix} y_1 \\ y_2 \\ \vdots \\ y_n \end{bmatrix}, \quad (13.5)$$

that is, $C = V^{-1}Y$, where V is the $n \times n$ matrix on the left in (13.5), the so-called *Vandermonde matrix* generated by the x-data. That the inverse exists is a consequence of the fact that the x_i's are distinct. (This may be easily verified using techniques from the next chapter.)

In the remainder of the section we develop a program that calculates the coefficients $c_0, c_1, \ldots, c_m$ of the polynomial of best fit and graphs the polynomial. Typical numerical input data is given in Figure 13.2. The coordinates of the

	A	B	C	D	E
3		degree	left	right	inc
4		5	2	7	.01
5	x data	y data	coeff	x values	y values
6	1	3			
7	2	5			
8	3	6			
9	4	8			

FIGURE 13.2: Spreadsheet for `PolyFitData`.

data points are entered in columns A and B, and the degree of the polynomial is entered in cell B4. The left and right endpoints of the graphing interval are entered in cells C4 and D4, respectively, and the graphing increment is entered in cell E4.

The main procedure `PolyFitData` reads the x and y data from columns A and B and then invokes `PolyCoeff` to calculate the coefficient matrix `C`. The latter is passed to `PolyPoints`, which generates points for the graph of the polynomial. The procedure `ClearColContent` from Section 9.2 removes any previously generated data.

```
Sub PolyFitData()
    Dim m As Integer, left As Double, right As Double, inc As Double
    Dim XYdata() As Double, C() As Double, error As Boolean
    If ChartObjects.Count > 0 Then ChartObjects.Delete    'clear charts
    Call ClearColContent(6, 5, 6)                            'and data
    m = Range("B4").Value                           'degree of polynomial
    left = Range("D4").Value            'left endpoint of graphing interval
    right = Range("E4").Value          'right endpoint of graphing interval
    inc = Range("F4").Value
    XYdata = MatrixIn(6, 1)                                    'read data
    C = PolyCoeff(m, XYdata, error)         'calculate the coefficients
    If error Then Exit Sub
    Call MatrixOut(C, 6, 4)                          'print coefficients
    Call PolyPoints(C, left, right, inc)         'points for the graph
    Call GraphPolynomial:  Call GraphDataPoints()
End Sub
```

The procedure `PolyCoeff` constructs the matrices S and T. It does this using the procedures `SumPowers`, which calculates the required sums of powers of the

x-data, and `SumProducts`, which calculates the required sums of powers of the x-data times the y-data. The matrix S is then inverted using the function `InvertMat` and the result is multiplied by the column matrix T.

```
Function PolyCoeff(m As Integer, XYdata() As Double, error As Boolean) _
        As Double()
    Dim C() As Double, S() As Double, T() As Double, SInv() As Double
    Dim i As Integer, j As Integer
    ReDim S(1 To m + 1, 1 To m + 1)          'memory for the matrices S,T
    ReDim T(1 To m + 1, 1 To 1)
    ReDim SInv(1 To m + 1, 1 To m + 1)       'memory for S inverse
    For i = 0 To m                           'calculate S entries
        For j = 0 To m
            S(i + 1, j + 1) = SumPowers(XYdata, i + j)
        Next j
    Next i
    For i = 0 To m
        T(i + 1, 1) = SumProducts(XYdata, i)       'calculate T entries
    Next i
    SInv = InvertMat(S, error)
    If Not error Then
        C = MultMat(SInv, T, error)      'coefficients of polynomial in C
        PolyCoeff = C
    End If
End Function
```

The functions `SumPowers` and `SumProducts` are based on the formulas in (13.3).

```
Function SumPowers(XYdata() As Double, power As Integer) As Double
    Dim i As Integer, sum As Double              'initialized to 0
    For i = 1 To UBound(XYdata, 1)
        sum = sum + XYdata(i, 1) ^ power
    Next i
    SumPowers = sum
End Function

Function SumProducts(XYdata() As Double, power) As Double
    Dim i As Integer, sum As Double              'initialized to 0
    For i = 1 To UBound(XYdata, 1)
        sum = sum + XYdata(i, 1) ^ power * XYdata(i, 2)
    Next i
    SumProducts = sum
End Function
```

The function `PolyPoints` takes the coefficient matrix C and calculates the values $y = c_0 + c_1 x + \cdots + c_m x^m$ for values of x determined by the given interval and increment. The x and y values are placed in columns E and F.

```
Sub PolyPoints(C() As Double, left, right, inc)
    Dim x As Double, y As Double, i As Integer, j As Integer
    x = left: j = 0                              'initial values
```

```
        Do While x <= right
            y = C(1, 1)
            For i = 1 To UBound(C, 1) - 1
                y = y + C(i + 1, 1) * x ^ i
            Next i
            Cells(6 + j, 5).Value = x: Cells(6 + j, 6).Value = y
            x = x + inc: j = j + 1
        Loop
    End Sub
```

The procedures `GraphPolynomial` and `GraphDataPoints` use the VBA graphing facility to graph the polynomial and the original data points, respectively.

```
Sub GraphPolynomial()
    Dim lrow As Long
    lrow = Cells(Rows.Count, 5).End(xlUp).row           'last row of data
    With Sheet1 _
        .ChartObjects.Add(left:=400, Width:=600, Top:=50, Height:=400)
        .Chart.SetSourceData Source:=Range("F" & 6 & ":" & "F" & lrow)
        .Chart.ChartType = xlLine
        .Chart.HasTitle = True
        .Chart.ChartTitle.Text = "graph"
        .Chart.SeriesCollection(1).XValues = _
        Range("E" & 6 & ":" & "E" & lrow)
    End With
End Sub

Sub GraphDataPoints()
    Dim lrow As Long
    lrow = Cells(Rows.Count, 1).End(xlUp).row
    With Sheet1 _
        .ChartObjects.Add(left:=1060, Width:=375, Top:=50, Height:=400)
        .Chart.SetSourceData Source:=Range("B" & 6 & ":" & "B" & lrow)
        .Chart.ChartType = xlXYScatter
        .Chart.HasTitle = True
        .Chart.ChartTitle.Text = "graph"
        .Chart.SeriesCollection(1).XValues = _
        Range("A" & 6 & ":" & "A" & lrow)
    End With
End Sub
```

---

*13.8   The Barnsley Fern

An *affine transformation* is a function that takes n dimensional points X into n dimensional points Y such that lines and parallelism are preserved but distances and angles may change. It may be shown that affine transformations are of the form $Y = AX + B$, where A is an $n \times n$ matrix and B is an $n \times 1$ matrix and the points X and Y are written as $n \times 1$ matrices. A *random iterated affine system* is a finite collection of affine transformations

$$Y = A_1 X + B_1, \ Y = A_2 X + B_2, \ldots, \ Y = A_k X + B_k$$

and a probability distribution $p_1, p_2, \ldots, p_k$ (positive numbers with sum 1) employed as follows:

(a) Choose a point X, say the origin.

(b) Select an index j with probability p_j.

(c) Calculate $Y = A_j X + B_j$ for that j.

(d) Copy Y to X.

(e) Repeat (b)–(d) the desired number of times.

The program `ChaosTriangleGame` of Section 9.9 is based on such a system, the equations changing each time a different vertex is randomly chosen. The program `AffineIterations` in the present section generalizes `ChaosTriangleGame` by allowing the user to enter arbitrarily many affine transformations on the spreadsheet. Figure 13.3 gives an example of input just below the output graph

	D	E	F	H	J
38	1	.95	.005	−.002	.85
39		−.005	.93	.5	
41	2	.035	−.11	−.05	.07
42		.27	.01	.005	
44	3	−.04	.11	.083	.07
45		.27	.01	.06	
47	4	0	0	0	.01
48		0	.24	−.4	

FIGURE 13.3: Spreadsheet for AffineIterations.

(not shown) for the case of four affine transformations, that is, four pairs of matrices A, B, and the probabilities of a pair being selected. The matrices A are entered in columns E and F, the matrices B in column H, and the probabilities in column J. Column D provides reference numbers for the data sets. Running the program for this particular input produces the so-called *Barnsley fern* (see Figure 13.4). Like the Sierpinski triangle, which was the output of `ChaosTriangleGame`, the fern is an example of a *fractal*, a pattern that looks the same at any magnification. The fern was introduced by Michael Barnsley in his book *Fractals Everywhere* (1988).

FIGURE 13.4: Idealized Barnsley fern.

The main procedure `AffineIterations` sets the number of iterations of the algorithm, calls `ComputeValues` to calculate and print the values of Y for the chart, and graphs the result.

```
Sub AffineIterations()
    Dim XRange As String, YRange As String, NumIterations As Integer
    If ChartObjects.Count > 0 Then ChartObjects.Delete
    Call ClearColContent(8, 1, 2)
    NumIterations = 10000                    'specify number of iterations
    Call ComputeValues(NumIterations)              'calculate points
    XRange = "A" & 8 & ":" & "A" & 8 + NumIterations
    YRange = "B" & 8 & ":" & "B" & 8 + NumIterations
    With Sheet1 _
        .ChartObjects.Add(Left:=200, Width:=800, Top:=10, Height:=500)
        .Chart.ChartType = xlXYScatter
        .Chart.HasTitle = True
        .Chart.ChartTitle.Text = ""
        .Chart.SetSourceData Source:=Range(YRange)
        .Chart.SeriesCollection(1).XValues = Range(XRange)
    End With
End Sub
```

The procedure `ComputeValues` first copies the probabilities from column J into array `P` and then runs a For Next loop to generate the points for the figure. All arrays are automatically initialized to zero. At each iteration the function `RndOut(P)` of Section 10.1 generates a random equation number 1–4 according the probability array P. This determines the row of the corresponding affine transformation. For each iteration the point Y is calculated and printed.

```
Sub ComputeValues(NumIterations As Integer)
    Dim j As Integer, row As Integer, P() As Double, error  As Boolean
    Dim A() As Double, B() As Double
    Dim Y() As Double, X(1 To 2, 1 To 1) As Double
    P = Probabilities()                          'get the probabilities
    For j = 1 To NumIterations
        row = 38 + 3 * (RndOut(P) - 1)
        A = MatrixIn(row, 5)
        B = MatrixIn(row, 8)
        Y = MultMat(A, X, error)
        Y = AddSubMat(Y, B, 1, error)
        Cells(j + 7, 1).Value = Y(1, 1)          'print x value of point
        Cells(j + 7, 2).Value = Y(2, 1)          'print y value of point
        X(1, 1) = Y(1, 1): X(2, 1) = Y(2, 1):
    Next j
End Sub
```

The function **Probabilities** returns the probabilities in a matrix array P. It first finds the last number in column D; this is the number of probabilities entered in the spreadsheet.

```
Function Probabilities() As Double()
    Dim j As Integer, row As Integer, col As Integer
    Dim P() As Double, lrow As Long
    lrow = Cells(Rows.Count, 4).End(xlUp).row         'row of last entry
    ReDim P(1 To Cells(lrow, 4).Value)               'array for probabilities
    row = 38: col = 10
    For j = 1 To UBound(P)
        P(j) = Cells(row + 3*(j-1), col).Value
    Next j
    Probabilities = P
End Function
```

The reader may wish to experiment with the program by varying the input parameters of the Barnsley fern. Mutations of the fern may be found online.

---

*13.9   A Matrix Calculator

The module in this section evaluates algebraic matrix expressions such as 3(A^2C - 2D3E^(-5))Z, where the capital letters are symbols for matrices entered by the user into the spreadsheet. Thus it is possible to calculate an expression containing up to 26 different matrix variables provided the user has the fortitude and patience to enter the matrices. The size of a matrix is limited only by the constraints of the spreadsheet and computer memory.

We shall call the portion of the columns containing the user-entered matrices the *matrix queue*. These start in cell C5 and the letters denoting the matrices in cell B5. The program assumes matrices in the queue are separated by a blank row. The answer appears one row below these. Figure 13.5 shows sample input and output.

	A	B	C	D
3	expression	$2A + B$	matrix	queue
5	matrices	A	1	2
6			3	4
8		B	1	1
9			1	1
11		answer	3	5
12			7	9

FIGURE 13.5: Spreadsheet for matrix calculator.

The main procedure `MatrixCalculator` first clears any answer from a previous run then reads the expression to be evaluated into `expr`. The variable `Idx` points to the current symbol in `expr`; it ranges from 1, the beginning of the expression, to `Len(expr)`. After an error check, the procedure calls `InsertAsterisks`, described below. The function `MatEval(0)` doles out the calculation tasks. The 0 places the program in the lowest precedence, matrix addition. Any other operation will be performed first. If `MatEval` returns with `error = False`, then the matrix answer Z is printed.

```
Public expr As String, Idx As Integer, error As Boolean
Public errorType As Integer
Sub MatrixCalculator()
    Dim Z() As Double, lRow As Long
    Call ClearLastAnswer
    Range("B4").Value = ""                      'erase last error msg, if any
    Range("B3").NumberFormat = "@"                      'format as text
    expr = RemoveWhiteSpace(Range("B3").Value)
    error = False: errorType = 0
    Call ErrorCheck
    If error Then Range("B4").Value = "error type " _
        & errorType & " in expression": Exit Sub
    Call InsertAsterisks
    Idx = 1                              'points to 1st character in expr
    Z = MatEval(0)
    If error Then Range("B4").Value = "error": Exit Sub
    lRow = Cells(Rows.Count, 3).End(xlUp).row 'last nonblank row in queue
    Call MatrixOut(Z, lRow + 2, 3)              'print matrix in this row
    Cells(lRow + 2, 2).Value = "Answer"              'and label it
End Sub
```

The original expression entered by the user is modified by the function `InsertAsterisks`, which inserts asterisks between various symbols to indicate either matrix or scalar multiplication. For example, the expression `3(A^2C - 2D3E^(-5))B` transforms into `3*(A^2*C - 2*D*3*E^(-5))*B`.

```
Sub InsertAsterisks()
    Dim char As String, nextchar As String, k As Integer: k = 1
    Do While k <= Len(Expr)
        char = Midd(Expr, k, 1): nextchar = Midd(Expr, k + 1, 1)
        If char Like "[A-Z]" And nextchar Like "[A-Z]" Or _
           char Like "[A-Z]" And InStr("0123456789.", nextchar) > 0 Or _
           char Like "[A-Z]" And nextchar = "(" Or _
           InStr("0123456789.", char) > 0 And nextchar Like "[A-Z]" Or _
           InStr("0123456789.", char) > 0 And nextchar = "(" Or _
           char = ")" And nextchar = "(" Or _
           char = ")" And nextchar Like "[A-Z]" Or _
           char = ")" And InStr("0123456789.", nextchar) > 0 Then
            Expr = InsertString(Expr, k, "*")
        End If
        k = k + 1
    Loop
End Sub
```

The procedure `MatEval` distributes the various calculation tasks to the matrix calculation procedures. It is similar to the function `ComplexEval` of Section 8.4 in that it calls itself to perform subtasks. The procedure consists of a Do While loop that advances the index `Idx`, which points to the symbols in the input expression `Expr`. The loop ends when either `Idx` exceeds its maximum value `Len(Expr)` or an error is encountered. The loop begins by extracting the current symbol from `Expr`. A Select Case statement decides which action to take based on the value of the symbol. If the symbol is a number it goes to the first case; if it is a capital letter it goes to the second case. The remaining cases treat the other symbols using the procedures `AddSubMat`, `ScalarMult`, `MultMat`, and `PowerMat` described in earlier sections.

```
Function MatEval(mode As Integer) As Double()
    Dim char As String, LonelyOp As Boolean, Z() As Double, W() As Double
    If error Then Exit Function
    Do While Idx <= EndIdx And Not error
        char = Mid(Expr, Idx, 1)                'extract current symbol
        If char Like "[0-9]" Or char = "." Then char = "#"         'number
        If char Like "[A-Z]" Then char = "@"             matrix letter
        Select Case char
            Case Is = "#"                                'number found
                Z = GetScalar()   'get no. in a 1 x 1 matrix, advance Idx
            Case Is = "@"                               'matrix letter found
                Z = GetMatrix(Mid(Expr, Idx, 1))   'get matrix from queue
                Idx = Idx + 1                           'move past letter
            Case Is = "+"
                If mode = 1 Then Exit Do       'MatEval was called earlier
```

```
              LonelyOp = (Idx = 1 Or Midd(Expr, Idx - 1, 1) = "(")
              Idx = Idx + 1                       'move past + sign
              W = MatEval(0)                      'evaluate what's next
              If LonelyOp Then                    'for the special case +W
                 Z = CopyMat(W)                   'just return the matrix
              Else                        'otherwise perform the addition
                 Z = AddSubMat(Z, W, 1, error)
              End If
          Case Is = "-"
              If mode = 1 Then Exit Do          'MatEval called earlier
              LonelyOp = (Idx = 1 Or Midd(Expr, Idx - 1, 1) = "(")
              Idx = Idx + 1                         'skip past -
              W = MatEval(1)                      'evaluate what's next
              If LonelyOp Then                    'for the special case -W
                 Z = ScalarMult(W, -1)             'multiply W by -1
              Else                                   'otherwise
                 Z = AddSubMat(Z, W, -1, error)    'do the subtraction
              End If
          Case Is = "*"
              Idx = Idx + 1           'move past multiplication operator
              W = MatEval(1)                      'evaluate what's next
              Z = MultMat(Z, W, error)           'do the multiplication
          Case Is = "^"
              Idx = Idx + 1                       'move past operator
              If Mid(Expr, Idx, 1) = "(" Then Idx = Idx + 1 'skip paren
              W = GetScalar()        'get the exponent in a 1 x 1 matrix
              If Mid(Expr, Idx, 1) = ")" Then Idx = Idx + 1 'skip paren
              Z = PowerMat(Z, CInt(W(1, 1)), error)
          Case Is = "("
              Idx = Idx + 1                               'skip"("
              Z = MatEval(0)
              Idx = Idx + 1                               'skip")"
          Case Is = ")": Exit Do
          Case Else: error = True: Exit Function
        End Select
      Loop
      MatEval = Z
  End Function
```

The function `GetScalar` extracts a number string from `Expr`. Since matrix multiplication and scalar multiplication are treated similarly, the function returns the scalar as a 1×1 matrix.

```
Function GetScalar() As Double()        'extract number string from expr
    Dim Z(1 To 1, 1 To 1) As Double, x As String, char As String
    Dim k As Integer, i As Integer, sign As Double
    sign = 1
    If Mid(Expr, Idx, 1) = "-" Then sign = -1: Idx = Idx + 1
    For i = Idx To EndIdx
        char = Mid(Expr, i, 1)
        k = Asc(char)
```

```
        If (k < 48 Or k > 57) And (char <> ".") Then Exit For
        x = x & char
    Next i
    Idx = i                              'Idx now after number
    Z(1, 1) = CDbl(x) * sign
    GetScalar = Z
End Function
```

The function `GetMatrix(MLetter)` retrieves the matrix labelled `MLetter` in column 2.

```
Function GetMatrix(MLetter As String) As Double()
    Dim m As Integer, lRow As Long
    lRow = Cells(Rows.Count,2).End(xlUp).row 'last nonblank row in col 2
    For m = 5 To lRow            'find entered matrix with symbol MLetter
        If Cells(m, 2).Value = MLetter Then Exit For  'found it in row m
    Next m
    GetMatrix = MatrixIn(m, 3)
End Function
```

The procedure `ErrorCheck`, although not foolproof, catches the main types of errors. It is similar in principle to the procedure of the same name in Section 8.4 in that each symbol has a restricted set of legal immediate predecessors and successors. Here are the rules:

1. An operator `+-` must be followed by a capital letter, a left parenthesis, a period, or a digit.

2. The operator `^` must be followed by a left parenthesis or a positive digit.

3. The operator `^` must be preceded by a right parenthesis, a capital letter, or a number.

4. A right parenthesis or capital letter must be followed by a right or left parenthesis, one of the operators `+-^`, a digit, a period, a capital letter, or nothing.

5. A left parenthesis cannot be followed by `^`.

6. There must be the same number of left parentheses as right parentheses.

```
Sub ErrorCheck()
    Dim s As String, sprev As String, snext As String, i As Integer
    Dim e(1 To 6) As Boolean
    For i = 1 To Len(expr)
        s = Midd(expr, i, 1)
        sprev = Midd(expr, i - 1, 1): snext = Midd(expr, i + 1, 1)
        e(1) = InStr("+-", s) > 0 And snext Like "[!A-Z]" And _
                snext <> "(" And snext <> "." And snext Like "[!0-9]"
        e(2) = (s = "^") And snext <> "(" And snext Like "[!0-9]"
```

```
        e(3) = (s = "^") And sprev <> ")" And sprev Like "[!A-Z]" _
               And sprev <> "." And sprev Like "[!0-9]"
        e(4) = (s = ")" Or s Like "[A-Z]") And i < L And _
               InStr("+-^", snext) = 0 And _
               snext Like "[!A-Z]" And snext <> "(" And snext <> ")" _
               And snext Like "[!0-9]" And snext <> "."
        e(5) = s = "(" And snext = "^"
        error = e(1) Or e(2) Or e(3) Or e(4) Or e(5)
        If error Then Exit For
    Next i
    e(6) = Not IsMatch(expr, "(", ")")
    For i = 1 To 6
        If e(i) Then errorType = i: error = True: Exit Sub
    Next i
End Sub
```

The procedure `ClearLastAnswer` finds the last nonempty row in column 2, the answer row, then finds the last nonempty column in the answer row, and finally finds the last nonempty row in column 3. These values are used to clear the answer from a previous run.

```
Sub ClearLastAnswer()
Dim lRow As Long, AnswerRow As Long, lCol As Long
    AnswerRow = Cells(Rows.Count, 2).End(xlUp).row
    If Cells(AnswerRow, 2).Value = "Answer" Then
        lCol = Cells(AnswerRow, Columns.Count).End(xlToLeft).Column
        lRow = Cells(Rows.Count, 3).End(xlUp).row
        Range(Cells(AnswerRow, 2), Cells(lRow, lCol)).ClearContents
    End If
End Sub
```

13.10 Exercises

1. Calculate the inverse by hand if it exists (the inverse, not your hand). Check your answers using `MatrixCalculator`.

   (a) $\begin{bmatrix} 1 & -1 & 0 \\ 0 & 1 & -1 \\ 1 & 0 & 1 \end{bmatrix}$ (b) $\begin{bmatrix} 1 & 2 & 3 \\ 2 & 3 & 1 \\ 3 & 1 & 2 \end{bmatrix}$ (c) $\begin{bmatrix} 1 & 3 & 5 \\ 7 & 9 & 11 \\ 13 & 15 & 17 \end{bmatrix}$ (d) $\begin{bmatrix} 0 & 3 & 5 \\ 7 & 9 & 11 \\ 13 & 15 & 17 \end{bmatrix}$

2. Calculate the production matrix by hand for each of the following IO and demand matrices and check your answer with `IOModel`.

   (a) $\begin{bmatrix} .1 & .2 & .3 \\ .2 & .3 & .1 \\ .3 & .2 & .1 \end{bmatrix}, \begin{bmatrix} 1 \\ 7 \\ 3 \end{bmatrix}$ (b) $\begin{bmatrix} .5 & .1 & .7 \\ .4 & .3 & .5 \\ .6 & .1 & .2 \end{bmatrix}, \begin{bmatrix} 2 \\ 5 \\ 4 \end{bmatrix}$

3. Use the program MatrixCalculator to show that $(AB)^{-1} = B^{-1}A^{-1}$.

4. Use the programs MultMat and Transpose to show that $(AB)^T = B^T A^T$.

5. Use the programs InvertMat and Transpose to show that $(A^T)^{-1} = (A^{-1})^T$.

6. Use the program MatrixCalculator to experiment with powers of matrices of the following forms, where the asterisks represent arbitrary numbers.

$$
\begin{bmatrix} 0 & * & * & * \\ 0 & 0 & * & * \\ 0 & 0 & 0 & * \\ 0 & 0 & 0 & 0 \end{bmatrix},
\begin{bmatrix} 0 & 0 & 0 & 0 \\ * & 0 & 0 & 0 \\ * & * & 0 & 0 \\ * & * & * & 0 \end{bmatrix},
\begin{bmatrix} * & * & * & 0 \\ * & * & 0 & 0 \\ * & 0 & 0 & 0 \\ 0 & 0 & 0 & 0 \end{bmatrix}, \text{ and }
\begin{bmatrix} 0 & 0 & 0 & 0 \\ 0 & 0 & 0 & * \\ 0 & 0 & * & * \\ 0 & * & * & * \end{bmatrix}.
$$

What conclusions do you draw?

7. (Permutation matrices) An $n \times n$ *permutation matrix* is a matrix obtained by permuting (rearranging) the columns of the $n \times n$ identity matrix. For example, the permutation matrix P below is obtained by switching columns 1 and 3:

$$
I = \begin{bmatrix} 1 & 0 & 0 \\ 0 & 1 & 0 \\ 0 & 0 & 1 \end{bmatrix} \longrightarrow P = \begin{bmatrix} 0 & 0 & 1 \\ 0 & 1 & 0 \\ 1 & 0 & 0 \end{bmatrix}.
$$

Multiplying an $n \times n$ matrix A on the right by a permutation matrix will permute the columns of A in exactly the same way. Thus in the leftmost matrix below columns 1 and 3 are switched since these columns are switched in the identity matrix.

$$
\begin{bmatrix} a & d & h \\ b & e & i \\ c & f & j \end{bmatrix}
\begin{bmatrix} 0 & 0 & 1 \\ 0 & 1 & 0 \\ 1 & 0 & 0 \end{bmatrix} =
\begin{bmatrix} h & d & a \\ i & e & b \\ j & f & c \end{bmatrix}.
$$

A permutation matrix may be described by a *permutation*, that is, a function that describes the rule used for permuting the columns. Such functions are typically described by two rows of integers. For example, the above matrix P above may be described by the permutation

$$
\begin{pmatrix} 1 & 2 & 3 \\ 3 & 2 & 1 \end{pmatrix},
$$

which indicates that column 1 goes to column 3, column 2 stays the same, and column 3 goes to column 1. The function may be written more succinctly as 3,2,1. Write a program PermuteMatCol that takes as input a matrix A and a permutation expressed in the succinct notation just described and returns A with its columns permuted. Do this directly without appealing to the module MultMat.

8. Use the function `Transpose` to write a program `PermuteMatRow` which permutes the rows in a fashion analogous to column permutation.

9. A *submatrix* of a matrix A is obtained by deleting specified rows and columns of A. Write a function `DeleteRows` that takes a user-entered matrix and a string of row numbers separated by commas and returns the submatrix obtained by deleting those rows. Write another function `DeleteCols` that is the analog of `DeleteRows`. Combine them in a function `SubMatrix` that takes a user-entered matrix, a string of row numbers separated by commas, and a string of column numbers separated by commas, returns the submatrix obtained by deleting those rows and columns. For example, for the matrix A below, the statement `SubMatrix(A, "2,3", "1,4")` should return the matrix B.

$$A = \begin{bmatrix} 1 & 2 & 3 & 4 \\ 5 & 6 & 7 & 8 \\ 9 & 10 & 11 & 12 \\ 13 & 14 & 15 & 16 \end{bmatrix} \xrightarrow[\text{delete columns 1,4}]{\text{delete rows 2,3}} \begin{bmatrix} 2 & 3 \\ 14 & 15 \end{bmatrix} = B.$$

10. (Random integer inverses). The program `RndIntMat` of Exercise 9 in Chapter 11 generates integer matrices with integer inverses, since no type 2 operations were used. Use this program to write a program `RndIntMatInverse` that displays the matrix and calculates its inverse. Check the result with `MatMult`.

11. (Binomial integer inverses). It may be shown that an $n \times n$ matrix with integer entries the *binomial coefficients* $a_{ij} = \binom{n+j-1}{i-1}$ (see Section 19.6) has an integer inverse matrix (Ericksen). Write a program `BinIntMatInverse` that displays the matrix and calculates its inverse. Use the VBA function `WorksheetFunction.Combin(n+j-1,i-1)` to populate the matrix.

*12. Refer to Section 8.4 for this exercise. Write a program `ComplexMatrixCalc` that is the complex number analog of `MatrixCalc`

Chapter 14

Determinants

The determinant of a square matrix A is a certain unique number calculated from the entries of A. In this chapter we describe several ways to find that number and give some properties and applications of determinants.

14.1 Definition and Properties

The standard notation for the determinant of an $n \times n$ matrix $A = [a_{ij}]$ is

$$\det(A), \quad |A|, \quad \text{or} \quad \begin{vmatrix} a_{11} & a_{12} & \cdots & a_{1n} \\ a_{21} & a_{22} & \cdots & a_{2n} \\ \vdots & \vdots & \ddots & \vdots \\ a_{n1} & a_{m2} & \cdots & a_{nn} \end{vmatrix}.$$

The number n is called the *order* of the determinant. A determinant of order 2 is evaluated by multiplying along the diagonals and subtracting:

$$\begin{vmatrix} a & b \\ c & d \end{vmatrix} = ad - bc. \tag{14.1}$$

To find the value of higher order determinants one has the choice of several methods. We begin with the method called *expansion along the first row*. The method is recursive in that it depends on the value of lower order determinants, ultimately reducing to the calculation (14.1). Here is the definition for determinants of order 3:

$$\begin{vmatrix} a_{11} & a_{12} & a_{13} \\ a_{21} & a_{22} & a_{23} \\ a_{31} & a_{32} & a_{33} \end{vmatrix} = a_{11} \begin{vmatrix} a_{22} & a_{23} \\ a_{32} & a_{33} \end{vmatrix} - a_{12} \begin{vmatrix} a_{21} & a_{23} \\ a_{31} & a_{33} \end{vmatrix} + a_{13} \begin{vmatrix} a_{21} & a_{22} \\ a_{31} & a_{32} \end{vmatrix}.$$

Note that the terms in the expansion are the entries of row 1 with alternating signs multiplied by second order determinants. These are obtained from the original matrix by deleting row 1 and the successive columns of the row 1 entries. The calculation is completed by using (14.1). Here is an example:

DOI: 10.1201/9781003351689-14

$$\begin{vmatrix} 1 & 2 & 3 \\ 4 & 5 & 6 \\ 7 & 8 & 9 \end{vmatrix} = 1 \cdot \begin{vmatrix} 5 & 6 \\ 8 & 9 \end{vmatrix} - 2 \cdot \begin{vmatrix} 4 & 6 \\ 7 & 9 \end{vmatrix} + 3 \cdot \begin{vmatrix} 4 & 5 \\ 7 & 8 \end{vmatrix} = -3 - 2(-6) + 3(-3) = 0.$$

The method for evaluating higher order determinants is entirely analogous. Moreover, one can expand along any row or column. To describe the general method we need the following notions: The *minor* of the (i, j) entry of a matrix A is the determinant of the matrix obtained by deleting row i and column j of A:

$$M_{ij} = \begin{vmatrix} a_{11} \cdots a_{1j} \cdots a_{1n} \\ \vdots \quad \vdots \quad \vdots \\ a_{i1} \cdots a_{ij} \cdots a_{in} \\ \vdots \quad \vdots \quad \vdots \\ a_{n1} \cdots a_{nj} \cdots a_{nn} \end{vmatrix}.$$

The *cofactor* of the (i, j) entry of A is defined as

$$C_{ij} = (-1)^{i+j} M_{ij}$$

The definition for expansion along row i is then given by

$$\begin{vmatrix} a_{11} & a_{12} & \cdots & a_{1n} \\ \vdots & \vdots & \ddots & \vdots \\ a_{i1} & a_{i2} & \cdots & a_{in} \\ \vdots & \vdots & \ddots & \vdots \\ a_{n1} & a_{n2} & \cdots & a_{nn} \end{vmatrix} = a_{i1} C_{i1} + a_{i2} C_{i2} + \cdots + a_{in} C_{in}.$$

The expansion along column i is obtained by reversing the order of the subscripts. The method of evaluating determinants along a row or column is called the *Laplace expansion*.

Example 14.1. An *upper triangular matrix* is a square matrix with zeros below the main diagonal. Evaluating along the first column in each case shows that the determinant of such a matrix is simply the product of the entries along the main diagonal. Here's an example:

$$\begin{vmatrix} 1 & 2 & 3 & 4 \\ 0 & 5 & 6 & 7 \\ 0 & 0 & 8 & 9 \\ 0 & 0 & 0 & 10 \end{vmatrix} = 1 \cdot \begin{vmatrix} 5 & 6 & 7 \\ 0 & 8 & 9 \\ 0 & 0 & 10 \end{vmatrix} = 1 \cdot 5 \cdot \begin{vmatrix} 8 & 9 \\ 0 & 10 \end{vmatrix} = 1 \cdot 5 \cdot 8 \cdot 10 = 400. \quad \lozenge$$

Determinants have many interesting properties, some of which are useful in reducing calculations. Here are three important properties. Others appear in the exercises.

- *Switching a pair of rows (or columns) changes the sign of the determinant.*

  For example,

  $$\begin{vmatrix} 2 & -1 & 3 \\ 4 & 8 & 9 \\ 3 & 2 & 7 \end{vmatrix} \underset{=\!=\!=\!=\!=}{R_1 \leftrightarrow R_3} - \begin{vmatrix} 3 & 2 & 7 \\ 4 & 8 & 9 \\ 2 & -1 & 3 \end{vmatrix}.$$

- *Rows or columns may be factored.*

  For example,

  $$\begin{vmatrix} 3 & -9 & 12 \\ 4 & 8 & 22 \\ 35 & -20 & 15 \end{vmatrix} = 3 \cdot 2 \cdot 5 \cdot \begin{vmatrix} 1 & -3 & 4 \\ 2 & 4 & 11 \\ 7 & -4 & 3 \end{vmatrix}.$$

  Here, 3 was factored from the first row, 2 from the second, and 5 from the last. Notice that this is very different from factoring a matrix, where a factor is taken from *all* rows.

- *Adding a multiple of one row to another does not change the value of the determinant.*

  This is useful because it allows the introduction of zeros into columns, making the resulting determinant easier to evaluate. The analogous property for columns holds as well.

Example 14.2. Here's an example that uses the second and third properties above to evaluate a determinant.

$$\begin{vmatrix} 2 & 6 & 0 \\ 2 & 9 & 3 \\ 3 & 14 & -5 \end{vmatrix} = 2 \begin{vmatrix} 1 & 3 & 0 \\ 2 & 9 & 3 \\ 3 & 14 & -5 \end{vmatrix} \underset{=\!=\!=\!=\!=}{\substack{-2R_1 + R_2 \\ -3R_1 + R_3}} 2 \begin{vmatrix} 1 & 3 & 0 \\ 0 & 3 & 3 \\ 0 & 5 & -5 \end{vmatrix}$$

$$= 2 \cdot 3 \cdot 5 \begin{vmatrix} 1 & 3 & 0 \\ 0 & 1 & 1 \\ 0 & 1 & -1 \end{vmatrix} \underset{=\!=\!=\!=\!=}{-1R_2 + R_3} 30 \cdot \begin{vmatrix} 1 & 3 & 0 \\ 0 & 1 & 1 \\ 0 & 0 & -2 \end{vmatrix} = -60. \qquad \Diamond$$

The last calculation comes from the fact, noted earlier, that the value of an upper triangular matrix is the product of the entries along the diagonal.

14.2 Determinants Using VBA

The function `DetEchelon(A)` uses the properties in the preceding section to evaluate the determinant of a user-entered square matrix A. It does so by converting the matrix A into reduced row echelon form R (more than is actually needed), keeping track of the number SW of row switches (type 1 operations) used in the conversion as well as the product P of the matrix entries whose reciprocals were used to multiply rows (type 2 operations) during the pivoting process. Since type 3 operations do not change the value of a determinant, $\det(A)$ is the product of the three quantities $\det(R)$, $(-1)^{SW}$ (undoing the row switches), and P (undoing the effect of the type 2 operations). Since R is an upper triangular matrix, $\det(R)$ is the product of its diagonal elements.

The function first calls the procedure RowEchelon(A,P,SW) to find the above mentioned components. A For Next loop then finds the product d of the diagonal elements of the echelon form. Determinant(A) returns the value d*P*(-1)^SW of the determinant.

```
Function DetEchelon(A() As Double) As Double
    Dim P As Double, SW As Integer, d As Double, size As Integer
    Dim i As Integer, R() As Double
    size = UBound(A, 1)                          'size of determinant
    R = RowEchelon(A, P, SW)                      'echelon form of A
    d = R(1, 1)                                   'first diagonal entry
    For i = 2 To size                'get product of diagonal entries
        d = d * R(i, i)
    Next i
    DetEchelon = d * (-1) ^ SW * P
End Function
```

Here's an alternate program that uses recursion. It has the advantages of being essentially self-contained and simpler in concept in that it does not rely on the procedure RowEchlelon. On the other hand, for large determinants the recursive version is significantly slower and could result in stack overflow. The idea here is to recurse on the submatrices obtained by deleting a row and column.

```
Function DetRecursive(A() As Double) As Double
    Dim B() As Double, D As Double, size As Integer, k As Integer
    size = UBound(A)
    If size = 2 Then
        D = A(1, 1) * A(2, 2) - A(1, 2) * A(2, 1) 'def of 2x2 determinant
    Else                               'expand determinant along row 1
        For k = 1 To size
            B = SubMatrix(A, 1, k)                 'omit row 1, col k of A
            D = D + (-1) ^ (k + 1) * A(1, k) * Determinant(B)    'recurse
        Next k
    End If
    Determinant = D
End Function
```

The function SubMatrix takes a square matrix, a row, and a column and returns the matrix obtained by removing the row and column. It is a short version of the more general program considered in Exercise 9 of Chapter 13.

```
Function SubMatrix(Mat() As Double, r As Integer, c As Integer)
    Dim m As Integer, n As Integer, i As Integer, j As Integer
    Dim size As Integer, SubMat() As Double
    size = UBound(Mat)
    ReDim SubMat(1 To size - 1, 1 To size - 1)
    For m = 1 To size      'transfer Mat to SubMat except row r and col c
        If m <> r Then
            i = i + 1: j = 1
            For n = 1 To size
                If n <> c Then SubMat(i, j) = Mat(m, n): j = j + 1
            Next n
        End If
    Next m
    SubMatrix = SubMat
End Function
```

The reader may wish to compare the speeds of the two versions by calculating determinants of the form

$$
\begin{vmatrix}
0 & \cdots & 0 & 0 & 1 \\
0 & \cdots & 0 & 2 & 1 \\
0 & \cdots & 3 & 0 & 1 \\
\vdots & \ddots & \vdots & \vdots & \vdots \\
n & \cdots & 0 & 0 & 1
\end{vmatrix}
$$

for large n. For example for $n = 10$ the echelon version returns the value -3628800 almost instantaneously while the recursive version is considerably slower. For larger values of n the recursive version is not practical on standard computers.

14.3 Cramer's Rule

Consider a system of n linear equations in n unknowns:

$$
\begin{aligned}
a_{11}x_1 + a_{12}x_2 + \cdots + a_{1n}x_n &= b_1 \\
a_{21}x_1 + a_{22}x_2 + \cdots + a_{2n}x_n &= b_2 \\
&\vdots \\
a_{n1}x_1 + a_{n2}x_2 + \cdots + a_{nn}x_n &= b_n.
\end{aligned}
\tag{14.2}
$$

As usual, we write the system in matrix form as $AX = B$. Cramer's rule asserts that if the determinant of A is not zero, then the system has a unique solution $(x_1, \ldots, x_n)$ given by the rule

$$x_k = \frac{|A_k|}{|A|}, \quad \text{where} \quad A_k = \begin{bmatrix} a_{11} & \cdots & a_{1,k-1} & b_1 & a_{1,k+1} & \cdots a_{1n} \\ a_{21} & \cdots & a_{2,k-1} & b_2 & a_{2,k+1} & \cdots a_{2n} \\ \vdots & & \vdots & \vdots & \vdots & \vdots \\ a_{n1} & \cdots & a_{n,k-1} & b_n & a_{n,k+1} & \cdots a_{nn} \end{bmatrix}.$$

Note that the matrix A_k is obtained by replacing column k of A with the column matrix B. For example, in the system of Example 11.4 we have

$$A = \begin{bmatrix} 2 & 2 & 3 \\ 4 & 5 & 6 \\ 7 & 8 & 9 \end{bmatrix}, \quad B = \begin{bmatrix} 9 \\ 12 \\ 15 \end{bmatrix},$$

$$A_1 = \begin{bmatrix} 9 & 2 & 3 \\ 12 & 5 & 6 \\ 15 & 8 & 9 \end{bmatrix}, \quad A_2 = \begin{bmatrix} 2 & 9 & 3 \\ 4 & 12 & 6 \\ 7 & 15 & 9 \end{bmatrix}, \text{ and } A_3 = \begin{bmatrix} 2 & 2 & 9 \\ 4 & 5 & 12 \\ 7 & 8 & 15 \end{bmatrix}.$$

The solution according to Cramer's rule is

$$x_1 = \frac{|A_1|}{|A|} = 0, \quad x_2 = \frac{|A_2|}{|A|} = -6, \quad x_3 = \frac{|A_3|}{|A|} = 7.$$

It should be emphasized that Cramer's rule applies only to the case where the determinant of the coefficient matrix is nonzero. If this condition fails then the system may have infinitely solutions or no solution.

Cramer's rule is readily implemented in VBA using the determinant function developed in the preceding section. The module `Cramer` assumes that the user has entered the coefficient matrix of the system with top left corner in C3. The right side of the system is entered one column to the right. Figure 14.1 gives an example of input and output of the program. The coefficient matrix of the 7×7 system was entered in columns C–I, the right side in column K, column J as a separator. The solution (rounded) appears in column N.

	C	D	E	F	G	H	I	K	M	N
3	4	3	−4	7	3	−1	7	−1	$x_1 =$	−2.28
4	2	−1	1	−9	4	3	4	2	$x_2 =$	0.76
5	1	−2	3	1	−5	−4	9	−5	$x_3 =$	−0.09
6	3	6	8	9	6	2	3	3	$x_4 =$	−0.20
7	2	3	5	−4	7	−3	2	8	$x_5 =$	1.17
8	2	4	1	8	9	9	5	7	$x_6 =$	−0.28
9	−4	6	−6	5	−3	2	−7	6	$x_7 =$	0.44

FIGURE 14.1: Spreadsheet for CramerRule.

The procedure `CramersRule` below, which may be easily implemented by a command button, reads the data into the matrix arrays `A` and `B` and then calls the function `Cramer` to calculate the solutions, if possible.

```
Sub CramersRule()
    Dim k As Integer, nrows As Integer, ncols As Integer, err As Boolean
    Dim A() As Double, B() As Double, d As Double, sol() As Double
    A = MatrixIn(3, 3)                          'read coefficient matrix
    nrows = UBound(A, 1): ncols = UBound(A, 2)
    If ncols <> nrows Then MsgBox "System not square.": Exit Sub
    B = MatrixIn(3, 4 + nrows)       'read col matrix; one col separation
    sol = Cramer(A, B, err)                         'array of solutions
    If err Then MsgBox "No solution.": Exit Sub
    For k = 1 To ncols                                'print solutions
        Cells(2 + k, 6 + ncols).Value = "x" & k & " = "        'labels
        Cells(2 + k, 7 + ncols).Value = sol(k)
    Next k
End Sub
```

The function `Cramer` calculates the denominator and numerators as per Cramer's rule. The function `Numerator` forms the matrix A_k as described in Cramer's rule and returns its determinant.

```
Function Cramer(A() As Double, B() As Double, err As Boolean) As Double()
    Dim k As Integer, d As Double, sol() As Double
    ReDim sol(1 To UBound(A, 1))
    d = Determinant(A)                  'determinant of coefficient matrix
    err = (d = 0)
    If err Then Exit Function
    For k = 1 To UBound(A, 2)                      'implement Cramer's rule
        sol(k) = Numerator(A, B, k) / d                      'solutions
    Next k
    Cramer = sol
End Function
```

```
Function Numerator(A() As Double, B() As Double, k As integer) As Double
    Dim n As Integer, i As Integer, j As Integer, Ak() As Double
    n = UBound(A, 1)                    'number of rows of augmented matrix
    ReDim Ak(1 To n, 1 To n)                   'memory for numerator array Ak
    For i = 1 To n
        For j = 1 To n
            Ak(i,j) = A(i, j)                   'same as coefficient matrix
            If j = k Then Ak(i, j) = B(i,1)        'replace kth col by B
        Next j
    Next i
    Numerator = Determinant(Ak)                      'return determinant
End Function
```

Since there are two versions of `Determinant`, there are, accordingly, two versions of `CramersRule`: `CramerRuleEchelon` and `CramerRuleRecursive`. As expected, for large systems the recursive version is considerably slower.

14.4 Collinear and Coplanar Points Using Determinants

The equation of line in the (x, y) coordinate plane has the form $ax+by+c = 0$, where not both a and b are zero. Points (x_1, y_1), (x_2, y_2), (x_3, y_3) are said to be *co-linear* if they lie on the same line, that is, if there exist a, b, c, with a and b not both zero, such that

$$ax_1 + by_1 + c = 0$$
$$ax_2 + by_2 + c = 0$$
$$ax_3 + by_3 + c = 0$$

This may be viewed as a system of linear equations in the variables a, b, c. The determinant of the coefficient matrix of the system must therefore be zero, otherwise Cramer's rule would imply that $a = b = 0$, contradicting the above requirement. Thus the points can be on the same line if and only if

$$\begin{vmatrix} x_1 & y_1 & 1 \\ x_2 & y_2 & 1 \\ x_3 & y_3 & 1 \end{vmatrix} = 0.$$

If we think of (x_2, y_2) and (x_3, y_3) as given points and (x_1, y_1) as a variable point (x, y), then we can express the equation of a line passing through the given points as

$$\begin{vmatrix} x & y & 1 \\ x_2 & y_2 & 1 \\ x_3 & y_3 & 1 \end{vmatrix} = 0.$$

Expanding along the first row yields the alternate form

$$\begin{vmatrix} y_2 & 1 \\ y_3 & 1 \end{vmatrix} x - \begin{vmatrix} x_2 & 1 \\ x_3 & 1 \end{vmatrix} y + \begin{vmatrix} x_2 & y_2 \\ x_3 & y_3 \end{vmatrix} = 0. \tag{14.3}$$

A similar analysis applies to planes in the x, y, z spatial coordinate system. Such a plane has equation of the form $ax + by + cz + d = 0$, where the constants $a, b,$ and c are not all zero. Points (x_1, y_1, z_1), $(x_2, y_2, z_2), (x_3, y_3, z_3)$, and (x_4, y_4, z_4) in space are said to be *co-planar* if they lie on a plane, that is, if there exist a, b, c, d with $a, b,$ and c not all zero such that

$$ax_1 + by_1 + cz_1 + d = 0$$
$$ax_2 + by_2 + cz_2 + d = 0$$
$$ax_3 + by_3 + cz_3 + d = 0$$
$$ax_4 + by_4 + cz_4 + d = 0$$

This may be viewed as a system of linear equations in the variables a, b, c, d.

Arguing as in the case of three points on a line we see that the points are co-planar if and only if

$$\begin{vmatrix} x_1 & y_1 & z_1 & 1 \\ x_2 & y_2 & z_2 & 1 \\ x_3 & y_3 & z_3 & 1 \\ x_4 & y_4 & z_4 & 1 \end{vmatrix} = 0.$$

Thus if (x_2, y_2, z_2), (x_3, y_3, z_3), and (x_4, y_4, z_4) are given points and (x_1, y_1, z_1) is a variable point (x, y, z) then the equation of a plane passing through the given points is

$$\begin{vmatrix} x & y & z & 1 \\ x_2 & y_2 & z_2 & 1 \\ x_3 & y_3 & z_3 & 1 \\ x_4 & y_4 & z_4 & 1 \end{vmatrix} = 0.$$

Expanding along the first row yields the alternate form

$$\begin{vmatrix} y_2 & z_2 & 1 \\ y_3 & z_3 & 1 \\ y_4 & z_4 & 1 \end{vmatrix} x - \begin{vmatrix} x_2 & z_2 & 1 \\ x_3 & z_3 & 1 \\ x_4 & z_4 & 1 \end{vmatrix} y + \begin{vmatrix} x_2 & y_2 & 1 \\ x_3 & y_3 & 1 \\ x_4 & y_4 & 1 \end{vmatrix} z = \begin{vmatrix} x_1 & x_2 & y_2 \\ x_2 & x_3 & y_3 \\ x_3 & x_4 & y_4 \end{vmatrix}. \tag{14.4}$$

14.5   Areas and Volumes Using Determinants

Let $P = (p_1, p_2)$, $Q = (q_1, q_2)$, and $R = (r_1, r_2)$ be points in a planar coordinate system. Define quantities

$$a_1 = q_1 - p_1, \quad a_2 = q_2 - p_2, \quad b_1 = r_1 - p_1, \quad b_2 = r_2 - p_2$$

It may be shown that the area A of the triangle PQR is given by the formula

$$\pm \frac{1}{2} \begin{vmatrix} a_1 & a_2 \\ b_1 & b_2 \end{vmatrix},$$

whichever sign gives a positive value.

A similar analysis applies to tetrahedrons, which are defined by four points in space,

$$P = (p_1, p_2, p_2), \quad Q = (q_1, q_2, q_2), \quad R = (r_1, r_2, r_2), \text{ and } S = (s_1, s_2, s_2),$$

not all in the same plane (see Figure 14.2). Define quantities

$$a_1 = q_1 - p_1, \quad a_2 = q_2 - p_2, \quad a_3 = q_3 - p_3$$
$$b_1 = r_1 - p_1, \quad b_2 = r_2 - p_2, \quad b_3 = r_3 - p_3$$
$$c_1 = s_1 - p_1, \quad c_2 = s_2 - p_2, \quad c_3 = s_3 - p_3.$$

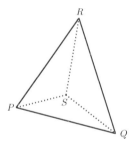

FIGURE 14.2: Tetrahedron $PQRS$.

It may be shown that the volume V of the tetrahedron is given by the formula

$$\pm \frac{1}{6} \begin{vmatrix} a_1 & a_2 & a_3 \\ b_1 & b_2 & b_3 \\ c_1 & c_2 & c_3 \end{vmatrix},$$

whichever sign gives a positive value.

14.6   Circumscribing a Triangle

A circle in the xy plane with center (h, k) and radius r is given by the equation

$$(x - h)^2 + (y - k)^2 = r^2.$$

Given three non-collinear points $P = (x_1, y_1)$, $Q = (x_2, y_2)$, and $R = (x_3, y_3)$ (the vertices of a triangle), there exists a unique circle containing these points and therefore circumscribing the triangle. This is verified by solving the following system of equations for the variables h, k, r:

$$(x_1 - h)^2 + (y_1 - k)^2 = r^2$$
$$(x_2 - h)^2 + (y_2 - k)^2 = r^2$$
$$(x_3 - h)^2 + (y_3 - k)^2 = r^2.$$

These are *nonlinear* equations, but the system may be linearized by expanding the equations and writing the system as

$$2x_1 h + 2y_1 k + \ell = x_1^2 + y_1^2$$
$$2x_2 h + 2y_2 k + \ell = x_2^2 + y_2^2$$
$$2x_3 h + 2y_3 k + \ell = x_3^2 + y_3^2,$$

where $\ell = r^2 - h^2 - k^2$. The system is linear in the variables h, k, ℓ and so may be solved for these variables by Cramer's rule. The radius r may then be recovered from ℓ, h, k. The center (h, k) of the circle is called the *circumcenter* of the triangle; it is the also the intersection of the perpendicular bisectors of the sides (see Figure 14.3).

The function `Circumcenter(P,Q,R,h,k,r,error)` implements this process. It takes as input the three points written as 1×2 arrays and calculates h, k and r.

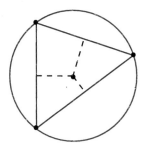

FIGURE 14.3: Circle through 3 points.

```
Sub Circumcenter(P() As Double, Q() As Double, R() As Double,
        h As Double, k As Double, radius As Double, error As Boolean)
    Dim A(1 To 3, 1 To 3) As Double, B(1 To 3, 1 To 1) As Double
    Dim sol() As Double
        'get the matrix A for Cramer
    A(1, 1) = 2 * P(1): A(1, 2) = 2 * P(2): A(1, 3) = 1
    A(2, 1) = 2 * Q(1): A(2, 2) = 2 * Q(2): A(2, 3) = 1
    A(3, 1) = 2 * R(1): A(3, 2) = 2 * R(2): A(3, 3) = 1
        'get the matrix B for Cramer
    B(1, 1) = (P(1)) ^ 2 + (P(2)) ^ 2
    B(2, 1) = (Q(1)) ^ 2 + (Q(2)) ^ 2
    B(3, 1) = (R(1)) ^ 2 + (R(2)) ^ 2
    sol = Cramer(A, B, error)                           'run Cramer
    h = sol(1): k = sol(2): radius = (h ^ 2 + k ^ 2 + sol(3)) ^ (1 / 2)
End Sub
```

The following procedure uses the graph facility of VBA to graph the triangle and the lines from the circumcenter to the vertices. It should be compared with the program `GraphIncenter` in Section 9.5.

```
Sub GraphCircumcenter()
    Dim XRange As String, YRange As String
    Dim P() As Double, Q() As Double, R() As Double
    Dim h As Double, k As Double, radius As Double
    Dim sol() As Double, Grid As TGrid, error As Boolean
            'clear old chart and data
    If ChartObjects.Count > 0 Then ChartObjects.Delete
    Call ClearColumns(8, 1, 2)       'clear columns 1,2 starting at row 8
    Grid = MakeGrid(3, 4, 27, 28, 2.3, 15, 24)
    Call InstallCoord(Grid, Grid.B, Grid.L)
```

```
Call GetVertexCoord(Grid, P, Q, R)
Call Circumcenter(P, Q, R, h, k, radius, error)
If error Then Exit Sub
        'list the X and Y data in columns 1 and 2
Cells(8, 1).Value = P(1):    Cells(8, 2).Value = P(2)
Cells(9, 1).Value = Q(1):    Cells(9, 2).Value = Q(2)
Cells(10, 1).Value = R(1):   Cells(10, 2).Value = R(2)
Cells(11, 1).Value = P(1):   Cells(11, 2).Value = P(2)
Cells(12, 1).Value = h:      Cells(12, 2).Value = k
Cells(13, 1).Value = Q(1):   Cells(13, 2).Value = Q(2)
Cells(14, 1).Value = h:      Cells(14, 2).Value = k
Cells(15, 1).Value = R(1):   Cells(15, 2).Value = R(2)
XRange = "A" & 8 & ":" & "A" & 15
YRange = "B" & 8 & ":" & "B" & 15
With Sheet1 _
    .ChartObjects.Add(Left:=580, Width:=460, Top:=15, Height:=400)
    .Chart.SetSourceData Source:=Range(YRange)                'y values
    .Chart.ChartType = xlXYScatterLines
    .Chart.HasTitle = True
    .Chart.ChartTitle.Text = "Circumcenter"
    .Chart.SeriesCollection(1).XValues = Range(XRange)        'x values
End With
End Sub
```

14.7   Exercises

1. Use the determinant program to find the determinants of matrices A of the form

$$\begin{bmatrix} a & 0 & 0 & 0 \\ * & b & 0 & 0 \\ * & * & c & 0 \\ * & * & * & d \end{bmatrix} \quad \begin{bmatrix} * & * & * & d \\ * & * & c & 0 \\ * & b & 0 & 0 \\ a & 0 & 0 & 0 \end{bmatrix} \quad \begin{bmatrix} 0 & 0 & 0 & d \\ 0 & 0 & c & * \\ 0 & b & * & * \\ a & * & * & * \end{bmatrix},$$

where the asterisks represent arbitrary numbers. What conclusions do you draw?

2. Use the functions MultMat and DetEchelon or DetRecursive to check that $|AB| = |A|\,|B|$. Is it true that $|A + B| = |A| + |B|$?.

3. Use the functions Transpose and DetEchelon or DetRecursive to check that $|A^T| = |A|$.

4. Use the programs InvertMat and DetEchelon or DetRecursive to check that $|A^{-1}| = 1/|A|$.

5. The $n \times n$ *Vandermonde matrix* $V = V(x_1, \ldots, x_n)$ is defined as

$$V = \begin{bmatrix} 1 & x_1 & x_1^2 & \cdots & x_1^{n-1} \\ 1 & x_2 & x_2^2 & \cdots & x_2^{n-1} \\ 1 & x_3 & x_3^2 & \cdots & x_3^{n-1} \\ \vdots & \vdots & \vdots & \cdots & \vdots \\ 1 & x_n & x_n^2 & \cdots & x_n^{n-1} \end{bmatrix}.$$

Write a program `VanderRecursive` using `DetRecursive` that displays the Vandermonde matrix V and evaluates $|V|$ for user entered x_j. Use properties of determinants to show that

$$|V| = \prod_{1 \le i < j \le n} (x_j - x_i).$$

Incorporate this into the program as a check.

6. Write a program `LineEqn` that takes as input two points (x_1, y_1) and (x_2, y_2) and returns the equation of the line through the points by using determinants as in (14.3)

7. Write a program `PlaneEqn` that takes as input three points (x_1, y_1, z_1), (x_2, y_2, z_2) and (x_3, y_3, z_3) and returns the equation of the plane through the points by using determinants as in (14.4)

8. Write an analog `RowOperationsDet` of the program `RowOperations` that calculates the determinant of a square matrix at each step.

9. (Adjugate of a Matrix). The *cofactor matrix* of an $n \times n$ A is the $n \times n$ matrix $C = (C_{ij})$. The *adjugate* Adj(A) of A is the matrix C^T. Write a program `Adjugate` that returns the adjugate of a user entered matrix A. Use the program to show that the product $A \cdot \text{Adj}(A)$ is the $n \times n$ diagonal matrix with $\det(A)$ along the diagonal. Thus if $\det(A) \ne 0$, then A^{-1} exists and equals Adj(A) multiplied by $1/\det(A)$.

*10. Refer to Section 8.4 for this and the next exercise. Write a program `CxDeterminant` that is the complex number analog of `Determinant`.

*11. Write a program `CxCramer` that is the complex number analog of `Cramer`.

*12. Refer to Section 8.5 for this exercise. Write a program `PolyDeterminant` that is the polynomial analog of `Determinant`.

*13. Let A be an $n \times n$ real matrix. An *eigenvalue* of A is a number x such that $AX = xX$ or, equivalently, $(A - xI)X = 0$ for some $n \times 1$ matrix X whose entries are not all zero. Eigenvalues have important applications in physics, engineering, and data analysis, to mention just a few. If x is an eigenvalue of A then it must be the case that $\det(A - xI) = 0$, otherwise Cramer's rule would imply that the entries of X are all zero, contradicting the definition of eigenvalue. The determinant $\det(A - xI)$ is a polynomial in x and is called the *characteristic polynomial* of A. Using the result of the previous exercise, write a program `CharPoly` that reads a matrix A and returns its characteristic polynomial.

Part III

Logic

Chapter 15

Propositional Logic

We have already had some acquaintance with logic in the form of the VBA logical operators And, Or, and Not. In this chapter we introduce mathematical versions of these operators as well as others. The operators act on arbitrary statements and thus place the notion of logical operator in the broader context of formal propositional logic.

15.1   Compound Statements

A *statement* or *proposition* is a declarative sentence which is either true or false but not both. For example, consider the sentences

- The square roots of four are ± 2.

- Every real number is positive.

- How are you?

- Find the prime decomposition of 72.

- This statement is false.

Clearly, the first sentence is true and the second false. The third is a question and the fourth a command and as such do not have truth values. The fifth sentence is a *logical paradox*: if we assume the sentence is true, then we must conclude that it is false, and vice versa. Thus only the first two sentences are statements in the above sense.

We shall use the letters p, q, r,... to designate what are called *simple statements*. A *compound statement* is a statement constructed from other statements by using the following *logical operations*:[1]

• pq	p and q
• $p + q$	p or q
• p'	not p
• $p \rightarrow q$	p implies q
• $p \leftrightarrow q$	p if and only if q

---

[1] The notation $p \wedge q$, $p \vee q$, and $\sim p$ for the first three operations is also used. Our choice is governed by the need to enter compound statements into a spreadsheet.

DOI: 10.1201/9781003351689-15

The statements pq and $p + q$ are called, respectively, the *conjunction* and *disjunction* of p and q, and p' is called the *negation* of p. The connective $\rightarrow$ is called the *conditional* and $\leftrightarrow$ the *biconditional*. In the statement $p \rightarrow q$, p is called the *antecedent* and q the *consequent*.

A statement, simple or compound, has *truth values* true or false, symbolized in this chapter by the integers 1 and 0, respectively. The truth values of a statement may be conveniently displayed by a *truth table*. Truth tables for the above compound statements are shown in the following figure. The first two columns list in conventional format all possible truth values of the pair p, q. The remaining columns give the corresponding truth values of the various compound statements.

Truth Tables for the Logical Operators

p	q	pq	$p + q$	p'	$p \rightarrow q$	$p \leftrightarrow q$
1	1	1	1	0	1	1
1	0	0	1	0	0	0
0	1	0	1	1	1	0
0	0	0	0	1	1	1

Note that $p \rightarrow q$ is true if p is false. This is called the *principle of explosion* ("from falsehood anything follows"), and may be made reasonable by considering nonsense statements such as "If elephants can fly, then $1 = 2$." Note also that $p \leftrightarrow q$ is true precisely when p and q have the same truth values.

Here is an example of how truth tables may be used to determine the truth values of a complicated compound statement. The extra columns are included to facilitate calculations.

Truth Table for $(p' + q)(p + q')$

p	q	p'	q'	$p' + q$	$p + q'$	$(p' + q)(p + q')$
1	1	0	0	1	1	1
1	0	0	1	0	1	0
0	1	1	0	1	0	0
0	0	1	1	1	1	1

There is a *hierarchy* or *precedence* of logical operations analogous to that for arithmetic operators. For example, in the statement $pq' + p'r$ the negations are evaluated first, then the conjunctions, and then the disjunction. A similar hierarchy occurs in the statement $pq' \rightarrow p'r$, where the conditional is evaluated last. Parentheses may be needed in more complex statements. For example, in the statement $(pq' + p'r) \leftrightarrow (pq' \rightarrow p'r)$ the parentheses force the biconditional to be evaluated last; removing the parentheses would produce a radically different statement. The program Stmt2TruthTable developed in Section 15.3 observes precedence rules.

15.2   Equivalent Statements and Laws of Logic

Two statements a and b are said to be *equivalent*, written $a \equiv b$, if they have precisely the same truth values. For example, referring to the truth tables of the last section we see that the statements $p' + q$ and $p \to q$ have identical columns, hence are equivalent. Similarly, $(p' + q)(p + q')$ and $p \leftrightarrow q$ are equivalent.

An alternate way to test equivalence of statements a and b is to check if the biconditional $a \leftrightarrow b$ is always true, that is, if its truth value is always 1. Such a statement is called a *tautology*. If c is a tautology we write $c \equiv 1$. Thus, by the observations of the preceding paragraph,

$$(p' + q) \leftrightarrow (p \to q) \equiv 1 \quad \text{and} \quad (p' + q)(p + q') \leftrightarrow (p \leftrightarrow q) \equiv 1.$$

At the opposite extreme is a statement c whose truth value is always 0, written $c \equiv 0$. Such a statement is called a *contradiction*. The negation of a tautology is a contradiction and vice-versa.

Equivalences are frequently used to describe laws of logic. The following list presents the main laws; others are given in the exercises. The module `Stmt2TruthTable`, described in the next section, may be used to verify these laws.

Laws of Logic

- Double Negation: $p'' \equiv p.$
- Commutative Laws: $p + q \equiv q + p,\ pq \equiv qp.$
- Associative Laws: $(p + q) + r \equiv p + (q + r),\ (pq)r \equiv p(qr).$
- DeMorgan's Laws: $(p + q)' \equiv p'q',\ (pq)' \equiv p' + q'.$
- Contrapositive: $(p \to q)' \equiv q' \to p'.$
- Distributive Laws: $p(q + r) \equiv pq + pr,\ p + qr \equiv (p + q)(p + r).$
- Absorption Laws: $pq + p \equiv p,\ (p + q)p \equiv p.$
- Idempotence: $pp \equiv p,\ p + p \equiv p.$
- Law of the Excluded Middle: $p + p' \equiv 1.$
- Law of Noncontradiction $pp' \equiv 0.$
- Identity Laws $p + 1 \equiv 1,\ p1 \equiv p,\ p + 0 \equiv p,\ p0 \equiv 0.$

The associative laws allow one to dispense with parentheses in statements such as $(p + q) + r$, so that one may simply write $p + q + r$. Also, DeMorgan's Laws and the distributive laws extend to more than two statements. For example, one has

$$(p + q + r + s)' \equiv p'q'r's'$$

and

$$p(q + r + s) \equiv pq + pr + ps.$$

Laws of logic may be used to simplify some compound statements. For example, consider the statement

$$pr + r' \equiv p + r' \tag{15.1}$$

This may be verified directly by assigning truth values or by using the program Stmt2TruthTable. Instead, we give a proof using some of the above laws. First we calculate the negation of the left side of (15.1):

$$
\begin{aligned}
(pr + r')' &\equiv (p' + r')r & \text{(DeMorgan's law and double negation)} \\
&\equiv p'r + r'r & \text{(distributive law)} \\
&\equiv p'r + 0 & \text{(law of noncontradiction)} \\
&\equiv p'r & \text{(identity law)}
\end{aligned}
$$

On the other hand, the negation of the right side of (15.1) is, by DeMorgan's law and double negation,

$$(p + r')' \equiv p'r.$$

Thus both sides of (15.1) have the same negations and so must be equivalent. The example shows that negating a compound statement sometimes reveals a simplification that might otherwise go unnoticed.

15.3   Truth Tables with VBA

The program described in this section takes as input a logical statement and generates the corresponding truth table. Figure 15.1 shows sample input and output, truncated to save space. The variables are entered starting in

	B	C	D	E	F	H
5	p	q	r	s	t	$(p + qr' + rs') \rightarrow (p' + qst')$
6	1	1	1	1	1	0
7	1	1	1	1	0	1
8	1	1	1	0	1	0
9	1	1	1	0	0	0
10	1	1	0	1	1	0
11	1	1	0	1	0	1

FIGURE 15.1: Truncated example for Stmt2TruthTable.

row 5 starting at column B; a single column separates the variables from the statement. The program reads the information in row 5 and generates truth values for the variables (single letters) and the statement. The symbol $\rightarrow$ for the conditional is entered by concatenating the symbols $-$ and $>$ to produce $->$. Similarly, the biconditional symbol $\leftrightarrow$ is entered as $<->$.

The main procedure `Stmt2TruthTable` uses a Do While loop to create a string `Vars` from the entered variables. The input statement `Stmt` is then retrieved. For the above example, `Vars= "pqrst"` and `Stmt= "(p+qr'+rs') -> (p' + qst')"`. To simplify the coding, asterisks are inserted between the character pairs)q,)(, p(, 'p, '(, and pq, so that, for example, `Stmt` becomes `"(p+q*r'+r*s')->(p'+q*s*t')"`. A For Next loop processes each row of the truth table. The function `InsertVarValues` takes `Stmt` and replaces the variables by their values in the row to produce `VarStmt`. In our example, for $i = 1$ the function returns the string `VarStmt = "(1+1*1'+1*1')->(1'+1*1*1')"`. The function `Eval` calculates the truth value of `VarStmt`.

```
Sub Stmt2TruthTable()
    Dim Stmt As String, VarStmt As String, VarVals As String
    Dim Vars As String, NVars As Integer, i As Integer
    Do While Not IsEmpty(Cells(5, i + 2))       'get variable name string
        Vars = Vars & Cells(5, i + 2).Value
        i = i + 1
    Loop
    NVars = Len(Vars)
    Stmt = Cells(5, 3 + NVars).Value                      'get statement
    Stmt = RemoveWhiteSpace(Stmt)                   'remove extra spaces
    Stmt = InsertAsterisks(Stmt)                     'for ease of coding
    For i = 1 To 2 ^ NVars                    'insert 1's and 0's into Stmt
        VarStmt = InsertVarValues(Stmt, Vars, i)        'producing VarStmt
        'evaluate statement for this row of variable values
        Cells(5 + i, 3 + NVars).Value = Eval(VarStmt, 1, 0, err)
    Next i
End Sub


Function InsertAsterisks(expr As String) As String
    Dim k As Integer, ch As String, nextch As String, insert As Boolean
    k = 1
    Do While k <= Len(expr)
        ch = Midd(expr, k, 1)                           'current character
        nextch = Midd(expr, k + 1, 1)                          'next one
        If ch = ")" And IsLower(nextch) Or _
           ch = ")" And nextch = "(" Or _
           IsLower(ch) And nextch = "(" Or _
           ch = "'" And IsLower(nextch) Or _
           ch = "'" And nextch = "(" Or _
           IsLower(ch) And IsLower(nextch) Then
            expr = InsertString(expr, k, "*")
        End If
        k = k + 1
    Loop
    InsertAsterisks = expr
End Function
```

The function `InsertVarValues` is passed a row number i of the evolving truth table, computes and prints the truth values of the variables for that row, and replaces the variable names by these values.

```
Function InsertVarValues(Stmt As String, Vars As String, i As Integer) _
         As String
    Dim j As Integer, VarStmt As String, y As String
    VarStmt = Stmt                           'copy original statement
    For j = 1 To Len(Vars)               'print variable values in ith row
        y = TruthVal(i, j, Len(Vars))             'get i,j truth value
        Cells(i + 5, j + 1).Value = y                'print it in table
        VarStmt = Replace(VarStmt,Midd(Vars, j, 1), y) 'put into jth var
    Next j
    InsertVarValues = VarStmt
End Function


Function TruthVal(i As Integer, j As Integer, N As Integer) As Integer
    Dim x As Integer
    x = Int((i - 1) * 2 ^ (j - N))
    TruthVal = (1 + (-1) ^ x) / 2
End Function
```

The procedure `TruthVal`, which generates the truth values of the variables $p, q, \ldots$, is based on the algebraic function

$$(i - 1)/2^{n-j}, \quad 1 \le i \le 2^n, \quad 1 \le j \le n,$$

where n is the number of variables, i is a row number, and j is a column number. For example, if $n = 3$ then $0 \le i - 1 \le 7$ and the preceding formula generates the following fractions (the denominator values for $j = 1, 2, 3$ are combined for notational convenience):

$$\frac{i-1}{2^{n-j}} = \frac{0}{4,2,1}, \ \frac{1}{4,2,1}, \ \frac{2}{4,2,1}, \ \frac{3}{4,2,1}, \ \frac{4}{4,2,1}, \ \frac{5}{4,2,1}, \ \frac{6}{4,2,1}, \ \frac{7}{4,2,1}.$$

Taking the integer parts x of these fractions yields the corresponding sequences

$$x = 0,0,0, \ \ 0,0,1, \ \ 0,1,2, \ \ 0,1,3, \ \ 1,2,4, \ \ 1,2,5, \ \ 1,3,6, \ \ 1,3,7.$$

(combined in accordance with the above notational scheme). These values are then converted by the function $y = (1 + (-1)^x)/2$ to produce eight rows with the truth values

$$y = 1,1,1, \ \ 1,1,0, \ \ 1,0,1, \ \ 1,0,0, \ \ 0,1,1, \ \ 0,1,0, \ \ 0,0,1, \ \ 0,0,0.$$

The y values are inserted into the copy `VarStmt` of `Stmt` using the VBA function `Replace` of Section 6.5.

The engine of the program is the function `Eval`, which takes an expression like `"(1+1*1'+1*1')->(1'+1*1')"` and finds its value. The procedure is entirely similar in concept to earlier calculators such as `ComplexEval` (Section 8.4), calling itself to perform the calculations using the procedures `Negation`, `Conjunction`, `Disjunction`, `Conditional`, and `Biconditional`.

```
Function Eval(expr As String, idx As Integer, mode As Integer, _
                err As Boolean) As String
    Dim z As String, w As String, char As String
    Do While idx <= Len(expr) And Not err
        char = Mid(expr, idx, 1)                'character at position idx
        If InStr("01", char) > 0 Then           'if char = 0 or 1
           char = "#" 'integer                  'go to first case
        Select Case char
          Case Is = "#"
                z = Mid(expr, idx, 1): idx = idx + 1
          Case Is = "+"
                If mode > 0 Then Exit Do        'wait for higher mode ops
                idx = idx + 1
                w = Eval(expr, idx, 0, err)
                If err Then Exit Do
                z = Disjunction(z, w)
          Case Is = "*"
                idx = idx + 1
                w = Eval(expr, idx, 1, err)             'get next factor
                If err Then Exit Do
                z = Conjunction(z, w)
          Case Is = "-"                                 'conditional
                idx = idx + 2                           'skip ">"
                w = Eval(expr, idx, 1, err)             'get next factor
                z = Conditional(z, w)
          Case Is = "<"                                 'biconditional
                idx = idx + 3                           'skip "->"
                w = Eval(expr, idx, 1, err)             'get next factor
                z = Biconditional(z, w)
          Case Is = "'"
                idx = idx + 1                           'skip "'"
                z = Negation(z)
            Case Is = "("
                idx = idx + 1                           'skip "("
                z = Eval(expr, idx, 0, err)
                If err Then Exit Do
                idx = idx + 1                           'skip")"
            Case Is = ")": Exit Do
            Case Else: err = True: Exit Do
        End Select
    Loop
    Eval = z                                    'return calculation
End Function

Function Negation(p As String) As String
    Negation = 1 - CInt(p)
End Function

Function Conjunction(p As String, q As String) As String
    Conjunction = CStr(p * q)
End Function
```

```
Function Disjunction(p As String, q As String) As String
    Dim x As Integer
    x = CInt(p) + CInt(q)
    If x = 2 Then x = 1
    Disjunction = CStr(x)
End Function


Function Conditional(p As String, q As String) As String
    Dim cond As String
    cond = "1"                                        'default
    If p = "1" And q = "0" Then cond = "0"      'only time cond is False
    Conditional = cond
End Function


Function Biconditional(p As String, q As String) As String
    Dim bicond As String
    bicond = 1
    If p <> q Then bicond = "0"                 'only time bicond is False
    Biconditional = bicond
End Function
```

15.4 Statement from a Truth Table

Given an arbitrary truth table, it is always possible to construct a statement that fits the table. For example, suppose you want a statement with variables p, q, r that is true precisely for the values (a) $p, q, r = 1, 1, 1$, (b) $p, q, r = 1, 0, 1$, and (c) $p, q, r = 1, 0, 0$. Such a statement may be constructed by observing that pqr is true for precisely the values given in (a), $pq'r$ is true for precisely the values given in (b), and $pq'r'$ is true for precisely the values given in (c). It follows that the disjunction $pqr + pq'r + pq'r'$ is a statement of the required type. Such a statement is called a *disjunction of basic conjunctions* and is said to be in *disjunctive normal form*.

In this section we describe a module TruthTable2DisjNormal which takes as input variables p, q, ... and a given column of truth values and prints a statement with precisely these values above the column, essentially reversing Stmt2TruthTable. The ones are placed where the statement is desired to be true; blanks are interpreted as zeros. Figure 15.2 illustrates spreadsheet input for the above example; running the program produces the statement $pqr + pq'r + pq'r'$ in cell E5.

The program works as follows: After entering the variable names in row 5, the user presses the first command button to generate the truth table columns headed by the variable names. The user then enters ones at the places where the value True is desired. It is not necessary to enter the zeros. Pressing the

	B	C	D	E
5	p	q	r	
6	1	1	1	1
7	1	1	0	
8	1	0	1	1
9	1	0	0	1
10	0	1	1	
11	0	1	0	
12	0	0	1	
13	0	0	0	

FIGURE 15.2: Spreadsheet for `TruthTable2DisjNormal`.

second command button constructs the statement, which is then placed in row 5.

```
Sub CommandButton1_Click()
    Call MakeTruthColumns
End Sub

Private Sub CommandButton2_Click()
    Call FormStatement
End Sub
```

The procedure MakeTruthColumns uses the formula `TruthVal`, described earlier, for the truth value of an entry.

```
Sub MakeTruthColumns()
    Dim i As Integer, j As Integer, x As Integer, y As Integer
    Dim N As Integer, row As Integer, col As Integer
    row = 7: col = 2  'variables start here
    Do While Not IsEmpty(Cells(row, N + col))      'get no. of variables
        N = N + 1
    Loop
    For i = 1 To 2 ^ N                          'form the variable columns
        For j = 1 To N                          '(N = number of variables)
            Cells(i + row, col + j - 1).Value = TruthVal(i, j, N)
        Next j
    Next i
End Sub
```

The function `FormStatement` runs down the column under the desired statement, selecting the appropriate basic conjunction if a 1 appears in the column, ignoring blanks.

```
Sub FormStatement()
    Dim i As Integer, j As Integer, Stmt As String, L As Integer
    Dim N As Integer, Vars As String, row As Integer, col As Integer
    row = 5: col = 2                          'variables start here
    Do While Not IsEmpty(Cells(row, N + col))      'get variables
        Vars = Vars & Cells(row, N + col).Value      'variables string
        N = N + 1
    Loop
    'if a 1 is in statement column, append the conjunction to disjunction
    For i = 1 To 2 ^ N
        If Cells(i + row, col + N + 1).Value = 1 Then
            Stmt = Stmt & MakeConj(i, Vars, row, col) & "+"
        End If
    Next i
        'chop off last + sign
    Cells(5, 3 + Len(Vars)).Value = Midd(Stmt, 1, Len(Stmt) - 1)
End Sub


Function MakeConj(i, Vars, row, col) As String
    Dim j As Integer, conj As String, var As String, N As Integer
    N = Len(Vars)
    For j = 1 To N                           'form the conjunction in row i
        var = Mid(Vars, j, 1)                     'get jth variable

        If Cells(i + row, col + j - 1).Value = 1 Then    'if variable = 1
            conj = conj & var                         'include variable as is
        Else
            conj = conj & var & "'"           'otherwise include its negation
        End If
    Next j
    MakeConj = conj
End Function
```

15.5   Valid Arguments

An *argument* is a sequence of logical steps that starts with given propositions $p_1, p_2, \ldots p_n$, called *premises*, and ends with statement q, called the *conclusion*. An argument is said to be *valid* if the implication

$$p_1 p_2 \cdots p_n \to q \tag{15.2}$$

is a tautology. Since an implication is false only when the antecedent is true and the consequent false, it follows that a valid argument is one for which the conclusion q is true whenever all of the statements $p_1, p_2, \cdots, p_n$ are true. Figure 15.3 shows some common forms of valid arguments displayed in standard form: the premises placed above a line and the conclusion below.

(a) $\dfrac{\begin{array}{c}p \to q\\ q \to r\end{array}}{p \to r}$ (b) $\dfrac{\begin{array}{c}p+q\\ p'\end{array}}{q}$ (c) $\dfrac{\begin{array}{c}p \to q\\ p\end{array}}{q}$ (d) $\dfrac{\begin{array}{c}p \to q\\ q'\end{array}}{p'}$

FIGURE 15.3: Valid arguments.

The following examples illustrate these arguments. It should be noted that in spite of their validity, in none of these arguments is it necessary that the premises or the conclusion be true. What matters is the logical necessity of the conclusion given the premises, not the actual truth of the statements.

(a) *If $1 \neq 2$, then birds cannot bellow.*
If birds cannot bellow, then elephants can fly.

   If $1 \neq 2$, then elephants can fly.

(b) *Either $1 = 2$ or elephants can fly.*
$1 \neq 2$.

   Elephants can fly.

(c) *If $1 = 2$, then elephants can fly.*
$1 = 2$.

   Elephants can fly.

(d) *If $1 = 2$, then elephants can fly.*
Elephants can't fly.

   $1 \neq 2$.

Note that these arguments are valid even though most elephants cannot fly.

Arguments (a)–(d) are known, respectively, as *hypothetical syllogism, disjunctive syllogism, modus ponens*, and *modus tollens*. Here is a calculation verifying the validity of the disjunctive syllogism (b):

$$
\begin{aligned}
(p+q)p' \to q &\equiv [(p+q)p']' + q & \text{(equivalent form)}\\
&\equiv [(pp') + qp']' + q & \text{(distributive law)}\\
&\equiv [0 + qp']' + q & \text{(law of noncontradiction)}\\
&\equiv (qp')' + q & \text{(identity law)}\\
&\equiv q' + p + q & \text{(DeMorgan's law)}
\end{aligned}
$$

As the last statement is a tautology so is the first, and the argument is valid. Similar calculations establish the validity of the other arguments. Alternately, one can use truth tables.

Sometimes a modus ponens argument is in the form illustrated by the following examples:

All birds can fly.	*No man is an island.*
Socrates is a bird.	*Socrates is a man.*
Socrates can fly.	*Socrates is not an island.*

These may be recast in standard form by replacing the first premise in the first argument by "If x is a bird, then x can fly" and the first premise in the second argument by "If x is a man, then x is not an island."

An invalid argument may be phrased in the manner of a valid argument, with premises p_1, p_2, ... p_n and conclusion q, but in this case the implication (15.2) is no longer a tautology.

Here are some of examples of invalid arguments even though in each case the conclusion is true.

(e) *If elephants cannot fly, then $1 \neq 2$.*
    $1 \neq 2$.

    Elephants cannot fly.

(f) *Either $1 \neq 2$ or elephants cannot fly.*
    $1 \neq 2$.

    Elephants cannot fly.

(g) *If $1 \neq 2$ then elephants cannot fly.*
    $1 = 2$.

    Elephants cannot fly.

15.6 Exercises

1. Show by taking negations that $(r + q' + rq') \equiv (r + q')$

2. Suppose p, q, and r are statements such that only p is true. Which of the following are true?
   (a) $pq + pq' + r$
   (b) $pq + p' + r'$
   (c) $(p'q + pr')(pq' + p'r)$
   (d) $p' + (pq + pr)$

3. Construct truth tables by hand and check your answers with Stmt2TruthTable.
   (a) $p'q' + p + q$
   (b) $p'q + q'p$
   (c) $(p' + q')(p + q')(p' + q)$

(d) $p \rightarrow (p + q')$

(e) $(p + p') \rightarrow p'$

(f) $p(p \rightarrow q) \rightarrow q$

4. Use `Stmt2TruthTable` to determine which of the following statements are tautologies.

   (a) $(p + p')(q' + r)$

   (b) $(p + p') + (q + r')$

   (c) $(pq + p'q) + (pq' + p'q')$

   (d) $[p + (q' + r)][p + (q' + r')]$

   (e) $[(p + q)(p' + q)][(p + q')(p' + q')]$

   (f) $[p + (q' + r)] \rightarrow [p + (q' + r')]$

   (g) $[(p \rightarrow q)(q \rightarrow r)] \rightarrow (p \rightarrow r)$

   (h) $(p + q)(q + r)(r + p) \leftrightarrow [pq + (qr + rq)]$

   (i) $(p \rightarrow q) \leftrightarrow [q(p + r) + (p + qr)]$

5. Use `Stmt2TruthTable` to verify the following equivalences:

   (a) $pq + pq'r \equiv pq + pr$

   (b) $(p + q)(p + q' + r) \equiv (p + q)(p + r)$

   (c) $pq + pq' \equiv p$

   (d) $(p + q)(p + q') \equiv p$

   (e) $p + p'q \equiv p + q$

   (f) $p(p' + q) \equiv pq$

6. Use laws of logic to simplify

   (a) $p'q' + (p + q')'$

   (b) $(pq' + qr)(qr' + pr)$

   (c) $(pq' + p'q)pr((p' + q') + (pq' + p'q))$

   (d) $(((pq')' + rp')'(pq'r'))'$

7. The XOR ("exclusive or") in logic is defined by p XOR q if and only if either p or q is true but not both. Use truth tables to show that p XOR q is equivalent to the statement $pq' + p'q$. Give meaning to the statement p XOR q XOR r.

8. The Sheffer stroke[2] of p and q, denoted by $p \mid q$ or p NAND q, is defined as $p' + q'$. Thus $p \mid q$ is false exactly when p and q are both true. Use truth tables to verify the following:

   (a) $p' \equiv p \mid p$.

   (b) $pq \equiv (p \mid q) \mid (p \mid q)$.

   (c) $p + q \equiv (p \mid p) \mid (q \mid q)$.

   Check your answers with `Stmt2TruthTable`.

9. The Peirce arrow $p \downarrow q$ of statements p, q (also written as p NOR q) is

---

[2]Named after Henry M. Sheffer, who in 1913 showed that the standard logical operators may be defined solely in term of the stroke.

defined as $p'q'$ and is therefore true if and only if both statements are false. Show that negation, conjunction, and disjunction of p, q may be expressed solely in terms of the Peirce arrow.

10. Show that one can express all the logical connectives (in the sense of equivalence) in terms of the operators

    (a) "not" and "or"   (b) "not" and "and"   (c) "not" and "if then"

11. Write the general form of the arguments in the elephant examples (e)–(g).

12. Use truth tables to check by hand the validity of the following arguments. Verify your answers with `Stmt2TruthTable`.

    (a) p' (b) q (c) $q \to r$ (d) $p' \to q$ (e) $p \to q$
        $\overline{p \to q}$ $\overline{p \to q}$ q pq $q \to r$
                                        $\overline{p \to r}$ $\overline{q'}$ p
                                                                                         $\overline{r}$

13. Use `Stmt2TruthTable` to check the validity of the following arguments.

    (a) $p + q$ (b) $r' + s'$ (c) $q' + r$ (d) $(p \to q) \to (r \to s)$
         $p \to r$ $p \to r$ $p' \to q$ pr
         $q \to s$ $q \to s$ p $\overline{s}$
         $\overline{r + s}$ $\overline{p' + q'}$ $\overline{r}$

    (e) $p \to [q \to (r \to s)]$
         p
         $\overline{s}$

14. Use the laws of logic to show that the hypothetical syllogism, disjunctive syllogism, modus ponens, and modus tollens are equivalent, respectively, to the tautologies

    (a) $pr' + qr' + p' + r$ (b) $p'q' + p + q$ (c) $pq' + p' + q$ (d) $pq' + p' + q$.

15. Since every logical statement has a truth table it follows from the discussion in Section 15.4 that each such statement is equivalent to a disjunction of basic conjunctions, that is, may be put into disjunctive normal form. Combine the programs `Stmt2TruthTable` and `TruthTable2DisjNormal` into a single program `Stmt2DisjNormal` that takes a statement and produces its disjunctive normal form.

16. Write a program `ValidArgument` that takes a column of premises and a conclusion and determines whether the argument is valid. Test it on the arguments above.

*17. In this exercise we discuss an analog of the disjunctive normal form of a statement. A *basic disjunction* in the variables $p, q, \ldots$ is a disjunction formed from the variables and their negations. For example, the basic disjunctions in the variables p, q are $p + q$, $p' + q$, $p + q'$, and $p' + q'$. A statement is said to be in *conjunctive normal form* if it is a conjunction of basic disjunctions. For example, the statement $(p + q' + r)(p' + q + r')$ is in conjunctive normal form for the variables p, q, r. Notice that the first factor is false for the values $(p, q, r) = (0, 1, 0)$ and the second factor is false for the values $(p, q, r) = (1, 0, 1)$. Thus the statement $(p + q' + r)(p' + q + r')$ is false if either $(p, q, r) = (0, 1, 0)$ or $(p, q, r) = (1, 0, 1)$, and is true otherwise. As with the disjunctive normal form it is always possible to construct a statement in conjunctive normal form from a given a truth table: simply take the conjunctions of the basic disjunctions corresponding to the row in the statement column where a zero lies.

Write a program a module `TruthTable2ConjNormal` that takes as input variables p, q, ... and a given column of truth values and prints a conjunctive normal statement with precisely these values. The program should takes as input the variable names and a column of zeros placed where the statement is desired to be false.

*18. Combine the programs `Stmt2TruthTable` and `TruthTable2ConjNormal` into a single program `Stmt2ConjNormal` that takes a statement and produces its conjunctive normal form.

Chapter 16

Switching Circuits

16.1 Introduction

The propositional algebra described in the preceding chapter has an interesting application to *switching circuits*. These are circuits that contain devices such as switches, relays, or transistors that are connected by wires and have two states, on or off. For definiteness, we shall assume that the devices are switches. Figure 16.1 shows circuits in two states. In the circuit on the right current flows and the battery lights the bulb; in the circuit on the left current does not flow. Hereafter, we omit the battery and light and ignore any electrical

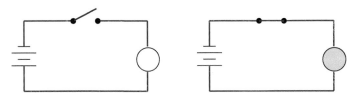

FIGURE 16.1: Switching circuit.

characteristics of the circuit, focusing only on whether or not current flows.

We shall denote switches in a diagram by lower case letters $p, q, r, \ldots$, these having values 0 or 1. A switch p is *closed* if $p = 1$, allowing current to flow, and *open* if $p = 0$, thus blocking current. The value of a switch is called its *state*. Two switches that have identical states, that is, that open and close together (say, by the same handle), will be denoted by the same letter. Switches with opposite states will be denoted by, for example, p and p'. We show that a circuit may be rendered into a compound statement, thus allowing the possibility of applying laws of logic to produce an equivalent simplified circuit. In the last section we show how circuit statements may be generated from diagrams using VBA.

DOI: 10.1201/9781003351689-16

16.2   Series and Parallel Circuits

The circuit on the left in Figure 16.2, called a *series circuit*, conducts current if and only if $p = 1$ and $q = 1$. The circuit on the right, called a *parallel circuit*, conducts current if and only if either $p = 1$ or $q = 1$. Thus we may identify

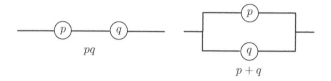

FIGURE 16.2: Series and parallel circuits.

former with the conjunction pq and the latter with the disjunction $p + q$.

A circuit is built from series and parallel circuits and thus may be represented by a logical expression. Figure 16.3 shows some examples.

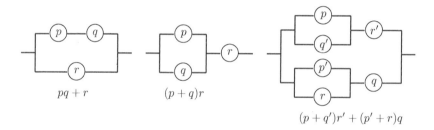

FIGURE 16.3: Switching circuits.

16.3   Equivalent Circuits

Two circuits are said to be *equivalent* if whenever current flows in one it flows in the other. This is the same as saying that the statements representing the circuits are logically equivalent. The equivalent circuits in Figure 16.4 illustrate three laws of logic.

Laws of logic can be used to simplify some circuits. For example, the statement for the circuit in Figure 16.5 is $(pq + r' + s' + t')(rst + pq)$. Applying

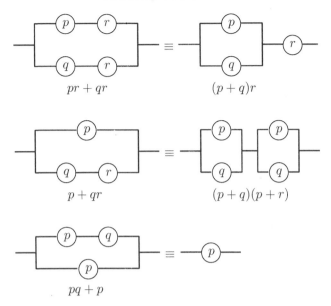

FIGURE 16.4: Distributive and absorption laws for circuits.

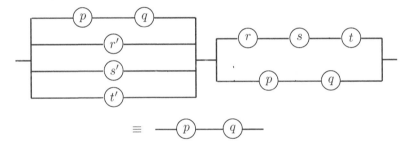

FIGURE 16.5: Simplification example.

the first distributive law yields

$$(pq + r' + s' + t')(rst + pq) \equiv pqrst + r'rst + s'rst + t'rst + (pq + r' + s' + t')pq$$
$$\equiv pqrst + 0 + 0 + 0 + (pq + r' + s' + t')pq$$
$$\equiv pq(rst + r' + s' + t') \equiv pq,$$

the last equivalence because the term in parentheses is a tautology. Thus the circuit in the figure may be reduced to a simple two-switch series circuit.

16.4   Circuit Expressions with VBA

In this section we develop a module that generates a compound statement from a user-drawn circuit in a grid. To draw such a circuit one can highlight the desired cells of the grid using the Shift key together with the arrow keys and then click the Good button in the Style section under the Home tab. This produces a light green swath on the grid with color index number 35. The Normal button in the same section may be used to remove the color. Figure 16.6 shows a typical circuit drawn in this manner.

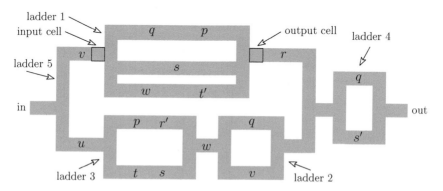

FIGURE 16.6: Characteristics of an admissible circuit.

Every circuit may be constructed in such a way as to be viewed as a group of connected rectangles which we shall call *ladders*. The horizontal cross pieces of a ladder are the "rungs" and the vertical segments embracing the rungs are the "rails." The circuit in Figure 16.6 has five ladders. The rungs of ladders 1–4 are straight, while the rungs of ladder 5 are themselves *inner* ladders. Ladder 5 is said to *embrace* ladders 1,2, and 3. Thus a circuit is a series of ladders embracing ladders, embracing ladders, etc. We shall call a ladder that does not embrace another ladder *simple* and one that embraces other ladders *complex*. Thus ladders 1–4 are simple and ladder 5 is complex.

To reduce the complexity of the code we impose the following rules for drawing a circuit: First, each ladder has an input cell and an output cell. These are to the immediate left and right of the ladder, respectively, and both must be on the same row of the spreadsheet.[1] Second, there can be no rung connecting the input and output cells. This allows input and output cells to be easily determined. Also, each ladder must have a distinguishable top left corner, a top right corner, a bottom left corner, and a bottom right corner. Finally, no letter can be on a rail. Circuits drawn in this way will be called *admissible*. All circuits may be rendered into equivalent admissible forms.

---

[1]For definiteness we are assuming current travels from left to right.

How the Program Works

The program first locates the corners, the input cell, and the output cell of each ladder, simple or complex. It then forms the conjunctions of the symbols in the rungs of simple ladders and places them in the cells immediately to the left of the left rail. For ladder 1 in Figure 16.6, this results in the diagram in Figure 16.7. Next, the program forms the disjunctions of these conjunctions

FIGURE 16.7: State of ladder 1 after conjunctions formed.

and places them in the input cells. Ladder 1 is now "empty" (Figure 16.8.) Next, the program collapses the empty simple ladders into horizontal segments

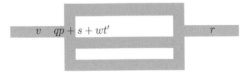

FIGURE 16.8: State of ladder 1 after disjunctions formed.

connecting the input and output cells. Figure 16.9 shows the steps in the collapsing until the final reduction. The program, detecting a single rung and no more rails forms the conjunction of the expressions in the rung and prints it in cell B2.

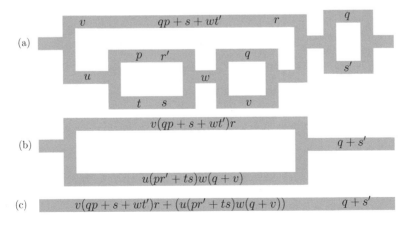

FIGURE 16.9: States of circuit after ladders are collapsed.

The Main Procedure

The module uses a new data type, `TLadder`, which collects the circuit characteristics described in the introduction. It also uses types `TGrid` and `TPos` (not shown). The main procedure of the module is `SwitchCircuitExpr`, which is invoked by a command button. After getting the circuit dimensions, a Do While loop performs the cycle of steps explained in the previous section, forming disjunctions (rails) of conjunctions (rungs) until only a single rung is left (detected by the procedure `SingleRung`). When this occurs the conjunctions along the rung are formed by the procedure `Finish` and printed. The circuit is then restored to its original configuration by the procedure `RestoreCircuit`. The user has the option of selecting the pause feature to observe the steps in the circuit reduction.

```
Private Type TLadder
    TLrow  As Integer: TLcol As Integer    'coordinates of top left corner
    TRrow  As Integer: TRcol As Integer    'coordinates of top right corner
    BLrow  As Integer: BLcol As Integer    'coordinates of bottom left corner
    BRrow  As Integer: BRcol As Integer    'coordinates of bottom right corner
    Inrow  As Integer: Incol As Integer    'coordinates of input to a ladder
    Outrow As Integer: Outcol As Integer   'coordinates of output from a ladder
    IsSimple As Boolean                    'True if ladder is simple
End Type


Public step As String, idxmax As Integer
        'dimensions of smallest rectangle containing circuit:
Public CTop As Integer, CBot As Integer, CLeft As Integer, CRight As Integer


Sub SwitchCircuitExpr()
    Dim Ladder(1 To 50) As TLadder, Grid As TGrid
    Grid = MakeGrid(5, 4, 45, 45, 2.2, 10.5, 24)
    Call GetCircuitBounds(Grid)                    'CTop, CBot, CLeft, CRight
    Call RecordCircuit        'save circuit in Sheet2 for later restoration
    step = Range("B3").Value                       'step through program?
    Do Until SingleRung                 'stop when a single rung is left
        Call GetLadderSpecs(Ladder)              'detect circuit geometry
        Call FormConjunctions(Ladder)            'conjunctions along rungs
        Call Pause                                        'user's option
        Call FormDisjunctions(Ladder)            'conjunctions along rails
        Call Pause
    Loop
    Cells(2, 2).Value = Finish()    'conjunction of expr's along single rung
    Call RestoreCircuit
```

The procedure `GetCircuitBounds` scans the grid for the highest (`CTop`), lowest (`CBot`), leftmost (`CLeft`) and rightmost (`CRight`) positions of the circuit. This is simply to condense the scanning area.

```
    Private Sub GetCircuitBounds(Grid As TGrid)
        Dim i As Integer, j As Integer
```

```
        CTop = Grid.B: CBot = Grid.T                    'worst candidates
        For i = Grid.T To Grid.B
            For j = Grid.L To Grid.R
                If Cells(i, j).Interior.ColorIndex = 35 Then
                    If i < CTop Then CTop = i       'better candidate for top
                    If i > CBot Then CBot = i       'better candidate for bottom
                End If
            Next j
        Next i
        CLeft = Grid.R: CRight = Grid.L                 'worst candidates
        For j = Grid.L To Grid.R
            For i = CTop To CBot
                If Cells(i, j).Interior.ColorIndex = 35 Or _
                Not IsEmpty(Cells(i, j)) Then
                    If j > CRight Then CRight = j 'better candidate for right
                    If j < CLeft Then CLeft = j     'better candidate for left
                End If
            Next i
        Next j
    End Sub
```

The function `SingleRung` scans the circuit to determine if there is more than one rung.

```
Function SingleRung() As Boolean
    Dim m As Integer, n As Integer, num As Integer
    For n = CLeft To CRight                          'run through columns
        num = 0                    'initialize number of rungs for each column
        For m = CTop To CBot
            If Cells(m, n).Interior.ColorIndex = 35 Then
                num = num + 1                        'found a rung in current column
            End If
        Next m
        If num > 1 Then Exit For                    'found more than one rung
    Next n
    SingleRung = (num = 1)
End Function
```

The procedure `RecordCircuit` prints a copy of the original circuit in Sheet2. The sister procedure `RestoreCircuit` uses the copy to replace the collapsed circuit at the end of the program.

```
    Private Sub RecordCircuit()
        Dim m As Integer, n As Integer
        Sheet2.Cells.Clear                              'wipe clean
        Sheet2.Cells.Interior.color = xlNone
        For m = CTop To CBot                            'copy circuit into Sheet2
            For n = CLeft To CRight
                If Cells(m, n).Interior.ColorIndex = 35 Then
                    Sheet2.Cells(m, n).Interior.ColorIndex = 35
                    Sheet2.Cells(m, n).Value = Cells(m, n).Value
                End If
```

```
            Next n
        Next m
    End Sub


    Private Sub RestoreCircuit()
        Dim m As Integer, n As Integer
        For m = CTop To CBot
            For n = CLeft To CRight
                'remove color from added rungs in Sheet1 and copy from Sheet2
                Cells(m, n).Interior.ColorIndex = 0
                If Sheet2.Cells(m, n).Interior.ColorIndex = 35 Then
                    Cells(m, n).Interior.ColorIndex = 35
                    Cells(m, n).Value = Sheet2.Cells(m, n).Value
                ElseIf Not IsEmpty(Cells(m, n)) Then
                    Cells(m, n).Value = ""
                End If
            Next n
        Next m
    End Sub
```

Detecting Ladder Characteristics

The procedure `GetLadderSpecs` detects ladder corners and inputs. As mentioned earlier, these are used to place conjunctions and disjunctions at the appropriate places in the circuit.

```
    Private Sub GetLadderSpecs(Ladder() As TLadder)
        Call LabelCorners                            'label ladder corners
        Call GetLadderCorners(Ladder)                 'get their positions
        Call Pause
        Call RemoveLabels(Ladder)                        'no longer needed
        Call GetLadderInOut(Ladder)   'positions of ladder inputs and outputs
        Call DenoteSimpleLadders(Ladder)       'these have unencumbered rungs
    End Sub
```

The procedure `LabelCorners` runs through the circuit, separately labeling the corners of each ladder. By the rules imposed on circuit construction, these are easily detectable by looking at green cell patterns.

```
    Private Sub LabelCorners()
        Dim i As Integer, j As Integer
        For i = CTop To CBot
            For j = CLeft To CRight
                If Cells(i, j).Interior.ColorIndex = 35 Then
                    Call LabelTopLeft(i, j): Call LabelTopRight(i, j)
                    Call LabelBottomLeft(i, j): Call LabelBottomRight(i, j)
                End If
            Next j
        Next i
    End Sub
```

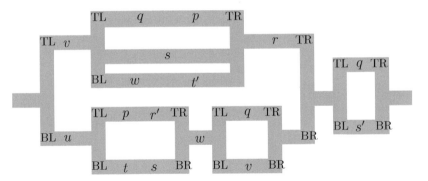

FIGURE 16.10: State of circuit after corners are labeled.

```
''''''''''''''''''''''''' detect patterns '''''''''''''''''''''''''
Sub LabelTopLeft(i As Integer, j As Integer)
    Dim IsTopLeft As Boolean
    IsTopLeft = Cells(i - 1, j).Interior.ColorIndex < 0 And _
                Cells(i, j - 1).Interior.ColorIndex < 0 And _
                Cells(i + 1, j).Interior.ColorIndex = 35 And _
                Cells(i, j + 1).Interior.ColorIndex = 35
    If IsTopLeft Then Cells(i, j).Value = "TL"
End Sub

Sub LabelTopRight(i As Integer, j As Integer)
    Dim IsTopRight As Boolean
    IsTopRight = Cells(i - 1, j).Interior.ColorIndex < 0 And _
                 Cells(i, j + 1).Interior.ColorIndex < 0 And _
                 Cells(i, j - 1).Interior.ColorIndex = 35 And _
                 Cells(i + 1, j).Interior.ColorIndex = 35
    If IsTopRight Then Cells(i, j).Value = "TR"
End Sub

Sub LabelBottomLeft(i As Integer, j As Integer)
    Dim IsBottomLeft As Boolean
    IsBottomLeft = Cells(i, j - 1).Interior.ColorIndex < 0 And _
                   Cells(i + 1, j).Interior.ColorIndex < 0 And _
                   Cells(i, j + 1).Interior.ColorIndex = 35 And _
                   Cells(i - 1, j).Interior.ColorIndex = 35
    If IsBottomLeft Then Cells(i, j).Value = "BL"
End Sub

Sub LabelBottomRight(i As Integer, j As Integer)
    Dim IsBottomRight As Boolean
    IsBottomRight = Cells(i, j + 1).Interior.ColorIndex < 0 And _
                    Cells(i + 1, j).Interior.ColorIndex < 0 And _
                    Cells(i - 1, j).Interior.ColorIndex = 35 And _
                    Cells(i, j - 1).Interior.ColorIndex = 35
    If IsBottomRight Then Cells(i, j).Value = "BR"
End Sub
```

The procedure `GetLadderCorners` fills an array `Ladder` of type `TLadder` with the positions of the ladder corners. These are denoted in the code by `TLr` (top left row) `TLc` (top left column), etc. The procedure starts with the top left corner and proceeds clockwise.

```
Private Sub GetLadderCorners(Ladder() As TLadder)
    Dim i As Integer, j As Integer, idx As Integer
    Dim TLr As Integer, TLc As Integer, TRr As Integer, TRc As Integer
    Dim BRr As Integer, BRc As Integer, BLr As Integer, BLc As Integer
    For i = CTop To CBot
        For j = CLeft To CRight
            If Not Cells(i, j).Value = "TL" Then Go To continue
            idx = idx + 1                        'new index for ladder array
            TLr = i: TLc = j                              'found top left
            Ladder(idx).TLrow = TLr: Ladder(idx).TLcol = TLc     'assign
            TRc = GetTRc(TLr, TLc): TRr = TLr 'go right to get top right
            Ladder(idx).TRrow = TRr: Ladder(idx).TRcol = TRc     'assign
            BRr = GetBRr(TRr, TRc): BRc = TRc 'go down to get bot. right
            Ladder(idx).BRrow = BRr: Ladder(idx).BRcol = BRc     'assign
            BLc = TLc: BLr = BRr   'bottom left col = top left col, etc.
            Ladder(idx).BLrow = BLr: Ladder(idx).BLcol = BLc     'assign
continue:
        Next j
    Next i
    idxmax = idx                                  'number of ladders
End Sub

Function GetTRc(TLr As Integer, TLc As Integer) As Integer
    Dim m As Integer: m = TLc
    Do While Cells(TLr, m).Value <> "TR"     'proceed right from top left
        m = m + 1
    Loop
    GetTRc = m                                'found top right
End Function

Function GetBRr(TRr As Integer, TRc As Integer) As Integer
    Dim m As Integer: m = TRr
    Do While Cells(m, TRc).Value <> "BR"     'proceed down from top right
        m = m + 1
    Loop
    GetBRr = m                                'found bottom right
End Function
```

The corner locations are now in the array `Ladder` so the corner labels are no longer needed.

```
Private Sub RemoveLabels(Ladder() As TLadder)
    Dim i As Integer
    For i = 1 To idxmax
        Cells(Ladder(i).TLrow, Ladder(i).TLcol).Value = ""
        Cells(Ladder(i).TRrow, Ladder(i).TRcol).Value = ""
```

```
        Cells(Ladder(i).BLrow, Ladder(i).BLcol).Value = ""
        Cells(Ladder(i).BRrow, Ladder(i).BRcol).Value = ""
    Next i
End Sub
```

The procedure `GetLadderInOut` gets the remaining ladder information by going down the rails of the ladder until it finds the appropriate circuit pattern.

```
Sub GetLadderInOut(Ladder() As TLadder)
    Dim idx As Integer, i As Integer
    For idx = 1 To idxmax
        For i = Ladder(idx).TLrow To Ladder(idx).BLrow
            If Cells(i,Ladder(idx).TLcol-1).Interior.ColorIndex = 35 Then
                Ladder(idx).Inrow = i                    'found input
                Ladder(idx).Incol = Ladder(idx).TLcol - 1
            End If
            If Cells(i,Ladder(idx).TRcol+1).Interior.ColorIndex = 35 Then
                Ladder(idx).Outrow = i                   'found output
                Ladder(idx).Outcol = Ladder(idx).TRcol + 1
            End If
        Next i
    Next idx
End Sub
```

The final step in getting the ladder specifications is to denote the simple ladders, that is, those without inner ladders. This is done by checking if the part of the rung inside the rails is straight. If it is not then another ladder exists in the rails.

```
Private Sub DenoteSimpleLadders(Ladder() As TLadder)
Dim i As Integer, j As Integer, idx As Integer, simple As Boolean
    For i = 1 To idxmax
        Ladder(i).IsSimple = True
        For j = Ladder(i).Incol+2 To Ladder(i).Outcol-2    'inside rails
            If Cells(Ladder(i).Inrow-1, j).Interior.ColorIndex = 35 And _
               Cells(Ladder(i).Inrow+1, j).Interior.ColorIndex = 35 Then
                Ladder(i).IsSimple = False             'rung not straight:
                Exit For                              'an inner ladder exits
            End If
        Next j
    Next i
End Sub
```

Forming the Conjunctions

The procedure `FormConjunctions` runs through the simple ladders, forming the conjunction of symbols along each rung and placing them to the left of the rung. This makes for easy detection during the next step of forming disjunctions. Figure 16.11 shows the outcome of the procedure.

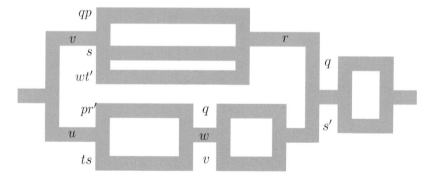

FIGURE 16.11: State of circuit after rung conjunctions are formed.

```
Sub FormConjunctions(Ladder() As TLadder)
    Dim idx As Integer, i As Integer, j As Integer
    For idx = 1 To idxmax   'check each ladder
        If Not Ladder(idx).IsSimple Then GoTo continueidx
            'go across each rung of ladder forming conjunctions
            'place them to the left of the rung
        For i = Ladder(idx).TLrow To Ladder(idx).BLrow
            For j = Ladder(idx).TLcol + 1 To Ladder(idx).TRcol - 1
                If IsEmpty(Cells(i, j)) Then GoTo continuej
                Cells(i, Ladder(idx).TLcol - 1).Value = _
                Cells(i, Ladder(idx).TLcol - 1).Value & Cells(i, j).Value
                Cells(i, j).Value = ""
continuej:
            Next j
        Next i
continueidx:
    Next idx
End Sub
```

Forming the Disjunctions

The procedure `FormDisjunctions` runs through the simple ladders, forming the disjunction of symbols along the left rail and placing them in the input cell. The conjunctions along the left rail are removed by `RemoveConjunctions`.

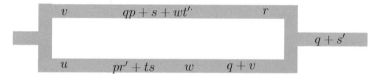

FIGURE 16.12: State of circuit after disjunctions are formed and ladders collapsed. Process is then repeated.

```
Private Sub FormDisjunctions(Ladder() As TLadder)
    Dim i As Integer, j As Integer, disj As String
    For i = 1 To idxmax                            'check each ladder
        If Not Ladder(i).IsSimple Then GoTo continue
            'go down the cells to the left of the left rail
            'and form the disjunctions
        For j = Ladder(i).TLrow To Ladder(i).BLrow
            If Not IsEmpty(Cells(j, Ladder(i).TLcol - 1)) Then
                disj = disj & "+" & Cells(j, Ladder(i).TLcol - 1).Value
            End If
        Next j
        Cells(Ladder(i).Inrow, Ladder(i).Incol).Value _
            = "(" & Mid(disj, 2) & ")"
        Call RemoveConjunctions(i, Ladder)
        Call CollapseLadder(i, Ladder)
continue:
    Next i
End Sub

Private Sub RemoveConjunctions(idx As Integer, Ladder() As TLadder)
    Dim i As Integer
    For i = Ladder(idx).TLrow To Ladder(idx).BLrow
        If i <> Ladder(idx).Inrow Then
            Cells(i, Ladder(idx).TLcol - 1).Value = ""
        End If
    Next i
End Sub
```

Collapsing an Empty Simple Ladder

The procedure CollapseLadder removes the color from the rails and rungs of the ladder, and removes symbols from the rungs as well. It then connects the input and output cell with a horizontal cross piece.

```
Private Sub CollapseLadder(i As Integer, Ladder() As TLadder)
Range(Cells(Ladder(i).TLrow, Ladder(i).TLcol), _
    Cells(Ladder(i).BRrow, Ladder(i).BRcol)).Interior.ColorIndex = 0
Range(Cells(Ladder(i).TLrow, Ladder(i).TLcol), _
    Cells(Ladder(i).BRrow, Ladder(i).BRcol)).Value = ""
Range(Cells(Ladder(i).Inrow, Ladder(i).Incol), _
    Cells(Ladder(i).Outrow, Ladder(i).Outcol)).Interior.ColorIndex = 35
End Sub
```

The Final Step

The procedure Finish forms and returns the conjunction of the expressions along the single rung

```
Function Finish() As String
Dim m As Integer, n As Integer, answer As String
```

```
For m = CTop To CBot
    For n = CLeft To CRight
        expr = Cells(m, n).Value
        If Not IsEmpty(expr) Then answer = answer & expr
    Next n
Next m
Finish = answer
End Function
```

The program can handle circuits with great complexity, limited only by the size of the grid (which can be made quite large) and, of course, computer memory. If it is desired to know the switch positions that result in current flow, the expression may be run through the program `Stmt2TruthTable`.

16.5   Exercises

1. Draw equivalent circuits that illustrate the equivalences

   (a) $(pr + r') \equiv (p + r')$ (b) $(r + q' + rq') \equiv (r + q')$

2. Describe in words a circuit with four switches p, q, r, s such that closing any one of them turns the light on.

3. Draw circuits that realize the expressions or their simplified equivalents.

   (a) $p + (q + r)'$ (b) $p(qr)'$ (c) $pq'r + p(s + t)$

4. Construct an interesting circuit and run it through the program `SwitchCircuitExpr`. Then run the resulting statement through the program `TruthTable2DisjNormal` of Section 15.4 and draw the circuit for the resulting statement.

5. Use logical equivalence to simplify the circuits pictured below.

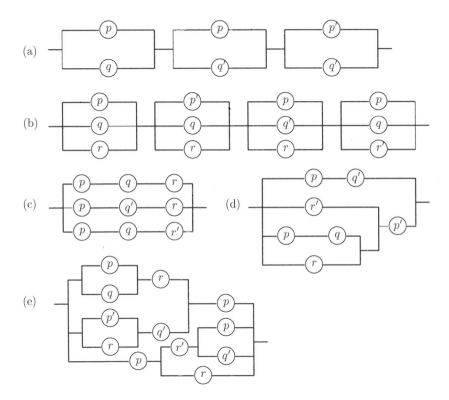

(a)

(b)

(c)

(d)

(e)

Chapter 17

Gates and Logic Circuits

A (*logic*) *gate* is a small transistor circuit that has several inputs and one output. Input and output signals correspond to the values 1 (high voltage) or 0 (no voltage). Gates may be wired together to form what is called a *logic circuit*. We show that a logic circuit may be represented by a truth table and hence by a compound statement. This opens up the possibility of using propositional algebra to simplify logic circuits. We give examples of logic circuits and show how their associated logic expressions may be generated by VBA.

17.1 The Gates NOT, AND, OR

These are the basic gates. Their circuit symbols and definitions (the input-output tables) are shown in Figure 17.1. The logic statement corresponding to a table (viewed as a truth table) is also indicated.

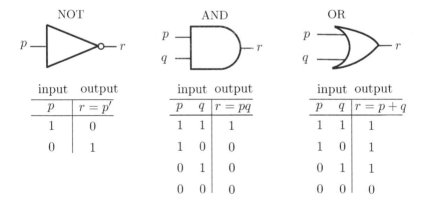

NOT input	output
p	$r = p'$
1	0
0	1

AND input		output
p	q	$r = pq$
1	1	1
1	0	0
0	1	0
0	0	0

OR input		output
p	q	$r = p + q$
1	1	1
1	0	1
0	1	1
0	0	0

FIGURE 17.1: Logic gates.

Gates may be wired together to produce logic circuits with several inputs and one output. The output values may be found by taking each set of input

DOI: 10.1201/9781003351689-17

values and tracing along the circuit from start to finish. A similar technique may be used to find the logical expression associated with a circuit. An example is given in Figure 17.2.

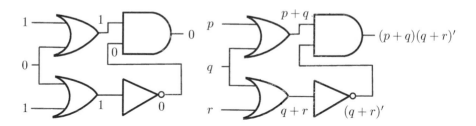

FIGURE 17.2: Input/output.

Wiring together several AND gates produces an AND gate with multiple inputs. A similar construction works for OR gates. Figure 17.3 gives the wiring for three inputs.

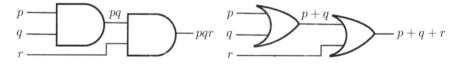

FIGURE 17.3: Multiple AND, OR gates.

17.2 The Gates XOR, NAND, NOR

The circuit symbols and definitions (the input-output tables) for these gates are shown in Figure 17.4. The output for the XOR (*exclusive or*) gate is 1 if and only if the inputs are unequal. The NAND and NOR gates are simply the negations of the AND and OR gates.

Two circuits A, B are said to be *equivalent*, written A≡ B, if they have equivalent logical statements. This simply means that identical inputs produce identical outputs. Figure 17.5 shows that the XOR gate is equivalent to a circuit consisting of AND, OR, and NOT gates.

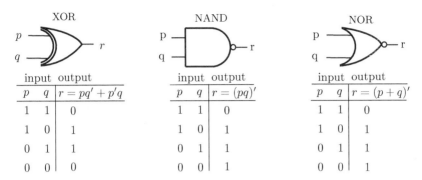

FIGURE 17.4: XOR, NAND, NOR gates.

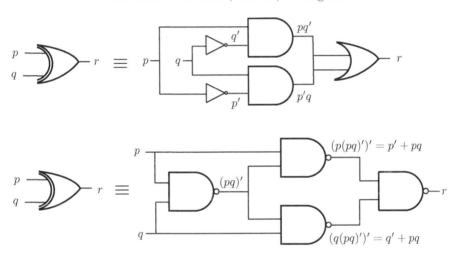

FIGURE 17.5: XOR gate equivalence.

17.3   Logic Circuit Expressions with VBA

In this section we develop a program that reads a logic circuit drawn by the user and returns the corresponding logic expression. To draw such a circuit one can highlight the desired cells of the grid using the Shift key together with the arrow keys and then click the Good button in the Style section under the Home tab. This produces a light green swath on the grid with color index number 35. The Normal button in the same section may be used to remove the color. Figure 17.6 shows a typical circuit drawn in this manner.

The gates are denoted on the spreadsheet by capital letters N (NOT), A (AND), O (OR), X (XOR) (we omit NAND and NOR gates). The cells containing the gate letters must have no background color and "wires" must be separated, not overlap and not touch the grid border. Figure 17.6 shows a typical circuit with inputs p, q, r, all entered by the user. Running the program produces the expression at the extreme right. The program populates the green

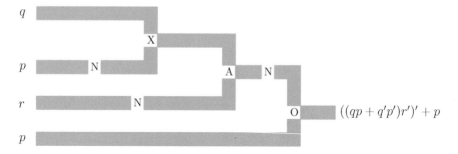

FIGURE 17.6: Logic circuit input/output.

cells with copies of the input and output expressions, removing the green color in the process. Figure 17.7 shows how the program proceeds from the initial circuit to the final expression by propagating intermediate results through the circuit, removing green along the way to keep track of progress. The

FIGURE 17.7: Logic circuit input/output.

program runs fairly quickly and accommodates circuits of essentially arbitrary complexity, limited only by the size of the grid. Here is main procedure.

```
Sub LogicCircuitExpr()
    Dim Grid As TGrid, str As String, outrow As Integer, outcol As Integer
    Grid = MakeGrid(4, 4, 50, 55, 2.1, 10.5, 24)
    Call GetCircuitOutput(Grid)                    'run the circuit
    Call GetOutPosition(outrow, outcol, Grid)  'rightmost entry position
    str = Format(Cells(outrow, outcol).Value)       'format entry there
    Cells(outrow, outcol).Value = ""                    'delete it
    Range("B3").Value = str
    MsgBox "clear"
    Call CleanCircuit(Grid)
End Sub
```

The procedure **GetOutPosition** finds the rightmost nonempty cell of the grid that holds the final (unformatted) logic expression.

```
Private Sub GetOutPosition(row As Integer, col As Integer, Grid As TGrid)
    Dim m As Integer, n As Integer
    For n = Grid.R To Grid.L Step -1
        For m = Grid.T To Grid.B
            If Not IsEmpty(Cells(m, n)) Then
                row = m: col = n: GoTo lastline
            End If
        Next m
    Next n
lastline:
End Sub
```

Here is the heart of the module. It consists of a Do While loop that continues as long as there are still empty green cells to be filled with logic expressions.

```
Private Sub GetCircuitOutput(Grid As TGrid)
    Do While NumEmptyGreen(Grid) > 0
        'send inputs and outputs scurrying along green trails
        Call PopulateCircuit(Grid)
        'get the inputs, evaluate gates, print outputs
        Call EvalGates(Grid)
    Loop
End Sub
```

The procedure `NumEmptyGreen` scans the grid, counting how many green cells remain to be populated with logic expressions.

```
Private Function NumEmptyGreen(Grid As TGrid) As Integer
    Dim m As Integer, n As Integer, num As Integer
    For m = Grid.T To Grid.B
        For n = Grid.L To Grid.R
            If CIdx(m, n) = 35 And IsEmpty(Cells(m, n)) Then
                num = num + 1
            End If
        Next n
    Next m
    NumEmptyGreen = num
End Function
```

The procedure `PopulateCircuit` scans the grid for logic expressions and moves these throughout the circuit. For this it calls the procedure `PopulateTrail`, which uses recursion to propagate the current logic expression `entry` along a branch of the circuit to the next gate or to the final output.

```
Private Sub PopulateCircuit(Grid As TGrid)
    Dim m As Integer, n As Integer, entry As String
    For m = Grid.T To Grid.B          'scan grid for expression characters
        For n = Grid.L To Grid.R
            If IsEmpty(Cells(m, n)) Then GoTo continue
            entry = Cells(m, n).Value                    'current entry
            If Asc(entry) < 65 Or Asc(entry) > 90 Then   'if not cap then
```

```
              Call PopulateTrail(m, n, entry)   'send entry along branch
           End If
continue:
         Next n
      Next m
   End Sub


   Private Sub PopulateTrail(m As Integer, n As Integer, entry As String)
      Dim roff As Integer, coff As Integer, i As Integer, j As Integer
      For roff = -1 To 1
         For coff = -1 To 1
            i = m + roff: j = n + coff
            If roff ^ 2 + coff ^ 2 = 1 And CIdx(i, j) = 35 And _
               IsEmpty(Cells(i, j)) Then
               Cells(i, j).Value = entry        'put in empty green cell
               Call PopulateTrail(i, j, entry)
            End If
         Next coff
      Next roff
   End Sub
```

The procedure `EvalGates` scans the grid for the gate letters N,A,O,X. Having found a gate, it uses the procedure `GateReady` to check whether there are sufficient inputs available. If so, it retrieves the inputs with the procedure `GateInputs`, passes these to the function `GateOuput`, which returns the required output depending on the type of gate. This value is then printed one cell to the right of the gate letter.

```
   Private Sub EvalGates(Grid As TGrid)
      Dim m As Integer, n As Integer, gate As String
      Dim in1 As String, in2 As String
      For m = Grid.T To Grid.B
         For n = Grid.L To Grid.R
               If IsEmpty(Cells(m, n)) Then GoTo continue
               If Not IsEmpty(Cells(m, n + 1)) Then GoTo continue
               gate = Cells(m, n).Value
               If 65 <= Asc(gate) And Asc(gate) <= 90 _
                     And GateReady(m, n) Then
                 Call GetGateInputs(m, n, in1, in2)
                 Cells(m, n + 1).Value = GateOutput(in1, in2, gate)
                 If pause Then MsgBox "continue"   'step through the gates?
               End If
continue:
         Next n
      Next m
   End Sub
```

The Boolean procedure `GateReady(m,n)` checks the neighbors on the input side of a gate cell at row m and column n and returns True if none of these is green, meaning that the gate has the required inputs and so is ready to be evaluated.

```
Private Function GateReady(m As Integer, n As Integer) As Integer
    Dim roff As Integer, coff As Integer, i As Integer, j As Integer
    Dim ready As Boolean
    ready = True
    For roff = -1 To 1
        For coff = -1 To 0
            i = m + roff: j = n + coff
            If roff ^ 2 + coff ^ 2 = 1 And CIdx(i, j) = 35 And _
                IsEmpty(Cells(i, j)) Then
                    ready = False   'found green empty neighbor:inputs lacking
                    GoTo lastline                      'gate not ready--try later
            End If
        Next coff
    Next roff
lastline:
    GateReady = ready
End Function
```

The procedure `GetGateInputs` checks the neighbors to the left of a gate cell for inputs and places them in the variables `in1,in2`.

```
Sub GetGateInputs(m As Integer, n As Integer, in1 As String, in2 As String)
    Dim roff As Integer, coff As Integer
    If Not IsEmpty(Cells(m, n - 1)) Then
        in1 = Cells(m, n - 1).Value: in2 = ""
    ElseIf Not IsEmpty(Cells(m - 1, n)) And _
            Not IsEmpty(Cells(m + 1, n)) Then
        in1 = Cells(m + 1, n).Value: in2 = Cells(m - 1, n).Value
    End If
End Sub
```

The procedure `GateOutput` forms the required expression from the inputs and returns a formatted version.

```
Function GateOutput(in1 As String, in2 As String, gate As String) As String
    Dim out As String
    in1 = "(" & in1 & ")"            'surround expressions with parentheses
    in2 = "(" & in2 & ")"
    Select Case gate
        Case Is = "N": out = in1 & "'"
        Case Is = "A": out = in1 & in2
        Case Is = "O": out = "(" & in1 & "+" & in2 & ")"
        Case Is = "X": out = "(" & in1 & in2 & "'" & _
                             "+" & in2 & in1 & "'" & ")"
    End Select
    out = Format(out)                          'remove extraneous parens
    GateOutput = "(" & out & ")"        'still might need them here though
End Function
```

The function `Format` takes the string returned by `GateOutput` and removes unnecessary parentheses. The procedure `GetMatchingParens` finds the matching parentheses, which are then removed by four functions.

```
Function Format(str As String) As String
    Dim lp() As Integer, rp() As Integer
    'memory for left paren positions and matching right paren positions
    ReDim lp(1 To Len(str) + 1): ReDim rp(1 To Len(str) + 1)
    Call GetMatchingParens(str, lp, rp)
    str = RemoveParens1(str, lp, rp)      'conversion: ((...)) -> (...)
    str = RemoveParens2(str, lp, rp)      'conversion: ((...)') -> (...)'
    str = RemoveParens3(str)                    'conversion: (p) -> p
    str = RemoveParens4(str)                    'conversion: (p') -> p'
    str = Replace(str, "'", "")
    Format = str
End Function
```

The procedure `GetMatchingParens(str,lp,rp)` fills the left parenthesis array `lp` and the right parenthesis array `rp` with the positions in `str` of matching parentheses, as demonstrated in Figure 17.8.

$$lp(1)\ lp(2)\ lp(3)\ rp(3)\quad lp(4)\ lp(5)\ rp(5)\qquad rp(4)\ rp(2)\ rp(1)$$
$$(\quad(\quad(\ p\ )\ +\ (\quad(\ q'\ )\ +\ r\ )'\quad)\quad)$$

FIGURE 17.8: Matching parentheses.

```
Sub GetMatchingParens(str As String, lp() As Integer, rp() As Integer)
    Dim i As Integer, j As Integer, k As Integer: k = 1
    Dim NumLeft As Integer, NumRight As Integer
    For i = 1 To Len(str)                        'find left paren positions
        If Mid(str, i, 1) = "(" Then                'found a left paren at i
            NumLeft = 0
            NumRight = 0
            For j = i To Len(str)      'count left, right parens until equal
                If Mid(str, j, 1) = "(" Then NumLeft = NumLeft + 1
                If Mid(str, j, 1) = ")" Then NumRight = NumRight + 1
                    'if a left-right paren match at j,
                If NumLeft = NumRight Then
                    lp(k) = i: rp(k) = j   'record matching pair positions
                    k = k + 1: Exit For    'go to next left paren at new i
                End If
            Next j
        End If
    Next i
End Sub
```

The procedure `RemoveParens1(str,lp,rp)` find patterns $((\cdots))$ and deletes the outer parentheses. For example, it transforms $(((p) + ((q') + r)'))$ into $((p) + ((q') + r)')$.

```
Function RemoveParens1(str As String, lp() As Integer, rp() As Integer) _
        As String
    Dim k As Integer
    For k = 1 To Len(str)              'find double parens and delete one
        If lp(k) <> 0 And rp(k) <> 0 Then
            If lp(k + 1) = lp(k) + 1 And rp(k + 1) = rp(k) - 1 Then
                'found ((...)) so  delete outers
                str = RemoveInsertString(str, lp(k), " ")
                str = RemoveInsertString(str, rp(k), " ")
            End If
        End If
    Next k
    'don't remove whitespace yet: messes up paren positions
    RemoveParens1 = str
End Function
```

The procedure `RemoveParens2(str,lp,rp)` find patterns $((\cdots)')$ and deletes the outer parentheses. For example, it transforms $((p) + ((q') + r)')$ into $(p) + ((q') + r)'$.

```
Function RemoveParens2(str As String, lp() As Integer, rp() As Integer) _
        As String
    Dim k As Integer
    For k = 1 To Len(str)   'find double parens with prime and delete one
        If lp(k + 1) = lp(k) + 1 And rp(k + 1) = rp(k) - 2 And _
                'found ((...)') so delete outers
            Midd(str, rp(k) - 1, 1) = "'" Then
                str = RemoveInsertString(str, lp(k), " ")
                str = RemoveInsertString(str, rp(k), " ")
            End If
    Next k
    RemoveParens2 = RemoveWhiteSpace(str)
End Function
```

The procedure `RemoveParens3(str)` removes parens around a single character For example, it transforms $(p) + ((q') + r)'$ into $p + ((q') + r)'$.

```
Function RemoveParens3(str As String) As String
    Dim i As Integer
    For i = 1 To Len(str)              'delete parens around single char
        If Midd(str, i, 1) = "(" And Midd(str, i + 2, 1) = ")" Then
            str = RemoveInsertString(str, i, " ")
            str = RemoveInsertString(str, i + 2, " ")
        End If
    Next i
    RemoveParens3 = RemoveWhiteSpace(str)
End Function
```

The procedure `RemoveParens4(str)` removes parens around a single character with a prime. For example, $p + ((q') + r)'$ becomes $p + (q' + r)'$.

```
Function RemoveParens4(str As String) As String
    Dim i As Integer
    For i = 1 To Len(str)    'delete parens around single char with prime
        If Midd(str, i, 1) = "(" And Midd(str, i + 2, 1) = "'" And _
        Midd(str, i + 3, 1) = ")" Then
            str = RemoveInsertString(str, i, " ")
            str = RemoveInsertString(str, i + 3, " ")
        End If
    Next i
    RemoveParens4 = RemoveWhiteSpace(str)
End Function
```

Removes unnecessary outer parentheses:

```
Function RemoveOuterParens(str As String) As String
    If Midd(str, 1, 1) = "(" And Midd(str, Len(str), 1) = ")" Then
        str = Mid(str, 2, Len(str) - 2)
    End If
    RemoveOuterParens = str
End Function
```

This procedure `CleanCircuit` restores the original circuit, removing the litter left on the trails by the program.

```
Private Sub CleanCircuit(Grid As TGrid)
    Dim m As Integer, n As Integer
    For m = Grid.T To Grid.B
        For n = Grid.L To Grid.R
            If CIdx(m, n) = 35 Then Cells(m, n).Value = ""
        Next n
    Next m
End Sub
```

17.4   Half Adders

XOR gates are used to build computer circuits that perform arithmetic of *binary numbers*. A binary (or base 2) number is a sequence of 1's and 0's, called *binary digits* or *bits*, that stand for digits multiplying powers of 2, just like ordinary (base 10) numbers are sequences of digits 0–9 that multiply powers of 10. For example, the binary number 1011 represents the base 10 number $1 \times 2^3 + 0 \times 2^2 + 1 \times 2^1 + 1 \times 2^0 = 11$. (Number bases are discussed in detail in Section 23.2.)

Binary numbers are added just like base 10 numbers. For example, adding the one-digit binary numbers 1 and 1 produces the binary number 10, which is the decimal number 2. For the purpose of the programs that follow, the zero

c	c	c	c
1	0	0	0

$$\begin{array}{c} 1 \\ \downarrow\,1 \\ \hline 1\ 0 \\ s \end{array} \qquad \begin{array}{c} 1 \\ 0 \\ \hline 1 \\ s \end{array} \qquad \begin{array}{c} 0 \\ 1 \\ \hline 1 \\ s \end{array} \qquad \begin{array}{c} 0 \\ 0 \\ \hline 0 \\ s \end{array}$$

FIGURE 17.9: Four additions of two 1-bit numbers.

in 10 is called the *sum*, denoted by s, and the 1 is the *carry*, denoted by c. Figure 17.9 gives the complete scheme for adding two 1-digit binary numbers.

The half-adder adds two binary digits. The input-output definition of the half-adder is given in Figure 17.10. The binary number output may be obtained by juxtaposing the digits in the carry and sum columns.

p	q	c	s	binary no.
1	1	1	0	10
1	0	0	1	01
0	1	0	1	01
0	0	0	0	00

FIGURE 17.10: Half-adder definition.

The circuit for the half-adder is given in Figure 17.11

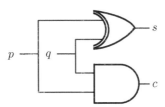

FIGURE 17.11: Half-adder circuit.

17.5   VBA Simulation of a Half Adder

The program in this section takes inputs 1 or 0 and produces the half adder outputs. The grid circuit is shown in Figure 17.12, where p and q represent the inputs. The user enters 1's and 0's here. These bits are then propagated

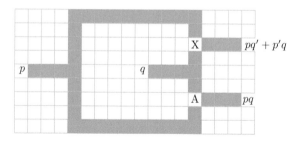

FIGURE 17.12: Half-adder spreadsheet circuit.

along the green trails, the color eliminated along the way. When the bits reach the gates they are evaluated and sent along the trails to the right, eventually ending their journeys. Figure 17.13, shows the circuit with inputs 1 and 1.

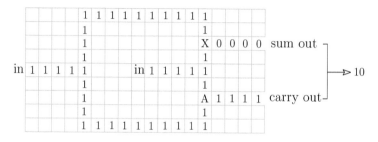

FIGURE 17.13: Half-adder propagation.

The program is essentially the same as the module in Section 17.3 with the following exceptions: (1) `LogicCircuitExpr` is replaced by `HalfAdder` below, (2) `GateOutput` is modified as shown below, (3) the procedures `GetOutPosition` and `Format` are omitted. Here is the code needed:

```
Sub HalfAdder()
    Dim Grid As TGrid, str As String
    Grid = SetGrid(5, 4, 30, 30, 2.1, 10.5): Call AddBorder(Grid, 24)
    Call GetCircuitOutput(Grid)                     'run the circuit
    MsgBox "restore circuit"
    Call CleanCircuit(Grid)
End Sub

Function GateOutput(in1 As String, in2 As String, gate As String) As String
    Dim out As String
    If gate = "A" Then out = CInt(i1) * CInt(in2)
    If gate = "X" Then out = Abs(CInt(in1) - CInt(in2))
    GateOutput = out
End Function
```

17.6 Full Adders

A *full adder* is like a half adder but has a carry input as well. This is added to the other two inputs to produce the binary sum of three binary digits. Figure 17.14 gives the definition of a 1-bit full adder. Again, juxtaposing the digits in the carry out and sum columns yields the binary number. Figure 17.15

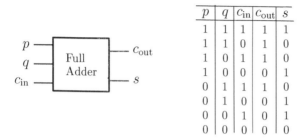

p	q	c_{in}	c_{out}	s
1	1	1	1	1
1	1	0	1	0
1	0	1	1	0
1	0	0	0	1
0	1	1	1	0
0	1	0	0	1
0	0	1	0	1
0	0	0	0	0

FIGURE 17.14: Full-adder definition.

shows the circuit of the full adder as a combination of two half-adders and an OR gate.

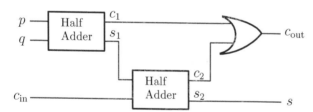

FIGURE 17.15: Full-adder circuit.

By combining 1-bit full adders in series one may add two binary numbers of any length. For 4-bit numbers the circuit is based on the following standard addition algorithm.

```
carries:  1  1  1  1          c_out  c_3  c_2  c_1

          1  1  0  1                 p_3  p_2  p_1  p_0
          1  0  1  1                 q_3  q_2  q_1  q_0
sums:     1  0  0  0                 s_3  s_2  s_1  s_0
```

FIGURE 17.16: Addition of two 4-bit numbers.

Notice that the concrete example would result in an overflow for an 4-bit machine.

Here is the circuit for a 4-bit adder in terms of 1-bit full adders. There is a c_{in} input in case this is the second part of an 8-bit adder formed from two 4-bit adders. Similarly, there is a c_{out} output in case this is the first part of an 8-bit adder. With this scheme one can wire adders together to form adders with arbitrarily many bits, subject only to memory restrictions.

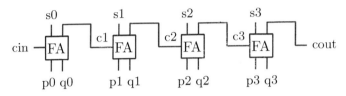

FIGURE 17.17: Circuit for addition of two 4-bit numbers.

17.7   VBA Simulation of a 4-bit Adder

In this section we develop a program that simulates a 4-bit adder. This is accomplished by wiring together four full adder circuits as defined in Figure 17.15. Figure 17.18 shows the spreadsheet circuit. The c's and s's in the

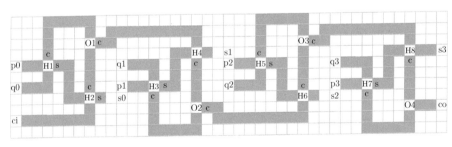

FIGURE 17.18: Spreadsheet circuit for addition of two 4-bit numbers.

diagram are the outputs of the intermediate gates and are omitted in the actual spreadsheet. The symbols $p0, p1, p2, p3$ and $q0, q1, q2, q3$ represent the digits of input binary numbers $p3p3p1p0$ and $q3q2q1q0$. The symbol ci is the input carry from the carry output of another (fictitious) 4-bit adder. The symbols $s0, s1, s2, s3$ represent the final outputs. These are concatenated to produce the binary sum of the inputs.

Here's how the program works. The user enters the carry input and the binary numbers $p3p2p1p0$ and $q3q2q1q0$ into cells B4, B5 and B6, respectively. The digits of the numbers are extracted and placed in the required spreadsheet locations. Running the program causes the bits to propagate along the green trails leading to the half-adders and OR gates. These are evaluated and the output printed to their right. The final sum is printed in cell B7. Figure 17.19 shows the process in midstream for the first part of the circuit. The module

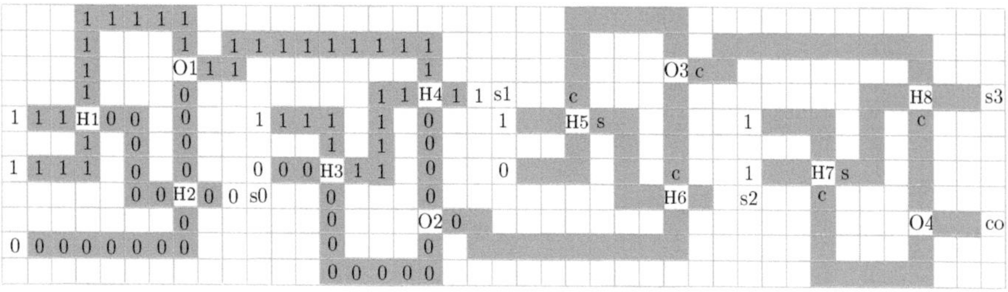

FIGURE 17.19: Propagation of bits in a 4-adder.

has numerous Public variables that are used locate the positions of the inputs, outputs, gates and half-adders. The suffixes r and c are used in these for row and column designations. The prefixes p, q are used in the locations of the input digits, and s in the location of the output digits. For example, $H1c$ is the column number of the first half adder, $p3r$ the row number of the first digit in the first input number, and $s0c$ the column number of the last digit in the output number. The letters ci and co refer to input and output carry digits, respectively.

```
Public cir As Integer, cic As Integer, cor As Integer, coc As Integer
Public p0r As Integer, p0c As Integer, p1r As Integer, p1c As Integer
Public p2r As Integer, p2c As Integer, p3r As Integer, p3c As Integer
Public q0r As Integer, q0c As Integer, q1r As Integer, q1c As Integer
Public q2r As Integer, q2c As Integer, q3r As Integer, q3c As Integer
Public s0r As Integer, s0c As Integer, s1r As Integer, s1c As Integer
Public s2r As Integer, s2c As Integer, s3r As Integer, s3c As Integer
Public H1r As Integer, H1c As Integer, H2r As Integer, H2c As Integer
Public H3r As Integer, H3c As Integer, H4r As Integer, H4c As Integer
Public H5r As Integer, H5c As Integer, H6r As Integer, H6c As Integer
Public H7r As Integer, H7c As Integer, H8r As Integer, H8c As Integer
Public O1r As Integer, O1c As Integer, O2r As Integer, O2c As Integer
Public O3r As Integer, O3c As Integer, O4r As Integer, O0c As Integer
Public Binp As String, Binq As String          'binary numbers input
Public ci As String, co As String     'carry in and carry out variables
Public delayval As Integer    'delay value to slow down visual display
```

The module uses the procedures `CleanCircuit`, `GetCircuitOutput`, and `PopulateCircuit` of earlier sections. Here is the procedure that launches the circuit activity.

```
Sub FullAdder4()
    Dim Grid As TGrid, str As String, outrow As Integer, outcol As Integer
    Grid = MakeGrid(4, 4, 16, 62, 2.1, 10.5, -1)
    delayval = Range("B3").Value                'slow down the visuals?
```

```
        ci = Range("B4").Value                  'input carry, usually set to 0
        Binp = Range("B5").Value
        Binq = Range("B6").Value                          'binary inputs
        Call GetLocations(Grid)    'locations of gates, inputs, outputs, etc.
        Call DistributeBits
        Call GetCircuitOutput(Grid)                       'run the circuit
        Call OutputSum
        Call CleanCircuit(Grid)
    End Sub
```

The sub `GetLocations` finds the row, column numbers of input digits, gates, etc.

```
Private Sub GetLocations(Grid As TGrid)
    Dim m As Integer, n As Integer
    For m = Grid.T To Grid.B  'scan grid to get positions of the letters
        For n = Grid.L To Grid.R    'of the half-adders, variables, etc.
            If IsEmpty(Cells(m, n)) Then GoTo continue
                Select Case Cells(m, n).Value
                    Case Is = "H1": H1r = m:   H1c = n
                    Case Is = "H2": H2r = m:   H2c = n
                    Case Is = "H3": H3r = m:   H3c = n
                    Case Is = "H4": H4r = m:   H4c = n
                    Case Is = "H5": H5r = m:   H5c = n
                    Case Is = "H6": H6r = m:   H6c = n
                    Case Is = "H7": H7r = m:   H7c = n
                    Case Is = "H8": H8r = m:   H8c = n
                    Case Is = "O1": O1r = m:   O1c = n
                    Case Is = "O2": O2r = m:   O2c = n
                    Case Is = "O3": O3r = m:   O3c = n
                    Case Is = "O4": O4r = m:   O4c = n
                    Case Is = "p0": p0r = m:   p0c = n
                    Case Is = "p1": p1r = m:   p1c = n
                    Case Is = "p2": p2r = m:   p2c = n
                    Case Is = "p3": p3r = m:   p3c = n
                    Case Is = "q0": q0r = m:   q0c = n
                    Case Is = "q1": q1r = m:   q1c = n
                    Case Is = "q2": q2r = m:   q2c = n
                    Case Is = "q3": q3r = m:   q3c = n
                    Case Is = "s0": s0r = m:   s0c = n
                    Case Is = "s1": s1r = m:   s1c = n
                    Case Is = "s2": s2r = m:   s2c = n
                    Case Is = "s3": s3r = m:   s3c = n
                    Case Is = "co": cor = m:   coc = n
                    Case Is = "ci": cir = m:   cic = n
                End Select
continue:
            Next n
        Next m
End Sub
```

The procedure `DistributeBits` extracts the digits of the binary input numbers `Binp,Binq` entered in cells A5 and A6 and places them in their designated locations on the spreadsheet.

```
Sub DistributeBits()
    Dim p3 As String, p2 As String, p1 As String, p0 As String
    Dim q3 As String, q2 As String, q1 As String, q0 As String
    p3 = Mid(Binp, 1, 1): p2 = Mid(Binp, 2, 1)        'get digits of Binp
    p1 = Mid(Binp, 3, 1): p0 = Mid(Binp, 4, 1)
    q3 = Mid(Binq, 1, 1): q2 = Mid(Binq, 2, 1)        'get digits of Binq
    q1 = Mid(Binq, 3, 1): q0 = Mid(Binq, 4, 1)
    Cells(p0r, p0c).Value = p0: Cells(q0r, q0c).Value = q0 'place digits
    Cells(p1r, p1c).Value = p1: Cells(q1r, q1c).Value = q1       'in the
    Cells(p2r, p2c).Value = p2: Cells(q2r, q2c).Value = q2  'appropriate
    Cells(p3r, p3c).Value = p3: Cells(q3r, q3c).Value = q3      'locations
    Cells(cir, cic).Value = ci
End Sub
```

The procedure `GateReady` returns True if there are two inputs for the gate at cell position m, n, these positions being one offset away from the gate location.

```
Private Function GateReady(m As Integer, n As Integer, gate As String) _
        As Boolean
    Dim in1r As Integer, in1c As Integer, inr2 As Integer, in2c As Integer
    Select Case gate             'get row, column numbers of gate inputs
        Case Is = "H1": in1r = m:    in1c = n - 1: in2r = m + 1: in2c = n
        Case Is = "H2": in1r = m:    in1c = n - 1: in2r = m + 1: in2c = n
        Case Is = "H3": in1r = m:    in1c = n - 1: in2r = m - 1: in2c = n
        Case Is = "H4": in1r = m:    in1c = n - 1: in2r = m - 1: in2c = n
        Case Is = "H5": in1r = m:    in1c = n - 1: in2r = m + 1: in2c = n
        Case Is = "H6": in1r = m:    in1c = n - 1: in2r = m + 1: in2c = n
        Case Is = "H7": in1r = m:    in1c = n - 1: in2r = m - 1: in2c = n
        Case Is = "H8": in1r = m:    in1c = n - 1: in2r = m - 1: in2c = n
        Case Is = "O1": in1r = m - 1: in1c = n   : in2r = m + 1: in2c = n
        Case Is = "O2": in1r = m - 1: in1c = n   : in2r = m + 1: in2c = n
        Case Is = "O3": in1r = m - 1: in1c = n   : in2r = m + 1: in2c = n
        Case Is = "O4": in1r = m - 1: in1c = n   : in2r = m + 1: in2c = n
    End Select
    GateReady = Not(IsEmpty(Cells(in1r,in1c)) Or IsEmpty(Cells(in2r,in2c)))
End Function
```

If the gate at cell position (m, n) is ready, then the procedure `GetGateInputs` retrieves the inputs for the gate.

```
Sub GetGateInputs(m As Integer, n As Integer, in1 As String, _
                in2 As String, gate As String)
    Dim gatetype As String
    gatetype = Mid(gate, 1, 1)                        'first letter is H or O
    If gatetype = "O" Then
        in1 = Cells(m - 1, n).Value: in2 = Cells(m + 1, n).Value
    ElseIf gatetype = "H" Then
```

```
        in1 = Cells(m, n - 1).Value
        Select Case gate
           Case Is = "H1":  in2 = Cells(m + 1, n).Value
           Case Is = "H2":  in2 = Cells(m + 1, n).Value
           Case Is = "H3":  in2 = Cells(m - 1, n).Value
           Case Is = "H4":  in2 = Cells(m - 1, n).Value
           Case Is = "H5":  in2 = Cells(m + 1, n).Value
           Case Is = "H6":  in2 = Cells(m + 1, n).Value
           Case Is = "H7":  in2 = Cells(m - 1, n).Value
           Case Is = "H8":  in2 = Cells(m - 1, n).Value
        End Select
     End If
End Sub
```

The procedure GateOutputs takes the inputs of a gate at cell position (m, n) and prints the output in the appropriate position outside the gate.

```
Sub GateOutputs(m As Integer, n As Integer, in1 As String, in2 As String, _
              gate As String)
    Dim gatetype As String, OutO As String, OutC As String, OutS As String
    gatetype = Mid(gate, 1, 1)                    'first letter is H or O
    OutO = CInt(in1) + CInt(in2)
    OutC = CInt(in1) * CInt(in2)
    OutS = Abs(CInt(in1) - CInt(in2))
    If gatetype = "O" Then
       Cells(m, n + 1).Value = OutO
    ElseIf gatetype = "H" Then
       Cells(m, n + 1) = OutS
       Select Case gate
          Case Is = "H1": Cells(m - 1, n).Value = OutC
          Case Is = "H2": Cells(m - 1, n).Value = OutC
          Case Is = "H3": Cells(m + 1, n).Value = OutC
          Case Is = "H4": Cells(m + 1, n).Value = OutC
          Case Is = "H5": Cells(m - 1, n).Value = OutC
          Case Is = "H6": Cells(m - 1, n).Value = OutC
          Case Is = "H7": Cells(m + 1, n).Value = OutC
          Case Is = "H8": Cells(m + 1, n).Value = OutC
       End Select
    End If
End Sub
```

The last procedure reads the cells containing the digits referring to each output sum and carry, concatenates them, and prints the result.

```
Sub OutputSum()
    Dim s3 As String, s2 As String, s1 As String, s0 As String
    s0 = Cells(s0r, s0c - 2).Value: s1 = Cells(s1r, s1c - 2).Value
    s2 = Cells(s2r, s2c - 2).Value: s3 = Cells(s3r, s3c - 2).Value
    co = Cells(cor, coc - 2).Value
    Range("B7").Value = co & s3 & s2 & s1 & s0
End Sub
```

17.8   Exercises

1. Construct a circuit for each of the Boolean expressions:

   (a) $q + pr$ (b) $q(p + r)$ (c) $p'q + q'p + r$

2. Find the output Boolean expression in each of the circuits.

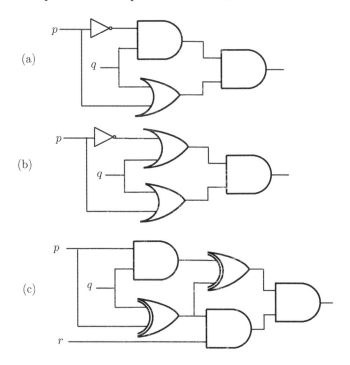

3. Write a program `FullAdder8` that simulates an 8-bit adder. You will need to wire together two circuits like the one used in `FullAdder4`. Figure 17.20 shows the circuit split into two pieces. Attach these at the cells labeled s3, p4, q4.

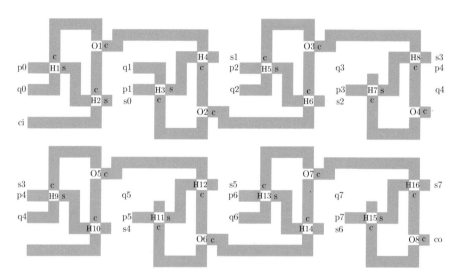

FIGURE 17.20: Spreadsheet circuit for addition of two 8-bit numbers.

Part IV

Combinatorics

Chapter 18

Sets

The notion of set is familiar to everyone. Indeed, in the preceding chapters we have frequently used the concept implicitly. In this chapter we take a closer look at the notion, discussing set operations, Venn diagrams, and other related ideas.

18.1 Introduction

A *set* is a collection of objects called the *members* or *elements* of the set. Abstract sets are usually denoted by capital letters and members of sets by small letters. We write $x \in A$ if x is a member of the set A and $x \notin A$ otherwise. The *empty set*, denoted by $\emptyset$, is the set with no members. While this may seem like an odd notion, it is as important in set theory as the number zero is in arithmetic, and indeed plays a similar role.[1]

Sets with just a few members may described by listing the elements between braces. For example, the set of even integers greater than 1 and less than 9 may be described by the notation $\{2, 4, 6, 8\}$. The listing technique may also be used for some infinite sets. For example, the set of all positive even integers may be written as $\{2, 4, 6, \ldots\}$. The three dots are called an *ellipsis* and indicate an omission of information that may be inferred from context, although admittedly this device could introduce some ambiguity.

Another way to describe sets is by *set-builder notation*. This has the form

$$\{x \mid P(x)\},$$

which is read "the set of all x such that $P(x)$." Here $P(x)$ is a well-defined property that x must satisfy in order to belong to the set. For example, the set of all even positive integers may be described in set builder notation as

$$\{n \mid n = 2m \text{ for some positive integer } m\}.$$

A set A is a *subset* of a set B, written $A \subseteq B$, if every member of A is a

---

[1]There is, of course, an obvious connection: zero is the number of elements in $\emptyset$.

member of B. If A is not a subset of B, that is, if there is a member of A that is not in B, we write $A \not\subseteq B$. For example,

$$\{1,2,3\} \subseteq \{1,2,3,4\} \quad \text{and} \quad \{1,2,3\} \not\subseteq \{2,3,4,5\}.$$

Note that $\emptyset \subseteq A$ for all sets A. (Can you find a member of $\emptyset$ not in A?)

Two sets A and B are said to be *equal*, written $A = B$, if $A \subseteq B$ and $B \subseteq A$, that is, if they have precisely the same members. For example,

$$\{1,2,2,3,3,3\} = \{1,2,3\} \quad \text{and} \quad \{x \mid x^2 = 4\} = \{2,-2\}.$$

The first example shows that members of a set need not be listed more than once.

Relationships among sets may sometimes be illustrated to good effect by a schematic device called a *Venn diagram*. Here, sets are depicted as closed regions in a rectangle. Figure 18.1 is a Venn diagram illustrating various membership and inclusion relations. The rectangle depicts a *universal set U*, that is, a set containing all members in a particular discussion. Universal sets, though not unique, are frequently left unspecified, as their choice is usually clear from context. For example, in ordinary algebra the universal set is the set of all real numbers.

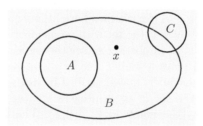

FIGURE 18.1: $A \subseteq B$, $C \not\subseteq B$, $x \in B$, $x \notin A$.

18.2   The Union of Two Sets

Sets may be combined in various ways, resulting in an algebra of sets analogous to the algebra of propositions discussed in Chapter 15. The standard operations on sets are *union, intersection,* and *complement*. In this section we treat the first of these.

The *union* of sets A and B is defined with set builder notation as

$$A + B = \{x \mid x \in A \ \text{ or } \ x \in B\}.$$

Thus for something to be in the union of two sets it must be in at least one of them. For example, if

$$A = \{1, 2, 3, 4, 5\} \text{ and } B = \{4, 5, 6, 7\},$$

then

$$A + B = \{1, 2, 3, 4, 5, 6, 7\}.$$

The more traditional notation for the union is $A \cup B$. We have chosen the notation $A + B$ to emphasize the connection between set theory and the propositional calculus and also to be able to enter set operations in a spreadsheet.

The function Union below takes as input two strings setA and setB that represent sets and returns a string that represents the union. For example, the statement Union("1,2,3","3,4,5") returns the string "1,2,3,4,5". The input strings are concatenated with a comma separator to form the output string NewSet. The function DeleteDupStr, discussed in Section 6.8, removes duplicate members from NewSet.

```
Function Union(setA As String, setB As String) As String
    Dim NewSet As String
    NewSet = setA & "," & setB
    Union = DeleteDupStr(NewSet, ",")        'remove duplcates from union
End Function
```

18.3   The Intersection of Two Sets

The *intersection* of sets A and B, denoted by AB, is defined by

$$AB = \{x \mid x \in A \text{ and } x \in B\}.$$

Thus for something to be in the intersection of two sets it must be in both of them. For example, if

$$A = \{1, 2, 3, 4, 5\} \text{ and } B = \{4, 5, 6, 7\},$$

then

$$AB = \{4, 5\}.$$

An alternate notation for intersection is $A \cap B$. We have chosen the notation AB for reasons similar to those regarding union.

The function Intersection is analogous to Union. It takes strings representing sets A and B as input and returns a string representing AB. Thus the statement Intersect("1,2,3,4,5,6", "2,3,4,7") returns the string "2,3,4". The procedure uses the function MidDelim discussed in Section 6.2 to find members of the sets. It also uses the procedure `Card` (short for "Cardinality") to find the number of distinct elements of a set. It does so by converting the comma-delimited set string to an array and then adding 1 to the upper index of the array (whose indices start at 0).

```
Function Intersection(ByVal setA As String, ByVal setB As String) As String
    Dim i As Integer, j As Integer, NewSet As String, memberA As String
    setA = DeleteDupStr(setA, ",")
    setB = DeleteDupStr(setB, ",")
    For i = 1 To Card(setA)              'run through members of setA
        memberA = MidDelim(setA, i, ",")      'get a member from setA
        For j = 1 To Card(setB)          'check if it is also in setB
            If memberA = MidDelim(setB, j, ",") Then
                NewSet = NewSet & memberA & "," 'attach common member
                Exit For
            End If
        Next j
    Next i
    If Len(NewSet) = 0 Then
        Intersection = ""
    Else
        Intersection = Mid(NewSet, 1, Len(NewSet) - 1)   'avoid last comma
    End If
End Function

Function Card(ByVal setA As String) As Integer
    Dim count As Integer, setArray() As String
    If setA = "" Then GoTo lastline
    setArray = Split(setA, ",")
    count = UBound(setArray) + 1
lastline:
    Card = count
End Function
```

18.4   The Complement of a Set

The *complement* A' of a set A is the set

$$A' = \{x \in U \mid x \notin A\}.$$

Here U is a universal set so the complement depends on which universal set is chosen. Again, either this is specified or is clear from context.

More generally, the set AB' is called the *relative complement of B in A* and consists of those members of A that are not in B. For example, if

$$A = \{1, 2, 3, 4, 5\} \text{ and } B = \{4, 5, 6, 7\}$$

then
$$AB' = \{1, 2, 3\}, \text{ and } A'B = \{6, 7\}.$$
Figure 18.2 illustrates the operation.

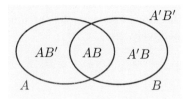

FIGURE 18.2: Venn diagram for two sets.

The function `Complement(A,B)` takes strings representing sets A and B as input and returns a string representing AB', the relative complement of B in A. For example, the instruction `Complement("1,2,3,4,5,6","2,3,4,7,8")` returns the string `"1,5,6"`.

```
Function Complement(setA As String, setB As String) As String
    Dim i As Integer, j As Integer, NewSet As String, memberA As String
    Dim FoundCommon As Boolean
    setA = DeleteDupStr(setA, ",")
    setB = DeleteDupStr(setB, ",")
    For i = 1 To Card(setA)                     'run through members of setA
        FoundCommon = False
        memberA = MidDelim(setA, i, ",")                'get a member of A
        For j = 1 To Card(setB)         'check if memberA is also in setB
            If memberA = MidDelim(setB, j, ",") Then        'if so, then
                FoundCommon = True                          'acknowledge
                Exit For
            End If
        Next j
        If Not FoundCommon Then                     'if not in B, then
            NewSet = NewSet & memberA & ","     'keep members of A not in B
        End If
    Next i
    If Len(NewSet) = 0 Then
        Complement = ""
    Else
        Complement = Mid(NewSet, 1, Len(NewSet) - 1)    'avoid last comma
    End If
End Function
```

18.5   Extensions to Three or More Sets

The notions of union and intersection extend to arbitrarily many sets. The definitions are essentially the same. For example, the union and intersection of three sets are defined, respectively, as

$$A + B + C = \{x \mid x \in A \ \text{ or } \ x \in B \ \text{ or } \ x \in C\}.$$

and

$$ABC = \{x \mid x \in A \ \text{ and } \ x \in B \ \text{ and } \ x \in C\}.$$

One can even define the union and intersection of infinitely many sets $A_1, A_2, \ldots$:

$$A_1 + A_2 + \cdots = \{x \mid x \in A_n \ \text{ for some } n\}.$$

and

$$A_1 A_2 \cdots = \{x \mid x \in A_n \ \text{ for every } n\}.$$

Such sets occur frequently in probability theory (see Chapter 20).

A *partition* of a set S is a collection of nonempty sets $S_1, S_2, \ldots S_n$ whose union is S and no two of the sets intersect. Figure 18.3 shows a Venn diagram with three sets partitioned into component parts. Venn diagrams for more than three sets are usually impractical (but see Exercise 2).

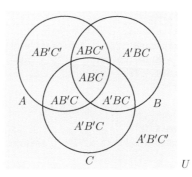

FIGURE 18.3: Venn diagram for three sets.

18.6   Calculating Sets with VBA

The program in this section takes a set expression such as $(AC + D'B)'$, where the capital letters are symbols for sets entered into the spreadsheet by the user, and prints out the members of the set. An example is given in the spreadsheet depiction Figure 18.4. The program assumes that the universal set U is the first set entered. The module is similar to `MatrixCalculator` (Section 13.9) but easier to code since the operations are simpler and strings rather than matrices are retrieved from the queue.

	A	B	C
3	Expression	$(AB'C + D'E' + A'BE)'$	
5	Sets	U	$1, 2, 3, 4, 5, 6, a, b, c$
6		A	$1, 2, 3$
7		B	$3, 4, 5$
8		C	$1, 2, 3, 4, 5, a, b$
9		D	$1, 2$
10		E	$2, 3, 4, 5, 6$
12		Answer	$3, 6$

FIGURE 18.4: Input-output example for SetCalculator.

The main procedure `SetCalculator` begins by reading from the spreadsheet the set expression and the universal set into the string variables `Expr` and `U`, respectively. Whitespace in the former and duplicate entries in the latter are then removed. The procedure inserts asterisks between pairs of letters so that the program can easily detect intersections. The code for this is similar to the analogous procedure in `MatrixCalculator`, hence the code is omitted. The main procedure then calls the calculation controller `SetEval`, which returns the desired set.

```
Public Expr As String, Idx As Integer, U As String, Error As Boolean
Sub SetCalculator()
    Dim setX As String, ansrow As Integer
    Range("B4").Value = ""                       'erase last error msg, if any
    U = DeleteDupStr(Range("C5").Value, ",")
    Expr = RemoveWhiteSpace(Range("B3").Value)
    InsertAsterisks()                            'used to detect intersection
    Idx = 1                                      'initializations
    Error = False
    setX = SetEval(0)
        'last non blank row in set symbol col:
    ansrow = 2 + Cells(Rows.count, 2).End(xlUp).row
    If Error Then
        Range("B4").Value = "Error"
    ElseIf setX = "" Then
        Cells(ansrow, 3).Value = "empty"
        Cells(ansrow, 2).Value = "Answer"
    Else
        Cells(ansrow, 3).Value = setX
        Cells(ansrow, 2).Value = "Answer"
    End If
End Sub
```

The procedure SetEval doles out the tasks union, intersection, and complement. Hierarchy is enforced using the variable mode.

```
Function SetEval(mode As Integer) As String
    Dim symbol As String, setX As String, setY As String
    Do While Idx <= EndIdx
        symbol = Mid(Expr, Idx, 1)
        If IsCap(symbol) Then symbol = "@"              'found a set name
        Select Case symbol
           Case Is = "@"                                'letter found
               setX = GetSet(Mid(Expr, Idx, 1))    'get corresponding set
               Idx = Idx + 1
           Case Is = "+"
               If mode = 1 Then Exit Do      'SetEval() was called earlier
               Idx = Idx + 1
               setY = SetEval(0)                       'evaluate sets after +
               If Not Error Then setX = Union(setX, setY)
           Case Is = "*"
               Idx = Idx + 1
               setY = SetEval(1)                       'evaluate sets after *
               If Not Error Then setX = Intersection(setX, setY)
           Case Is = "'"
               Idx = Idx + 1
               setX = Complement(U, setX)
           Case Is = "("
               Idx = Idx + 1                                    'skip"("
               setX = SetEval(0)
               Idx = Idx + 1                                    'skip")"
           Case Is = ")": Exit Do
           Case Else: Error = True: Exit Function
        End Select
        Ctr = Ctr + 1
    Loop
    SetEval = setX
End Function
```

The function GetSet(SetName) retrieves the set in column 3 named SetName in column 2.

```
Function GetSet(SetName As String) As String
    Dim m As Integer, lastrow As Integer, setX As String
    lastrow = Cells(Rows.count, 2).End(xlUp).row
    For m = 5 To lastrow                   'find entered set named SetName
        If Cells(m, 2).Value = SetName Then Exit For    'it's in row m
    Next m
    setX = DeleteDupStr(Cells(m, 3).Value, ",")
    GetSet = RemoveWhiteSpace(setX)  'return set
End Function
```

18.7 Venn Diagram Components with VBA

The program `VennParts` takes as input a column of user-entered sets in column B with labels $A, B, C, \ldots$ in column A and prints all parts of the corresponding Venn diagram and their elements. Figure 18.5 illustrates the result for three sets. The user has entered the data in B4 and rows 6–8 of columns A and B. Running the program produces the data in rows 10–17 of columns A and B.

	A	B			A	B
4	universal set	$1, 2, 3, 4, 5, 6, 7, 8, 9$		12	$AB'C$	empty
5	sets			13	$A'B'C$	$7, 8$
6	A	$1, 2, 3$		14	ABC'	3
7	B	$3, 4, 5, 9$		15	$A'BC'$	$4, 9$
8	C	$5, 6, 7, 8$		16	$AB'C'$	$1, 2$
10	ABC	empty		17	$A'B'C'$	empty
11	$A'BC$	5				

FIGURE 18.5: Components of a Venn diagram.

The main procedure `VennParts` allocates memory for the given sets $A, B, C, \ldots$ and their complements and then calls various procedures that do the work.

```
Public NumSets As Long, setlabels As String
Sub VennParts()
    Dim i As Integer, sets() As String, complements() As String
    Dim MaxSets As Integer, U As String, binary() As Integer
    MaxSets = 10                        'this gives max 1024 parts
    ReDim sets(1 To MaxSets)                    'array for the sets
    ReDim complements(1 To MaxSets)      'array for their complements
    U = Range("B4").Value                          'universal set
    Call GetSetLabels                            'set labels string
    NumSets = Len(setlabels)
    If NumSets > MaxSets Then MsgBox "Too many sets": Exit Sub
    Call GetSets(sets)
    Call GetComplements(U, sets, complements)
    Call GenerateVennParts(sets, complements)
End Sub
```

The procedure `GetSetLabels` forms a Public string `setlabels` from the set labels in column 1.

```
Sub GetSetLabels()
    Dim i As Integer
```

```
    setlabels = ""                              'initialize Public variable
    Do While Not IsEmpty(Cells(6 + i, 1))  'labels start at row 6, col 1
        setlabels = setlabels & Cells(6 + i, 1).Value      'label string
        i = i + 1
    Loop
End Sub
```

The procedure `GetSets` places the set elements entered in column 2 in the array `sets` and `GetComplements` places their complements with respect to the universal set U in the array `complements`.

```
Sub GetSets(sets() As String)
    Dim i As Integer
    i = 1
    Do While Not IsEmpty(Cells(5 + i, 2))
        sets(i) = Cells(5 + i, 2).Value
        i = i + 1
    Loop
End Sub


Sub GetComplements(U, sets() As String, complements() As String)
    Dim i As Integer
    For i = 1 To NumSets
        complements(i) = Complement(U, sets(i))
    Next i
End Sub
```

The procedure `GenerateVennParts` generates the set labels for the Venn parts as well as the sets corresponding to the labels. It does this by generating binary sequences, that is, sequences of ones and zeros precisely like those in truth tables (but in a different order). These sequences determine whether to include a set or its complement: a zero means include the set, a one to include its complement. For example, for three sets A, B, C the binary sequences generated are

$$000, 100, 010, 110, 001, 101, 011, 111$$

These tell the program to generate the sets

$$ABC, A'BC, AB'C, A'B'C, ABC', A'BC', AB'C', A'B'C'.$$

```
Sub GenerateVennParts(sets() As String, complements() As String)
    Dim i As Integer, j As Integer, binary() As Integer, setpart As String
    ReDim binary(1 To NumSets)
    For i = 1 To 2 ^ NumSets     'make binary sequences of length NumSets
        For j = 1 To NumSets     'get ith sequence; store in array binary
            binary(j) = ZeroOne(i, j)           'get jth digit of sequence
        Next j
        Cells(i + 6 + NumSets, 1) = VennPartLabel(binary)       'get label
        setpart = VennPartSet(binary, sets, complements)        'get set
        If setpart = "" Then setpart = "empty"
```

```
            Cells(i + 6 + NumSets, 2) = setpart
      Next i
End Sub

Function ZeroOne(i As Integer, j As Integer) As Integer
      Dim n As Integer
      n = Int((i - 1) * 2 ^ (1 - j))   'similar to truth value calculation
      ZeroOne = (1 - (-1) ^ n) / 2
End Function
```

The procedure `VennPartLabel` includes in `partlabel` the jth set label in the string `setlabel` if `binary(j)= 0`; otherwise it includes the label for its complement. Analogously, the procedure `VennPartSet` includes the jth set in the string `setpart` if `binary(j)= 0,`; otherwise it includes its complement.

```
Function VennPartLabel(binary() As Integer) As String
      Dim j As Integer, partlabel As String
      For j = 1 To NumSets
            partlabel = partlabel & Mid(setlabels, j, 1)          'default
            If binary(j) = 1 Then partlabel = partlabel & "'"
      Next j
VennPartLabel = partlabel
End Function

Function VennPartSet(binary() As Integer, sets() As String, _
            complements() As String) As String
      Dim j As Integer, setpart As String
      If binary(1) = 0 Then          'get the first set for the intersection
        setpart = sets(1)                         'thus initializing setpart
      Else
        setpart = complements(1)
      End If
      For j = 2 To NumSets                            'get the remaining sets
          If binary(j) = 0 Then
              setpart = Intersection(setpart, sets(j))
          Else
              setpart = Intersection(setpart, complements(j))
          End If
      Next j
      VennPartSet = setpart
End Function
```

18.8   Laws of Set Equality

The set operations introduced in earlier sections obey laws similar to the laws of logic discussed in Section 15.2. As we shall see, this is no coincidence. Indeed, one might guess at a connection based on the fact that union, intersection, and complement are defined using the logical operators "or", "and", "not", respectively.

Laws of sets naturally fall into two classes: the laws of equality and the laws of inclusion. We treat the former in this section, the latter in the next.

Laws of Set Equality

- Double Complement: $A'' = A$.
- Commutative Laws: $A + B = B + A,\ AB = BA$.
- Associative Laws: $(A + B) + C = A + (B + C),$
  $(AB)C = A(BC)$.
- DeMorgan's Laws: $(A + B)' = A'B',$
  $(AB)' = A' + B'$.
- Distributive Laws: $A(B + C) = AB + AC,$
  $A + BC = (A + B)(A + C)$.
- Absorption Laws: $AB + A = A,\ (A + B)A = A$.
- Idempotence: $AA = A,\ A + A = A$.
- Law of the Excluded Middle: $A + A' = U$.
- Law of Noncontradiction $AA' = \emptyset$.
- Identity Laws $A + U = U,\ AU = A,$
  $A + \emptyset = A,\ A\emptyset = \emptyset$.

The reader will notice that the above properties may be formally obtained from the laws of logic in Section 15.2 by replacing $\equiv$ by $=$. To see why this is the case, let's give an argument to establish the validity of the first distributive law. The heart of the argument is to show that an arbitrary member of each side of the equality is a member of the other. We proceed as follows: An element x is in $A(B + C)$ if and only if $x \in A$ and $x \in B + C$, which is the case if and only if $x \in A$ and $(x \in B$ or $x \in C)$. By the first distributive law of logical equivalence, the last statement is equivalent to

$$\big(x \in A \text{ and } x \in B\big) \text{ or } \big(x \in A \text{ and } x \in C\big),$$

which is the statement $x \in AB + AC$. We have used the first distributive law of logical equivalence to prove the first distributive law of sets. The other laws may be proved in a similar manner.

The preceding discussion suggests that the validity of equality between two general set expressions may be established with the program Stmt2TruthTable. In Exercise 10 the reader is asked to carry this out in the more general setting of set inclusion.

18.9 Laws of Set Inclusion

Here are the main properties governing set inclusion. As we shall see in the example below and in the exercises, the properties may be used to discover interesting logical connections among statements.

Laws of Set Inclusion

- Reflexivity: $A \subseteq A$.
- Antisymmetry: $A \subseteq B$ and $B \subseteq A$ implies $A = B$.
- Transitivity: $A \subseteq B$ and $B \subseteq C$ implies $A \subseteq C$.
- Union Property: $A \subseteq C$ and $B \subseteq C$ implies $A + B \subseteq C$.
- Intersection Property: $C \subseteq A$ and $C \subseteq B$ implies $C \subseteq AB$.
- Inclusion Equivalences The statements $A \subseteq B$, $B' \subseteq A'$, $AB = A$, and $A + B = B$ are equivalent.

Let's prove one the inclusion equivalence in the last property that says that $A \subseteq B$ is equivalent to $AB = B$. Suppose first that $A \subseteq B$. To prove that $AB = A$, first choose a member x of the left side. Then x is in both A and B and so x is in the right side. Now choose a member x of the right side. Then, of course $x \in A$, and since $A \subseteq B$, $x \in B$ as well. Therefore, x is in the right side. We have proved that $A \subseteq B$ implies $AB = A$. The converse is proved in a similar manner.

Set inclusion may be used to deduce certain consequences in a logical problem. Here is a typical example:

Example 18.1. Determine what conclusions may be drawn from the following statements:

(a) A person who is not well-read is not educated.
(b) A well-read person never makes a mistake.
(c) Philosophers are well-educated.
(d) Every philosopher makes mistakes.

We first render the problem into symbols by letting

R = the set of well-read people.
E = the set of educated people.
P = the set of philosophers, and
M = the set of people who make mistakes.

From (a)–(d) we obtain the inclusions

$$R' \subseteq E', \quad R \subseteq M', \quad P \subseteq E, \quad P \subseteq M,$$

which lead to

$$P \subseteq E \subseteq R \subseteq M' \subseteq P'.$$

However, the inclusion $P \subseteq P'$ is possible only if P is empty. Thus we are led to the sad conclusion that are no philosophers. $\diamond$

18.10 Cartesian Products, Relations, and Functions

An *ordered pair* is a sequence (a, b) consisting of two elements. The terminology reflects the fact that the order in which the elements are listed is important: (a, b) is not the same as (b, a) unless $a = b$. The *Cartesian product* of sets A and B, denoted by $A \times B$, is the set of all ordered pairs (a, b), where $a \in A$ and $b \in B$. In set builder notation

$$A \times B = \big\{(a, b) \mid a \in A \text{ and } b \in B\big\}.$$

For example, if $A = \{a, b, c\}$ and $B = \{1, 2\}$, then

$$A \times B = \big\{(a, 1),\, (a, 2),\, (b, 1),\, (b, 2),\, (c, 1),\, (c, 2)\big\}.$$

The definition of Cartesian product extends to more than two sets:

$$A_1 \times A_2 \times \cdots \times A_n = \big\{(a_1, a_2, \ldots, a_n) \mid a_1 \in A_1, \ldots, a_n \in A_n\big\}.$$

The symbol $(a_1, a_2, \ldots, a_n)$ is called an *n-tuple*.

If A and B are nonempty sets, then any subset R of $A \times B$ is called a *relation from A to B*. Membership $(a, b) \in R$ is frequently written as aRb and read as "a is related to b." For example, let A be a set of parents and B a set of children and define a relation aRb to mean that "b is an offspring of a." If x has three daughters x_1, x_2, x_3 and y has two sons y_1, y_2, then

$$R = \{(x, x_1), (x, x_2), (x, x_3), (y, y_1), (y, y_2)\}$$

An *equivalence relation* on a set A is a relation R from A to A that has the following properties

- aRa for all $a \in A$ (*reflexive property*)

- If aRb then bRa (*symmetric property*)

- If aRb and bRc then aRc (*transitive property*).

For an example, let S be a set partitioned by the subsets $S_1, S_2, \ldots, S_n$ (Section 18.5). For $x, y \in S$ define xRy to mean that x and y are in the same member of the partition. The reader may easily check that R is an equivalence relation from S to S. All equivalence relations R arise in this way. The partition members here are the sets $[x]$, called *equivalence classes*, which contain precisely those members y for which xRy.

Heretofore, we have used the notion of function informally. It is now possible place the concept in the precise setting of set theory. Let A and B be nonempty sets. A *function F from A to B*, denoted by $F : A \to B$, is a relation with the property that for each $x \in A$ there exists a unique $y \in B$ such that $(x, y) \in F$. In this case we write $y = F(x)$. We may then think of F as a rule that transforms a member of A into a member of B, thus retrieving the standard informal notion of function.

If $G : A \to B$ and $F : B \to C$ are functions with $G(x) \in B$ for all $x \in A$, then the *composition of F and G*, denoted by $F \circ G : A \to C$ is defined by $(F \circ G)(x) = F(G(x))$. If $C = A$ and $(F \circ G)(x) = x$ for all $x \in A$ and $(G \circ F)(y) = y$ for all $y \in B$, then G is called the *inverse* of F. We shall see these notions later in the context of permutations.

18.11   Exercises

1. In Figure 18.3 shade the following sets:

   (a) $AB' + A'C$ (b) $AB' + AC + B'C$ (c) $(A' + B)(A + C)$

2. Fill in the regions in the following Venn diagram for 4 sets with suitable set labels.

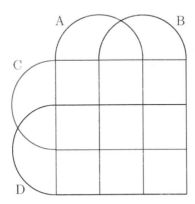

3. Render the following statements into inclusions and draw conclusions.

   (a) An athlete that does not train hard is not a good athlete.
   (b) Good athletes win meets.
   (c) An athlete that trains hard and win meets will win a medal.

4. Render the following statements into inclusions.

(a) There are no women premed students who are not good at physics.
(b) Students who do not take calculus are not good at physics.
(c) Students who take calculus are not good at biology.

Which is the following may be deduced from this information?

(1) No premed student is good at physics.
(2) All premed students take calculus and biology.
(3) All premed students are good at physics.
(4) Women premed students are not good at biology.

5. Render the following statements into inclusions.

   (a) Every member of the Smith family eats broccoli.
   (b) Every member of the Jones family eats kale.
   (c) No member of either the Smith family or the Jones family eats both broccoli and kale.
   (d) Allison eats broccoli.

   Which is the following may be deduced from this information?

   (1) Allison is a member of the Smith family.
   (2) Allison is not a member of the Jones family.

6. Simplify as much as possible.

   (a) $(A + B)(A' + C)(B' + C')$
   (b) $A + A'B$
   (c) $AB' + ABC + AB$
   (d) $A + B + C + A'B'C'$
   (e) $(A + B + C)(A + AB + ABC)$
   (f) $(A'B' + A'B + AB')'$

7. Write a function `SetsEqual(setA,setB)` that returns True if the sets are equal and False otherwise. The members of each set are entered by the user.

8. Write a program `SetOperations` that reads a collection of sets entered in column one and prints their union and intersection.

*9. Write a program `Cartesian` that generates the Cartesian product of arbitrarily many sets entered by the user.

*10. Write a program `SetInclusion(expr1, expr2)` that takes as input two abstract set expressions and decides whether `expr1` included in `expr2` and vice versa. For example, if `expr1 = AB+C` and `expr1 = AC+B` the program should return False. (Suggestion: Give the sets in `expr1` all possible truth values 1,0 (meaning that an arbitrary x is or isn't in a set). For each combination of values that results in `expr1 = 1`, give those variables in `expr2` that are not in `expr1` all possible truth values. If in this process there is an overall combination that result in `expr2 = 0`, then `expr1` is not contained in `expr2`.

Chapter 19

Counting

In this chapter we consider methods that allow one to calculate the number of ways various tasks can be carried out. The computational techniques here have important applications in geometry, network analysis, algebra, computer science and, as we shall see in the next several chapters, probability theory.

19.1 The Addition Principle

The number of (distinct) elements in a set A is called the *cardinality* of A and is denoted by $n(A)$. For example, $n(\{a, b, c, d, e\}) = 5$ and $n(\emptyset) = 0$. The *addition principle* asserts that if A and B are sets with no common elements (that is, $AB = \emptyset$), then $n(A+B) = n(A)+n(B)$. The principle can be extended in the obvious way to more than two sets:

If no pair of sets $A_1, A_2, \ldots, A_k$ have elements in common then

$$n(A_1 + A_2 + \cdots + A_k) = n(A_1) + n(A_2) + \cdots + n(A_k).$$

The following example illustrates how the addition principle may be used in certain counting problems.

Example 19.1. A survey of 300 people questioning their participation in regular exercise activities **A**, **B**, and **C** revealed the following:

- 100 participate in activity **A**
- 100 participate in activity **B**
- 120 participate in activity **C**
- 70 participate in **B** but not in **A**
- 80 participate in **A** but not in **C**
- 40 participate in **B** but not in **A** or **C**
- 10 participate in all three.

DOI: 10.1201/9781003351689-19

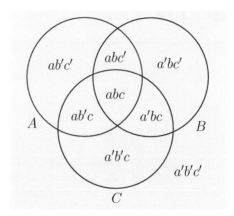

FIGURE 19.1: Cardinalities of subsets in Figure 18.3.

We wish to find the number of people in the survey that participate in exercises **B** and **C** but not **A**.

To this end let A, B, and C denote the sets of people that participate in activities **A**, **B**, and **C**, respectively, and let U denote the population surveyed. Using shorthand such as $a = n(A)$, $a'b = n(A'B)$, etc. we may summarize the given information as

$$a = b = 100, \quad c = 120, \quad a'b = 70, \quad ac' = 80, \quad a'bc' = 40, \quad abc = 10, \quad \text{and } u = 300.$$

The diagram in Figure 19.1 may then be used to derive additional information. For example, from the diagram we see that $a'bc + a'bc' = a'b$. This, together with the given information $a'bc' = 40$ and $a'b = 70$, implies that $a'bc = 30$, the solution to our problem.   ◇

---

*19.2   Venn Solutions with VBA

In this section we develop a program that solves a 3-set Venn diagram problem like the one in Example 19.1. As in that example, the cardinalities of sets are labeled with corresponding small letters. We shall call these *Venn numbers*. The program assumes that the Venn number labels have been entered in column B and their numerical codes 1–27 in column A, as shown in Figure 19.2. The Venn data is entered in column C. The figure shows the input data from Example 19.1 (in boldface) entered in rows 7, 9, 11, 16, 17, 25, 31, and 33 of column C. The program calculates the remaining Venn numbers

	A	B	C
7	1	a	**100**
8	2	a'	200
9	3	b	**100**
10	4	b'	200
11	5	c	**120**
12	6	c'	180
13	7	ab	30
14	8	ab'	70
15	9	ac	20

	A	B	C
16	10	ac'	**80**
17	11	$a'b$	**70**
18	12	$a'b'$	130
19	13	$a'c$	100
20	14	$a'c'$	100
21	15	bc	40
22	16	bc'	60
23	17	$b'c$	80
24	18	$b'c'$	120

	A	B	C
25	19	abc	**10**
26	20	$ab'c$	10
27	21	abc'	20
28	22	$ab'c'$	60
29	23	$a'bc$	30
30	24	$a'b'c$	70
31	25	$a'bc'$	**40**
32	26	$a'b'c'$	60
33	27	u	**300**

FIGURE 19.2: Input-output for Venn example.

TABLE 19.1: Venn Equations and Their Codes.

$u = a + a'$	27,1,2	$c' = bc' + b'c'$	6,16,18
$u = b + b'$	27,3,4	$ab = abc + abc'$	7,19,21
$u = c + c'$	27,5,6	$a'b = a'bc + a'bc'$	11,23,25
$a = ab + ab'$	1,7,8	$ab' = ab'c + ab'c'$	8,20,22
$a = ac + ac'$	1,9,10	$a'b' = a'b'c + a'b'c'$	12,24,26
$a' = a'b + a'b'$	2,11,12	$ac = abc + ab'c$	9,19,20
$a' = a'c + a'c'$	2,13,14	$a'c = a'bc + a'b'c$	13,23,24
$b = ab + a'b$	3,7,11	$ac' = abc' + ab'c'$	10,21,22
$b = bc + bc'$	3,15,16	$a'c' = a'bc' + a'b'c'$	14,25,26
$b' = b'c + b'c'$	4,8,12	$bc = abc + a'bc$	15,19,23
$b' = b'c + b'c'$	4,17,18	$b'c = ab'c + a'b'c$	17,20,24
$c = ac + a'c$	5,9,13	$bc' = abc' + a'bc'$	16,21,25
$c = bc + b'c$	5,15,17	$b'c' = ab'c' + a'b'c'$	18,22,26
$c' = ac' + a'c'$	6,10,14		

(shown in the figure in non-boldface) and prints these in the appropriate rows of column C.

The program works as follows: The unknown Venn numbers are calculated using the 27 equations that are listed in Table 19.1. Each equation has a three digit code to its right that refers to the components of the equation. For example, the equation $a = ab + ab'$ has components a, ab, and ab'. Since the labels in column A for these components are 1, 7, and 8, respectively, the equation receives the code 1,7,8. The program assumes that the equation codes have been entered in Sheet2 as a 27×3 matrix with the $(1,1)$ entry in cell B2.

The main procedure `Venn3Solver` begins by allocating the array E for the 27 equation codes and the array V for the 27 Venn numbers that will eventually appear in column C. It then calls `GetEqnCodes`, which retrieves the equation codes from Sheet2 and stores them in the array E. For example, for the fourth equation $a = ab + ab'$ in the above list (coded 1,7,8),

$$E(4,1) = 1, \quad E(4,2) = 7, \quad E(4,3) = 8.$$

The procedure `GetEqnCodes` is simply an integer version of `MatrixIn` and is not listed. A Do While loop reads the initial Venn numbers entered by the user as well as any new numbers that were generated by the function `CalculateVennNums` on earlier iterations. The loop terminates after `count` reaches 27 iterations (the number of Venn parts) or if `status= -1`, a signal that the entered data is inconsistent.

```
Sub Venn3Solver()
    Dim V(0 To 27) As Integer              'array for Venn numbers
    Dim Eq(1 To 27, 1 To 3) As Integer     'array for equation codes
    Dim count As Integer
    Dim status As Integer
    Call GetEqnCodes(Eq)                   'get codes from spreadsheet
    Do
        count = count + 1
        Call GetNumbers(V)                 'get entered spreadsheet numbers
        status = CalculateVennNums(V, Eq)      'find the remaining data
        If status = -1 Then MsgBox "Inconsistent Data": Exit Sub
        Call PrintVennNums(V)
    Loop While status = 0 And count < 28
End Sub
```

In the first iteration of the Do While loop, the procedure `GetNumbers` retrieves the Venn numbers entered by the user. If Venn number i is missing, then $V(i)$ is set to -1. For our example, the retrieved numbers are

$$V(1) = 100, \quad V(3) = 100, \quad V(5) = 120, \quad V(10) = 80,$$
$$V(11) = 70, \quad V(19) = 10, \quad V(25) = 40, \quad V(27) = 300.$$

Additional entries of V are calculated (and the -1's removed) during successive iterations as long as the variable `status` remains zero.

```
Sub GetNumbers(V() As Integer)
    Dim i As Integer
    For i = 1 To 27                        'load Venn data
        V(i) = -1                          'default: no data
        If Not IsEmpty(Cells(6 + i, 3)) Then
            V(i) = Cells(6 + i, 3).Value           'found data
        End If
    Next i
End Sub
```

The main calculations are done by `CalculateVennNums`, which attempts to find the missing Venn numbers by solving the equations

$$V(E(i,1)) = V(E(i,2)) + V(E(i,3)), \quad i = 1, 2, 3. \tag{19.1}$$

To do so it must first determine which terms of an equation are missing a value. This is done by the procedure `Unknown`, which tests whether any of the terms

$$V(E(i,j)), \quad j = 1, 2, 3,$$

has the value -1, and if so returns the corresponding index j.

```
Function Unknown(V() As Integer, E() As Integer, i As Integer) As Integer
    Dim j As Integer, count As Integer, status As Integer
    For j = 1 To 3                    'count how many missing values in eqn i
        If V(E(i, j)) < 0 Then
            count = count + 1: status = j        'missing value:V(E(i, j))
        End If
    Next j
    If count > 1 Then status = 0         'not ready to solve equation yet
    If count = 0 And V(E(i, 1)) <> V(E(i, 2)) + V(E(i, 3)) Then _
        status = -1                                      'data error
    Unknown = status
End Function
```

If `Unknown` returns an index j, then `CalculateVennNums` solves the equation (19.1) for the term `V(E(i, j))` and prints the calculated Venn number in column C. If more than one term is missing a value, then `Unknown` returns 0. In this case the Do While Loop in `VennCalculator` performs another iteration in an attempt to fill in more gaps in column C. If there are inconsistent data entries, the function `Unknown` returns -1. The entire process continues until it is no longer possible to fill more gaps, indicated by the variable `status` no longer equalling 0 or the number of iterations exceeding its limit. The last procedure below prints the Venn numbers.

```
Function CalculateVennNums(V() As Integer, E() As Integer) As Integer
    Dim i As Integer, status As Integer
    status = 1 'default
    i = 1
    Do While i <= 27 And status > -1
        Select Case Unknown(V, E, i)
            Case Is = -1: status = -1
            Case Is = 0:  status = 0
            Case Is = 1:  V(E(i, 1)) = V(E(i, 2)) + V(E(i, 3))
            Case Is = 2:  V(E(i, 2)) = V(E(i, 1)) - V(E(i, 3))
            Case Is = 3:  V(E(i, 3)) = V(E(i, 1)) - V(E(i, 2))
        End Select
        i = i + 1
    Loop
    CalculateVennNums = status
End Function

Sub PrintVennNums(V() As Integer)
    Dim i As Integer
    For i = 1 To 27                          'print the Venn numbers
```

```
            If V(i) > 0 Then Cells(6 + i, 3).Value = V(i)
        Next i
    End Sub
```

19.3   The Multiplication Principle

The *multiplication principle* is the analog of the addition principle and may be formulated as follows:

If $A_1, A_2, \ldots, A_k$ are arbitrary finite sets, then

$$n(A_1 \times A_2 \times \cdots \times A_n) = n(A_1)n(A_2)\cdots n(A_n).$$

Here $A_1 \times A_2 \times \cdots \times A_n$ denotes the Cartesian product of the sets (see Section 18.10). The reader should verify the principle for the case $n = 2$ by explicitly listing the pairs of elements from the sets $A_1 = \{a, b, c\}$ and $A_2 = \{1, 2, 3\}$. The above formula implies the following rule, also called the multiplication principle:

> *When performing a task requiring n steps, if there are m_1 ways to complete step 1, and if for each of these there are m_2 ways to complete step 2, etc., then there are $m_1 m_2 \cdots m_n$ ways to perform the task.*

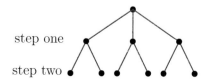

step one

step two

FIGURE 19.3: The multiplication principle.

Applications of the principle are given in the next section and in Chapter 20.

19.4   Permutations

Consider the problem of counting the number of three letter sequences that can be formed from the letters of the word "computer," where no letter is used more than once. We can describe the task of forming such a sequence as a three-step process, namely that of selecting the letters one at a time. Since letters may be used only once, each selection reduces the number of available choices by one. Thus there are 8 choices for the first letter, 7 for the second, and 6 for the third. By the multiplication principle, the total number of non-repeating three-letter sequences is therefore $8 \cdot 7 \cdot 6 = 336$.

The sequences described in this example are called *permutations*. More generally, let n and r be positive integers with $1 \le r \le n$. A *permutation of n items taken r at a time* is an ordered list of r items (an r-tuple) chosen from the n items. An argument similar to that in the preceding paragraph shows that the number of such r-tuples is

$$n(n-1)(n-2) \cdots (n-r+1) = \frac{n!}{(n-r)!},$$

where we have used the *factorial* notation

$$m! = m(m-1)(m-2) \cdots 2 \cdot 1, \quad 0! = 1! = 1. \tag{19.2}$$

In particular, the number of permutations of n items taken n at a time, which is simply the number of arrangements of n objects, is $n!$

Example 19.2. In how many ways can a group of 4 women and 3 men line up for a photograph if no two adjacent people are of the same gender?

The lineup must be of the form WMWMWMW. Since there are 4! arrangements of the women and 3! arrangements of the men, the multiplication principle implies that the total number of lineups is $(4!)(3!) = 144$. ◊

19.5   Generating Permutations with VBA

The function `Permutations` in this section is given a string and returns an array of all its permutations. For example, if the string is abc it will return the array of strings

$$\{abc,\ acb,\ bac,\ bca,\ cba,\ cab\}$$

The procedure first uses the factorial function `Factorial` of Section 8.1 to get the dimension of the array and then calls `Split1` (Section 6.8) to convert the string into an array. The recursive procedure `GetPerms` generates the permutations.

```
Function Permutations(str As String) As String()
    Dim S() As String, P() As String
    ReDim P(1 To Fact(Len(str)))          'memory for permutations
    S = Split1(str, "")               'convert to array for convenience
    Call GetPerms(S, 1, P, 0)                   'start the recursion
    Permutations = P                  'return the array of permutations
End Function
```

The function `GetPerms` is passed an array `S` of symbols to be permuted and returns an array `P` of the permutations of the symbols. It keeps track of the current number `N` of permutations already generated.

```
Sub GetPerms(S() As String, idx As Integer, P() As String, N As Long)
    Dim i As Integer, k As Long
    If idx = UBound(S) Then                          'base case
        k = N + 1                                    'spot for new permutation
        P(k) = Join(S, "")                           'merge array into string
        N = k                            'update current number of permutations
    Else
        For i = idx To UBound(S)
            Call swap(S(idx), S(i))          'interchange values of variables
            Call GetPerms(S, idx + 1, P, N)      'permute string at idx+1
            Call swap(S(idx), S(i))
        Next i
    End If
End Sub
```

The following figure shows how the program works for the case of three letters abc. The down arrows indicate going into recursion, the up arrows coming out of recursion. The pairs of numbers labeling the up and down arrows are the values of *idx* and *i*, respectively, and indicate the swap $S(idx) \leftrightarrow S(1)$.

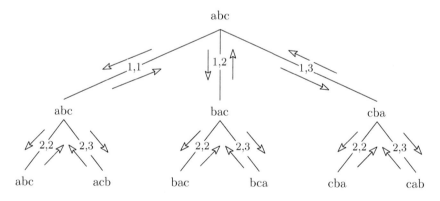

19.6 Combinations

While a permutation is an ordered list, a combination is an *unordered* list. More precisely, if n and r are positive integers with $1 \leq r \leq n$, then a *combination of n items taken r at a time* is a set of r items chosen from the n items. For example, the combinations of the five letters a, b, c, d, and e taken three at a time (written, without set braces or comma separators) are

$$abc, \ abd, \ abe, \ acd, \ ace, \ ade, \ bcd, \ bce, \ bde, \text{ and } cde.$$

To make a connection with permutations, let c and p denote, respectively, the number of combinations and the number of permutations of n items taken r at a time. Since each such combination gives rise to $r!$ permutations we have the relation $c(r!) = p$ and so $c = p/r!$. Using (19.2) we conclude that the number of combinations n things taken r at a time is the *binomial coefficient*

$$\binom{n}{r} = \frac{n!}{(n-r)!r!} \qquad (19.3)$$

Here are two examples combining the multiplication and addition principles.

Example 19.3. A bag contains 5 red, 4 yellow, and 3 green marbles. In how many ways is it possible to randomly draw from the bag 4 marbles consisting of exactly 2 reds and no more than 2 greens?

We have the following decision scheme:

    Case 1: No green marbles.
        Step 1: Choose 2 reds: $\binom{5}{2} = 60$ possibilities.
        Step 2: Choose 2 yellows: $\binom{4}{2} = 12$ possibilities.

    Case 2: Exactly 1 green marble.
        Step 1: Choose the green: 3 possibilities.
        Step 2: Choose 2 reds: $\binom{5}{2} = 60$ possibilities.
        Step 3: Choose 1 yellow: 4 possibilities.

    Case 3: Exactly 2 green marbles.
        Step 1: Choose 2 greens: $\binom{3}{2} = 3$ possibilities.
        Step 2: Choose 2 reds: $\binom{5}{2} = 60$ possibilities.

By the addition and multiplication principles there are a total of $60 \cdot 12 + 3 \cdot 60 \cdot 4 + 3 \cdot 60 = 1620$ possibilities. ◊

Example 19.4. How many different arrangements of the letters in the word "mathematics" are possible?

There are duplicate letters, so the answer $11!$ is incorrect. We proceed as follows:

    Step 1: Select positions for the 2 m's: $\binom{11}{2} = 55$ choices.

    Step 2: Select positions for the 2 a's: $\binom{9}{2} = 36$ choices.

    Step 3: Select positions for the 2 t's: $\binom{7}{2} = 21$ choices.

    Step 5: Fill remaining 5 spots with the letters h, e, i, c, s: $5! = 120$ choices.

Thus there are $55 \cdot 36 \cdot 21 \cdot 120 = 4,989,600$ different arrangements. ◊

The binomial coefficients arise naturally in the important *Binomial Theorem*:

Let a and b be real numbers and n a positive integer. Then

$$(a+b)^n = \sum_{k=0}^{n} \binom{n}{k} a^k b^{n-k}.$$

Here is a combinatorial argument that shows why the formula holds. Expanding $(a+b)^n$ results in a sum of products each having n factors consisting of a's and b's. For each $k = 0, 1, \ldots n$ collect together all products that have exactly k a's. Each such product may be written as $a^k b^{n-k}$. There are as many of these as there are ways to choose k spots out of n. Thus for each k there are $\binom{n}{k}$ terms of the form $a^k b^{n-k}$, proving the theorem. Another proof, somewhat more rigorous, uses the prnciple of mathematical induction discussed in Section 23.5.

For an application we show that a set S with n members has 2^n subsets (including S and $\emptyset$). Indeed, since $\binom{n}{r}$ gives the number of subsets of size r, the total number of subsets of S (including S and $\emptyset$) is

$$\binom{n}{0} + \binom{n}{1} + \cdots + \binom{n}{n}$$

By the Binomial Theorem this is $(1+1)^n = 2^n$.

19.7   Generating Combinations with VBA

In this section we develop a program that generates all subsets of a given size taken from a given set. The main procedure `Combinations` is passed a string of comma-separated characters and the desired subset size and returns an array of all subsets of that size. For example, if the input string is a, b, c, d, e and length 4 subsets are desired, then the program generates the array of strings

$$a,b,c,d \quad a,b,c,e \quad a,b,d,e \quad a,c,d,e \quad b,c,d,e.$$

The basic idea of the program is to generate all possible sequences of 1's and 0's with length equal to the size of the given set. These are used to select the subsets, 1 indicating that the element in that position is to be included, 0 indicating that it is not. For example if the entered set is the string a, b, c, d, e, then 10101 stands for the subset a, c, e and 11011 stands for the subset a, b, d, e. If size 4 subsets are desired, then the former string is ignored and the latter retained. Here is the command button that launches the program.

```
Sub CommandButton1_Click()
    Dim Combo() As String, SetStr As String, SubSize As Integer
    Call ClearColumns(8, 1, 1)
    SetStr = Range("B5").Value              'get the set S as a string
    SubSize = Range("B6").Value   'get desired size of the subsets of S
    Combo = Combinations(SetStr, SubSize)
    For k = 1 To UBound(Combo)
        Cells(7 + k, 1).Value = Combo(k)
    Next k
End Sub
```

Binary sequences are generated by nested For Next loops using the function ZeroOne(i,j) of Section 18.7 The function NumOnes returns the number of ones in a sequence. If that number equals the subset size, then the function SelectSubset is called to extract a subset according to the rule in the preceding paragraph. The subset is stored in the next available spot in the list of combinations.

```
Function Combinations(SetStr As String, SubSize As Integer) As String()
    Dim Combo() As String, bin As String, SetSize As Integer
    Dim i As Integer, j As Integer, k As Integer: k = 1
    SetSize = Card(SetStr)                          'size of S
    If SubSize = 0 Or SubSize > SetSize Then Exit Function
    ReDim Combo(1 To 2 ^ SetSize)                   'array of combinations
    For i = 1 To 2 ^ SetSize                        'generate binary sequences
        bin = ""                                    'clear last binary sequence
        For j = 1 To SetSize                        'build ith binary sequence
            bin = bin & ZeroOne(i, j)
        Next j
        If NumOnes(bin) = SubSize Then
            Combo(k) = SelectSubset(bin, SetStr)           'store subset
            k = k + 1
        End If
    Next i
    Combinations = Combo                    'return list of combinations
End Function
```

The function NumOnes(bin) simply removes all zeros from bin and returns the length of the resulting string.

```
Function NumOnes(bin As String) AsInteger
    NumOnes = Len(Replace(bin, "0", ""))
End Function
```

The function SelectSubset takes as input the binary selector sequence bin and the string SetS. A For Next loop runs through the members of the binary sequence extracting the ith member of SetS precisely when the ith character in bin is one. The subset string SubS is built from these members by concatenation and then returned by the function.

```
Function SelectSubset(bin As String, SetString As String) As String
    Dim i As Integer, SubSet As String
    For i = 1 To Len(bin)
        If Mid(binary, i, 1) = 1 Then              'include ith char if 1
        SubSet = SubSet & MidDelim(SetString, i, ",") & ","
        End If
    Next i
    SelectSubset = Mid(SubSet, 1, Len(SubSet) - 1)   'remove last comma
End Function
```

19.8 Traveling Salesman Problem

The problem asks for the shortest route a salesperson can take if he must visit each of several towns exactly once, starting and ending at the same town. If we designate a particularly town as the starting point and there are $n + 1$ towns, then there are a total of $n!$ possible routes: n possible towns to visit from the first town, for each of these $n - 1$ towns to visit from the second town, etc. This suggests the following "brute force" solution: compute the length of each of the $n!$ possible routes and then select a shortest route (there may be more than one). While this solution is practical for only a few towns,[1] it has the advantage of being easy to implement.

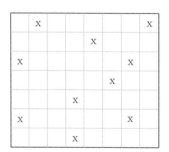 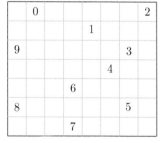

FIGURE 19.4: A shortest route for the travelling salesman.

The module in this section takes the brute force approach. The user places x's on a spreadsheet grid for the towns. The program then finds a shortest route, which we shall call a *best path*, the distance between towns taken as the number of cells one must traverse in horizontal and vertical directions to get from one town to another. Figure 19.4 shows the state of the spreadsheet for 10 towns before and after the program has run. The program labels the towns $0 - 9$ in an order that gives a shortest route, in this case comprised of 32 cells.

---

[1]More than 11 towns is generally not feasible on a typical computer.

The choice of the origin town is immaterial; one could start with any town, as paths are circuits.

The main procedure `TravelSales` begins by setting up the grid that displays the towns and allocating memory for the position array `Loc` of town locations, the current path array `CurrPath`, and the current best path array `LeastPath`. It then calls `InitialData`, which scans the grid for towns, recording their locations in `Loc`, and initializes `CurrPath` and `LeastPath`. The position of the ith town in the scan (which is left to right, top to bottom) is recorded as `Loc(i)`. The first town encountered in the scan (at location `Loc(0)`) is designated as town 0. The sequence $0, 1, 2, \ldots, k, 0$ then represents the path through the towns in the order obtained from the scan. An arbitrary path is represented by a sequence $0, n_1, n_2, \ldots, n_k, 0$, where $n_1, n_2, \cdots, n_k$ is a permutation of $1, 2, \ldots, k$. The zeros are redundant, so a path may be represented simply by $n_1, n_2, \ldots, n_k$. The distance from town i to town j is stored as entry (i, j) of a $k + 1$ by $k + 1$ matrix D. The matrix is constructed by the function `GetDistances`.

```
Public MaxTowns As Integer, NumTowns As Integer, LeastLength As Integer
Sub TravelSales()
      Dim Loc() As TPos, Grid As TGrid, D() As Integer
      Dim CurrPath() As Integer, LeastPath() As Integer
      Grid = MakeGrid(3, 3, 26, 26, 3, 13, 24)
      MaxTowns = 11
      ReDim Loc(0 To MaxTowns - 1)              'array for town locations
      ReDim CurrPath(1 To MaxTowns - 1)          'array for current route
      ReDim LeastPath(1 To MaxTowns - 1)     'array for current best route
      ReDim D(0 To MaxTowns - 1, 0 To MaxTowns - 1)
      Call InitialData(Grid, Loc, CurrPath, LeastPath)      'scan for towns
      If NumTowns = 0 Or NumTowns > MaxTowns Then Exit Sub
      Call GetDistances(Loc, D)                  'distances between towns
      LeastLength = GetPathLength(LeastPath, D)            'initial route
      Call permute(CurrPath, LeastPath, D, 1)              'get new route
      Range("AC8").Value = LeastLength       'print distance of best route
      Call PrintPath(Loc, LeastPath)                   'label best route
End Sub
```

The procedure `InitialData` retrieves the positions of the towns on the spreadsheet. These are placed in the array `Loc`. It also takes as the current path and best path the order of the towns visited in the scan. These are updated by the program later.

```
Private Sub InitialData(Grid As TGrid, Loc() As TPos,
                  CurrPath() As Integer, LeastPath() As Integer)
      Dim i As Integer, j As Integer, m As Integer
      For i = Grid.T To Grid.B     'scan grid and read in town positions
          For j = Grid.L To Grid.R
              If Not IsEmpty(Cells(i,j).Value) And m <= MaxTowns - 1 Then
                   'location of mth town in scan:
                   Loc(m).row = i: Loc(m).col = j
                   If m > 0 Then CurrPath(m) = m:  LeastPath(m) = m
```

```
              m = m + 1
          End If
        Next j
      Next i
      NumTowns = m
  End Sub
```

The distances between the towns are calculated from the position arrays by the procedure `GetDistances`. The distance $D(i, j)$ between the ith and jth towns encountered in the initial scan is the length of the elbow path connecting them. It is calculated using the VBA function `Abs`. The distances are printed in matrix form on the spreadsheet.

```
Private Sub GetDistances(Loc() As TPos, D() As Integer)
    Dim i As Integer, j As Integer
    For i = 0 To NumTowns - 1
        For j = 0 To NumTowns - 1
            D(i, j) = Abs(Loc(i).row - Loc(j).row) + _
                    Abs(Loc(i).col - Loc(j).col)
            Cells(31 + i, j + 2).Value = D(i, j)      'print as matrix
        Next j
    Next i
End Sub
```

As mentioned earlier, the module examines all possible routes through the towns, avoiding towns previously visited. The routes are symbolized by permutations $n_1, n_2, \ldots, n_k$ of the nonzero indices $1, 2, \ldots, k$ of the position arrays. The first such route was supplied by the procedure `InitialData`, which simply forms the path $1, 2, \ldots, k$. The remaining routes are generated by `permute`, which is similar to the recursive procedure `GetPerms` (Section 19.4) but instead of printing out a permutation at the end of the recursion, the program calls `CheckPath` to calculate the length of the route with the distance matrix D, and then updates `LeastPath` and `LeastLength`, if warranted.

```
Sub permute(CurrPath() As Integer, LeastPath() As Integer, _
          D() As Integer, idx As Integer)
    Dim i As Integer
    If idx = NumTowns - 1 Then            'recursion terminating case
        Call CheckPath(CurrPath, LeastPath, D)
    Else
        For i = idx To NumTowns - 1      'permute the towns except town 0
            Call swap(CurrPath(idx), CurrPath(i))        'interchange
            Call permute(CurrPath, LeastPath, D, idx + 1)
            Call swap(CurrPath(idx), CurrPath(i))
        Next i
    End If
End Sub
```

The procedure `CheckPath` calls `GetPathLength(CurrPath,D)`, which returns the length of the current path. If this is less than the current best length, then the path is stored in the current best path array `LeastPath`.

```
Sub CheckPath(CurrPath() As Integer, LeastPath() As Integer, _
              D() As Integer)
    Dim length As Integer
    length = GetPathLength(CurrPath, D)
    If length < LeastLength Then   'if length < current best least length
        LeastLength = length                              'update
        Call StorePath(CurrPath, LeastPath)      'and store shorter path
    End If
End Sub
```

The function `GetPathLength` sums the distances from town 0 to the first and last towns in the route described in `Path` and then adds to the sum the intermediate distances between the towns.

```
Function GetPathLength(CurrPath() As Integer, D() As Integer)
    Dim length As Integer, i As Integer
    'distance from town 0 to towns Path(1) and Path(NumTowns - 1))
    length = D(0, CurrPath(1)) + D(0, CurrPath(NumTowns - 1))
    For i = 1 To NumTowns - 2 'get distances for towns in between
        length = length + D(CurrPath(i), CurrPath(i + 1))
    Next i
    GetPathLength = length
End Function
```

```
Sub StorePath(CurrPath() As Integer, LeastPath() As Integer)
    Dim i As Integer
    For i = 1 To NumTowns - 1
        LeastPath(i) = CurrPath(i)
    Next i
End Sub
```

```
Private Sub PrintPath(Loc() As TPos,  LeastPath() As Integer)
    Dim i As Integer
    Cells(Loc(0).row, Loc(0).col) = 0
    For i = 1 To NumTowns - 1
        Cells(Loc(LeastPath(i)).row, Loc(LeastPath(i)).col).Value = i
    Next i
End Sub
```

---

*19.9   Permutation Algebra

A permutation $p = (n_1, n_2, \ldots, n_k)$ of the integers $1, 2, \ldots, k$ may be viewed as a function whose value at j is n_j. To emphasize the function interpretation and make calculations more transparent, the permutation is sometimes displayed as

$$p = \begin{pmatrix} 1 & 2 & \cdots & k \\ n_1 & n_2 & \cdots & n_k \end{pmatrix}$$

For example, in this notation the permutation $(4, 3, 5, 1, 2)$ becomes

$$\begin{pmatrix} 1 & 2 & 3 & 4 & 5 \\ 4 & 3 & 5 & 1 & 2 \end{pmatrix}$$

While the latter notation more cumbersome, it has the advantage of increased clarity when performing operations on permutations.

The set of permutations of the integers $1, 2, \ldots, k$ may be given an algebraic structure. The fundamental operation of the system is *product* of two permutations p and q, denoted by the juxtaposition of the permutations and defined as illustrated by the following example:

$$pq = \begin{pmatrix} 1 & 2 & 3 & 4 & 5 \\ 4 & 3 & 5 & 1 & 2 \end{pmatrix} \begin{pmatrix} 1 & 2 & 3 & 4 & 5 \\ 3 & 5 & 1 & 2 & 4 \end{pmatrix} = \begin{pmatrix} 1 & 2 & 3 & 4 & 5 \\ 5 & 2 & 4 & 3 & 1 \end{pmatrix} \quad (19.4)$$

The permutation on the right was obtained by applying the transformation rules of the permutations on the left in succession, starting with the second permutation q. Thus we have

$$1 \overset{q}{\to} 3 \overset{p}{\to} 5, \ 2 \overset{q}{\to} 5 \overset{p}{\to} 2, \ 3 \overset{q}{\to} 1 \overset{p}{\to} 4, \ 4 \overset{q}{\to} 2 \overset{p}{\to} 3, \ 5 \overset{q}{\to} 4 \overset{p}{\to} 1 \ .$$

Note that multiplication of permutations is simply composition of functions. Here is another example:

$$\begin{pmatrix} 1 & 2 & 3 & 4 & 5 \\ 3 & 5 & 2 & 1 & 4 \end{pmatrix} \begin{pmatrix} 1 & 2 & 3 & 4 & 5 \\ 1 & 2 & 3 & 4 & 5 \end{pmatrix} = \begin{pmatrix} 1 & 2 & 3 & 4 & 5 \\ 3 & 5 & 2 & 1 & 4 \end{pmatrix}$$

We call the second permutation on the left side of the equation the *identity permutation*. The analogous equation with the order of the permutations on the left reversed also holds.

The function `MultPerm` returns the product of two permutations. The input permutations are written as strings `permA` and `permB` as is the output permutation `permC = permA permB`. For example, referring to (19.4), `MultPerm("4,3,5,1,2","3,5,1,2,4")` returns the string `"5,2,4,3,1"`.

```
Function MultPerm(permA As String, permB As String) As String
    Dim permC As String, i As Integer, A() As String, B() As String
    A = Split1(permA, ","): B = Split(permB, ",")      'convert to arrays
    For i = 1 To UBound(A)                    'split1 returns arrays with idx 1
        permC = permC & A(CInt(B(i)) & ","            'form composition
    Next i
    MultPerm = Mid(permC, 1, Len(permC) - 1)          'kill last comma
End Function
```

The *inverse* of a permutation is denoted with a superscript -1 and is obtained by reversing the transformation rules. For example,

$$\begin{pmatrix} 1 & 2 & 3 & 4 & 5 \\ 3 & 5 & 2 & 1 & 4 \end{pmatrix}^{-1} = \begin{pmatrix} 1 & 2 & 3 & 4 & 5 \\ 4 & 3 & 1 & 5 & 2 \end{pmatrix}$$

Notice that

$$\begin{pmatrix} 1 & 2 & 3 & 4 & 5 \\ 3 & 5 & 2 & 1 & 4 \end{pmatrix}^{-1} \begin{pmatrix} 1 & 2 & 3 & 4 & 5 \\ 3 & 5 & 2 & 1 & 4 \end{pmatrix} = \begin{pmatrix} 1 & 2 & 3 & 4 & 5 \\ 1 & 2 & 3 & 4 & 5 \end{pmatrix},$$

the identity permutation. Interchanging the permutations on the left yields the same result.

The following function takes a string `perm` returns the inverse as another string.

```
Function InvPerm(perm As String) As String
    Dim iperm() As String, k As Integer, L As Integer, _
        splitperm() As String
    splitperm = Split(perm, ",")
    L = UBound(splitperm)     '0 To L
    ReDim iperm(1 To L + 1)                       'array for inverse
    For k = 1 To L + 1                            'get inverse
        iperm(CInt(splitperm(k - 1))) = CStr(k)   'definition of inverse
    Next k
    InvPerm = Join(iperm, ",")                    'return string version
End Function
```

A permutation raised to a positive integer power n is, by definition, the product of the permutation times itself $n - 1$ times. A permutation raised to a negative integer power n is the inverse of the permutation raised to the positive power $-n$. A permutation raised to the zero power is by definition the identity permutation. The following function raises a permutation to a power n, which can be positive, negative of zero.

```
Function PowerPerm(perm As String, n As Integer) As String
    Dim outperm As String, iperm As String, k As Integer, L As Integer
    L = Len(perm)
    If n = 1 Then PowerPerm = perm: Exit Function    'return the input
    If n = 0 Then
        For k = 1 To L
            outperm = outperm & "," & CStr(k)        'create identity perm.
        Next k
    End If
    If n < 0 Then
        iperm = InvPerm(perm)
        n = -n                               'replace n by its negative
        perm = iperm            'replace iperm by perm to get power of iperm
    End If
    If n > 1 Then
        outperm = perm
        For k = 2 To n
            outperm = MultPerm(perm, outperm)    'get perm^2, perm^3, etc.
        Next k
    End If
    PowerPerm = outperm
End Function
```

The set of all permutations of $1, 2, \ldots, k$ together with the operations of multiplication and inversion defined in the previous paragraphs is denoted by S_k and is called the *permutation group of order k!*. Permutation groups have applications in geometry, combinatorics and other branches of mathematics, as well as in physics and chemistry. They have even been used to good effect in solving Rubic's cube.

*19.10 Permutation Cycles

A *cyclic permutation* or *cycle* is a permutation that transforms its members in a cyclic fashion. For example, for the permutation

$$\begin{pmatrix} 1 & 2 & 3 & 4 & 5 \\ 4 & 1 & 5 & 3 & 2 \end{pmatrix}$$

we have $1 \to 4 \to 3 \to 5 \to 2 \to 1$. The permutation is then written as $(1\,4\,3\,5\,2)$ Every permutation may be decomposed into a product of cycles. For example, the cycles of

$$\begin{pmatrix} 1 & 2 & 3 & 4 & 5 & 6 \\ 6 & 5 & 1 & 4 & 2 & 3 \end{pmatrix}$$

are $(1\,6\,3)$, $(2\,5)$, and (4); the permutation is then written as

$$\begin{pmatrix} 1 & 2 & 3 & 4 & 5 & 6 \\ 6 & 5 & 1 & 4 & 2 & 3 \end{pmatrix} = (1\,6\,3)(2\,5)(4)$$

The program `PermCycle` below prints out the cycles of a permutation. The user enters the permutation in cell B6 as a comma separated string or lets `RndIntList` do this. The string is then converted into a string array Q. The program then calculates

$$Q(1),\ Q(Q(1)),\ Q(Q(Q(1))), \ldots$$

setting these array members to the null string "" as they are used up. The process continues as far as possible, that is, until a null string string is encountered. If $Q(2)$ is not the null string the program calculates the cycle

$$Q(2),\ Q(Q(2)),\ Q(Q(Q(2))), \ldots$$

This continues until all of the entries of Q are null strings. Here is the code:

```
Sub PermCycles()
    Dim perm As String, m As Integer
    Dim L As Integer, cycle As String
    Dim P() As Integer, Q() As String, prod As String
```

```
        If Range("B3").Value = "y" Then      'if random permutation is desired
            L = Range("B4").Value                     'then get its length
            If L = 0 Then MsgBox "no length in B4": Exit Sub
            ReDim P(1 To L)
            Call RndIntList(P, 1, L)  'populate array P with random integers
            Q = Convert2StringArray(P)
            Range("B6").Value = Join(Q, ",")
            GoTo cycles                              'skip the manual stuff
        End If
        perm = Range("B6").Value                 'read in user-permutation
        If perm = "" Then Exit Sub
        Q = Split1(perm, ",") 'make Split start at index 1 like permutations
        L = UBound(Q)
cycles:
        For m = 1 To L
            cycle = GetCycle(Q, m)                    'cycle starting at m
            If cycle <> "" Then                      'if cycle ended
                prod = prod & "(" & cycle & ")"      'update cycle product
            End If
        Next m
        Range("B7").Value = prod
        End Sub
```

The function `GetCycle` implements the scheme discussed above.

```
Function GetCycle(Q() As String, m As Integer) As String
    Dim k As Integer, j As Integer, cycle As String
    If Q(m) = "" Then GoTo lastline      'perm member already in a cycle
    cycle = m                                 'start of new cycle
    k = Q(m)                                  'next member of cycle
    Q(m) = ""                        'remove it from Q: no longer needed
      'continue running through perm cyclically
    Do While k <> m And Q(k) <> ""
        cycle = cycle & "," & k
        j = Q(k)                                  'next in cycle
        Q(k) = ""                       'remove from Q: no longer needed
        k = j
    Loop
lastline:
    GetCycle = cycle
End Function
```

The function `Convert2StringArray` takes the integer array generated by `RndIntList` and converts it into a string array, since permutations are written as strings.

```
Function Convert2StringArray(P() As Integer) As String()
    Dim i As Integer, Q() As String, L As Integer
    L = UBound(P)
    ReDim Q(1 To L)
    For i = 1 To L
```

```
      Q(i) = CStr(P(i))
   Next i
   Convert2StringArray = Q
End Function
```

19.11   Exercises

1. A restauranteur needs to hire a pastry chef, a roast chef, a vegetable chef, and three fry chefs. If there are 10 applicants equally qualified for the positions, in how many ways can the positions be filled?

2. A bag contains 5 red, 4 yellow, and 3 green marbles. In how many ways is it possible to draw 5 marbles at random from the bag with exactly 2 reds and no more than 1 green?

3. How many different 12-letter arrangements of the letters of the word "arrangements" are there?

4. Work the following problem by hand and then check your answer with **Venn3Solver**. A sample of 100 people resulted in the following data:

   (a) 60 play tennis
   (b) 45 play basketball
   (c) 45 play soccer
   (d) 15 play basketball and soccer
   (e) 25 play basketball and tennis
   (f) 20 play soccer and tennis
   (g) 5 play all three

   How many

   (1) play no sports?
   (2) play exactly one sport?
   (3) play exactly two sports?

5. Work the following problem by hand and then check your answer with **Venn3Solver**. A sample of the music tastes of 100 people resulted in the following data:

   (a) 52 like classical
   (b) 45 like jazz
   (c) 60 like rock
   (d) 25 like classical and jazz
   (e) 30 like classical and rock
   (f) 28 like jazz and rock
   (g) 6 liked none of these

How many people like all three genre's.

6. Work the following problem by hand and then check your answer with `Venn3Solver`. In a certain group of people it was found that

(a) 70 read Spanish
(b) 40 read French
(c) 40 read German
(d) 20 read French and Spanish
(e) 15 read French and German
(f) 25 read Spanish and German
(g) 5 read French, Spanish and German
(h) every person in the group reads in least one of these languages.

How many people are in the group?

7. Work the following problem by hand and then check your answer with `Venn3Solver`. In a certain group of children it was found that

(a) 51 liked apples
(b) 49 liked bananas
(c) 60 liked cantaloupes
(d) 36 apples and cantaloupes
(e) 34 apples and bananas
(f) 32 cantaloupes and bananas
(g) 24 liked all three.
(h) 1 didn't like any
How many children liked exactly two fruits?

8. Write a program `PermutationsNonRecursive` that generates all permutation of the letters of a string in the following manner: If the entered string is `"xzyvu"` then the first letter `"x"` is printed in column 1; the next letter `"z"` is used with the entry column 1 to form the two permutations `"zx"` and `"xz"`, which are printed in column 2; the next letter `"y"` is used with the entries in column 2 to form the six permutations `"yzx","zyx","zxy","yxz","xyz","xzy"`, and so on. At each stage the new letter is inserted in all of the positions of each permutation obtained in the preceding stage.

9. Revise `PrintPath` so that it lays out a colored path along the best route. Use colored L-shaped paths to connect one town to the next.

*10. Write a program `Venn4Solver` that handles four sets. Use the program to solve the following problem: A survey of 300 high school graduates found that of the four sports soccer, lacrosse, basketball, and baseball

- 61 had played soccer
- 68 had played lacrosse
- 64 had played basketball

- 66 had played baseball

- 44 had played soccer and lacrosse

- 30 had played lacrosse and basketball

- 15 had played lacrosse and basketball but not soccer

- 29 had played soccer and lacrosse but not basketball

- 9 had played lacrosse but neither soccer nor basketball

- 24 had played soccer, lacrosse, and baseball

- 12 had played soccer, basketball, and baseball

- 20 had played lacrosse, basketball, and baseball

- 13 had played basketball and baseball but neither soccer nor lacrosse

- 2 had played lacrosse and baseball but neither soccer nor basketball

- 13 had played basketball and baseball but neither soccer nor lacrosse

- 1 had played soccer, and baseball but neither lacrosse nor basketball

- 8 had played soccer and basketball but neither lacrosse nor baseball

- 14 had played soccer and lacrosse but neither basketball nor baseball

- 9 had played all four sports.

How many

(1) play no sports? (2) play exactly one sport? (3) play exactly two sports?

*11. Write a program **PermCalc** that reads from the spreadsheet a user-entered string of permutations such as

$$(3,4,2,1,5)(2,1,5,3,4)\char`^(-3)(3,5,1,4,2)\char`^2$$

and prints its value. Use the functions of Section 19.9.

Part V

Probability

Chapter 20

Probability

A *probability* is a numerical measure of the likelihood of some event. The term *event* refers to an outcome of a *random experiment*. Experiments in this context are simply repeatable activities that result in well-defined, observable consequences. Familiar examples are the toss of a coin, the roll of a die, or the drawing of a card from a deck. In this chapter we discuss the basics of probability theory and show how VBA may be used to simulate probability experiments.

20.1 Sample Spaces, Probabilities, and Events

The set of all outcomes of a random experiment is called the *sample space* of the experiment and is typically denoted by Ω. For example, tossing a single die and noting the number of dots appearing on the top face is an experiment whose outcomes are the integers 1 through 6. The sample space of the experiment is the set $\Omega = \{1, 2, 3, 4, 5, 6\}$. In probability theory one starts by assigning probabilities to the outcomes of the experiment, that is, to the members of the sample space. The probability of an the outcome ω is denoted by $\mathbb{P}(\omega)$. If the sample space consists of the outcomes $\{\omega_1, \omega_2, \ldots\}$ we shall call the set of probabilities $\mathbb{P}(\omega_1), \mathbb{P}(\omega_2), \ldots$ the *probability distribution of the experiment*. The only condition on these numbers is that they be nonnegative and sum to 1. In the die example, if the die is fair then each outcome is equally likely hence the probability distribution is

$$\mathbb{P}(1) = \mathbb{P}(2) = \mathbb{P}(3) = \mathbb{P}(4) = \mathbb{P}(5) = \mathbb{P}(6) = 1/6.$$

Another example of a random experiment is the flip of a fair coin. If the coin is flipped twice, then the sample space Ω may be written as $\{HH, HT, TH, TT\}$, where, for example, the notation HT signifies that the coin came up heads on the first flip and tails on the second. Since the coin is fair we would expect these outcomes to be equally likely and therefore assign the probability 1/4 to each. Thus the probability distribution of the experiment is

$$\mathbb{P}(HH) = \mathbb{P}(HT) = \mathbb{P}(TH) = \mathbb{P}(TT) = 1/4.$$

DOI: 10.1201/9781003351689-20

An *event* in a random experiment is a set of outcomes, that is, a subset of Ω.[1] The probability of an event $A \subseteq \Omega$ is denoted by $\mathbb{P}(A)$ and is defined as the sum of the probabilities $\mathbb{P}(\omega)$ of the outcomes ω comprising A. This may be described symbolically with summation notation:

$$\mathbb{P}(A) = \sum_{\omega \in A} \mathbb{P}(\omega).$$

If all outcomes of a random experiment are equally likely, as was the case in the coin flip and dice toss examples, then the values $\mathbb{P}(\omega)$ are all the same and so must equal $1/n(\Omega)$, where the notation $n(A)$ means the number (cardinality) of outcomes in a set A (see Chapter 19). It follows that

$$\mathbb{P}(A) = \underbrace{\frac{1}{n(\Omega)} + \frac{1}{n(\Omega)} + \cdots + \frac{1}{n(\Omega)}}_{n(A) \text{ terms}} = \frac{n(A)}{n(\Omega)},$$

The determination of probabilities in this case therefore reduces to the calculation of cardinalities of events. For example, the probability that a single toss of a fair die produces an outcome of at least 4 is $3/6 = .5$, since there are three equally likely ways this can happen. For another example, consider a voting population of 100 people of which 45 prefer candidate **A** and 55 prefer candidate **B**. The probability that a person chosen at random from the population prefers candidate **A** is .45.

If A and B are events in a probability experiment, then, of course, so are the sets $A + B$, AB, and A'. In probability theory these are frequently referred to as "A or B," "A and B," and "not A," respectively, Event probabilities have the following properties:

(a) If $AB = \emptyset$, then $\mathbb{P}(A + B) = \mathbb{P}(A) + \mathbb{P}(B)$.

(b) $\mathbb{P}(A') = 1 - \mathbb{P}(A)$.

(c) $\mathbb{P}(A + B) = \mathbb{P}(A) + \mathbb{P}(B) - \mathbb{P}(AB)$.

To verify (a), suppose that A consists of the outcomes $a_1, \ldots, a_k$ and B the outcomes $b_1, \ldots, b_m$. Since the events don't overlap

$$\mathbb{P}(A + B) = \mathbb{P}(a_1) + \cdots + \mathbb{P}(a_k) + \mathbb{P}(b_1) + \cdots + \mathbb{P}(b_m) = \mathbb{P}(A) + \mathbb{P}(B).$$

For (b), simply apply part (a) to $B = A'$ and use the fact that $A + B = \Omega$ and $\mathbb{P}(\Omega) = 1$. The proof of (c) may be seen to be similar to that of (a) by noting that subtracting $\mathbb{P}(AB)$ compensates for the overlapping members of $A + B$.

Events $A_1, \ldots, A_n$ are said to be *mutually exclusive* if $A_j A_j = \emptyset$ for $i \neq j$. Property (a) may be extended to such events:

$$\mathbb{P}(A_1 + A_2 + \cdots + A_n) = \mathbb{P}(A_1) + \mathbb{P}(A_2) + \cdots + \mathbb{P}(A_n).$$

The property is called the *addition rule for probabilities*.

---

[1]Sets are discussed in Chapter 18. The reader who has not read the chapter should refer to the relevant concepts there as needed.

20.2 Throwing a Pair of Fair Dice

In this and the following two sections we give some common examples of random experiments. Others appear in the exercises.

Consider the experiment of throwing of a pair of distinguishable fair dice, say a red die and a green die. The outcomes of the toss, that is, the face up numbers of dots on each die may be symbolized by the ordered pairs (i, j), where i is the number of dots that are face up on the red die and j is the number of dots face up on the green die.

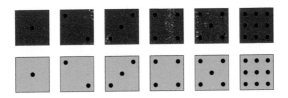

The sample space then consists of all ordered pairs (i, j), where $1 \leq i, j \leq 6$. As there are 36 equally likely outcomes, the probability of each toss is $1/36$.

We could also consider the numbers $2, \ldots, 12$ as outcomes, these being the sums of the dots on the top faces. For example, the outcome 7 is represented by the event $\{(1, 6), (2, 5), (3, 4), (4, 3), (5, 2), (6, 1)\}$ and so has probability $6/36 = 1/6$. Figure 20.1 gives theoretical probabilities of the outcomes $2, \ldots, 12$ in chart form.

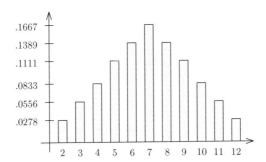

FIGURE 20.1: Theoretical probabilities for fair dice throw.

The following program simulates the experiment by "tossing" the dice many times and calculating the relative frequency of the outcomes $2, \ldots, 12$. Relative frequencies are sometimes called *empirical probabilities* in contrast to the *theoretical probabilities* obtained by logical deduction. The use of empirical probabilities is sometimes necessary as theoretical probabilities may be difficult to deduce. The justification for the use of empirical probabilities is the law of large numbers, discussed in the next chapter.

The procedure below tabulates the frequency of the sums red + green, where red is the number on the red die and green the number on the green die. The numbers are chosen with the function RndInt (Section 10.1). A For Next loop carries out the throws NumThrows times, a value entered by the user in cell B5. The array freq counts the number of times a particular sum red + green comes up. For example, if a toss comes up red = 4 and green = 3 then the array entry freq(7) is incremented by one. A For Next loop prints the sums 2–12 and their relative frequencies (frequency divided by NumThrows). The procedure then graphs the probabilities.

```
Sub DiceRelativeFreq()
    Dim freq(2 To 12) As Long, red As Integer, green As Integer
    Dim k As Long, NumThrows As Long
    Range("B7:C38").NumberFormat = ".000000"              'format
    NumThrows = Range("B5").Value        'user entered no. of dice throws
    For k = 1 To NumThrows               'toss the dice this many times
        red = RndInt(1, 6)                        'throw red die
        green = RndInt(1, 6)                       'throw red die
        freq(red + green) = freq(red + green) + 1   'increment freq.
    Next k
    For k = 2 To 12          'print sums and their relative frequencies
        Cells(6 + k, 1).Value = k
        Cells(6 + k, 2).Value = freq(k) / NumThrows      'form averages
    Next k
    With Sheet1 _
        .ChartObjects.Add(Left:=250, Width:=300, Top:=80, Height:=200)
        .Chart.SetSourceData Source:=Range("B8:B18")        'y values
        .Chart.ChartType = xlColumnClustered              'chart type
        .Chart.HasTitle = True
        .Chart.ChartTitle.Text = "Relative Frequencies"
        .Chart.SeriesCollection(1).XValues = Range("A8:A18")  'x values
    End With
End Sub
```

20.3 Poker Hands

A poker deck consists of 52 cards with 13 *denominations*, labeled 2–10, and Jack, Queen, King, and Ace. Each denomination has four *suits*: hearts, diamonds, clubs, and spades. A poker hand consists of five cards chosen randomly from the deck. The total number of poker hands is therefore

$$\binom{52}{5} = 2,598,960.$$

There are many potentially winning hands. We calculate the probabilities of four of these, rounding in each case.

- *Single pair*: To find the number of such hands first select a denomination for the pair (13 choices), then select a pair from the four cards in the denomination, ($\binom{4}{2} = 6$ choices), then select the remaining three cards, avoiding cards in previously chosen denominations ($48 \cdot 44 \cdot \cdot 40/3! = 14,080$ choices). By the multiplication principle the total number of hands with a single pair is therefore $13 \cdot 6 \cdot 14,080 = 1,098,240$ hence the probability of getting a single pair is

$$\frac{1,098,240}{2,598,960} = 0.4225690376$$

- *Two distinct pairs*: The number of such hands is obtained as follows: First choose two denominations for the pairs ($\binom{13}{2} = 78$ choices), then choose two cards from each denomination ($\left(\binom{4}{2}\right)^2 = 36$ choices), and finally choose the remaining card avoiding the selected denominations (44 choices). The number of possible hands with two distinct pairs is therefore

$$78 \cdot 36 \cdot 44 = 123,552$$

and so the probability of such a hand is

$$\frac{123,552}{2,598,960} = 0.0475390156$$

- *Three of a kind* (the remaining cards not a pair): To find the number of such hands first choose a denomination for the triple (13 choices). Then select three cards from that denomination (4 choices). Finally, choose the remaining two cards, avoiding the denomination for the triple as well as pairs ($\frac{48 \cdot 44}{2} = 1056$) choices. (The divisor 2 is needed since, for example, the choice $5, 8$ is the same as the choice $8, 5$.) Thus the number of poker hands with three of a kind is $13 \cdot 4 \cdot 1056 = 54,912$ and so the probability of such a hand is

$$\frac{54,912}{2,598,960} = .0211284514$$

- *Full house* (3 cards of one denomination and 2 cards of another, for example, three kings and two jacks): For the number of such hands first choose denominations for the triple and pair ($\binom{13}{2} = 78$ choices). Then select 2 cards from one denomination ($\binom{4}{2} = 6$ choices) and 3 cards from the other denomination (4 choices). Thus the number of full house poker hands is $78 \cdot 6 \cdot 4 \cdot 2 = 3744$ so the probability for such a hand is

$$\frac{3744}{2,598,960} = .001440576$$

- *Four of a kind*: Choose a denomination (13 choices) for the quadruple then choose 1 card from the remaining 48, giving $13 \cdot 48 = 624$ choices. Thus the probability for such a hand is

$$\frac{624}{2,598,960} = .000240096$$

The following program simulates these probabilities by "drawing" many hands, counting the number of these that result in one of the hands and dividing by the number of trials.

```
Public Num1Pairs As Long, Num2Pairs As Long, NumTriples As Long
Public NumFull As Long, NumQuads As Long, NumTrials As Long
Sub PokerHand()
    Dim deck(1 To 52) As Integer, hand(1 To 5) As Integer, k As Long
    Num1Pairs = 0: Num2Pairs = 0: NumTriples = 0: NumFull = 0
    NumQuads = 0: NumTrials = Range("B3").Value           'initialize
    For k = 1 To NumTrials                              'simulate hands
        Call MakeDeck(deck)                    'make a new deck each time
        Call DrawHand(hand, deck)
        Call Eval(hand)                                'get no. pairs, etc.
    Next k
    Range("B5").Value = Num1Pairs / NumTrials        'relative frequencies
    Range("B6").Value = Num2Pairs / NumTrials
    Range("B7").Value = NumTriples / NumTrials
    Range("B8").Value = NumFull / NumTrials
    Range("B9").Value = NumQuads / NumTrials
End Sub

Sub MakeDeck(deck() As Integer)
    Dim i As Integer, j As Integer
    For i = 1 To 52
          'use either one"
        'deck(i) = 1 + (i - 1) Mod 13 '1,2,3,4,5,6,7,8,9,10,11,12,13,...
        deck(i) = 1 + Int((i - 1) / 4)            '1,1,1,1,2,2,2,2,...
    Next i
End Sub
```

The procedure **DrawHand** simulates the random drawing of five cards.

```
Sub DrawHand(hand() As Integer, deck() As Integer)
    Dim k As Integer, i As Integer
    Randomize
    For i = 0 To 4
        k = 1 + Int(Rnd * (52 - i))          'randomly choose kth card from
        hand(i + 1) = deck(k)                             'remaining deck
        deck(k) = deck(52 - i)               'replace by current last card
    Next i
End Sub
```

The procedure **Eval** finds the number of pairs, triples, and full houses by making a frequency chart of the denominations.

```
Sub Eval(hand() As Integer)
    Dim i As Integer, DenFreq(1 To 13) As Integer
    Dim NumPairs As Integer, NumTrips As Integer
    For i = 1 To 5                      'run through hand, tallying frequency of
        DenFreq(hand(i)) = DenFreq(hand(i)) + 1    'the card denomination
```

```
    Next i
    For i = 1 To 13              'check for pairs, triples, and quadruples
        If DenFreq(i) = 2 Then NumPairs = NumPairs + 1
        If DenFreq(i) = 3 Then NumTrips = NumTrips + 1
        If DenFreq(i) = 4 Then NumQuads = NumQuads + 1
    Next i
                    'update count of pairs, etc.
    If NumPairs = 0 And NumTrips = 1 Then NumTriples = NumTriples + 1
    If NumPairs = 1 And NumTrips = 0 Then Num1Pairs = Num1Pairs + 1
    If NumPairs = 2 Then Num2Pairs = Num2Pairs + 1
    If NumPairs = 1 And NumTrips = 1 Then NumFull = NumFull + 1
End Sub
```

20.4   Drawing Balls from an Urn

A ball is randomly drawn from an urn containing five red balls and six yellow balls. The likelihood of drawing a particular ball is the same for all balls, namely $1/11$.

Since there are five ways to draw a red ball, the probability of doing so is $5/11$. Similarly, the probability of drawing a yellow ball is $6/11$. We now draw five balls, replacing each ball before drawing the next and ask What is the probability of drawing the particular sequence RRRYY, that is, first three red balls and then two yellow balls? To answer this we use the multiplication principle described in Section 19.3. Since there are five red balls, each may be chosen in any one of 5 ways. Similarly, since there are six yellow balls, each of these may be chosen in any one of six ways. By the multiplication principle, the sequence RRRYY may be drawn in any one of $5 \cdot 5 \cdot 5 \cdot 6 \cdot 6$ ways. Since there are $11 \cdot 11 \cdot 11 \cdot 11 \cdot 11$ possible sequences, each equally likely to appear, the probability of the sequence RRRYY is

$$\frac{5 \cdot 5 \cdot 5 \cdot 6 \cdot 6}{11 \cdot 11 \cdot 11 \cdot 11 \cdot 11} = \left(\frac{5}{11}\right)^3 \left(\frac{6}{11}\right)^2.$$

If we set

$$p = \frac{5}{11} \quad \text{and} \quad q = \frac{6}{11} \ (= 1 - p)$$

these being, respectively, the probabilities of drawing a red ball and yellow ball, then we may write the result symbolically as $\mathbb{P}(\text{RRRYY}) = p^3 q^2$. Notice that the order of the R's and Y's is immaterial; only the number of each matters. Thus one also has $\mathbb{P}(\text{YRRYR}) = p^3 q^2$.

Now we ask: if we draw five balls with replacement, what is the probability of getting three red balls and two yellow balls in any order? The event in question now consists of *all* sequences XXXXX with 3 R's and 2 Y's, not just the particular sequence RRRYY. For the solution, note that there are $\binom{5}{3}$ such sequences, corresponding to the number of ways of choosing subsets of size 3 from the set of X's. As noted above, each such sequence has probability $p^3 q^2$. Therefore, the desired probability is the sum of these, namely $\binom{5}{3} p^3 q^2$.

More generally, suppose that the urn contains r red balls and y yellow balls. If a sample of size n is drawn with replacement, then an argument similar to the one in the preceding paragraph shows that the probability of drawing exactly k red balls in this sample is

$$\binom{n}{k} p^k q^{s-k}, \quad 0 \le k \le n, \tag{20.1}$$

where

$$p = \frac{r}{r+y} \quad \text{and} \quad q = \frac{y}{r+y} \ (= 1 - p), \quad 0 \le k \le n.$$

Note that the expression in (20.1) is the kth term (starting at 0) of the binomial expansion of $(p + q)^n$ (see Section 19.6). The probabilities in (20.1) form the so-called *binomial distribution*. Figure 20.2 graphs the binomial probabilities for $n = 6$ and three values of p.

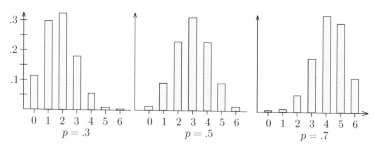

FIGURE 20.2: Binomial probability distribution.

The urn experiment described above is an example of what is called *sampling with replacement*. The analogous experiment of drawing the balls without returning them to the urn is called *sampling without replacement* and leads to the so-called *hypergeometric distribution*. For an application, let the balls represent individuals in a population of size N from which a sample of size n is randomly selected and then polled as to preferences. For example, red balls may represent individuals in the sample who favor candidate **A** for political office, and yellow balls those who favor candidate **B**. Pollsters use the sample to estimate the fraction of people in the general population who favor a particular candidate. Since a person is not questioned twice as to their preference, the experiment is sampling without replacement.

The urn problem may be cast in the following more general setting: Call each drawing a *trial* and call the selection of a red ball a *success* and the selection of a yellow ball a *failure*. Equation (20.1) then gives the probability of k successes in n trials. Note that because the urn configuration is restored after each drawing the trials are *independent*, that is, the probabilities for each trial are the same and do not depend on results of previous trials. Any experiment consisting of n independent trials, each of which has two outcomes, arbitrarily called *success* and *failure*, with respective probabilities p and $q := 1 - p$ has the binomial probability distribution (20.1).

20.5 Simulation of the Binomial Distribution

An interesting visual realization of the binomial distribution was developed by the English statistician Francis Galton. His device, called a Galton board, is an upright board with rows of equally spaced pegs in the shape of a triangle. A ball is dropped above the top peg and meanders down the triangle. At each peg the ball goes either left or right with equal probability. At the end of its journey it falls into one of the bins under the bottom row of pegs. If the rows of pegs represent the trials and we arbitrarily call a movement to the left a success and a movement to the right a failure then the ball's motion through the pegs is governed by a binomial probability law and after many balls the bins take on the approximate shape of a binomial probability distribution. Figure 20.3 illustrates a red ball's progress from the top peg to a bin.

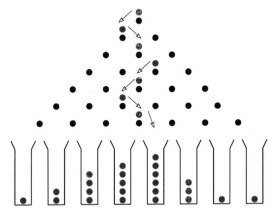

FIGURE 20.3: Red ball's progress and the result of many balls.

The module `GaltonsBoard` in ths section displays a red cell (ball) bouncing off black cells (the pegs) until it reaches the last row, when it is placed in the column underneath it. The columns grow downward to allow bins of arbitrary size (Figure 20.4). The main procedure `Galton` begins by retrieving user-entered

FIGURE 20.4: Grid activity.

information: the number of desired peg rows, numbered 1 to N, the probability `ProbLeft` of left motion of the ball (in case the board is slightly tilted), the number `NumBalls` of balls sent down the board in any one run, and the delay value `DelVal` of the ball's motion. The number N is stored in Sheet3 since it is used to make grids of suitable size. If the user changes this value, the grid will adjust automatically. After the grid is set the procedure calls `DrawGalton` to draw the pegs and then calls `RollBalls` to send balls cascading down the board.

```
Public startrow As Integer, startcol As Integer
Public ProbLeft As Single, NumBalls As Integer, DelVal
Sub Galton()
    Dim Grid As TGrid, height As Integer, breadth As Integer, N As Integer
    Dim T As Integer, L As Integer, B As Integer, R As Integer
    N = Range("B3").Value                 'peg rows numbered 1 to N
    Sheet3.Cells(9, 1).Value = N
    ProbLeft = Range("B4").Value              'probability of left motion
    NumBalls = Range("B5").Value               'needed for ball bins
    DelVal = Range("B6").Value              'large value slows down ball
    height = 6 * N
    breadth = 4 * N + 2                   'grid specs
    T = 8: L = 4: B = T + height: R = L + breadth
    Grid = MakeGrid(T, L, B, R, 0.7, 5, -1)
    startrow = T + 1:                      'initial position of ball
    Call DrawGalton(N)
    Call RollBalls(N)
End Sub
```

The procedure `DrawGalton` draws the pegs as black cells. It begins by coloring the top cell at (`startrow,startcol`) then moves down the rows by two cells. Each row starts two cells to the left of the previous row. Black cells in a particular row are printed every 4th cell.

```
Sub DrawGalton(N As Integer)
    Dim nextrow As Integer, leftcol As Integer, rightcol As Integer
    Dim i As Integer, j As Integer
    Cells(startrow, startcol).Interior.ColorIndex = 1    'peg color black
    nextrow = startrow
    leftcol = startcol                                   'initial values
    For i = 1 To N   'draw the pegs
        nextrow = nextrow + 2                    'rows separated by 2 cells
        leftcol = leftcol - 2                    'next column 2 units to left
        For j = 0 To i  'run through the row, print black every 4th cell
            Cells(nextrow, leftcol + 4 * j).Interior.ColorIndex = 1
        Next j
    Next i
End Sub
```

The procedure `RollBalls` iterates `RollBall`. The latter causes a ball (red cell) to drop through the grid. The direction is determined by `ProbLeft` and the function `Rnd`.

```
Sub RollBalls(N As Integer)
    Dim i As Integer
    For i = 1 To NumBalls                               'this many trials
        Call RollBall(N)
    Next i
End Sub
```

```
Sub RollBall(N As Integer)
    Dim nextrow As Integer, nextcol As Integer, leftcol As Integer
    Dim i As Integer
    nextrow = startrow - 1                   'start ball above first peg
    nextcol = startcol
    Cells(nextrow, nextcol).Interior.ColorIndex = 3         'color ball
    Call Delay(DelVal)                             'delay for effect
    Cells(nextrow, nextcol).Interior.ColorIndex = 0    'now remove color
    For i = 1 To N                                 'run through the rows
        nextrow = nextrow + 2               'next ball position two below
        Randomize: x = Rnd()                      'get a random number
        If x < ProbLeft Then
            nextcol = nextcol - 2                      'ball goes left
        Else
            nextcol = nextcol + 2                      'ball goes right
        End If
        Cells(nextrow, nextcol).Interior.ColorIndex = 3      'color ball
        Call Delay(DelVal)                             'delay for effect
        Cells(nextrow, nextcol).Interior.ColorIndex = 0'now remove color
    Next i
    Call PlaceOnColumn(nextrow + 2, nextcol)       'add ball to its column
End Sub
```

Th procedure `PlaceOnColumn` searches for the bottom of the column and places the ball below it.

```
Private Sub PlaceOnColumn(row As Integer, col As Integer)
Dim i As Integer
    i = row
    Do While Cells(i, col).Interior.ColorIndex = 3      'find bottom of
        i = i + 1                                       'ball column
    Loop
    Cells(i, col).Interior.ColorIndex = 3           'place ball on bottom
End Sub
```

20.6   Conditional Probability

It may happen that additional information emerges in an ongoing random experiment that changes the probabilities of outcomes. For example, suppose in a group of 60 women and 40 men that 20 women and 15 men prefer classical music to jazz. The data may be pictured as in Figure 20.5. If we randomly

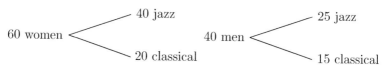

60 women   40 jazz   40 men   25 jazz

20 classical   15 classical

FIGURE 20.5: Preference sample.

select an individual from the group, then the probability that the person prefers classical music to jazz is $(20 + 15)/100 = 7/20$. The sample space in this case is the set of 100 people. However, if we incorporate the knowledge that the person selected is a woman, then sample space reduces to the set of 60 women, and the new probability is $20/60 = 1/3$. We can symbolize these calculations as follows: Let A denote the event that the person selected prefers classical music to jazz and B the event that the person selected is a women. The original probability is

$$\mathbb{P}(A) = \frac{n(A)}{n(\Omega)}$$

and the new probability is

$$\frac{n(AB)}{n(B)} = \frac{n(AB)/n(\Omega)}{n(B)/n(\Omega)} = \frac{\mathbb{P}(AB)}{\mathbb{P}(B)}.$$

The example suggests the following general definition: If A and B are events in an experiment and $\mathbb{P}(B) > 0$, then the *conditional probability of A given B* is defined as

$$\mathbb{P}(A \mid B) = \frac{\mathbb{P}(AB)}{\mathbb{P}(B)}. \qquad (20.2)$$

Note that $\mathbb{P}(A \mid B)$ is undefined if $\mathbb{P}(B) = 0$.

If we solve (20.2) for $\mathbb{P}(AB)$, we obtain

$$\mathbb{P}(AB) = \mathbb{P}(A \mid B)\mathbb{P}(B). \qquad (20.3)$$

For three events we apply the last result twice to obtain

$$\mathbb{P}(ABC) = \mathbb{P}(A \mid BC)\mathbb{P}(BC) = \mathbb{P}(A \mid BC)\mathbb{P}(B \mid C)\mathbb{P}(C).$$

Arguing repeatedly in this manner and we arrive at the *multiplication rule for conditional probabilities*:

If $A_1, A_2, \ldots, A_n$ are events with $\mathbb{P}(A_1 A_2 \cdots A_{n-1}) > 0$, then
$\mathbb{P}(A_1 A_2 \cdots A_n) = \mathbb{P}(A_1)\mathbb{P}(A_2 \mid A_1)\mathbb{P}(A_3 \mid A_1 A_2) \cdots \mathbb{P}(A_n \mid A_1 A_2 \cdots A_{n-1})$.

Example 20.1. A jar contains 5 red and 6 yellow marbles. Randomly draw 3 marbles in succession without replacement. Let R_1 denote the event that the first marble is red, R_2 the event that the second marble is red, and Y_3 the event that the third marble is yellow. By the multiplication rule the probability that the first two marbles are red and the third is yellow is

$$\mathbb{P}(R_1 R_2 Y_3) = \mathbb{P}(R_1)\mathbb{P}(R_2 \mid R_1)\mathbb{P}(Y_3 \mid R_1 R_2) = (5/11)(4/10)(6/9) \approx .12 \quad \Diamond$$

20.7 Bayes' Theorem

One of the most important applications of conditional probability is Bayes' Theorem. To set the stage, let A and B be events with positive probabilities, so that the conditional probabilities

$$\mathbb{P}(A \mid B) = \frac{\mathbb{P}(AB)}{\mathbb{P}(B)} \quad \text{and} \quad \mathbb{P}(B \mid A) = \frac{\mathbb{P}(BA)}{\mathbb{P}(A)}$$

are defined. Solving for $\mathbb{P}(BA) = \mathbb{P}(AB)$ in the second equation and substituting the result in the first we have

$$\mathbb{P}(A \mid B) = \frac{\mathbb{P}(B \mid A)\,\mathbb{P}(A)}{\mathbb{P}(B)} \qquad (20.4)$$

Since $B = BA + BA'$ we can use the addition and multiplication rules for probabilities to write the denominator as

$$\mathbb{P}(B) = \mathbb{P}(BA) + \mathbb{P}(BA') = \mathbb{P}(B \mid A)\mathbb{P}(A) + \mathbb{P}(B \mid A')\mathbb{P}(A').$$

Substituting this into (20.4) we arrive at *Bayes' Theorem*:

$$\mathbb{P}(A \mid B) = \frac{\mathbb{P}(B \mid A)\mathbb{P}(A)}{\mathbb{P}(B \mid A)\mathbb{P}(A) + \mathbb{P}(B \mid A')\mathbb{P}(A')}. \qquad (20.5)$$

The point here is that the conditional probability $\mathbb{P}(A \mid B)$, where the event B is the given information, is expressed in terms of "reversed" conditional probabilities, where now A is the given information.

Bayes' theorem may be extended to arbitrarily many events $A_1, A_2, \ldots A_n$, where

$$\Omega = A_1 + A_2 + \cdots + A_n \quad \text{and} \quad A_i A_j = \emptyset \ (i \neq j).$$

Indeed, by (20.4),

$$\mathbb{P}(A_k \mid B) = \frac{\mathbb{P}(B \mid A_k)\, \mathbb{P}(A_k)}{\mathbb{P}(B)} \tag{20.6}$$

Using the expansion

$$\mathbb{P}(B) = \mathbb{P}(BA_1) + \mathbb{P}(BA_1) + \cdots + \mathbb{P}(BA_n)$$
$$= \mathbb{P}(B \mid A_1)\mathbb{P}(A_1) + \mathbb{P}(B \mid A_2)\mathbb{P}(A_2) + \cdots + \mathbb{P}(B \mid A_n)\mathbb{P}(A_n).$$

we obtain the following general version of Bayes' theorem:

$$\mathbb{P}(A_k \mid B) = \frac{\mathbb{P}(B \mid A_k)\mathbb{P}(A_k)}{\mathbb{P}(B \mid A_1)\mathbb{P}(A_1) + \mathbb{P}(B \mid A_2)\mathbb{P}(A_2) + \cdots + \mathbb{P}(B \mid A_n)\mathbb{P}(A_n)}.$$

Example 20.2. A factory has three machines that manufacture knortin rods. The following table gives data regarding production and defects. What is the probability that a randomly chosen defective knortin rod comes from machine 3?

machine	% of production	% of defects
1	40	1
2	35	2
3	25	3

To solve the problem, let B denote the event that a defective rod was chosen and let A_k be the event that the rod came from machine k. We then have the probabilities $\mathbb{P}(A_1) = .4$, $\mathbb{P}(A_2) = .35$, $\mathbb{P}(A_3) = .25$, and $\mathbb{P}(B \mid A_k) = (.01)k$. Thus by Bayes' theorem,

$$\mathbb{P}(A_3 \mid B) = \frac{\mathbb{P}(B \mid A_3)\mathbb{P}(A_3)}{\mathbb{P}(B \mid A_1)\mathbb{P}(A_1) + \mathbb{P}(B \mid A_2)\mathbb{P}(A_2) + \mathbb{P}(B \mid A_3)\mathbb{P}(A_3)}$$

$$= \frac{(.03)(.25)}{(.01)(.4) + (.02)(.35) + (.03)(.25)}$$

$$= \frac{75}{40 + 70 + 75}$$

$$\approx .405 \qquad \qquad \diamond$$

20.8 Disease Testing

Suppose a population is tested for a certain disease. Say the test has a *sensitivity rate* of 98%, that is, 98% of people with the disease test positive; a *specificity rate* of 97%, that is, 97% of people without the disease test negative; and a *prevalence rate* of 1%, that is, 1% of the population has the disease. A person is randomly chosen from the population. If the person tests positive for the disease, what is the probability that the person actually has the disease? If the person tests negative for the disease, what is the probability that the person does not have the disease?

To answer the questions, let D be the event "the person has the disease" and P the event "the person tests positive." The data may then be expressed in the following form:

$$\mathbb{P}(D) = .01, \quad \mathbb{P}(P \mid D) = .98 \quad \text{and} \quad \mathbb{P}(P' \mid D') = .97,$$

It follows that

$$\mathbb{P}(P' \mid D) = 1 - \mathbb{P}(P \mid D) = .02 \quad \text{and} \quad \mathbb{P}(P \mid D') = 1 - \mathbb{P}(P' \mid D') = .03.$$

By Bayes' theorem, the desired probability in the first question (rounded to two places) is

$$\mathbb{P}(D \mid P) = \frac{\mathbb{P}(P \mid D)\mathbb{P}(D)}{\mathbb{P}(P \mid D)\mathbb{P}(D) + \mathbb{P}(P \mid D')\mathbb{P}(D')} = \frac{(.98)(.01)}{(.98)(.01) + (.03)(.99)} = .25$$

Thus only one out of four people that test positive actually have the disease.

The answer to the second question is

$$\frac{\mathbb{P}(P' \mid D')\mathbb{P}(D')}{\mathbb{P}(P' \mid D')\mathbb{P}(D') + \mathbb{P}(P' \mid D)\mathbb{P}(D)} = \frac{(.97)(.99)}{(.97)(.99) + (.02)(.01)} = .99.$$

Thus it virtually certain that if you test negative you don't have the disease. More generally, set

$$
\begin{aligned}
PR &= \mathbb{P}(D) && \text{prevalence} \\
SE &= \mathbb{P}(P \mid D) && \text{sensitivity} \\
SP &= \mathbb{P}(D' \mid P') && \text{specificity} \\
TP &= \mathbb{P}(D \mid P) && \text{probability of a true positive} \\
TN &= \mathbb{P}(D' \mid P') && \text{probability of a true negative}
\end{aligned}
$$

The first three are the given data. The remaining quantities are calculated in terms of this data by Bayes' theorem:

$$TP = \frac{\mathbb{P}(P \mid D)\mathbb{P}(D)}{\mathbb{P}(P \mid D)\mathbb{P}(D) + \mathbb{P}(P \mid D')\mathbb{P}(D')} = \frac{SE \cdot PR}{SE \cdot PR + (1 - SP) \cdot (1 - PR)}$$

$$TN = \frac{\mathbb{P}(P' \mid D')\mathbb{P}(D')}{\mathbb{P}(P' \mid D')\mathbb{P}(D') + \mathbb{P}(P' \mid D)\mathbb{P}(D)} = \frac{SP \cdot (1 - PR)}{SP \cdot (1 - PR) + (1 - SE) \cdot PR}$$

Note also that the probability of a false positive is $\mathbb{P}(D' \mid P) = 1 - TP$ and the probability of a false negative $\mathbb{P}(D \mid P') = 1 - TN$.

Here's a simple program that graphs the true positive and true negatives against the prevalence rate. The user enters the sensitivity and specificity in cells B4 and B5, respectively.

```
Sub TruePosTrueNeg()
    Dim NumPoints As Integer, Inc As Double, j As Integer
    Dim XRange As String, YRange As String, prevalence As Single
    Dim sensitivity As Single, specificity As Single
    If ChartObjects.Count > 0 Then ChartObjects.Delete  'clear old chart
    Call ClearColumns(8, 3, 5)                                'and data
    Inc = 0.0005: NumPoints = 40                        'graph parameters
    sensitivity = Range("B4").Value
    specificity = Range("B5").Value
    For j = 1 To NumPoints                        'calculate the formulas
        Cells(j + 7, 1).Value = prevalence
        Cells(j + 7, 2).Value = sensitivity * prevalence /
        (sensitivity * prevalence + (1 - specificity) * (1 - prevalence))
        Cells(j + 7, 3).Value = specificity * (1 - prevalence) /
        (specificity * (1 - prevalence) + (1 - sensitivity) * prevalence)
        prevalence = prevalence + Inc
    Next j
    XRange = "A" & 8 & ":" & "A" & 7 + NumPoints
    YRange = "B" & 8 & ":" & "B" & 7 + NumPoints
    Call MakeChart(xlXYScatter, XRange, YRange, 250, 400, 50, 350, _
        "True Positive")
    XRange = "A" & 8 & ":" & "A" & 7 + NumPoints
    YRange = "C" & 8 & ":" & "C" & 7 + NumPoints
    Call MakeChart(xlXYScatter, XRange, YRange, 650, 400, 50, 350, _
        "True Negative")
End Sub

Sub MakeChart(CType As Integer, XRange As String, YRange As String, _
            L As Integer, W As Integer, T As Integer, H As Integer, _
            title As String)
    With Sheet1 _
        .ChartObjects.Add(Left:=L, Width:=W, Top:=T, Height:=H)
        .Chart.SetSourceData Source:=Range(YRange)          'y values
        .Chart.ChartType = CType                          'chart type
        .Chart.HasTitle = True
        .Chart.ChartTitle.Text = title
        .Chart.SeriesCollection(1).XValues = Range(XRange)  'x values
        .Chart.Axes(xlCategory).HasTitle = True
        .Chart.Axes(xlCategory).AxisTitle.Text = "Prevalence"
        .Chart.Axes(xlValue, xlPrimary).HasTitle = True
        .Chart.Axes(xlValue, xlPrimary).AxisTitle.Characters.Text = _
            "Probabilities"
    End With
End Sub
```

```
Sub ClearColumns(row As Integer, col1 As Integer, col2 As Integer)
    Dim lrow As Long
    If IsEmpty(Cells(row, col1)) Then Exit Sub
    lrow = Cells(Rows.Count, col1).End(xlDown).row
    Range(Cells(row, col1), Cells(lrow, col2)).ClearContents
End Sub
```

20.9 Independence

Events A and B in a probability experiment are said to be *independent* if

$$\mathbb{P}(AB) = \mathbb{P}(A)\mathbb{P}(B).$$

If A and B have positive probabilities then independence implies that

$$\mathbb{P}(A \mid B) = \frac{\mathbb{P}(AB)}{\mathbb{P}(B)} = \frac{\mathbb{P}(A)\mathbb{P}(B)}{\mathbb{P}(B)} = \mathbb{P}(A) \text{ and}$$

$$\mathbb{P}(B \mid A) = \frac{\mathbb{P}(AB)}{\mathbb{P}(A)} = \frac{\mathbb{P}(A)\mathbb{P}(B)}{\mathbb{P}(A)} = \mathbb{P}(B).$$

Thus the information imparted by either event does not affect the probability of the other.

Example 20.3. In the dice throw experiment, let A be the event that the sum of the dice is 7, B the event that the sum of the dice is 8, and C the event that the first die is even. Then $\mathbb{P}(A) = 1/6$, $\mathbb{P}(B) = 5/36$, $\mathbb{P}(C) = 1/2$, and $\mathbb{P}(AC) = \mathbb{P}(AB) = 1/12$, so the events A and C are independent, but B and C are not. ◇

An arbitrary collection of events is said to be *independent* if for all choices $A_1, A_2, \ldots, A_n$ in the collection

$$\mathbb{P}(A_1 A_2 \cdots A_n) = \mathbb{P}(A_1)\mathbb{P}(A_2) \cdots \mathbb{P}(A_n).$$

Successive coin flips are independent, as a coin has no memory.[2] For example if you toss a coin 3 times in succession and A_j is the event that the jth toss comes up heads, then A_1, A_2, and A_3 are easily seen to be independent. This explains the use of the phrase "independent trials." Here is another such example:

Example 20.4. An urn contains 5 red and 6 yellow marbles. We randomly draw marbles one at a time, each time replacing the marble. Let R_1 denote the event that the first marble is red, R_2 the event that the second marble is red, and Y_3 the event that the third marble is yellow. Since the marbles were

---

[2] A common fallacy is that a long string of successive heads in the toss of a *fair* coin implies a high likelihood of tails in the next toss.

replaced, the events are independent. Thus the probability that the first two marbles are red and the third is yellow is

$$\mathbb{P}(R_1 R_2 Y_3) = \mathbb{P}(R_1)\mathbb{P}(R_2)\mathbb{P}(Y_3) = (5/11)(5/11)(6/11) \approx .113.$$

Note that, unlike Example 20.1, the original marble configuration in the urn is restored after each draw, ensuring independence. $\Diamond$

20.10 Exercises

1. You have 5 history books, 4 mathematics books, and 3 physics books that you randomly place next to each other on a shelf. What is the probability that all books within the same subject will be adjacent?

2. A group of 3 women and 3 men line up for a photograph. Find the probability that at least two people of the same gender are adjacent.

3. Hanna and Eric intend to have a ping pong tournament, the grand winner being the first one who wins 3 games. Based on previous games the probability of Eric winning a particular game is p, while that of Hanna is q. Find the probability that Hanna wins the tournament. Assume that there are no ties $(p + q = 1)$.

4. A "loaded" die has the property that the numbers 1 to 5 are equally likely but the number 6 is three times as likely to occur as any of the others. What is the probability that the die comes up even?

5. Balls are randomly thrown one at a time at a row of 30 open-topped jars numbered 1 to 30. Assuming that each ball lands in some jar, find the smallest number N of throws required so that there is a better than a 60% chance that two balls land in the same jar.

6. A hat contains six slips of paper numbered 1 through 6. A slip is drawn at random, the number is noted, the slip is replaced in the hat, and the procedure is repeated. What is the probability that after three draws the slip numbered 1 was drawn exactly twice, given that the sum of the numbers on the three draws is 8.

7. An urn contains 12 marbles: 3 reds, 4 greens, and 5 yellows. A handful of 6 marbles is drawn at random. Let A be the event that there are at least 3 green marbles and B the event that there is exactly 1 red. Find $\mathbb{P}(A|B)$. Are the events independent?

8. Write a program `DiceProbabilities` that calculates and graphs the theoretical probabilities in a dice throw by keeping track of the number of pairs (i, j) that add up to $k = 2, 3, \ldots, 12$ and dividing this number by 36.

9. Roll a fair die twice. Let A be the event that the first roll comes up odd, B the event that the second roll is odd, and C the event that the sum of the dice is odd. Show that any two of the events A, B, and C are independent but the events A, B, and C are not independent.

10. Write a program `Bayes` that calculates Bayes probabilities. Figure 20.6 depicts a possible spreadsheet with input in columns B and D and output (rounded to two places) in column F.

	A	B	C	D	E	F
6	$\mathbb{P}(A1) =$	.09	$\mathbb{P}(B \mid A1) =$	.04	$\mathbb{P}(A1 \mid B) =$	.11
7	$\mathbb{P}(A2) =$	.23	$\mathbb{P}(B \mid A2) =$	.02	$\mathbb{P}(A2 \mid B) =$	.14
8	$\mathbb{P}(A3) =$	.16	$\mathbb{P}(B \mid A3) =$	.06	$\mathbb{P}(A3 \mid B) =$	.29
9	$\mathbb{P}(A4) =$	.12	$\mathbb{P}(B \mid A4) =$	.01	$\mathbb{P}(A4 \mid B) =$	.04
10	$\mathbb{P}(A5) =$	.35	$\mathbb{P}(B \mid A5) =$	.03	$\mathbb{P}(A5 \mid B) =$	.32
11	$\mathbb{P}(A6) =$	.05	$\mathbb{P}(B \mid A6) =$	.07	$\mathbb{P}(A6 \mid B) =$	.11

FIGURE 20.6: Input/output for `Bayes`.

11. A manufacturer uses shipping companies A, B, C to ship their product, propeller wash, to airports. The following table gives the percentages of shipments from each company and the percentages of times a shipment is late. What is the probability that a randomly chosen late shipment comes from company B?

company	% of shipments	% of late shipments
A	35	1
B	15	2
C	50	3

12. A study of a group of single-sport athletes revealed the percentages of those participating in the sports and the percentage of injuries. Use the program `Bayes` to find the probability that a randomly chosen injured person plays (a) football, (b) soccer, (c) basketball, (d) baseball, (e) tennis, (f) golf.

sport	% in sport	% of injuries
football	10	30
soccer	8	25
basketball	28	20
baseball	22	15
tennis	15	8
golf	17	2

Chapter 21

Random Variables

A *random variable* is a function defined on the sample space Ω of an experiment. Random variables are typically denoted by capital letters. If X is a random variable on Ω, then the *values* of X are the numbers $X(\omega)$, $\omega \in \Omega$. For example, in the toss of a pair of dice, the sum S of the numbers that appear is a random variable with values $S(r, g) = r + g$, where r is the number that appears on the red die and g the number that appears on the green. In this chapter we give some common examples of random variables and calculate their mean and standard deviation. We also show how the mean may be approximated by sample averages generated by a program.

21.1 Examples of Random Variables

We use the notation $\{X = x\}$ to denote the event that a random variable X takes on the value x, and $\mathbb{P}(X = x)$ to denote the probability of the event. If $x_1, x_2, \ldots, x_m$ are the distinct values of X, then the numbers

$$\mathbb{P}(X = x_1), \; \mathbb{P}(X = x_2), \; \ldots, \mathbb{P}(X = x_m)$$

form what is called the *probability distribution of* X. For example, the random variable S mentioned in the introduction has probability distribution

$$\mathbb{P}(S = 2), \; \mathbb{P}(S = 3), \; \ldots, \; \mathbb{P}(S = 12).$$

Here are some additional examples:

Example 21.1. The following table which summarizes the grades (0 - 4) of a group of 100 students, where the first row gives the number of students having the grades listed in the second row.

no. of students	9	17	18	19	13	11	7	6
grades	2.2	2.3	2.5	2.7	2.9	3.1	3.4	3.5

If we select a student ω at random then we are also selecting the student's grade

DOI: 10.1201/9781003351689-21

$X(\omega)$, one of the eight numbers in the last row of the table. The probability distribution of X is therefore given by

$$\mathbb{P}(X = 2.2) = \frac{9}{100}, \ \mathbb{P}(X = 2.3) = \frac{17}{100}, \ \ldots, \mathbb{P}(X = 3.5) = \frac{6}{100}.$$

This process may be carried out for any data set.   ◇

Example 21.2. (Bernoulli random variable X with parameter p): X takes on two values, 1 and 0, with probabilities p and $1 - p$, respectively:

$$\mathbb{P}(X = 1) = p, \quad \mathbb{P}(X = 0) = 1 - p.$$

An example is the number of heads (one or zero) that come up on the toss of a single coin with probability of heads p.   ◇

Example 21.3. (Binomial random variable X with parameters n, p): X takes on the values $0, 1, \ldots, n$ with probabilities

$$\mathbb{P}(X = k) = \binom{n}{k} p^k q^{n-k}, \quad q = 1 - p, \quad k = 0, 1, \ldots, n.$$

An example is the number of heads that come up in the toss of n coins, where p is the probability of heads in a single toss.   ◇

The above random variables take on only finitely many values. However, there are many important random variables that take on infinitely many values. Here is an example.

Example 21.4. (Geometric random variable X with parameter p): X takes on the values $1, 2, 3 \ldots$ with probabilities

$$\mathbb{P}(X = k) = q^{k-1} p, \quad q = 1 - p, \quad k = 1, 2, 3, \ldots, n.$$

An example is the number coin tosses required until the first head appears, where p is the probability of a head on a single toss. Thus $X = 5$ if and only if the first four tosses result in tails and the fifth toss in a head, this occurring with probability $qqqqp = q^4 p$.   ◇

Random variables with values in a finite or infinite set of the form $\{x_1, x_2, x_3, \ldots\}$ are called *discrete*. Random variables that take on all values in an interval are said to be *continuous*. The latter typically arise from *continuous data* such as weight of a new born baby, inches of rainfall, or stock values. Probability distributions for continuous random variables are given by a "smooth histogram." The most important example of these is the following.

Example 21.5. *(Normal random variable X with parameters μ and σ):* The probability $\mathbb{P}(a < X < b)$ that X lies in an interval (a, b) is defined as the area under the bell-shaped curve with equation

$$y = \frac{1}{\sigma\sqrt{2\pi}} e^{-(\frac{x-\mu}{\sigma})^2/2}.$$

The maximum of the curve occurs at μ. The parameter σ controls the shape: the larger the value of σ the flatter the curve. In the figure $\mu = 4$ and $\sigma = 1$. The shaded region in the figure is the probability $\mathbb{P}(2.7 < X < 4.5)$.

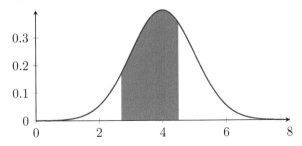

Normal random variables are important in statistics and are often used in situations involving random variables whose distributions are unknown. Their importance derives largely from the *Central Limit Theorem*, discussed later. ◊

21.2 Expected Value of a Random Variable

To motivate the definition of expected value we return to the student grades example in the last section. The class average, that is, the average of the grade data, is

$$[(2.2)9 + (2.3)17 + (2.5)18 + (2.7)19 + (2.9)13 + (3.1)11 + (3.4)7 + (3.5)6]/100$$

which is 2.72. If X denotes the grade of a student chosen at random then average may be written in terms of the probability distribution of X:

$$2.2 \cdot \mathbb{P}(X = 2.2) + 2.3 \cdot \mathbb{P}(X = 2.3) + \cdots + 3.5 \cdot \mathbb{P}(X = 3.5).$$

Now consider an arbitrary experiment with probabilities $\mathbb{P}(\omega_1), \ldots, \mathbb{P}(\omega_n)$. Suppose a random variable X takes on the distinct values $x_1, x_2, \ldots, x_m$. The *mean, expectation,* or *expected value* of X is defined as the sum

$$\mathbb{E}(X) = \sum_{k=1}^{m} x_k \mathbb{P}(X = x_k). \tag{21.1}$$

As the example suggests, the average of any collection of data may be described in this way.

We can write the expected value of a random variable X in another way. Recall that the probability of the event $\{X = x_k\}$ is the sum of the individual probabilities of the outcomes comprising the event:

$$\mathbb{P}(X = x_k) = \sum_{i:X(\omega_i)=x_k} \mathbb{P}(\omega_i)$$

where the notation means that the quantities $\mathbb{P}(\omega_i)$ are added over all indices i satisfying $X(\omega_i) = x_k$. Substituting this into (21.1) we obtain

$$\mathbb{E}(X) = \sum_{k=1}^{m} x_k \sum_{i:X(\omega_i)=x_k} \mathbb{P}(\omega_i)$$

For each k we can take x_k inside the second sum where it becomes $X(\omega_i)$. The resulting double sum is then the sum over all ω_i. Thus we have the alternate form

$$\mathbb{E}(X) = \sum_{i=1}^{n} X(\omega_i)\mathbb{P}(\omega_i). \tag{21.2}$$

Example 21.6. (Bernoulli random variable X). Recalling that

$$\mathbb{P}(X = 1) = p \text{ and } \mathbb{P}(X = 0) = 1 - p,$$

we see from (21.1) that

$$\mathbb{E}(X) = 1 \cdot \mathbb{P}(X = 1) + 0 \cdot \mathbb{P}(X = 0) = p. \qquad \Diamond$$

Example 21.7. The expected value of the sum S in a dice throw is, by (21.2),

$$\mathbb{E}(S) = \sum_{i=1}^{6}\sum_{j=1}^{6} S(i,j)\mathbb{P}(i,j) = \frac{1}{36}\sum_{i=1}^{6}\sum_{j=1}^{6}(i+j)$$

For a fixed i the inner sum on the right is

$$\sum_{j=1}^{6}(i+j) = 6i + 21.$$

Therefore,

$$\mathbb{E}(S) = \frac{1}{36}\sum_{i=1}^{6}(6i + 21) = \frac{6 \cdot 21 + 6 \cdot 21}{36} = 7.$$

Thus the average roll of the dice is seven. $\qquad \Diamond$

Example 21.8. (Binomial random variable X). Recalling that

$$P(X = k) = \binom{n}{k}p^k(1-p)^k, \quad k = 0, 1, \ldots, n,$$

we have

$$\mathbb{E}(X) = \sum_{k=1}^{n} k\mathbb{P}(X = k) = \sum_{k=1}^{n} k\binom{n}{k}p^k q^{n-k}.$$

Now write the general term in the sum on the right as

$$k\frac{n!}{k!\,(n-k)!}p^k q^{n-k} = np\frac{(n-1)!}{(k-1)!\,((n-1)-(k-1))!}p^{k-1}q^{(n-1)-(k-1)}.$$

Setting $m = n - 1$ and $j = k - 1$ we have

$$k\binom{n}{k}p^k q^{n-k} = np\frac{m!}{j!\,(m-j)!}p^j q^{m-j} = np\binom{m}{j}p^j q^{m-j}, \quad j = 0, 1, \ldots, n-1$$

and so

$$\mathbb{E}(X) = np \sum_{j=0}^{n-1} \binom{m}{j}p^j q^{m-j} = np(p+q)^{n-1} = np.$$

The second equality uses the Binomial Theorem discussed in Section 19.6. We can interpret this in the context of a ball drawing experiment. If an urn contains r red balls and y yellow balls and n balls are drawn in succession with replacement, then the average number of red balls drawn is $np = nr/(r+y)$. Thus if the urn contains 60% red balls and 40% yellow balls, the average percent of red balls in a sample is 60%, not unexpected. $\diamond$

Example 21.9. (Binomial stock model). Consider a stock whose price each day goes up by a factor $u > 1$ with probability p or down by a factor $d < 1$ with probability $q = 1 - p$. Let the initial price of the stock be S_0. Thus if the stock goes up the first day and down the next then its value after two days is udS_0 and the probability of this occurring is pq.

For fixed n we let X denote the number of times the stock went up in n days. Then X is a random variable with values $0, 1, \ldots, n$. The event that the stock went up exactly k times is $\{X = k\}$ and this can happen in any one of $\binom{n}{k}$ ways, which is the number of ways one can choose k u's. By independence, each of these has probability $p^k q^{n-k}$. Thus

$$\mathbb{P}(X = k) = \binom{n}{k}p^k q^{n-k}, \quad 0 \le k \le n,$$

so X is a binomial random variable with parameters p, n. After n days the price S of the stock is of the form $S = u^X d^{n-X} S_0$, signifying that the stock went up X times, down $n - X$ times. Thus S has values $d^n S_0$, $ud^{n-1}S_0$, $u^2 d^{n-2}S_0, \ldots, u^{n-1}dS_0$, $u^n S_0$. The probability of the event $S = u^k d^{n-k}S_0$ is just the probability of the outcome $X = k$:

$$\mathbb{P}(S = u^k d^{n-k}S_0) = \mathbb{P}(X = k) = \binom{n}{k}p^k q^{n-k}.$$

Therefore,

$$\mathbb{E}(S) = \sum_{k=0}^{n} \mathbb{P}(S = u^k d^{n-k}S_0)u^k d^{n-k}S_0$$

$$= S_0 \sum_{k=0}^{n} \binom{n}{k}p^k q^{n-k}u^k d^{n-k}$$

$$= S_0 \sum_{k=0}^{n} \binom{n}{k}(up)^k (dq)^{n-k}.$$

Using the Binomial Theorem (Section 19.6) on the last expression we see that the expected value of the stock after n days is $S_0(up + dq)^n$.

The binomial stock price model has been used extensively in the pricing of stock options. ◊

The mean of a random variable X that takes on infinitely many values $x_1, x_2, \ldots$ is defined as in the finite case except that the sum is infinite. We illustrate with the following example.

Example 21.10. (Expectation of a geometric random variable X). Recall that X has distribution

$$\mathbb{P}(X = k) = q^{k-1}p, \quad q = 1 - p, \quad k = 1, 2, 3, \ldots, n.$$

The expectation is then defined as

$$\mathbb{E}(X) = \sum_{k=1}^{\infty} kq^{k-1}p$$

where the infinite sum on the right is defined as the limit of the finite sums $\sum_{k=1}^{n} kq^{k-1}p$ as n gets larger and larger. It may be shown by standard methods that $\mathbb{E}(X) = 1/p$. Thus, on average, it takes two tosses of a fair coin to produce a head. ◊

Example 21.11. (Expectation of a normal random variable). It may be shown by standard calculus methods that the mean of the random variable X of Exercise 21.5 is μ. ◊

21.3   Properties of Expectation

In the dice experiment let R denote the number that comes up on the red die and G the number that comes up on the green die, so $S = R + G$. Then

$$\mathbb{E}(R) = \sum_{i=1}^{6} \frac{1}{6} R(i) = \frac{21}{6} = 3.5 = \mathbb{E}(G)$$

Recalling that $\mathbb{E}(S) = 7$ we see that

$$\mathbb{E}(R + G) = \mathbb{E}(R) + \mathbb{E}(G).$$

Note also that

$$\mathbb{E}(R \cdot G) = \sum_{k=2}^{12} k \cdot \mathbb{P}(RG = k) = \sum_{i=1}^{6} \sum_{j=1}^{6} ij \cdot \mathbb{P}(R = i, G = j),$$

Since the events $\{R = i\}$ and $\{G = j\}$ are independent, $\mathbb{P}(R = i, G = j) = \mathbb{P}(R = i)\mathbb{P}(G = j)$ and so the term on the right is

$$\sum_{i=1}^{6}\sum_{j=1}^{6} ij \cdot \mathbb{P}(G = j)\mathbb{P}(G = j) = \left(\sum_{i=1}^{6} i \cdot \mathbb{P}(R = i)\right) \cdot \left(\sum_{j=1}^{6} j \cdot \mathbb{P}(G = j)\right)$$

Thus

$$\mathbb{E}(R \cdot G) = \mathbb{E}(R) \cdot \mathbb{E}(G)$$

The above properties hold in general. Here is a summary of the most common properties of expectation, where X and Y are arbitrary random variables:

- $\mathbb{E}(X + Y) = \mathbb{E}(X) + \mathbb{E}(Y)$.
- $\mathbb{E}(X \cdot Y) = \mathbb{E}(X) \cdot \mathbb{E}(Y)$, provided that X and Y are *independent*, that is, (for discrete random variables) events of the form $\{X = x\}$ and $\{Y = y\}$ are independent for all x, y.
- $\mathbb{E}(cX) = c\mathbb{E}(X)$, where c is a constant.
- $\mathbb{E}(\mathbf{1}) = 1$, where $\mathbf{1}$ is the constant function whose values are 1.

Independence is needed in the second property. For example, in the dice experiment, the following quantities are dramatically unequal:

$$\mathbb{E}(R \cdot R) = (1^2 + 2^2 + 3^2 + 4^2 + 5^2 + 6^2)/6,$$

$$\mathbb{E}(R) \cdot \mathbb{E}(R) = (1 + 2 + 3 + 4 + 5 + 6)^2/36,$$

21.4 Variance and Standard Deviation of a Random Variable

Let X be a random variable with mean $m = \mathbb{E}(X)$. The *variance* $\mathbb{V}(X)$ and *standard deviation* $\sigma(X)$ of X are defined, respectively, as

$$\mathbb{V}(X) = \mathbb{E}[(X - m)^2] \quad \text{and} \quad \sigma(X) = \sqrt{\mathbb{V}(X)}.$$

Both quantities measure how much a random variable deviates from its mean: the smaller the variance or standard deviation the closer the data clusters around the mean. Here are the main properties of variance:

(a) $\mathbb{V}(X) = \mathbb{E}X^2 - m^2$.

(b) For real numbers a and b, $\mathbb{V}(aX + b) = a^2\mathbb{V}(X)$.

(c) If X and Y are independent, then $\mathbb{V}(X + Y) = \mathbb{V}(X) + \mathbb{V}(Y)$.

The verifications depend on properties of expectation developed in the previous section. For part (a) we use the expansion $(X - m)^2 = X^2 - 2mX + m^2$ to obtain

$$\mathbb{V}(X) = \mathbb{E}(X^2) - 2m\mathbb{E}(X) + m^2 = \mathbb{E}(X^2) - m^2.$$

Similarly, for (b) we have

$$\mathbb{E}[(aX + b)^2] = \mathbb{E}[a^2 X^2 + 2abX + b^2] = a^2\mathbb{E}(X^2) + 2abm + b^2$$

and

$$[\mathbb{E}(aX + b)]^2 = (am + b)^2 = a^2 m^2 + 2abm + b^2.$$

Subtracting these equations and using part (a) yields the desired result. To prove (c) note that

$$\mathbb{E}[(X + Y)^2] = \mathbb{E}(X^2 + 2XY + Y^2) = \mathbb{E}(X^2) + 2\mathbb{E}(XY) + \mathbb{E}(Y^2)$$

and

$$[\mathbb{E}(X + Y)]^2 = [\mathbb{E}(X) + \mathbb{E}(Y)]^2 = \mathbb{E}^2(X) + 2\mathbb{E}(X)\mathbb{E}(Y) + \mathbb{E}^2(Y).$$

Subtracting these equations and using independence and part (a) establishes the result.

Example 21.12. (Variance of a Bernoulli random variable X). Recalling that the mean of X is p we see that

$$\mathbb{V}(X) = \mathbb{E}(X^2) - p^2 = p(1 - p). \qquad \Diamond$$

Example 21.13. (Variance of a binomial random variable X). Since the mean of X is np we see that $\mathbb{V}(X) = \mathbb{E}(X^2) - (np)^2$. With a bit of work one can show that $\mathbb{E}(X^2) = (np)^2 + np(1 - p)$. Thus

$$\mathbb{V}(X) = np(1 - p). \qquad \Diamond$$

Example 21.14. (Variance of a binomial random variable X). Since the mean of X is np we see that $\mathbb{V}(X) = \mathbb{E}(X^2) - (np)^2$. With a bit of work one can show that $\mathbb{E}(X^2) = (np)^2 + np(1 - p)$. Thus

$$\mathbb{V}(X) = np(1 - p). \qquad \Diamond$$

Example 21.15. (Variance of a random variable). It may be shown with standard calculus methods that the variance of the random variable X of Example 21.5 is σ^2. $\qquad \Diamond$

21.5   The Law of Large Numbers

Consider an experiment that is repeated n times, the trials being independent. Let X_1 be a random variable describing the outcomes on the first trial, X_2 a random variable describing the outcomes on the second trial, etc. The *sample average* for n trials is the average $\overline{X}_n$ of the random variables $X_1, X_2, \ldots, X_n$:

$$\overline{X}_n = \frac{X_1 + X_2 + \cdots + X_n}{n}, \quad n = 1, 2, \ldots.$$

Suppose that the expectations of the random variables are equal with mean m. An important theoretical result called the *Law of Large Numbers* (LLN) asserts that as we take larger and larger n these sample averages get closer and closer to m, symbolically $\overline{X}_n \to m$. The LLN is the theoretical basis of the Monte Carlo method: repeated random sampling to calculate numerical quantities.

The law may be used to justify the use of relative frequencies as empirical probabilities: Let A be an event in the experiment and define a Bernoulli random variable X_n that has value 1 if A occurs on the nth trial, and 0 otherwise. Then $X_1 + X_2 + \cdots + X_n$ is the number of times the event A occurs in n trials. The corresponding sample average is the relative frequency of this outcome. The law of large numbers asserts that for large n the relative frequency of the event A occurring is very close to to the mean m of these variables. But $m = \mathbb{P}(A)$. Thus for large n the theoretical probability of the event A is approximately equal to the sample averages.

Mean Time For Two Heads

We use Monte Carlo simulation to calculate sample averages of the number N of coin tosses required for a pair of consecutive heads to appear. An example of such an outcome is THTHTHH, a member of the event $\{N = 7\}$. If p denotes the probability of a head on a single toss, then the probability of the outcome is $(1 - p)^3 p^4$. The function `MeanTimeForFirstHH` takes the probability `ProbH` of a head on a single toss and returns the average time required for two consecutive heads to appear. The trials are carried out by a For Next loop. For each trial a Do While loop uses `Rnd` to toss the coin until two heads appear together, signaled by the fact that `NumHeads` $= 2$. Heads are only counted after the last tail, since then one needs to start over.

```
Sub MeanTimeForFirstHH()
    Dim i As Long, NumTrials As Long, count As Long, mean As Double
    Dim  ProbH As Single, NumHeads As Integer
    NumTrials = 100000
    ProbH = Range("B5").Value                'get probability of a head
    For i = 1 To NumTrials
```

```
    count = 0: NumHeads = 0                           'initialize
    Do While NumHeads < 2          'keep tossing until 2 heads in a row
        Randomize
        If  Rnd > ProbH Then                      'if a tail comes up
            NumHeads = 0                          'start count over
        Else
            NumHeads = NumHeads + 1                   'count the head
        End If
        count = count + 1          'count number of throws for 2 heads
    Loop
    mean = mean + count / NumTrials                       'update mean
  Next i
  Range("B6").Value = mean
  Range("B7").Value = (1+ProbH)/ProbH^2             'print actual mean
End Sub
```

Bernoulli–Laplace model

Consider the following experiment. There are two urns. Urn one contains an even number of red balls and and urn two an equal number of yellow balls. At each trial a ball is chosen at random from each urn and the balls switched, so the number of balls in each urn is constant. The experiment is performed until there are an equal number of red and yellow balls in urn one (hence also in urn two). Suppose it takes N trials for this to happen. The program BernoulliLaplaceMeanTime calculates the average value of N

```
Sub BernoulliLaplaceMeanTime()
    Dim NumBalls As Integer, CurrNumRed1 As Integer, count As Long
    Dim color As Integer, Balls1() As Integer, Balls2() As Integer
    Dim TotalCount As Long, NumTrials As Long, n As Long
    Dim BallOne As Integer, BallTwo As Integer
    Dim red As Integer, yellow As Integer
    NumBalls = Range("B2").Value      'enter number of balls in each urn
    red = 1: yellow = 2
    ReDim Balls1(1 To NumBalls)              'urn 1 initially all red
    ReDim Balls2(1 To NumBalls)              'urn 2 initially all yellow
    NumTrials = Range("B3").Value
    For n = 1 To NumTrials                            'do the trials
        Call Init(Balls1, 1): Call Init(Balls2, 2)        'fill urns
        CurrNumRed1 = NumBalls  'urn 1 has this many red balls initially
        count = 0                             'reset after each trial
        Do While CurrNumRed1 > NumBalls / 2
            Idx1 = RandInt(1, NumBalls)                'choose ball in urn 1
            BallOne = Balls1(Idx1)
            Idx2 = RandInt(1, NumBalls)                'choose ball in urn 2
            BallTwo = Balls2(Idx2)
            If BallOne = red And BallTwo = yellow Then _
                CurrNumRed1 = CurrNumRed1 - 1
            If BallOne = yellow And BallTwo = red Then _
                CurrNumRed1 = CurrNumRed1 + 1
```

```
            Balls1(Idx1) = BallTwo: Balls2(Idx2) = BallOne  'switch balls
            count = count + 1
        Loop                           'count is now the time to reach 1/2 reds
            TotalCount = TotalCount + count
    Next n
    Range("B4").Value = TotalCount / NumTrials
    End Sub

Sub Init(balls() As Integer, color As Integer)
    Dim i As Integer
    For i = 1 To UBound(balls)
        balls(i) = color
    Next i
End Sub
```

21.6 Central Limit Theorem

The Central Limit Theorem (CLT) is one of the main pillars of statistics. It asserts that certain averages of a random variable taken over a large number of independent observations have probability distributions that are nearly normal. More precisely, suppose $X_1, X_2, \ldots$ is a sequence of independent random variables, each with mean μ and standard deviation σ. For example, X_j could be the number of heads that appear on the jth trial of a repeated coin toss experiment or the height of a person randomly chosen from a large population. As before, set

$$\overline{X}_n = \frac{X_1 + \cdots + X_n}{n}.$$

The CLT asserts that for large n

$$\mathbb{P}\left(\frac{\overline{X}_n - \mu_n}{\sigma_n} \leq x\right) \approx \mathbb{P}\left(\frac{X - \mu}{\sigma} \leq x\right),$$

where X is a normal random variable with parameters μ and σ and

$$\mu_n = \mathbb{E}(\overline{X}_n) = \mu, \quad \sigma_n = \sigma(\overline{X}_n) = \frac{\sigma}{\sqrt{n}}.$$

Setting

$$S_n = X_1 + \cdots + X_n = n\overline{X}_n$$

we may write the assertion as

$$\mathbb{P}\left(\frac{S_n - n\mu}{\sigma\sqrt{n}} \leq x\right) \approx \mathbb{P}\left(\frac{X - \mu}{\sigma} \leq x\right),$$

In the special case that the random variables X_j are Bernoulli with parameter $p \in (0,1)$, $\mu = p$ and $\sigma = \sqrt{p(1-p)}$ so the approximation becomes

$$\mathbb{P}\left(\frac{S_n - np}{\sqrt{np(1-p)}} \leq x\right) \approx \mathbb{P}\left(\frac{X - \mu}{\sigma} \leq x\right),$$

This result is known as the *DeMoivre-Laplace Theorem*. Notice that in this case S_n is a binomial random variable with parameters n, p

---

21.7   Exercises

1. Consider the experiment of throwing a pair of dice n times. Let S_1 denote the sum on the first toss, S_2 the sum on the second toss, etc. The sample averages are

$$\frac{S_1 + S_2 + \cdots + S_n}{n}.$$

The LLN asserts that as we take larger and larger n, these sample averages get closer and closer to the expected value of S, namely 7. Write a procedure `DiceSampleAvg` that simulates the experiment and prints the sample average over, say, $100,000$ trials. How close do the averages come to 7?

2. Write a program `MeanTimeForFirstH` that is the one-head analog of `MeanTimeForFirstHH`

3. Write a program `MeanTimeForFirstnH` that is the n-head analog of `MeanTimeForFirstHH`.

4. Write a program `MeanTimeForFirstHHorTT` that simulates the average number of tosses required until either two heads appear in a row or two tails appear in a row.

5. The *odds for an event* E are said to be r to 1 if E is r times as likely to occur as E', that is, $\mathbb{P}(E) = r\mathbb{P}(E')$. *Odds r to s* means the same thing as odds r/s to 1, and *odds r to s against* means the same as odds s to r for. A bet of one dollar on an event E with odds r to s is *fair* if the bettor wins s/r dollars if E occurs and loses one dollar if E' occurs. (If E occurs, the dollar wager is returned to the bettor.)

   (a) If the odds for E are r to s, show that $\mathbb{P}(E) = \dfrac{r}{r+s}$.

   (b) Verify that a fair bet of one dollar on E returns $1/\mathbb{P}(E)$ dollars (including the wager) if E occurs.

6. (Roulette). Pockets of a roulette wheel are numbered 1 to 36, of which 18 are red and 18 are black. There are also two green pockets numbered 0 and 00. The probability of a ball randomly falling into a black pocket is then $18/38 = 9/19$. Suppose you place a bet of one dollar on black and the house gives you odds r to 1 for black. (If you get on black you win r dollars, keeping the dollar bet, otherwise you lose one dollar. What are your expected winnings W? What values of r favor the house.

7. Consider the following experiment: A hat contains n slips of paper each with a positive number. A slip of paper is drawn, its number noted, and the slip is returned. The process is repeated k times, after which the numbers are added to form the sum S. Write a program `NumbersInHat` that takes a user-entered string `numberStr` of comma-separated positive integers such as `3,4,3,2,2,6,5` (the numbers in the hat) and a positive integer `numDraws` and calculates the sample averages that approximate $\mathbb{E}(S)$. For example, if `numberStr = "1,2,3,4,5,6"` and `numDraws= 2` then the experiment reduces to a two dice toss.

8. Consider the following experiment. Numbers between 0 and 1 are repeatedly chosen at random and added until the sum exceeds a user-entered number x. Write a program `SumExceedsX` that calculates the average number of times required for this to occur. What mathematical constant does the average seem to approximate when $x = 1$?

9. Write a program `SumExceedsN` analogous to `SumExceedsX` but where integers from 1 to 10 rather than decimals from 0 to 1 are chosen and the cutoff number is a user-entered integer N.

Chapter 22

Markov Chains

A *Markov chain* is a system that changes over time and at any instant can be in any one of m states, which we shall represent by the integers $1, 2, \ldots, m$. The state of the system at time k is denoted by s_k, which can have any one of the values $1, 2, \ldots, m$. Thus at time 0 the system is in the *initial state* s_0, at time 1 the system is in state s_1, etc. The *history of the system at time k* is the sequence $s_0, s_1, \ldots, s_k$. A probability law governs the transition from one state to the next. The key feature of a Markov chain is that the probability of going from state s_k to state s_{k+1} depends only on s_k; the previous states $s_0, s_1, \ldots, s_{k-1}$ of the system are irrelevant in this transition

Markov chains arise in many common circumstances. For example, an investor might decide to buy or sell a stock depending on its current value and not on how the stock reached that value (although that may be shortsighted!). A gambler might fashion a wager based on her current winnings, not on how she achieved the winnings. A robot is essentially Markovian, acting only on the information gleaned from its current position. Word completion in computer programs is another example: the current state is the last word (or last few words) entered. Words that could come next are assigned probabilities based on a large volume of textual data.

In this chapter we develop the basic mathematics of Markov chains and show how they may be modelled using VBA.

22.1 Transition Probabilities

The essential ingredient of a Markov chain is the probability of going from state i to state j during any time step. This is called a *transition probability* and is denoted p_{ij}. It may be interpreted as a conditional probability: the probability that the system will be in state j at the next time step, given that it is currently in state i. Transition probabilities may be displayed schematically by a *transition diagram*, as in Figure 22.1. While this method is visually informative, it is impractical if there are many states and, moreover, lacks computational utility. Instead, transition probabilities are frequently displayed

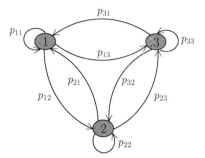

FIGURE 22.1: Transition diagram with states 1,2,3.

in a matrix, called the *transition matrix* of the system:

$$\begin{bmatrix} p_{11} & p_{12} & \cdots & p_{1m} \\ p_{21} & p_{22} & \cdots & p_{2m} \\ \vdots & \vdots & \ddots & \vdots \\ p_{m1} & p_{m2} & \cdots & p_{mm} \end{bmatrix}.$$

Note that each entry of the matrix is nonnegative. Also, since row i gives the probabilities of going from state i to all possible states, the sum of the probabilities in the row must add up to 1. A matrix with these properties is called a *stochastic matrix* Any stochastic matrix qualifies as the transition matrix for a Markov chain.

Here are some simple examples of Markov chains; more complex examples are considered in later sections.

Example 22.1. (Coin toss). Suppose a coin is tossed twice, the states of the system being heads or tails. If the probability of heads on a single toss is p, then the transition matrix is

$$\begin{bmatrix} p_{hh} & p_{ht} \\ p_{th} & p_{tt} \end{bmatrix} = \begin{bmatrix} p & 1-p \\ p & 1-p \end{bmatrix}$$

where, for example, p_{ht} is the probability that a tail comes up on the next toss given that a head just came up. Notice that the rows of the matrix are equal. This is the hallmark of independent trials. In general, a sequence of n independent trials has transition probabilities p_{ij} that do not depend on i. Thus the transition probability to the next state is not only independent of past states but is also independent of the current state. The transition matrix of such a system looks like

$$\begin{bmatrix} q_1 & q_2 & \cdots & q_m \\ q_1 & q_2 & \cdots & q_m \\ \vdots & \vdots & \cdots & \vdots \\ q_1 & q_2 & \cdots & q_m \end{bmatrix}$$

where the q_i are nonnegative and sum to one. $\Diamond$

Example 22.2. (Weather model). Let the state of the weather on any given day be represented by s: sunny, c: cloudy, r: rainy. Define weather transition probabilities p_{sc}, the probability that it will be cloudy tomorrow given that it's sunny today; p_{rs}, the probability that it will be sunny tomorrow given it's raining today; etc. The actual values could be determined from weather statistics. The transition matrix is

$$\begin{bmatrix} p_{ss} & p_{sc} & p_{sr} \\ p_{cs} & p_{cc} & p_{cr} \\ p_{rs} & p_{rc} & p_{rr} \end{bmatrix} \qquad \Diamond$$

Example 22.3. (Transmission of information). Suppose the news ("yes" or "no") regarding the success of a Mars landing is passed from one person to the next. The state of the system at any stage is the information received. Assume that at any stage the probability that a "yes" is passed to the next person as a "yes" is p and that the probability that a "no" is passed as a "no" is q, where $0 < p, q < 1$. The transition matrix is

$$\begin{bmatrix} p_{yy} & p_{yn} \\ p_{ny} & p_{nn} \end{bmatrix} = \begin{bmatrix} p & 1-p \\ 1-q & q \end{bmatrix},$$

where p_{yy} denotes the probability that "yes" is delivered as "yes," etc. $\qquad \Diamond$

Example 22.4. (Consumer preference). In order to better serve its customers the TV production company NutFlocks has collected data related to viewer preference. It has discovered that of its customers who recently watched an action movie, 70% selected another action movie, 20% selected a horror movie, and 10% selected a comedy. We can interpret this data as transition probabilities:

$$p_{aa} = .7, \quad p_{ah} = .2, \quad \text{and} \quad p_{ac} = .1$$

where, for example, p_{ah} is the probability that a customer chose a horror movie after watching an action movie. Similar data gives rise to the other transition probabilities yielding the following possible transition matrix:

$$\begin{bmatrix} p_{aa} & p_{ah} & p_{ac} \\ p_{ha} & p_{hh} & p_{hc} \\ p_{ca} & p_{ch} & p_{cc} \end{bmatrix} = \begin{bmatrix} .7 & .2 & .1 \\ .3 & .6 & .1 \\ .2 & .1 & .7 \end{bmatrix} \qquad \Diamond$$

22.2 Higher Order Transition Probabilities

Suppose a system with states $1, 2, 3$ has transition matrix

$$P = \begin{bmatrix} p_{11} & p_{12} & p_{13} \\ p_{21} & p_{22} & p_{23} \\ p_{31} & p_{32} & p_{33} \end{bmatrix}$$

and that the system is currently in state 2. The probability of being in state 3 in one time step is p_{23}. We are interested in finding the probability of being in state 3 after 2 steps. We denote this probability by $p_{23}^{(2)}$. (Caution: the superscript is *not* an exponent.) The transition from state 2 to state 3 in two steps can be achieved by the following one-step transitions:

$$2 \xrightarrow{p_{21}} 1 \xrightarrow{p_{13}} 3, \quad 2 \xrightarrow{p_{22}} 2 \xrightarrow{p_{23}} 3, \quad 2 \xrightarrow{p_{23}} 3 \xrightarrow{p_{33}} 3.$$

By the multiplication rule, these 2-step transitions have probabilities $p_{21}p_{13}$, $p_{22}p_{23}$, and $p_{23}p_{33}$, respectively. By the addition rule, the desired probability is the sum of these:

$$p_{23}^{(2)} = p_{21}p_{13} + p_{22}p_{23} + p_{23}p_{33}.$$

Note that this is the $(2,3)$ entry of the square P^2 of the transition matrix. By the same analysis, the probability $p_{ij}^{(n)}$ of going from state i to state j in n time steps is the (i,j) entry of P^n. The reasoning here applies to Markov chains with any number of states.

Example 22.5. (Transmission of information). Assume in Example 22.3 that there is a 95% chance that a "yes" will be relayed as a "yes" and a 95% chance that a "no" will be relayed as a "no." Given that the current answer is "yes," the probability that it will be "yes" 30 people from now is the $(1,1)$-entry of

$$\begin{bmatrix} .95 & .05 \\ .05 & .95 \end{bmatrix}^{30} = \begin{bmatrix} .52 & .48 \\ .48 & .52 \end{bmatrix},$$

only a slightly better than even chance. ◇

22.3   The Initial Distribution of a Markov Chain

Suppose that the initial (time zero) state of a Markov system is governed by a probability law. Specifically, assume that the probability of the system being initially in state i is p_i, $i = 1, 2, \ldots, m$, where $p_1 + p_2 + \cdots + p_m = 1$. We use the row matrix

$$\begin{bmatrix} p_1 & p_2 & \cdots & p_m \end{bmatrix}$$

to represent the probability distribution of the system at time 0. It is of interest to determine the probability distribution of the system at time n, that is, after n steps. Denote this distribution by the row matrix

$$\begin{bmatrix} p_1^{(n)} & p_2^{(n)} & \cdots & p_m^{(n)} \end{bmatrix},$$

Thus, for example, $p_1^{(1)}$ is the probability that the system will be in state 1 after 1 step, $p_1^{(2)}$ is the probability that the system will be in state 1 after 2 steps, etc. We set

$$[p_1^{(0)} \quad p_2^{(0)} \quad \cdots \quad p_m^{(0)}] = [p_1 \quad p_2 \quad \cdots \quad p_m].$$

To find $p_1^{(1)}$ we argue as follows: To get to state 1 in one step either the system was initially in state 1 and it transitioned in one step to state 1, or the system was initially in state 2 and it transitioned in one step to state 1, etc. By the multiplication rule, the probabilities of events are, respectively, $p_1 p_{11}$, $p_2 p_{21}$, etc. Since the events are mutually exclusive, the addition law implies that

$$p_1^{(1)} = p_1 p_{11} + p_2 p_{21} + \cdots + p_m p_{m1}.$$

Similarly, for any j,

$$p_j^{(1)} = p_1 p_{1j} + p_2 p_{2j} + \cdots + p_m p_{mj}.$$

Notice that this is the jth entry of the matrix product $[p_1 \quad p_2 \quad \cdots \quad p_m] \, P$. Thus we have

$$\left[p_1^{(1)} \quad p_2^{(1)} \quad \cdots \quad p_m^{(1)}\right] = [p_1 \quad p_2 \quad \cdots \quad p_m] \, P$$

Going from step 1 to step 2 produces a similar result:

$$\left[p_1^{(2)} \quad p_2^{(2)} \quad \cdots \quad p_m^{(2)}\right] = \left[p_1^{(1)} \quad p_2^{(1)} \quad \cdots \quad p_m^{(1)}\right] P = [p_1 \quad p_2 \quad \cdots \quad p_m] \, P^2$$

In general one has

$$\left[p_1^{(n)} \quad p_2^{(n)} \quad \cdots \quad p_m^{(n)}\right] = [p_1 \quad p_2 \quad \cdots \quad p_m] \, P^n \tag{22.1}$$

The $1 \times m$ matrix on the left in (22.1) is called the *n-step distribution vector*.

Example 22.6. (Weather model). Recall that in this model the state of the weather on any given day is represented by s: sunny, c: cloudy, r: rainy. Suppose that the transition matrix is

$$\begin{bmatrix} p_{ss} & p_{sc} & p_{sr} \\ p_{cs} & p_{cc} & p_{cr} \\ p_{rs} & p_{rc} & p_{rr} \end{bmatrix} = \begin{bmatrix} .9 & .05 & .05 \\ .2 & .6 & .2 \\ .1 & .1 & .8 \end{bmatrix},$$

so that it is more likely than not that the weather will be the same tomorrow. Given that it is sunny today, the probability that it will be sunny 3 days from now is the $(1,1)$ entry of

$$\begin{bmatrix} .9 & .05 & .05 \\ .2 & .6 & .2 \\ .1 & .1 & .8 \end{bmatrix}^3 \approx \begin{bmatrix} .8 & .1 & .1 \\ .4 & .3 & .3 \\ .25 & .15 & .5 \end{bmatrix}.$$

Suppose now that the forecast for today's weather is 80% sunny, 10% cloudy, and 10% rain. The probability distribution for today's weather is the row matrix

$$\begin{bmatrix} p_s & p_c & p_r \end{bmatrix} = \begin{bmatrix} .8 & .1 & .1 \end{bmatrix}.$$

From the calculation

$$\begin{bmatrix} .8 & .1 & .1 \end{bmatrix} \begin{bmatrix} .8 & .1 & .1 \\ .4 & .3 & .3 \\ .25 & .15 & .5 \end{bmatrix} \approx \begin{bmatrix} .7 & .1 & .2 \end{bmatrix} = \begin{bmatrix} p_s^{(3)} & p_c^{(3)} & p_r^{(3)} \end{bmatrix}$$

we see that the probability that it will be sunny 3 days from now is only .7. As the days progress, sunshine is less and less likely.   ◇

Example 22.7. (Genotypes). Humans and animals inherit characteristics that depend on the genetic code of their parents. We consider only the simplest of these traits: those that depend on a single gene. Each gene has two forms, *dominant* and *recessive*. We label the former by the letter G and the latter by the letter g. An offspring receives one gene from each parent; these combine to form the *genotypes* GG, Gg, gG and gg. A trait is said to be *dominant* if it is associated with the gene GG, *recessive* if it is associated with the gene gg, and *hybrid* if it is associated with the gene Gg or gG. The last two types are nearly indistinguishable and so may be assumed to be the same.

Now consider what happens when a GG individual (the "mother") breeds with other individuals ("fathers"). Suppose that mother GG is bred with father GG. The offspring will then be GG with probability $p_{GG,GG} = 1$. If mother GG is bred with father Gg then the offspring will be GG with probability $p_{Gg,GG} = .5$ and Gg with probability $p_{Gg,Gg} = .5$. If GG is bred with gg then the offspring will be Gg with probability $p_{gg,Gg} = .5$. If the states of the system are the various genotypes and the transitions are from father genotype to offspring genotype, then the transition probabilities are given by the matrix

$$P = \begin{bmatrix} p_{GG,GG} & p_{GG,Gg} & p_{GG,gg} \\ p_{Gg,GG} & p_{Gg,Gg} & p_{Gg,gg} \\ p_{gg,GG} & p_{gg,Gg} & p_{gg,gg} \end{bmatrix} = \begin{bmatrix} 1 & 0 & 0 \\ .5 & .5 & 0 \\ 0 & 1 & 0 \end{bmatrix}$$

If the genotypes of the initial population of fathers are equally distributed, that is, 1/3 of each type, then the genotypes of successive generations are given by

$$\begin{bmatrix} 1/3 & 1/3 & 1/3 \end{bmatrix} P^n$$

Using `PowerMat` from Section 13.5 for large powers of n we see that P^n is approximately

$$\begin{bmatrix} 1 & 0 & 0 \\ 1 & 0 & 0 \\ 1 & 0 & 0 \end{bmatrix}$$

hence $[1/3, 1/3, 1,3]P^n = [1, 0, 0]$. Thus mating a mother with fathers of uniform genotype distribution eventually results in a population of GG genotypes.   ◇

22.4 The Steady State Vector

The last observations in Examples 22.6 and 22.7 beg the following question: What is the long term behavior of the n-step distribution vector (22.1)? Specifically, as we take more and more steps do the entries of the vectors get closer to the entries of some *steady state vector* q? Symbolically,

$$\begin{bmatrix} p_1^{(n)} & p_2^{(n)} & \cdots & p_m^{(n)} \end{bmatrix} = \begin{bmatrix} p_1 & p_2 & \cdots & p_m \end{bmatrix} P^n \to q \text{ as } n \text{ gets larger?}$$

One can show that this happens if for some n the matrix P^n has no zero entries. To compute q in this case, multiply on the right by P to conclude that

$$\begin{bmatrix} p_1 & p_2 & \cdots & p_m \end{bmatrix} P^n P \to qP \text{ as } n \text{ gets larger.}$$

On the other hand we also have

$$\begin{bmatrix} p_1 & p_2 & \cdots & p_m \end{bmatrix} P^n P = \begin{bmatrix} p_1 & p_2 & \cdots & p_m \end{bmatrix} P^{n+1} \to q \text{ as } n \text{ gets larger}$$

Therefore, $qP = q$ or $q(P - I) = 0$, the zero vector. Taking transposes yields $(P - I)^T q^T = 0$. Also, since the components of each vector $\begin{bmatrix} p_1^{(n)} & p_2^{(n)} & \cdots & p_m^{(n)} \end{bmatrix}$ add to one, the same is true for q. We can combine the two observations to form the $(m + 1) \times m$ system

$$\begin{bmatrix} p_{11} - 1 & p_{21} & \cdots & p_{m1} \\ p_{12} & p_{22} - 1 & \cdots & p_{m2} \\ & & \ddots & \\ p_{1m} & p_{2m} & \cdots & p_{mm} - 1 \\ 1 & 1 & \cdots & 1 \end{bmatrix} \begin{bmatrix} q_1 \\ q_2 \\ \vdots \\ q_{m-1} \\ q_m \end{bmatrix} = \begin{bmatrix} 0 \\ 0 \\ \vdots \\ 0 \\ 1 \end{bmatrix}. \tag{22.2}$$

This may be solved for q using the methods of Chapter 11.

Example 22.8. (Weather model). In Example 22.6 the transition matrix was

$$P = \begin{bmatrix} .9 & .05 & .05 \\ .2 & .6 & .2 \\ .1 & .1 & .8 \end{bmatrix},$$

The augmented matrix of the system (22.2) in this case is

$$\begin{bmatrix} p_{11} - 1 & p_{21} & p_{31} & 0 \\ p_{12} & p_{22} - 1 & p_{32} & 0 \\ p_{13} & p_{23} & p_{33} - 1 & 0 \\ 1 & 1 & 1 & 1 \end{bmatrix} = \begin{bmatrix} -.1 & .2 & .1 & 0 \\ .05 & -.4 & .1 & 0 \\ .05 & .2 & -.2 & 0 \\ 1 & 1 & 1 & 1 \end{bmatrix}$$

Using `RowEchelon` we see that the steady state vector (rounded) is

$$q = \begin{bmatrix} .571 & .143 & .286 \end{bmatrix}. \qquad \diamond$$

22.5   Position Dependent Random Walk

Consider a single red cell moving in a spreadsheet grid. We take the state of the system as the pair (r, c), the current row, column position of the cell. Transition probabilities PL, PU, PR, and PD govern the movement of the cell as follows:

- $(r, c) \rightarrow (r, c - 1)$: move left with probability PL

- $(r, c) \rightarrow (r - 1, c)$: move up with probability PU

- $(r, c) \rightarrow (r, c + 1)$: move right with probability PR

- $(r, c) \rightarrow (r + 1, c)$: move down with probability PD.

The only requirements are that the probabilities be nonnegative, add up to one, and depend only on the current position (r, c). For example, one could take

$$PL = \frac{r + 2}{2(r + c)}, \quad PU = \frac{r - 2}{2(r + c)}, \quad PR = \frac{c - 2}{2(r + c)}, \quad PD = \frac{c + 2}{2(r + c)}.$$

The program in this section allows the user to enter formulas for transition probabilities PL, PU, and PR in cells B3 – B5 (preceded by an equal sign, as shown). It is not necessary to enter a formula for PD since this is calculated as $1 - (PL + PU + PR)$. After each movement of the cell, the entries in B1 and B2, which contain, respectively, the row and column of the cell, are updated and new transition probabilities are calculated from these values. A typical input is summarized in Figure 22.2, where we have used the above formulas to calculate the probabilities.

	A	B
1	row	17
2	col	22
3	PL	= (1/2)*(B1 + 2)/(B1 + B2)
4	PR	= (1/2)*(B1– 2)/(B1 + B2)
5	PU	= (1/2)*(B2 – 2)/(B1 + B2)

FIGURE 22.2: Input for `PositionRandWalk()`.

With the initial row and columns values in B1 and B2 as shown, the initial probabilities are

$$PL = \frac{17 + 2}{2(17 + 22)} \approx .244, \quad PU = \frac{17 - 2}{2(17 + 22)} \approx .192, \quad PR = \frac{22 - 2}{2(17 + 22)} \approx .256.$$

The main procedure of the module is `PositionRandomWalk`. It is virtually identical to the procedure `RandomWalk` of Section 10.4 except that code must be added to calculate the transition probabilities.

```
Sub PositionRandomWalk()
    Dim Steps As Integer, Speed As Integer, k As Integer
    Dim Grid As TGrid, Pos As TPos, NewPos As TPos
    Dim PL As Single, PU As Single, PR As Single
    Steps = 100: Speed = 12                      'motion parameters
    Grid = MakeGrid(8, 5, 45, 40, 2.2, 11, 24)
    Call ClearGrid(Grid,0)
    Call AddBorder(Grid, color)
    Pos.col = (Grid.L + Grid.R) / 2              'initial position of cell
    Pos.row = (Grid.T + Grid.B) / 2
    Range("B2").Value = Pos.row: Range("B3").Value = Pos.col
    Call PrintCell(Pos, 3)                       'print a red cell
    For k = 1 To Steps
        If Not GetTransitionProb(PL, PU, PR) Then Exit Sub
        NewPos = NextRandPos(PL, PL, PL, Pos)
        If (Cells(NewPos.row, NewPos.col).Interior.ColorIndex <= 0) Then
            Call MoveCell(Pos, NewPos)    'erase at Pos, print at NewPos
            Pos = NewPos                               'update
            Range("B2").Value = NewPos.row
            Range("B3").Value = NewPos.col
            Call Delay(Speed)            'delay before the next move
        End If
    Next k
End Sub
```

The function `GetTransitionProb` generates the probabilities *PL*, *PU*, and *PR* by evaluating the functions in cells B3–B5. The function returns `False` if the probabilities are negative or add up to a value larger than 1. The code invokes the VBA function `Evaluate` to make the calculations.

```
Function GetTransitionProb(PL As Single, PU As Single, PR As Single) _
        As Boolean
    PL = Evaluate("B4"): PR = Evaluate("B5"): PU = Evaluate("B6")
    If PL < 0 Or PR < 0 Or PU < 0 Or PL + PR + PU > 1 Then
        GetTransitionProb = False: Exit Function
    End If
    GetTransitionProb = True
End Function
```

22.6   Polya Urn Model

Consider an urn containing red balls and green balls. A ball is drawn at random from the urn, the color noted, the ball replaced, and a new ball of the same color is added to the urn. The process is repeated as many times as desired. The state of the system at a particular time is the number r of red balls and the number g of green balls in the urn, symbolized by (r, g). The probability of transition to the state $(r + 1, g)$ (a red ball was drawn) is $r/(r+g)$, the probability of drawing a red ball in an urn with r red balls and g green balls. Similarly, the transition probability to the state $(r, g + 1)$ (a green ball was drawn) is $g/(r+g)$. Figure 22.3 shows the process for three draws.

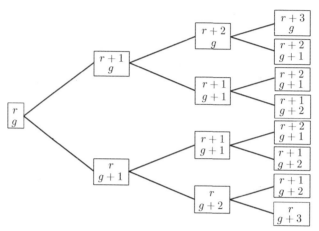

FIGURE 22.3: Polya urn model.

The module `PolyaUrn` in this section simulates the experiment. The urn is pictured in the spreadsheet as a bottomless grid with dimensions given by the UDT `TUrn`. The user enters the desired initial number of red and green balls in cells B3 and B4. The main procedure `DrawBall` calls `FillUrn` to place the balls, pictured in the grid as colored cells or, alternately, as colored letter O's, in the urn. A For Next loop executes the sequence of draws. The numbers `NumRed` and `NumGreen` of red and green balls are updated after each draw by the procedures `DrewRed` and `DrewGreen`. The VBA function `Rnd` is called to simulate the drawing a red ball or a green ball.

```
Private Type TUrn     'top, left, right, N balls                    'urn type
        T As Integer: L As Integer: R As Integer
End Type

Public NumRed As Integer, NumGreen As Integer
Sub DrawBalls()
        Dim x As Double, draws As Integer, i As Integer, Urn As TUrn
        Columns("C:AZ").ColumnWidth = 2.5              'compress columns
        draws = Range("B2").Value                      'desired no. of draws
        NumRed = Range("B3").Value                      'initial colors
        NumGreen = Range("B4").Value
        Urn.T = 3: Urn.L = 4: Urn.R = 30         '(bottomless) urn dimensions
        Call ClearColumns(Urn.T, Urn.L = 3, Urn.R)       'clear last urn
```

```
        Call FillUrn(Urn)                              'fill new urn
        MsgBox "Start the drawings."
        For i = 1 To draws
            Randomize: x = Rnd
            If x <= NumRed / (NumRed + NumGreen) Then
                Call DrewRed(Urn)                      'a red ball was drawn
            Else
                Call DrewGreen(Urn)                    'a green ball was drawn
            End If
        Next i
    End Sub
```

The procedure `DrewRed` adds another red ball to the urn, using the procedure `EmptyCell` to find a spot for the new red ball. `DrewGreen` performs the analogous task.

```
Private Sub DrewRed(Urn As TUrn)
    NumRed = NumRed + 1                          'update number of red balls
    Call PrintBall(EmptyCell(Urn), 3)
    Range("B8").Value = NumRed                   'print new number of reds
End Sub

Private Sub DrewGreen(Urn As TUrn)
    NumGreen = NumGreen + 1                      'update number of green balls
    Call PrintBall(EmptyCell(Urn), 4)
    Range("B9").Value = NumGreen                 'print new number of greens
End Sub

Private Sub PrintBall(Pos As TPos, color)
    ''''''''''''colored cell version ''''''''''''''''''''''''''
    Cells(Pos.row, Pos.col).Interior.ColorIndex = color
    Cells(Pos.row, Pos.col).BorderAround ColorIndex:=1

    ''''''colored 0 version (faster but less colorful)''''''
    'Cells(Pos.row, Pos.col).Font.ColorIndex = color
    'Cells(Pos.row, Pos.col).Font.FontStyle = "Bold"
    'Cells(Pos.row, Pos.col).Value = "0"
    ''''''''''''''''''''''''''''''''''''''''''''''''''''''''
End Sub
```

The function `EmptyCell` scans the urn and returns the first empty position.

```
Private Function EmptyCell(Urn As TUrn) As TPos
    Dim i As Integer, j As Integer, em As TPos
    i = Urn.T
    Do While 1 = 1
        For j = Urn.L To Urn.R
            If Cells(i, j).Interior.ColorIndex <= 0 Then GoTo line
                '0 version:
            'If IsEmpty(Cells(i,j)) Then GoTo line     'got an empty cell
        Next j
```

```
            i = i + 1
    Loop
line: em.row = i: em.col = j:  EmptyCell = em
End Function
```

The procedure `FillUrn` uses `PrintBall` to populates the grid with the initial number of red and green cells (or O's) representing the balls (Figure 22.4).

	A	B	C	D	E	F	G	H	I	J	K	L	M	O	P	Q
2	draws	20														
3	current red	44														
4	current green	39														
5	urn top	2														
6	urn bottom	8														
7	urn left	4														
8	urn right	16														

FIGURE 22.4: PolyaUrn spreadsheet; initial configuration of balls.

```
Private Sub FillUrn(Urn As TUrn)
    Dim R As Integer, G As Integer, Pos As TPos
    R = NumRed: G = NumGreen
    Pos.row = Urn.T: Pos.col = Urn.L
    Do While R > 0                              'fill with reds first
        Call PrintBall(Pos, 3)
        R = R - 1: Pos.col = Pos.col + 1
        If Pos.col > Urn.R Then Pos.col = Urn.L: Pos.row = Pos.row + 1
    Loop
    Do While G > 0                              'fill with greens
        Call PrintBall(Pos, 4)
        G = G - 1: Pos.col = Pos.col + 1
        If Pos.col > Urn.R Then Pos.col = Urn.L: Pos.row = Pos.row + 1
    Loop
End Sub

Private Sub PrintBall(Pos As TPos, color)
    Cells(Pos.row, Pos.col).Font.ColorIndex = color
    Cells(Pos.row, Pos.col).Font.FontStyle = "Bold"
    Cells(Pos.row, Pos.col).Value = "O"
End Sub
```

22.7   Ehrenfest Diffusion Model

Consider two urns labeled 1 and 2 that contain a total of N balls. A ball is chosen randomly from the N balls and returned to the opposite urn. We take the state of the system to be the number of balls in urn 1. If that number is $i = 0, 1, \ldots, N$, then the transition probabilities are

$$p_{i,i-1} = \frac{i}{N} \quad (i \geq 1), \quad \text{(ball was chosen from urn 1 and placed in urn 2)}.$$

and

$$p_{i,i+1} = \frac{N-i}{N} \quad (i < N), \quad \text{(ball was chosen from urn 2 and placed in urn 1)}.$$

The transition probabilities p_{ij} for all other values of i and j are zero. The model, named after physicists Tatiana and Paul Ehrenfest, has been used to describe the exchange of gas molecules between two containers. (Figure 22.5.)

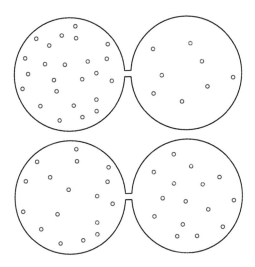

FIGURE 22.5: Ehrenfest model.

The module **EhrenfestUrns** simulates the diffusion model. The user enters the number of desired balls for each urn and the number of desired diffusion steps. An array of type **TUrn** holds the dimensions of the urns as well as the numbers of balls. The procedure **FillUrns** fills each urn with an initial number of balls, displayed as O's in the spreadsheet. The variables **Urn(1).N** and **Urn(2).N** keep track of the number of balls in the urns at any given moment. A Do While loop carries out the diffusion steps. The probability of drawing from a particular urn is proportional to the number of balls in the urn—the higher the number of balls in an urn the more likely it is that that urn will be chosen. Ultimately the system arrives at equilibrium, the number of balls in the urns differing by at most one. The VBA function **Rnd** carries out the drawings by respecting the proportion of balls in the urns, a method analogous to that of the Polya model.

```
Sub EhrenfestUrn()
    Dim x As Double, NumBalls As Integer, i As Integer, Urn() As TUrn
    Columns("C:AZ").ColumnWidth = 2.5                    'compress columns
    ReDim Urn(1 To 2
    Urn(1).N = Range("B5").Value           'initial no. of balls in urn 1
    Urn(2).N = Range("B6").Value           'initial no. of balls in urn 2
    Urn(1).T = 3: Urn(1).L = 4: Urn(1).R = 20              'urn dimensions
    Urn(2).T = 3: Urn(2).L = 24: Urn(2).R = 40
    Call ClearColumns(Urn(1).T, Urn(1).L, Urn(2).R)
    Call FillUrns(Urn)
    Do While Abs(Urn(1).N - Urn(2).N)) > 1 'continue until steady state
        i = i + 1
        Randomize: x = Rnd
        If x <= Urn(1).N / (Urn(1).N + Urn(2).N) Then
            Call DrewFromUrn(1, Urn)
        Else
            Call DrewFromUrn(2, Urn)
        End If
    Loop
End Sub


Private Sub FillUrns(Urn() As TUrn)
    Dim k As Integer, num As Integer, Pos As TPos
    For k = 1 To 2
        num = Urn(k).N
        Pos.row = Urn(k).T: Pos.col = Urn(k).L
        Do While num > 0
            Call PrintBall(Pos, 3)
            num = num - 1: Pos.col = Pos.col + 1
            If Pos.col > Urn(k).R Then _
                Pos.col = Urn(k).L: Pos.row = Pos.row + 1
        Loop
    Next k
End Sub
```

The procedure **DrewFromUrn** carries out the transfer of balls from one urn to the other.

```
Private Sub DrewFromUrn(whichUrn As Integer, Urn() As TUrn)
    If whichUrn = 1 Then                             'update urns
        Urn(2).N = Urn(2).N + 1: Urn(1).N = Urn(1).N - 1
        Call AddToUrn(Urn(2)): Call RemoveFromUrn(Urn(1))
    Else
        Urn(1).N = Urn(1).N + 1: Urn(2).N = Urn(2).N - 1
        Call AddToUrn(Urn(1)): Call RemoveFromUrn(Urn(2))
    End If
End Sub
```

The following procedures add or remove one ball from an urn.

```
Private Sub AddToUrn(Urn As TUrn)
    Dim i As Integer, j As Integer, lrow As Long, em As TPos
```

```
        lrow = Cells(Rows.Count, Urn.L).End(xlUp).row
        For i = Urn.T To lrow + 1           'find an empty cell to place ball
            For j = Urn.L To Urn.R
                If IsEmpty(Cells(i, j).Value) Then
                    em.row = i: em.col = j
                    Call PrintBall(em, 3)
                    Exit Sub
                End If
            Next j
        Next i
    End Sub

    Private Sub RemoveFromUrn(Urn As TUrn)
        Dim i As Integer, j As Integer, lrow As Long
        Dim LastRow As Integer, LastCol As Integer
        lrow = Cells(Rows.Count, Urn.L).End(xlUp).row
        For i = Urn.T To lrow               'find last nonempty cell
            For j = Urn.L To Urn.R
                If Cells(i, j).Value = "0" Then LastRow = i: LastCol = j
            Next j
        Next i
        Cells(LastRow, LastCol).Value = ""              'remove the ball
    End Sub
```

22.8   Stepping Stone Model

Consider an $m \times m$ grid, each cell of which has one of k colors, where k is a fixed integer. The state of the system at any particular time is the current color configuration of the grid. Since there are m^2 cells and k colors for each, the number of possible states is k^{m^2}. Each state may be represented by an $m \times m$ matrix containing the colors of the cells. The transition probabilities are determined by the cell's eight neighbors. The left and right boundaries of the grid are identified, as are the top and bottom boundaries, so that each cell will have the required 8 neighbors. Thus the grid is essentially a torus. (See Figure 22.6.)

At each stage of the process a cell is chosen at random. The cell's color changes randomly according to the color distribution of its neighbors. The transition probabilities are determined by how many neighbors there are of a particular color. For example, if a cell has 3 neighbors that are red, 4 neighbors that are yellow, and 1 neighbor that is green, then the probability that the cell gets the color red is 3/8, yellow 4/8, and green 1/8 (see Figure 22.7). The model has been used in the study of population genetics.

The main procedure SteppingStone first sets up the grid and then allocates memory for the arrays Colors and Pos. The entry Colors(i,j) is the color index

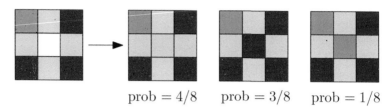

FIGURE 22.6: The neighbors of cell 0.

$$\text{prob} = 4/8 \qquad \text{prob} = 3/8 \qquad \text{prob} = 1/8$$

FIGURE 22.7: Transition probabilities.

of the cell at position `Pos(i,j).row`, `Pos(i,j).col`. The procedure `Initialize` uses the function `RndInt` of Section 10.1. to load the array `Colors` with random color numbers 1–56.

The array `Pos` transforms the grid into a torus so that cells which lie on the edge of the grid still have 8 neighbors. The procedure `MakeTorus` carries out the transformation.

A For Next loop carries out the transition from one color configuration to the next. The loop repeatedly calls the procedure `ColorNextCell`, which randomly selects a cell and changes its color according to the transition probabilities.

```
Sub SteppingStone()
    Dim Grid As TGrid, Colors() As Integer, Pos() As TPos
    Dim i As Long, NumDisplays As Long
    NumDisplays = Range("B2").Value                'number of displays
    Grid = MakeGrid(5, 5, 26, 26, 2.2, 11, 0)
    Colors = InitialColors(Grid)        'store, print random cell colors
    Pos = MakeTorus(Grid)               'transforms row T-1 to row B, etc.
    For i = 1 To NumDisplays
        Range("C2").Value = NumDisplays - i
        Call ColorNextCell(Pos, Colors, Grid)
    Next i
End Sub

Private Function InitialColors(G As TGrid) As Integer()
    Dim i As Integer, j As Integer, Colors() As Integer
    ReDim Colors(G.T To G.B, G.L To G.R)
    Range(Cells(G.T, G.L), Cells(G.B, G.R)).Borders.LineStyle _
        = xlContinuous
```

```
    For i = G.T To G.B
        For j = G.L To G.R
            Colors(i, j) = RndInt(1, 56)
            Cells(i, j).Interior.ColorIndex = Colors(i, j)
        Next j
    Next i
    InitialColors = Colors
End Function
```

The procedure `MakeTorus` constructs an array `P` of type `TPos` that transforms the grid into a torus. For this it uses the procedure `Convert2TorusPoint` of Section 7.2.

```
Private Function MakeTorus(G As TGrid) As TPos()
    Dim i As Integer, j As Integer, P() As TPos
    ReDim P(G.T - 1 To G.B + 1, G.L - 1 To G.R + 1)
    For i = G.T - 1 To G.B + 1
        For j = G.L - 1 To G.R + 1
            P(i, j).row = i: P(i, j).col = j
            Call Convert2TorusPoint(P(i, j).row, P(i, j).col, G)
        Next j
    Next i
MakeTorus = P
End Function
```

The procedure `ColorNextCell` is the heart of the module. Initially, a cell in the grid is chosen randomly by `RndInt`. The cell is passed the torus coordinates in the form of the array `Pos`. A For Next loop runs through the cell's neighbors and places their colors in the array `NborColors`. A second For Next loop runs through the color numbers 1–56 to obtain the frequency distribution of the neighbor colors. These are placed in the array `ColorFreq`. A third For Next loop converts the frequencies in `ColorFreq` to probabilities by dividing by the number of neighbors. The resulting probability distribution is passed to the procedure `RndOut` of Section 10.1, which returns a random color based on the distribution.

```
Private Sub ColorNextCell(Pos() As TPos, Colors() As Integer, _
        G As TGrid)
    Dim Nbor As TPos, k As Integer, i As Integer, j As Integer
    Dim color As Integer, NborColors(1 To 8) As Integer
    Dim roff As Integer, coff As Integer, ColorFreq(1 To 56) As Double
    i = RndInt(G.T, G.B): j = RndInt(G.L, G.R)        'get random indices
    k = 1                                  'first index of NborColors array
    For roff = -1 To 1  'run through neighbors of randomly selected cell
        For coff = -1 To 1                      'using row and column offsets
            If roff ^ 2 + coff ^ 2 <> 0 Then          'avoid 0,0 offset
                Nbor = Pos(i + roff, j + coff)        'position of neighbor
                NborColors(k) = Colors(Nbor.row, Nbor.col)
                k = k + 1
            End If
```

```
         Next coff
      Next roff
      For color = 1 To 56        'get freq. distribution of neighbor colors
         For k = 1 To 8                   'run through the neighbors
            If color = NborColors(k) Then          'kth neighbor color
               ColorFreq(color) = ColorFreq(color) + 1          'tally
            End If
         Next k
      Next color
      For color = 1 To 56        'convert freq. to to prob. distribution
         ColorFreq(color) = ColorFreq(color) / 8
      Next color
      Colors(i, j) = RndOut(ColorFreq)                 'get random color
      Call PrintCell(Pos(i,j), Colors(i, j))
   End Sub
```

22.9   Exercises

1. Let coins A and B have probabilities p_a and p_b of heads on a single toss. Initially, coin A is tossed until heads comes up, at which time coin B is tossed until heads comes up, then coin A again. Write a program SwitchCoinToss that simulates the experiment, printing out the sequence of A's and B's in one column and the sequence of H's and T's in an adjacent column.

2. (Direction dependent random walk). Consider again a single red cell moving in a spreadsheet grid. Define the system to be in state

   1 if the cell just moved right,   2 if the cell just moved left,
   3 if the cell just moved up,   4 if the cell just moved down.

   (This assumes that at time zero the cell was in motion, establishing an initial direction.) Denote by p_{ij} the transition probabilities that govern the motion of the cell. For example, p_{11} is the probability that the cell moves right, given that it just moved right, and p_{23} is the probability that it moves up, given that is just moved left. Write a program DirectionRandomWalk that simulates the experiment. You will have to enter the transition matrix of your choice in the spreadsheet.

3. Write a program PolyaUrn2 that modifies the module PolyaUrn as follows: Draw ball at random from the urn, note the color, replace it, and add to the urn s balls of the same color and o balls of the opposite color.

4. (Moran urn model). Simulate the following variation of the Pólya urn model:

   (1) Draw a ball randomly from an urn containing red and green balls.

   (2) Return the ball together with a ball of the same color.

   (3) Remove a ball randomly from the urn.

   The new feature is (3), which results in the number of balls in the urn not changing, but of course the color distribution does. What happens in the long run? The Moran urn model has been used to study genetic drift in populations.

5. (Hoppe urn model). Simulate the following variation of the Pólya urn scheme: An urn contains b black balls. At any draw, if a black ball is drawn replace the ball along with a ball of a non-black color randomly generated from the numbers $2, \cdots, 56$. If a non-black ball is chosen, return the ball and add another ball of the same color, as in the Pólya urn scheme. Thus the number of black balls remains constant. For a visually interesting effect place the non-black ball in the position of the black ball drawn and then add back the black ball at the bottom of the urn grid.

6. The version of SteppingStone changed one cell at a time. Make a version SteppingStoneSimultaneous that changes the whole grid before displaying the new colors. Suggestion: For each transition, copy the existing color configuration into a temporary array TempColors, carry out the transition rules using this array, copy back to Colors, and then print the colors. You may wish to to allow the user to enter the number of times the transitions are carried out before the next display Choosing a large number speeds up the "evolution" of the color configuration.

7. Find the steady state by hand for the following matrices P

   (a) $\begin{bmatrix} .9 & .1 \\ .4 & .6 \end{bmatrix}$ (b) $\begin{bmatrix} .7 & .3 \\ .5 & .5 \end{bmatrix}$ (c) $\begin{bmatrix} .2 & .8 \\ .1 & .9 \end{bmatrix}$ (d) $\begin{bmatrix} .07 & .93 \\ .84 & .16 \end{bmatrix}$

8. For this exercise refer to Example 22.4. Suppose NutFlocks determines that 65% of viewers prefer action movies, 30% prefer horror, and 15% prefer comedy.

   (a) Use the program MatrixCalculator of Chapter 13 to find the n-step distribution vector for $n = 10$.

   (b) Find the steady state vector.

Part VI

Properties of Numbers

Chapter 23

Divisibility and Prime Numbers

With this chapter we begin an examination of some of the deeper properties of positive integers, also called *natural numbers*. As will be seen, many of these properties have great theoretical importance as well as significant applications, particularly in cryptography. Our goal in this chapter is to explore some these properties and to show how VBA Excel may be used to automate computations and suggest patterns. Applications to cryptography are made in a later chapter.

23.1 The Division Algorithm

The division algorithm asserts that for any pair of integers a, b with $b \neq 0$ there exist unique integers q and r, called the *quotient* and *remainder*, respectively, such that

$$a = qb + r, \quad \text{where } 0 \leq r < |b|. \tag{23.1}$$

For example, for the pairs $(a, b) = (14, 3)$ and $(-14, -3)$ we have, respectively, $14 = \underline{4} \cdot 3 + \underline{2}$ and $-14 = \underline{5} \cdot (-3) + \underline{1}$, where we have underlined q and r. To find q and r we consider various cases:

- $a = b$: Then $a = 1 \cdot a + 0$, so take $q = 1$ and $r = 0$

- $0 < a < b$: Then $a = 0 \cdot b + a$, so take $q = 0$ and $r = a$

- $0 < b < a$: The multiples $b, 2b, 3b, \cdots$ eventually exceed a. If $(q + 1)b$ is the first to do so, then $qb \leq a$ and $(q + 1)b > a$. Set $r = a - qb$. From the first inequality $r \geq 0$ and from the second $r < b$.

- $b < 0 < a$: Apply positive version to a and $-b$ to get $a = q_1(-b) + r$, where $0 \leq r < -b = |b|$, and take $q = -q_1$.

- $a < 0 < b$: Apply positive version to $-a$ and b to obtain $-a = q_1 b + r_1$, where $0 \leq r_1 < b$ and so $a = -q_1 b - r_1$. If $r_1 = 0$ take $q = -q_1$ and $r = 0$. If $r_1 > 0$ write $a = -(q_1 + 1)b + (b - r_1)$ and take $q = -(q_1 + 1)$ and $r = b - r_1$.

- $a, b < 0$: Apply previous case to a and $-b$ to get $a = q_1(-b) + r$, where $0 \leq r < -b = |b|$, and take $q_1 = -q$ and

DOI: 10.1201/9781003351689-23

The procedures `Qtnt` and `Rmdr` below return q and r given a and b.

```
Function Qtnt(a As Long, b As Long) As Long
    Dim q As Long
    If b = 0 Then Exit Function
    If a = Int(a / b) * b Or 0 < b And 0 < a   Then
        q = Int(a / b)
    ElseIf b < 0 And 0 < a Then
        q = -Int(a / (-b))
    ElseIf a < 0 And 0 < b Then
        q = -Int((-a) / b) - 1
    ElseIf b < 0 And a < 0 Then
        q = Int((-a) / -(b)) + 1
    End If
    Qtnt = q
End Function


Function Rmdr(a As Long, b As Long) As Long
    Rmdr = a - Qtnt(a, b) * b
End Function
```

One could use Double instead of Long in the above code if larger integers a, b are desired.

23.2 Number Bases

The familiar base 10 expansion of a positive integer is illustrated by the example

$$123456 = \underline{1} \cdot 10^5 + \underline{2} \cdot 10^4 + \underline{3} \cdot 10^3 + \underline{4} \cdot 10^2 + \underline{5} \cdot 10^1 + \underline{6} \cdot 10^0,$$

where the digits have been underlined to emphasize their role in the expansion. There is nothing special about the number 10, however. Indeed, any positive integer greater than 1 may be used. For example, in base 7 the same number is written as

$$1022634_7 = \underline{1} \cdot 7^6 + \underline{2} \cdot 7^4 + \underline{2} \cdot 7^3 + \underline{6} \cdot 7^2 + \underline{3} \cdot 7^1 + \underline{4} \cdot 7^0.$$

Here we have used a subscript to designate the base; it is omitted for base 10. For bases larger than 10 we need extra symbols. We shall use capital letters:

$$A = 10, \quad B = 11, \quad C = 12, \quad \ldots, \quad Z = 35,$$

This enables us to express positive integers in bases as large as 36.[1] For example, 123456 may be written in base 14 as

$$32DC4_{14} = \underline{3} \cdot 14^4 + \underline{2} \cdot 14^3 + \underline{13} \cdot 14^2 + \underline{12} \cdot 14^1 + \underline{4} \cdot 14^0.$$

---

[1]Even larger bases are possible by using additional symbols such as small letters. We have resisted the temptation.

To convert an integer n from base 10 to base $b > 1$ one successively divides n and the resulting quotients by b. The remainders, in the reverse order that they were generated, form the digits of the base b representation. Here are the steps that convert 123456 to $32DC4_{14}$:

$$123456 = 8818 \cdot 14 + \underline{4}$$
$$8818 = 629 \cdot 14 + \underline{12}$$
$$629 = 44 \cdot 14 + \underline{13} \qquad 32DC4_{14}$$
$$44 = 3 \cdot 14 + \underline{2}$$
$$3 = 0 \cdot 14 + \underline{3}$$

The function `FromBase10` implements the algorithm. Given arguments n and b it returns the representation of n in base b. A For Next loop stores the 36 digit symbols $0 - 9$, $A - Z$ in the array `Digit`, where `Digit(i) = i` for $i = 0, 1, \ldots 9$ and `Digit(i) = Chr(55+i)` for $i = 10, 11, \ldots 35$. A Do Until loop keeps dividing n and successive quotients by b until a quotient is zero. The corresponding remainders are strung together to form the representation.

```
Function FromBase10(ByVal num_10 As String, base_b As Long) As String
    Dim x As Long, r As Long, num_b As String, Digit() As String
    Digit = GenerateDigits
    x = CLng(num_10)                          'start the division
    Do
        r = x Mod base_b:
        num_b = Digit(r) & num_b              'attach it to string
        x = x \ base_b                   'new x is quotient of last x
    Loop Until x = 0
    FromBase10 = num_b           'return base 10 input number in base_b
End Function

Function GenerateDigits() As String()
    Dim Digit(0 To 35) As String, i As Long
    For i = 0 To 35                    'generate digit symbols for output
        If i <= 9 Then Digit(i) = i                'characters 0 to 9
        If i > 9 Then Digit(i) = Chr(55 + i)       'characters A to Z
    Next i
GenerateDigits = Digit
End Function
```

The reverse procedure `ToBase10` takes a number in base b and returns the number in base 10.

```
Function ToBase10(ByVal num_b As String, base_b As Long) As String
    Dim i As Long, j As Long, d As String, Digit() As String
    Dim prod As String, num_10 As Long
    num_b = StrReverse(num_b)        'reverse digits since powers decrease
    Digit = GenerateDigits
```

```
For i = 0 To Len(num_b) - 1
    d = Mid(num_b, i + 1, 1)
    For j = 0 To 35                      'find number corresponding to digit
        If d = Digit(j) Then Exit For
    Next j
    num_10 = j * base_b ^ i + num_10
Next i
ToBase10 = CStr(num_10)            'return base_b input number in base 10
End Function
```

The following function combines the two procedures. It takes a pair of bases a, b and a number in base a and returns the number in base b.

```
Function BaseToBase(num_a As String, base_a As Long, base_b As Long) _
        As String                    'convert base_a  number num_a to base_b
    Dim num_10 As String, num_b As String
    num_10 = ToBase10(num_a, base_a)
    num_b = FromBase10(num_10, base_b)
    BaseToBase = FromBase10(num_10, base_b)
End Function
```

The function may be placed in a module by including the following command button procedure:

```
Private Sub CommandButton1_Click()
    Dim num_a As String, base_a As Long, num_b As String, base_b As Long
    base_a = Range("B4").Value                'enter base going from
    num_a = Range("C4").Value                   'number in base_a
    base_b = Range("B5").Value                  'enter base going to
    num_b = BaseToBase(num_a, base_a, base_b)
    Range("C5").Value = num_b
    Range("C7").Value = BaseToBase(num_b, base_b, base_a)
End Sub
```

*23.3　Wolfram's Binary Rules

The program in this section implements in VBA Stephan Wolfram's *Rule of 30* as well as other similar rules. The rule was introduced by Wolfram in 1983 as an example of a one-dimensional cellular automata, that is, a row of cells that change state according to simple rules. This is of interest since these rules frequently lead to unexpected complex patterns. The rule of 30 is of particular interest as it is chaotic and aperiodic.

Wolfram's rules for the change of state of a cell, in this case colors, depend on the color of the cell and those of its left and right neighbors. The rule of 30 is described in Figure 23.1. For example, in the first pattern the central cell is red as are its two neighbors. The rule then requires the cell color to change to white. In the fourth pattern the central cell is white as is its right neighbor, and its left neighbor is red. In this case the rule requires the cell color to change to red. The top pattern in Figure 23.3 shows the result of applying the rules 17 times with a single 1 in the first row. The bottom pattern in the figure shows the result of applying the rule over 60 times, with interesting patterns beginning to emerge.

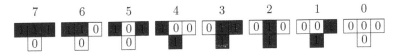

FIGURE 23.1: Rule 30.

The cells in Figure 23.1 have been labelled with 1's and 0's, 1 for red and 0 for white. In the first row these have been interpreted as binary numbers with values 7–0. Similarly, the second row is binary for the number 30, accounting for the rule's name. Changing the number changes the rule. Thus there are as many rules as there 8 digit binary numbers, namely $2^8 = 256$. The rule of 22 is given in Figure 23.2.

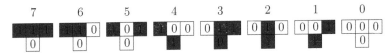

FIGURE 23.2: Rule 22.

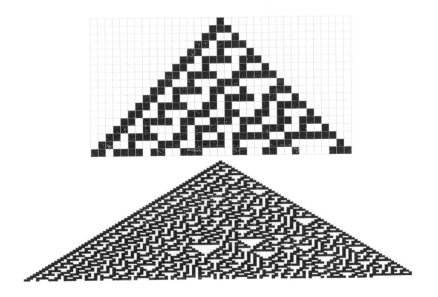

FIGURE 23.3: Rule 30 configuration.

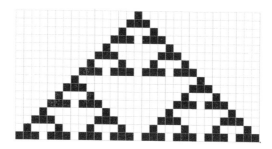

FIGURE 23.4: Rule 22 configuration.

The program `WolframRules` allows the user to enter any one of the 256 numbers 0–255. Running the program then produces the corresponding configuration. The number is entered in cell B5 and digits 1 are entered in row 6; the rest of the cells are assumed to be 0. Entering a single 1 produces the triangular shapes in the figures. Entering several 1's can produce interesting competing triangular shapes.

```
Public L As Integer, R As Integer, T As Integer, B As Integer
Sub WolframRules()
    Dim states(0 To 7) As Integer, i As Integer, rule As Integer
    L = 3: R = 150: T = 6: B = 90                        'grid position
    Range(Cells(T, L), Cells(B, R)).ClearFormats
    Range(Cells(, L), Cells(, R)).ColumnWidth = 1.7   'compress columns
    Range(Cells(T, L), Cells(B, R)).RowHeight = 8         'and rows
    rule = Range("B5").Value                   'get Wolfram rule number
    Call Binary(rule, states)             'convert rule to binary states
    Call FirstGen                             'initialize first row T
    For i = T + 1 To B
        Call NextGen(states, i)    'color row i using previous generation
    Next i
End Sub
```

The procedure `Binary` converts the rule number into binary form $d_7 d_6 \ldots d_1 d_0$ and stores the digits in the array `states` with `states(0)`$= d_0$, `states(1)`$= d_1$, etc. For example, if the rule number is 30 then `states(j) = 1` for $j = 1, 2, 3, 4$, and 0 otherwise. One could use the procedure `FromBase10` for this but `Binary` is more direct.

```
Sub Binary(rule As Integer, states() As Integer)
    Dim q1 As Integer, q As Integer
    q = rule
    For i = 0 To 7
        states(i) = q Mod 2                 'remainder on division by 2
        q1 = q \ 2                          'quotient of last quotient
        q = q1                              'new quotient
    Next i
End Sub
```

The procedure `FirstGen` simply colors red any cell in the top row T containing a digit 1.

```
Sub FirstGen()
    Dim j As Integer
    For j = L To R 'read first generation row T, color cells with a 1 red
        If Cells(T, j).Value = 1 Then Cells(T, j).Interior.ColorIndex = 3
    Next j
End Sub
```

The procedure `NextGen` is given a row and runs through the row's columns applying the binary rules.

```
Sub NextGen(states() As Integer, row As Integer)
    Dim j As Integer For j = L + 1 To R - 1
        Call ApplyRules(states, row, j)  'apply rules for the entire row
    Next j
End Sub
```

The heart of the module is the procedure `ApplyRules`. It is given the rule stored in the array `states` and also the row, column position of a new cell. The cell immediately above it and the left and right neighbors of this cell are assigned binary digits: 1 if the color of the cell is red, 0 otherwise. These form a three digit binary number `Lcell Mcell Rcell` with decimal value 0–7. The number is then converted into decimal form c. The number `states(c)` is either 0 or 1 depending on the rule, and the cell is colored accordingly. For example, for the configuration in Figure 23.1 labelled 2, `Lcell=0`, `Mcell=1`, `Rcell=0`. Since for rule 30 `states(2)=1`, the new cell gets the color red.

```
Sub ApplyRules(states() As Integer, row As Integer, col As Integer)
    Dim Lcell As Integer, Mcell As Integer, Rcell As Integer
    Dim c As Integer
    Lcell = 0: Mcell = 0: Rcell = 0                         'initialize
    If Cells(row - 1, col - 1).Interior.ColorIndex = 3 Then Lcell = 1
    If Cells(row - 1, col).Interior.ColorIndex = 3 Then Mcell = 1
    If Cells(row - 1, col + 1).Interior.ColorIndex = 3 Then Rcell = 1
    c = Lcell * 4 + Mcell * 2 + Rcell           'configuration number
    Cells(row, col).Interior.ColorIndex = 3 * states(c)  'red or nothing
End Sub
```

23.4 Greatest Common Divisor

A positive integer m is said to *divide* a positive integer a, if there exists a positive integer k such that $a = km$, that is, if division of a by m results in a remainder of zero. The integers m and k are called *divisors* or *factors* of a. Of course, a and 1 are always divisors of a. The remaining divisors, if any, are called *proper*. Note that no proper divisor of a can be larger than $a/2$. Indeed, if $a = km$ ($k \geq 2$) and $m > a/2$ then we would have $a = km > 2a/2$, impossible. If m divides a we write $m \mid a$; otherwise we write $m \nmid a$. Thus $4 \mid 12$ but $4 \nmid 14$.

The *greatest common divisor* (gcd) of a pair of positive integers a, b is the largest of all divisors common to both a and b. We shall use the notation $\gcd(a, b)$ for this number. For example,

$$\gcd(12, 7) = 1, \ \gcd(24, 30) = 6, \ \text{and} \ \gcd(987654321, 123456789) = 9,$$

the last example with the aid of the module GCD below.

The division algorithm gives us a way to calculate the gcd of an arbitrary pair of positive integers a, b. For this we may assume without loss of generality that $a > b$. Assuming this, we successively apply the division algorithm in the following manner, where $r_1 = a$ and $r_2 = b$:

$$r_1 = q_1 r_2 + r_3, \ r_2 = q_2 r_3 + r_4, \ r_3 = q_3 r_4 + r_5, \dots \ (r_1 > r_2 > r_3 > r_4 > r_5 \dots)$$

In general, we divide r_{n-2} by r_{n-1} to produce the quotient q_{n-2} and the remainder r_n:

$$r_{n-2} = q_{n-2} r_{n-1} + r_n, \ \ 0 \leq r_n < r_{n-1}. \tag{23.2}$$

Since the remainders are strictly decreasing and nonnegative, there is a smallest n for which $r_{n+1} = 0$. It may be shown that for this n, $r_n = \gcd(a, b)$. We illustrate this for $a = 74$, $b = 22$:

$$
\begin{array}{cccccccccccccccc}
r_1 & q_1 & r_2 & & r_3 & r_2 & q_2 & r_3 & & r_4 & r_3 & q_3 & r_4 & & r_5 & r_4 & q_4 & r_5 & & r_6 \\
74 & = & 3 & \cdot & 22 & + & 8, & & 22 & = & 2 & \cdot & 8 & + & 6, & & 8 & = & 1 & \cdot & 6 & + & 2, & & 6 & = & 3 & \cdot & 2 & + & 0
\end{array}
$$

so $\gcd(74, 22) = r_5 = 2$.

The above scheme may be easily implemented with VBA. Rather than do this, however, we give an algorithm that not only calculates the gcd of user-entered positive integers a, b, but also finds integers x and y such that

$$\gcd(a, b) = xa + yb.$$

This called *Bézout's equation*. For example,

$$\gcd(74, 22) = 2 = 3 \cdot 74 - 10 \cdot 22.$$

As we shall see, this is a very useful property of the gcd. Here's an obvious and sometimes useful consequence:

If $\gcd(a, b) = 1$ then there exist integers x and y such that $1 = ax + by$.

The converse of this is also true: if $1 = ax + by$ holds for some x and y, then $\gcd(a, b) = 1$. Indeed, any common factor of a and b must be a factor of 1 and so must equal 1.

To develop the aforementioned algorithm we begin by solving for r_n in (23.2) to obtain

$$r_n = r_{n-2} - q_{n-2}r_{n-1}, \quad n \geq 3. \tag{23.3}$$

Thus for $n = 3, 4, 5$ and $r_1 = a$, $r_2 = b$, we have

$$
\begin{aligned}
r_3 &= r_1 - q_1 r_2 \\
r_4 &= r_2 - q_2 r_3 = r_2 - q_2(r_1 - q_1 r_2) = -q_2 r_1 + (1 + q_1 q_2) r_2 \\
r_5 &= r_3 - q_3 r_4 = r_1 - q_1 r_2 - q_3(-q_2 r_1 + (1 + q_1 q_2) r_2) \\
&= (1 + q_2 q_3) r_1 - (q_1 + q_3 + q_1 q_2 q_3) r_2
\end{aligned} \tag{23.4}
$$

Continuing this process we see that for each $n \geq 3$ there exist integers x_n and y_n (not necessarily positive) such that $r_n = x_n r_1 + y_n r_2$. Recalling that for some n the remainder r_n is the gcd of $a = r_1$ and $b = r_2$, we conclude that the gcd may be expressed in the desired form $xa + yb$.

To calculate x and y define $y_1 = 0$, $y_2 = 1$ and notice that the coefficients y_n of r_2 in (23.4) follow the rules

$$
\begin{aligned}
y_1 - q_1 y_2 &= -q_1 = y_3 \\
y_2 - q_2 y_3 &= 1 - q_2(-q_1) = 1 + q_1 q_2 = y_4 \\
y_3 - q_3 y_4 &= (-q_1) - q_3(1 + q_1 q_2) = -(q_1 + q_3 + q_1 q_2 q_3) = y_5
\end{aligned}
$$

Writing this in reverse we obtain the general recursion scheme

$$y_1 = 0, \quad y_2 = 1, \quad y_n = y_{n-2} - q_{n-2} y_{n-1}, \quad q_{n-2} = r_{n-2} \backslash r_{n-1}, \quad n \geq 3. \tag{23.5}$$

Equations (23.3) and (23.5) form the basis of the *extended Euclidean algorithm*

The algorithm is implemented in the following procedure, which calculates $gcd(a, b)$ as well as x, and y.

```
Function GCD(a As Long, b As Long, x As Long, y As Long) As Long
    Dim r1 As Long, r2 As Long, r As Long, last_y As Long
    Dim q As Long, y1 As Long, y2 As Long, switched As Boolean, g As Long
    If a < b Then                        'algorithm wants a > b
        Call SwitchValues(a, b): switched = True
    End If
    y1 = 0: y2 = 1: r1 = a: r2 = b: y = y2
    Do
        q = r1 \ r2                      'integer division
        g = r2                           'possible gcd
        last_y = y                       'save
        y = y1 - q * y2                  'new y and r
        r = r1 - q * r2                  'cycle
        r1 = r2: r2 = r
        y1 = y2: y2 = y
    Loop Until r = 0
    y = last_y
```

```
      x = (g - b * y) / a
      If switched Then Call SwitchValues(x, y)      'x goes with a, y with b
      GCD = g
End Function

Sub SwitchValues(x As Long, y As Long)
      Dim z As Long
      z = x: x = y: y = z
End Sub
```

The greatest common divisor may be extended to more than two numbers. For example, $\gcd(a_1, a_2, a_3)$ is defined as the largest common divisor of a_1, a_2, and a_3. It follows easily that

$$\gcd(a_1, a_2, a_3) = \gcd(\gcd(a_1, a_2), a_3),$$

which the reader should verify. More generally, one has

$$\gcd(a_1, a_2, \ldots, a_n) = \gcd(\gcd(a_1, \ldots, a_{n-1}), a_n). \qquad (23.6)$$

The reader may wish to write a simple recursive program that uses this equation to find the gcd of a user-entered or randomly generated list of integers.

23.5   Prime Divisibility Property

Recall that an integer greater than 1 is said to be *prime* if it has no proper divisors, that is, if its only positive divisors are 1 and itself. An integer that is not prime is said to be *composite*. The prime divisibility property is one of the fundamental properties of prime numbers and will be used in a critical way in the next section.

If a prime divides a product of integers $a_1 a_2 \cdots a_n$ then it must divide one if these integers.

The proof requires the *Principle of Mathematical Induction* (which may also be used to establish (23.6)):

For each positive integer n, let $P(n)$ be a statement depending on n. Suppose that
(a) $P(1)$ *is true and*
(b) $P(n+1)$ *is true whenever $P(n)$ is true.*
Then $P(n)$ is true for all n. Furthermore, (b) may be replaced by
(c) $P(n+1)$ *is true whenever $P(k)$ is true for all $k \leq n$.*

One need not start the induction at 1; any positive integer may be used. The principle has been loosely described as the "domino principle": If dominoes are lined up vertically so that the $(n+1)$st domino will fall if the nth one falls, then, if the first domino is tipped, all the dominoes will fall.

The proof of the divisibility property uses induction starting at $n = 2$: For this case we need to show that if $p \mid a_1 a_2$, then either $p \mid a_1$ or $p \mid a_2$. Now, by hypothesis, $a_1 a_2 = pk$ for some integer k. If $p \mid a_1$ then we are done, so assume that $p \nmid a_1$. Then, because p is prime, p and a_1 have no common factors other than 1, that is, $\gcd(a_1, p) = 1$. By Bézout's equation (Section 23.4), $a_1 x + py = 1$ for some integers x and y. Multiplying the equation by a_2 we then have

$$a_2 = a_2(a_1 x + py) = a_1 a_2 x + p a_2 y = pk + p a_2 y = p(k + a_2 y).$$

Thus $p \mid a_2$, completing the first step of the induction argument.

Now assume that for some $n \geq 2$ the assertion holds for all products of length n. We wish to show it holds for products of length $n + 1$. To this end suppose that p divides $a_1 a_2 \cdots a_n a_{n+1}$. We may write the product as $b a_{n+1}$, where $b = a_1 a_2 \cdots a_n$. By the first paragraph, either $p \mid b$ or $p \mid a_{n+1}$. If the latter, then we're done. If the former, we use the induction hypothesis to conclude that p divides a_j for some $j = 1, 2, \ldots, n$.

23.6 The Fundamental Theorem of Arithmetic

The theorem stated in the heading, also called the *unique prime factorization theorem*, asserts that every integer $N \geq 2$ may be written uniquely as a product of prime numbers, called the *prime decomposition of N*. More precisely:

For any positive integer $N \geq 2$ there exists a unique set of primes $p_1 < p_2 \ldots < p_m$ and unique positive exponents e_1, e_2, ..., e_m such that $N = p_1^{e_1} p_2^{e_2} \cdots p_m^{e_m}$.

For example

$$16206750 = 2 \cdot 3^3 \cdot 5^3 \cdot 7^4 \quad \text{and} \quad 66666789 = 3^2 \cdot 7 \cdot 1058203.$$

To establish the existence of prime decompositions we use mathematical induction (the alternate version with (c)). The theorem is trivially true for $N = 2$. To see why the theorem holds in general, suppose we have established that all numbers $\leq N$ have a prime factorization. It then follows that $N + 1$ also has a prime factorization. This is clear if $N + 1$ is prime. On the other hand, if $N + 1$ is not prime then it may be written as a product ab, where a and b are less than $N + 1$, hence less than or equal to N. By our assumption, a

and b have prime factorizations. Multiplying these factorizations and collecting together like primes in increasing order produces the prime factorization for $N + 1$. From the principle of mathematical induction we now conclude that every N has at least one prime factorization.

To establish uniqueness of the prime factorization, suppose that N also has the prime factorization $N = q_1^{f_1} q_2^{f_2} \cdots q_m^{f_m}$, where the exponents f_j are positive and $q_1 < q_2 \ldots < q_n$. We then have

$$p_1^{e_1} p_2^{e_2} \cdots p_m^{e_m} = q_1^{f_1} q_2^{f_2} \cdots q_n^{f_n}. \tag{23.7}$$

We must show that $m = n$, $e_j = f_j$, and $p_j = q_j$ for all j. For this, we may assume without loss of generality that $m \geq n$. The argument we give uses the prime divisibility property. Since p_1 divides the left side of the above equation it divides the right side and so must divide q_j for some j, say $j = j_1$. Since q_{j_1} is prime, $p_1 = q_{j_1}$. Arguing similarly we have

$$p_1 = q_{j_1}, \; p_2 = q_{j_2}, \ldots, \; p_m = q_{j_m} \tag{23.8}$$

Since the p's are strictly increasing, the q's in this scheme are as well. By the prime divisibility property, all the q's in the original expansion are accounted for in (23.8) and it follows that $m = n$ and $p_k = q_k$ for all k. Thus from (23.7)

$$p_1^{e_1} p_2^{e_2} \cdots p_m^{e_m} = p_1^{f_1} p_2^{f_2} \cdots p_m^{f_m},$$

which is possible only if $e_k = f_k$ for all k. (If, say, $e_1 > f_1$, then canceling the factor $p_1^{f_1}$ would result in p_1 dividing the left side but not the right).

The Infinitude of Primes

We may use theorem to prove that there are infinitely many primes. This was already known to Euclid; we give a variation of his argument:

Suppose there were only finitely many primes, say $p_1, p_2, \ldots, p_n$. Form the positive integer $q = p_1 p_2 \cdots p_n + 1$ Clearly, q is larger than every p_j and so cannot be prime. Thus q must be composite and as such must have a prime factor p, say $q = mp$. Since the above list is exhaustive, $p = p_j$ for some j. For ease of notation we can assume that $j = 1$ (otherwise rename the list). We now have $p_1 p_2 \cdots p_n + 1 = mp_1$ and so $1 = p_1(p_2 \cdots p_n - m)$ But the last equation is impossible since the factors on the right are integers with $p_1 > 1$.

23.7   Prime Decomposition with VBA

The module in this section generates the prime number factorization of a positive integer $N \geq 2$. The program first allocates memory for the array `primes` in the factorization as well as the array `exps` for the exponents. Next, the program calls the procedure GetPrimes(N,primes,exps), which populates the arrays primes and exps. A Do While loop runs through these arrays, attaching powers of primes to the factorization string D.

```
Function PrimeDecomposition(N As Long) As String
    Dim primes() As Long, exps() As Integer
    Dim L As Integer, D As String, i As Integer
    ReDim primes(1 To N / 2)        'memory for primes in factorization
    ReDim exps(1 To N / 2)          'memory for exponents in factorization
    Call GetPrimes(N, primes, exps)
    i = 1
    Do While Not primes(i) = 0                  'build the factorization
        If exps(i) = 1 Then
            D = D & primes(i) & " " & "*" & " "
        ElseIf exps(i) > 1 Then
            D = D & primes(i) & "^" & exps(i) & " " & "*" & " "
        End If
        i = i + 1
    Loop
    L = Len(D)
    If L > 1 And Mid(D, L - 1, 1) = "*" Then
        D = Mid(D, 1, L - 2)   'omit last *
    End If
    PrimeDecomposition = D
End Function
```

The procedure GetPrimes is the workhorse of the module, calculating the prime powers in the factorization of N. The primes used to test N for divisibility are stored in the continually updated list TestPrime, obviating the need to initially generate a large list of primes. A Do While loop cycles through the list, determining if N is divisible by the prime. The integer x starts out with the value N and is repeatedly reduced by its prime factors. If x is found to be prime, which is determined by the procedure IsPrime, or if there are no more primes factors in N, then the loop quits, thoroughly exhausted.

```
Sub GetPrimes(N As Long, primes() As Integer, exps() As Integer)
    Dim i As Long, x As Long, TestPrimes() As Integer, exp As Integer
    ReDim TestPrimes(1 To N / 2)            'memory for primes testing N
    TestPrimes(1) = 2                       'first prime <= N
    i = 1: j = 1: x = N            'initializations for reduction process
    Do While TestPrimes(i) <> 0            'loop until no primes left in x
        If IsPrime(x) Then        'found a prime with exponent 1 in reduced N
            primes(j) = x: exps(j) = 1         'add it to list dividing N
            Exit Do                       'N completely reduced so quit
        End If
        exp = 0                           'still primes left in N
        Do While x Mod TestPrimes(i) = 0 'keep dividing x by TestPrime(i)
            x = x / TestPrimes(i)
            exp = exp + 1                  'keep track of how many divisions
        Loop
        If exp <> 0 Then          'if TestPrimes(i) divided x then add it
```

```
            primes(j) = TestPrimes(i)    'to the list of primes dividing N
            exps(j) = exp                    'add the exponent to the list
            j = j + 1            'next index in the list primes() dividing N
        End If
        Call AddPrime(TestPrimes, x, i)        'get new test prime <= x
        i = i + 1
    Loop
End Sub
```

The procedure `AddPrime` is passed the array `TestPrimes`, which holds the test primes obtained so far, and is also passed an integer x and the index of the last prime in the array. A new prime $k \leq x$ is added to the array if k is not divisible by any of the previous test primes.

```
Sub AddPrime(TestPrimes() As Integer, x As Long, idx As Long)
    Dim k As Long, j As Long
    For k = 3 To x                    'check numbers k <= x for primality
        For j = 1 To idx
            If k Mod TestPrimes(j) = 0 Then GoTo Next_k     'k not prime
        Next j
        TestPrimes(idx + 1) = k                    'k prime; add it to array
        Exit For                                        'done
Next_k:
    Next k
End Sub
```

The following function detects if a user-entered positive integer $N \geq 2$ is prime. It does so by checking whether N has a divisor $\leq \sqrt{N}$. It is unnecessary to check for larger divisors as these occur in pairs m, k with $N = mk$, and not both of these can be larger than $\sqrt{N}$.

```
Function IsPrime(N As Long) As Boolean        'return True if N is prime
    Dim i As Long, m As Long, PrimeNumber As Boolean
    PrimeNumber = True                                'default
    m = Int(N ^ (1 / 2))            'no need to test for larger factors
    For i = 2 To m                        'check if i a factor of N
        If N Mod i = 0 Then PrimeNumber = False: Exit For    'N not prime
    Next i
    IsPrime = PrimeNumber
End Function
```

To better appreciate how `GetPrimes` works, take $N = 3087 = 3^2 \cdot 7^3$. Since `TestPrime(1)`$= 2$ does not divide N and $x = N$ is not prime, nothing happens during the first iteration of the outer Do While loop until the second-to-last instruction when the new test prime 3 is added by `AddPrime` to the list. At this stage `TestPrime(2)`$= 3$. The outer Do While loop then executes a second iteration with $i = 2$, while x is still 3087 and $j = 1$. This time, however, the inner Do While loop engages and x is reduced to 7^3 by two divisions by 3. After this `primes(1)` is set to 3, `exps(1)` is set to 2, j is incremented by 1, and `AddPrime` adds 5 to the test primes so `TestPrime(3)`$= 5$. The outer Do

While loop then undergoes a third iteration with $x = 7^3$ and $j = 2$. Since $5 \nmid x$, the only thing that happens here is that the test prime 7 is added by AddPrime, so TestPrime(4)$= 7$. The outer Do While loop then executes a fourth iteration, still with $x = 7^3$ and $j = 2$. But now x can be reduced, this time with 3 divisions by 7, resulting in $x = 1$, primes(2)$= 7$, and exps(2)$= 3$. Next, AddPrime is called but does nothing, and i is incremented to 5. Since TestPrime(5)$= 0$, GetPrimes grinds to a screeching halt.

23.8   Exercises

1. The *least common multiple* (LCM) of a pair of positive integers a, b, denoted $\text{lcm}(a, b)$, is the smallest positive integer that is both a multiple of a and a multiple of b. For example, $\text{lcm}(3, 4) = 12$. Write a function LCM(a,b) that returns $\text{lcm}(a, b)$. Check that $\text{lcm}(a, b) \cdot \gcd(a, b) = ab$.

2. The *divisibility by seven rule* is an algorithm that determines whether or not an integer n with at least three digits is divisible by seven. It proceeds as follows: Remove the last digit d of n, forming a new number m, and then subtract $2d$ from m. Continue the process until a two digit number k appears. It may be shown that n is divisible by 7 if and only if k is divisible by 7. For example, 3346 is divisible by 7 but 2473 is not, as shown by the calculations

$$3346 \rightarrow 334 - 12 = 322 \rightarrow 32 - 4 = 28,$$
$$2473 \rightarrow 247 - 6 = 241 \rightarrow 24 - 2 = 22.$$

Write a program SevenRule that implements the algorithm, listing in a column all the steps.

3. The *divisibility by thirteen rule* is an algorithm that determines whether or not an integer n with at least three digits is divisible by 13. It proceeds as follows: Remove the last digit d of n forming a new number m, and then add $4d$ to m. Continue the process until a two digit number k appears. It may be shown that n is divisible by 13 if and only if k is divisible by 13. For example, 4615 is divisible by 13 but 3377 is not:

$$4615 \rightarrow 461 + 20 = 481 \rightarrow 48 + 4 = 52$$
$$3377 \rightarrow 337 + 28 = 365 \rightarrow 36 + 20 = 56.$$

Write a program ThirteenRule that implements the algorithm, listing in a column all the steps.

4. Use the program GCD to generate $\gcd(a, a+1)$ and $\gcd(a, a+2)$ for various values of a. What are your conclusions?

5. Use the program GCD to generate $\gcd(a + 2b, 2a + b)$ for various relatively prime a, b. What are your conclusions?

6. Write a program Primes-n-To-2n that lists all the primes between n and $2n$ for $n \leq$ some user entered integer N. (A theorem of Chebyshev asserts that there is always at least one such prime.) The following is a sample spreadsheet.

	A		B
2	N =	50	
3	n	$2n$	primes
4	1	2	2
5	2	4	2,3
6	3	6	2,3,5
53	50	100	53, 59, 61, 67, 71, 73, 79, 83, 89, 97

FIGURE 23.5: Spreadsheet for Primes-n-To-2n.

7. Two odd consecutive numbers that are both prime are called *twin primes*. For example, the first eight pairs of twin primes are

$$(3, 5), \ (11, 13), \ (17, 19), \ (29, 31), \ (41, 43), \ (59, 61), \ (71, 73), \ (101, 103).$$

As of this writing it is not known whether the number of such pairs is finite or infinite. Write a program TwinPrimes that prints all pairs of twin primes less than or equal to a user-entered number. Print the pairs in adjacent columns.

8. Write a program SquarePlusOne that prints out prime numbers of the form $n^2 + 1$, n a positive integer, that is, primes p for which $p - 1$ is a perfect square (e.g. 2, 5, 17, 37, 101). Print both the prime and n. As of this writing it is not known whether the number of such primes is finite or infinite.

9. Write a program NumberSpiral that generates in a grid a spiral of arbitrary length, as in Figure 23.6, where the prime numbers are colored red. M.

```
17-16-15-14-13 30
 |           |  |
18  5—4—3 12 29
 |  |    |  |  |
19  6  1—2 11 28
 |  |       |  |
20  7—8—9—10 27
 |              |
21-22-23-24-25-26
```

FIGURE 23.6: Number spiral.

Gardener and S. Ulam discovered that in large spirals there are many prominent lines containing a large number of primes. Large spirals may be achieved by using a grid with very small cells without the numbers and displaying a prime as a red cell and a non prime as an empty cell. Figure 23.7 was generated using 15000 numbers.

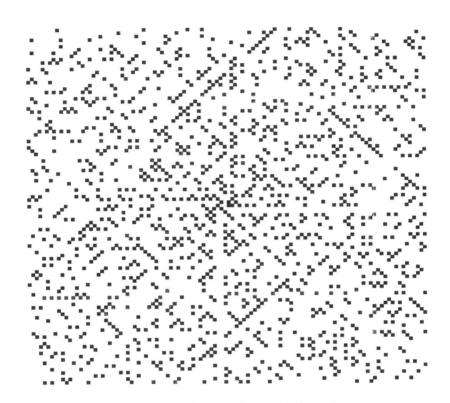

FIGURE 23.7: Number spiral for 15,000 numbers.

10. It is known that a prime p may be written as $a^2 + b^2$ (a and b positive integers) if and only if $p - 1$ is divisible by 4. Write a program `PrimesSumOfSquares` that prints out a column of such primes p less than a given integer N and integers a, b whose squares sum to p. ($N > 200$ takes awhile.)

Chapter 24

Congruence

In this chapter we explore another kind of arithmetic, one that is based on the notion of congruence. The theory was developed by the German mathematician Carl Friedrich Gauss in the late 18th century. It has since become an important tool in the study of divisibility and an integral component of cryptography. While the notion in its abstract form may be unfamiliar to the reader, the idea of congruence is implicit in ordinary "clock arithmetic."

24.1 Definition and Basic Properties

Let m, a, and b be integers with $m \geq 2$. We say that a *is congruent to b modulo m*, written

$$a \equiv b \pmod{m} \quad \text{or} \quad a \equiv_m b.$$

if $a - b$ is divisible by m, that is, if there exists an integer k such that $a - b = km$. If no such integer exists we write

$$a \not\equiv b \pmod{m} \quad \text{or} \quad a \not\equiv_m b.$$

For example,

$$17 \equiv 3 \pmod{7}, \quad -17 \equiv 1 \pmod{6}, \quad \text{and} \quad -17 \equiv -2 \pmod{5}$$

Note that if r is the remainder upon dividing a by m, that is,

$$a = qm + r, \quad 0 \leq r < m,$$

then $a - r = qm$ and so $a \equiv r \pmod{m}$. This shows that every integer a is congruent to one of the numbers $0, 1, \ldots, m - 1$ modulo m.

The 12-hour clock is based on congruence. For example, if it is 11 AM now, in 54 hours it will be

$$11 + 54 \equiv_{12} 5 \text{ PM.}$$

The 24-hour clock is related to congruence in a similar manner.

Here are three immediate facts regarding congruences. They assert that congruence is an equivalence relation (Section 18.10). Additional properties are given later.

DOI: 10.1201/9781003351689-24

(1) $a \equiv_m a$. (reflexivity)

(2) If $a \equiv_m b$, then $b \equiv_m a$. (symmetry)

(3) If $a \equiv_m b$ and $b \equiv_m c$, then $a \equiv_m c$ (transitivity)

For example, to verify (3) note that the hypotheses state that $a - b = jm$ and $b - c = km$ for some integers j and k. Adding these equations we obtain $a - c = (j + k)m$, which shows that $a \equiv_m c$.

24.2   Congruence Arithmetic

Congruences have arithmetic properties similar to those of ordinary arithmetic. Here are the most common:

If $a \equiv b \pmod{m}$ and $c \equiv d \pmod{m}$ and n is a positive integer, then

(1) $a + c \equiv b + d \pmod{m}$ (Addition Property).

(2) $a - c \equiv b - d \pmod{m}$ (Subtraction Property).

(3) $ac \equiv bd \pmod{m}$ (Multiplication Property).

(4) $a^n \equiv b^n \pmod{m}$, n a positive integer (Power Property).

To see why these hold we start with the hypothesis that $a - b = jm$ and $c - d = km$ for some integers j and k. For (1) simply note that

$$(a + c) - (b + d) = (a - b) + (c - d) = jm + km = (j + k)m.$$

A similar argument verifies (2). For (3) we have

$$ac - bd = a(c - d) + d(a - b) = akm + djm = (ak + dj)m.$$

Property (4) follows by iterating (3) (or by mathematical induction), taking $c = a$ and $d = b$. Alternatively, we could use the formula

$$a^n - b^n = (a - b)(a^{n-1} + a^{n-2}b + \cdots ab^{n-1}b^{n-1}),$$

which may be verified by multiplying out the right side and cancelling terms. Since by hypothesis $a - b$ is divisible by m, so is $a^n - b^n$.

Example 24.1. (Casting out nines). Since $10 \equiv 1 \pmod{9}$, it follows from (4) that $10^n \equiv 1 \pmod{9}$ for any natural number n. Now let a be a positive integer with decimal digit representation $a = a_k a_{k-1} \ldots a_1 a_0$, so that

$$a = a_0 + a_1 \cdot 10 + a_2 \cdot 10^2 + \cdots + a_k \cdot 10^k.$$

Let $a^* = a_0 + a_1 + a_2 + \cdots + a_k$, the sum of the digits of a. We claim that $a \equiv_9 a^*$. Indeed, since $10^n \equiv_9 1$ it follows from (3) that $a_n \cdot 10^n \equiv_9 a_n$ $(n = 1, 2, \ldots, k)$. Adding these congruences and using (1) verifies the assertion. For example, successively adding digits we see that

$$77777777 \equiv_9 56 \equiv_9 11 \equiv_9 2.$$

The result may be used as a check for multiplication. For example, using similar notation, property (3) and the result of the previous paragraph imply that

$$(ab)^* \equiv_9 ab \equiv_9 a^* b^*.$$

Thus if $(ab)^* \not\equiv_9 a^* b^*$, then the multiplication ab is incorrect. $\Diamond$

As in ordinary arithmetic, congruence arithmetic also has a cancellation law, but with an important caveat:

If $ac \equiv bc \pmod{m}$ *and* $\gcd(m, c) = 1$, *then* $a \equiv b \pmod{m}$. (24.1)

For example, in the congruence $12 \equiv 6 \pmod{3}$ we can cancel the common factor 2 in the equation to obtain the correct assertion $6 \equiv 3 \pmod{3}$, but cancelling the common factor 6 leads to the incorrect assertion $2 \equiv 1 \pmod{3}$.

To prove (24.1) we use Bézout's equation: $xc + ym = 1$ for some integers x and y. We then have

$$a - b = (a - b) \cdot 1 = (a - b)(xc + ym) = x(ac - bc) + y(a - b)m.$$

By hypothesis, the first term on the extreme right is divisible by m, and of course so is the second. Therefore, $a - b$ is divisible by m, as required.

24.3 Fermat's Little Theorem

The theorem asserts the following:

If a *is a positive integer, p is prime, and $p \nmid a$, then* $a^{p-1} \equiv 1 \pmod{p}$.

The theorem is a consequence of the cancellation property (24.1) and the following property, of interest in itself:

If a *is a positive integer and p is a prime, then* $a^p \equiv a \pmod{p}$.

The property provides the theoretical basis of the RSA cipher algorithm discussed in the next chapter. We prove it by induction on the positive integer a (Section 23.5). For $a = 1$ the assertion is trivial. Now suppose that $a^p \equiv$

$a \pmod{p}$ for some $a \geq 1$. We show that $(a+1)^p \equiv (a+1) \pmod{p}$, which will complete the induction. By the Binomial Theorem (Section 19.6)

$$(a+1)^p = a^p + pa^{p-1} + \frac{p(p-1)}{2} + \frac{p(p-1)(p-2)}{3!} + \cdots + p + 1.$$

Subtracting $a+1$ yields

$$(a+1)^p - (a+1) = (a^p - a) + pa^{p-1} + \frac{p(p-1)}{2} + \frac{p(p-1)(p-2)}{3!} + \cdots + p.$$

By hypothesis, p divides $a^p - a$, the first term on the right, and clearly p divides the remaining terms. Therefore, p divides $(a+1)^p - (a+1)$, completing the induction argument.

24.4   Remainder of a Product

The calculation of remainders by computer is obviously limited by storage capacity. For example, with Long numbers the statement 987654321 Mod 123456789 returns the correct answer 9 without hesitation, but overflow occurs if an extra digit is added to the first number. Larger numbers are possible using the Double type, but even this is inadequate for some applications. In this section and the next we consider two special cases which get around storage problems. The first, treated in the current section, allows the calculation of the remainder of a large number in factored form. The second, derived from the first and discussed in the next section, allows the calculation of the remainder of numbers raised to large powers. The procedures will be needed for the RSA algorithm in the next chapter but are of some intrinsic interest.

To start, fix a positive integer m and for any positive integer a let $R(a)$ denote the remainder when m divides a:

$$a = qm + R(a), \quad 0 \leq R(a) < m.$$

For ease of notation we drop the subscript m on $\equiv_m$. We claim that

$$R(ab) = R(R(a)R(b)). \tag{24.2}$$

For example, for $m = 3$

$$R(5 \cdot 8) = R(40) = 1 \text{ and } R(R(5)R(8)) = R(2 \cdot 2) = 1$$

To verify (24.2), note first that because $a \equiv R(a)$ and $b \equiv R(b)$, by the product property

$$ab \equiv R(a)R(b) \equiv R(R(a)R(b)).$$

Since $ab \equiv R(ab)$ it follows that

$$R(ab) \equiv R(R(a)R(b)).$$

Since each side of the last equivalence is less than m the congruence must actually be equality, verifying (24.2).

By iterating (24.2) (or using mathematical induction) we see that

$$R(a_1 a_2 \cdots a_n) = R(R(a_1)R(a_2) \cdots R(a_n)) \qquad (24.3)$$

Thus the problem of calculating the remainder of a product of possibly large factors $a_1, a_2, \ldots, a_n$ reduces to calculating the remainder of the product of the smaller numbers $R(a_1), R(a_2), \ldots, R(a_n)$.

The function `ProductRmdr` is based on equation (24.3). It takes as arguments the modulus m and an array of factors $a_1, a_2, \ldots, a_n$, and returns the remainder of the product, working with factors two at a time. An initial For Next loop runs through the array, replacing each factor a_j by its remainder r_j, the reduction justified by (24.3). A second For Next loop replaces each new remainder x by the remainder of x times its predecessor. To illustrate, the array $\{a_1, a_2, a_3, a_4\}$ undergoes the following transformations, where $r_i = R(a_i)$:

$$\{a_1, a_2, a_3, a_4\} \to \{r_1, r_2, r_3, r_4\}$$
$$\to \{r_1, R(r_1 r_2), r_3, r_4\}$$
$$\to \{r_1, R(r_1 r_2), (R(r_1 r_2)r_3), r_4\}$$
$$\to \left\{r_1, R(r_1 r_2), (R(r_1 r_2)r_3), R\left[(R(r_1 r_2)r_3)r_4\right]\right\}$$

The last entry in the array is

$$R\left[(R(R(a_1)R(a_2))R(a_3))R(a_4)\right],$$

which reduces to $R(a_1 a_2 a_3 a_4)$ by (24.2).

```
Function ProductRmdr(Factors() As Double, modulus As Double) As Double
    Dim i As Integer, number As Double, NumFactors As Integer
    NumFactors = UBound(Factors)
    For i = 1 To NumFactors                    'preliminary reduction
        Factors(i) = Rmdr(Factors(i), modulus)
    Next i
    For i = 1 To NumFactors - 1                'compute remainder pairwise
        Factors(i + 1) = Rmdr(Factors(i) * Factors(i + 1), modulus)
    Next i
    ProductRmdr = Factors(NumFactors)
End Function

Function Rmdr(number As Double, modulus As Double) As Double
    Rmdr = number - (Int(number / modulus) * modulus)
End Function
```

The reader may check that the function works well for large lists of factors not exceeding 15 digits. Using Long instead of Double decreases this to 9 digits.

24.5   Remainder of a Power

For the RSA algorithm in the next chapter we will need to find the remainder of powers c^e where c and e are (possibly large) positive integers. The algorithm that accomplishes this expresses the power e in base 2, allowing c^e to be expressed as a product of factors. The algorithm described in the preceding section may then be invoked to find the remainder of the power.

For a concrete example, suppose we take

$$e = 27 = 11011_2 = 2^4 + 2^3 + 2 + 1,$$

so that $c^e = c \cdot c^2 \cdot c^8 \cdot c^{16}$. By (24.2), we then have

$$R(c^e) = R\left(R(c) \cdot R(c^2) \cdot R(c^8) \cdot R(c^{16})\right).$$

By (24.2) again, the powers of c may be successively reduced modulo 2:

$$R(c), \quad R(c^2) = R\big(R(c) \cdot R(c)\big)), \quad R(c^4) = R\left(R(c^2) \cdot R(c^2)\right),$$
$$R(c^8) = R\left(R(c^4) \cdot R(c^4)\right), \quad R(c^{16}) = R\left(R(c^8) \cdot R(c^8)\right).$$

All but the third term (which corresponds to the 0 digit in 11011_2 are used as factors.

The function `PowerRmdr` implements the algorithm. It is passed the base, exponent, and modulus, and returns the remainder.

```
Function PowerRmdr(c As Double, e As Double, m As Double) As Double
    Dim i As Integer, j As Integer, r As Double
    Dim eBase2 As String, L As Integer, Factors() As Double
    eBase2 = FromBase10(e,2)                    'get base 2 representation
    ReDim Factors(1 To NumFactors(eBase2))        'NumFactors = no. of 1's
    L = Len(eBase2)                               'L = no. of digits
    r = c : i = 1                                 'initializations
    If Mid(eBase2, L, 1) = 1 Then                'if rightmost digit is 1
        Factors(1) = c                           'then first factor is c
        i = i + 1                                 'get ready for next factor
    End If
    For j = 1 To L - 1                           'run through digits of eBase2
        r = Rmdr(r ^ 2, m)                'get remainder of successive squares
        If Mid(eBase2, L - j, 1) = 1 Then            'if digit = 1 then
            Factors(i) = r: i = i + 1                   'add r to list
        End If
    Next j                                        'next digit
    PowerRmdr = ProductRmdr(Factors, m)          'product of factors
End Function


Function NumFactors(ByVal eBase2 As String) As Integer
    expbase2 = Replace(eBase2,"0", "1")
    NumFactors = Len(eBase2)
End Function
```

The function `NumFactors` counts the number of ones in the base 2 representation of the exponent. This is precisely the number of factors in c^e.

24.6 Solving Congruence Equations

It is sometimes possible to solve a congruence equation in an unknown variable. For example, the equation $3x \equiv 5 \pmod{14}$ has a unique solution $x = 11$ between 0 and 13; all other solutions are congruent to 11 modulo 14. Similarly, $x^2 \equiv 1 \pmod{12}$ has solutions $x = 1, 5, 7, 11$ between 0 and 11. On the other hand, $x^2 \equiv 2 \pmod{12}$ has no solutions.

One way to determine the solutions between 0 and $m - 1$ of such equations is to simply plug these numbers into the equation and see which ones work. The following module carries out these computations for polynomial equations with integer coefficients. The user enters the modulus m in cell B3 and a polynomial expression $a_n x^n + a_{n-1} x^{n-1} + \cdots + a_1 x + a_0 - b$ in cell B4. The program solves the congruence equation

$$a_n x^n + a_{n-1} x^{n-1} + \cdots + a_1 x + a_0 - b \equiv 0 \pmod{m} ,$$

or, equivalently,

$$a_n x^n + a_{n-1} x^{n-1} + \cdots + a_1 x + a_0 \equiv b \pmod{m} ,$$

by testing the values $x = 0, 1, \ldots, m - 1$. The user must enter the equation in the first form.

The program works like this: A For Next loop calls the VBA function `Replace` to place the numbers $j = 0, 1, \ldots, m - 1$ into the variable x in `expr`. The VBA procedure `Evaluate` evaluates the polynomial for these values. If a j-value produces a value of the polynomial that is divisible by m, then j is printed as a solution.

```
Sub CongruenceSolver()
    Dim expr As String, exprVal As String, m As Integer, j As Integer
    Dim y As Long, sol As String, L As Integer
    Range("B5").ClearContents
    m = Range("B3").Value
    expr = RemoveWhiteSpace(Range("B4").Value)
    For j = 0 To m - 1
        exprVal = Replace(expr, "x", j)                    'put j in x
        y = Evaluate(exprVal)
        If y Mod m = 0 Then sol = sol & j & ","   'add to solution list
    Next j
    L = Len(sol)
    Range("B5").Value = Mid(sol, 1, L - 1)              'kill last comma
End Sub
```

*24.7   Modular Arithmetic

It is possible to define addition and multiplication for the set of remainders modulo m, typically denoted by $\mathbb{Z}_m$:

$$\mathbb{Z}_m := \{0, 1, \ldots, m - 1\}.$$

For $a, b \in \mathbb{Z}_m$ we define $a +_m b$ to be the remainder when $a + b$ is divided by m. Similarly $a *_m b$ is defined as the remainder when ab is divided by m. The operations can be summarized in *congruence tables*, as shown for the cases $m = 4$ and $m = 5$.

$+_4$	0	1	2	3
0	0	1	2	3
1	1	2	3	0
2	2	3	0	0
3	3	0	1	2

$*_4$	0	1	2	3
0	0	0	0	0
1	0	1	2	3
2	0	2	0	2
3	0	3	2	1

$+_5$	0	1	2	3	4
0	0	1	2	3	4
1	1	2	3	4	0
2	2	3	4	0	1
3	3	4	0	1	2
4	4	0	1	2	3

$*_5$	0	1	2	3	4
0	0	0	0	0	0
1	0	1	2	3	4
2	0	2	4	1	3
3	0	3	1	4	2
4	0	4	3	2	1

Addition and multiplication mod m have properties similar to ordinary addition and multiplication. For example, it is easy to check that

$$a *_m (b +_m c) = a *_m b +_m a *_m c.$$

Also, integers mod m have negatives; that is, for each $a \in \mathbb{Z}_m$ there exists a unique $b \in \mathbb{Z}_m$ such that $a +_m b = 0$. This is clear if $a = 0$, and if $a > 0$ take $b = m - a$. The quantity b is called the *negative of a* and is denoted by $-a$. That every member of $\mathbb{Z}_m$ has a negative is reflected in the rows of the addition tables: each row has a zero. Using negatives, one may then define subtraction $a -_m b$ as $a +_m (-b)$

By contrast, it is not always possible to define division. For example, in the table for $*_4$, there is no b such that $2 *_4 b = 1$. This means that 1 cannot be divided by 2, that is, 2 does not have an *inverse*. On the other hand, $3 *_4 3 = 1$ so 1 can be divided by 3. This difference is accounted for by the fact that 3 and 4 are relatively prime while 2 and 4 are not. In general, the equation $a *_m x = 1$ can be solved for x if a and m are relatively prime. Indeed, in this case we have $ax + my = 1$ for some integers x and y and so $ax \equiv 1 \pmod{m}$. In particular, if m is prime then every member of $\mathbb{Z}_m$ except 0 has an inverse. This is reflected in the rows of the multiplication tables for primes: each row except the first contains a 1. The reader should check this for the table $*_5$ and compare with the table $*_4$.

Here is a program that generates addition and multiplication tables for a user-entered modulus m. The plus sign is printed in `Cells(PRow,PCol)`, the column labels start at `Cells(PRow + 1, PCol)` and the row labels start at `Cells(PRow, PCol+1)`. Similarly for the times sign.

```
Sub CongruenceTables()
    Dim m As Integer, i As Integer, j As Integer, PRow As Integer
    Dim PCol As Integer, TRow As Integer, TCol As Integer
    Columns("C:ZZ").ColumnWidth = 3.5              'compress columns
    Rows("3:50").RowHeight = 18
    Range("D3:ZZ1000").ClearContents               'clear previous tables
    m = Range("B3").Value
    PRow = 3: PCol = 4                             'position of + sign
    TRow = PRow + m + 2: TCol = PCol              'position of * sign
    Cells(PRow, PCol).Value = "+"                 'labels for the tables
    Cells(TRow, TCol).Value = "*"
    For i = 0 To m - 1
        For j = 0 To m - 1
            Cells(PRow + 1 + i, PCol).Value = i        'col labels for +
            Cells(PRow, PCol + 1 + j).Value = j        'row labels for +
            Cells(PRow + 1 + i, PCol + 1 + j).Value = (i + j) Mod m
            Cells(TRow + 1 + i, TCol).Value = i        'col labels for *
            Cells(TRow, TCol + 1 + j).Value = j        'row labels for *
            Cells(TRow + 1 + i, TCol + 1 + j).Value = (i * j) Mod m
        Next j
    Next i
End Sub
```

24.8 Exercises

1. Write a program `SquareRootModp` that takes as input a number M and a prime p and prints out all numbers $n \leq M$ that have an integer square root k modulo p, that is, $k^2 \equiv n \pmod{p}$. Print the k's next to the n's.

2. Write a program `AddSubMatMod` that returns the sum or difference of two integer matrices modulo m. Do the same for `ScalarMultMod`, and `MultMatMod`.

3. Write a program `InvertMatMod` that finds the inverse of an integer matrix modulo m for a prime m, if it exists. Use the method of Exercise 9 that calculates the adjugate of a matrix.

4. Write a program `CramerMod` that finds the solution modulo m of a system of equations with integer coefficients.

*5. Write a program `RowEchelonMod` that calculates the reduced row echelon form of a matrix modulo a prime p

*6. Write a program `CongCalc` in the sprit of `ComplexCalc` that does congruence arithmetic.

Chapter 25

Cryptography

Cryptography is the study of enciphering and deciphering messages. The practice has been used for centuries in conducting affairs of state to prevent unwanted actors from seeing sensitive material. It is no less important today when so many activities, commercial, government, military, etc., are conducted via the internet. For this reason it is essential to have secure ways of transmitting data. Cryptography plays an indispensable role in such transmissions.

Cryptography is based on the following idea: There are two functions $\mathcal{E}()$ and $\mathcal{D}()$, the *encrypter* and the *decrypter*, respectively. The functions have the crucial property that decryption reverses encryption:

$$\mathcal{D}\big(\mathcal{E}(M)\big) = M \quad \text{for all messages } M.$$

One party transmits a message M, also called *plaintext*, encrypted as $\mathcal{E}(M)$. Another party receives the encrypted message $C = \mathcal{E}(M)$, called *ciphertext*, and decrypts it by applying the function $\mathcal{D}$, thereby retrieving the original message M. Ideally, only intended parties are privy to $\mathcal{D}$ (or to a *key* from which $\mathcal{D}$ may be derived), thus assuring the desired privacy. An algorithm that implements either $\mathcal{E}$ or $\mathcal{D}$ is called a *cipher*. In this chapter we consider several ciphers, both classical and modern. We also develop programs that allow one to encode and decode messages using these ciphers. Some of the material relies on the matrix theory and number theory developed in earlier chapters.

25.1 Encoding the Alphabet

Many ciphers are based on the assignment of numbers to capital letters of the alphabet, facilitating a numerical approach to encrypting and decrypting messages. We shall assign to the letters A,B,C,..., Z the numbers 0,1,2, ..., 25, respectively, as shown in Figure 25.1. This choice is motivated by the fact

A	B	C	D	E	F	G	H	I	J	K	L	M	N	O	P	Q	R	S	T	U	V	W	X	Y	Z
0	1	2	3	4	5	6	7	8	9	10	11	12	13	14	15	16	17	18	19	20	21	22	23	24	25

FIGURE 25.1: Alpha values table.

DOI: 10.1201/9781003351689-25

that these numbers are the remainders modulo 26, allowing the introduction of mathematical techniques via the modulus function. We shall call the number that corresponds to a letter the *alpha value* or *alpha number* of that letter. In this regard the following functions will occasionally be useful:

```
Function Alph2Val(letter As String) As Integer
    Alph2Val = Asc(letter) - 65
End Function

Function Val2Alph(val As Integer) As String
    Val2Alph = Chr(val Mod 25 + 65)
End Function
```

25.2   Caesar's Shift Cipher

The shift cipher is an example of a *monoalphabetic substitution cipher*, where each character in the plaintext is replaced by another character to form the ciphertext. In the shift cipher each letter of the message is replaced by the letter appearing a fixed number of positions down the alphabet, as shown in the top table of Figure 25.2, where the shift is 3. For example, the message " Delete the

A	B	C	D	E	F	G	H	I	J	K	L	M	N	O	P	Q	R	S	T	U	V	W	X	Y	Z
D	E	F	G	H	I	J	K	L	M	N	O	P	Q	R	S	T	U	V	W	X	Y	Z	A	B	C

0	1	2	3	4	5	6	7	8	9	10	11	12	13	14	15	16	17	18	19	20	21	22	23	24	25
3	4	5	6	7	8	9	10	11	12	13	14	15	16	17	18	19	20	21	22	23	24	25	0	1	2

FIGURE 25.2: Table for Caesar cipher, shift by 3 positions.

files", rendered as DELETETHEFILES, encodes to GHOHWHWKHILOHV. The second table in the figure shows the corresponding alpha numbers of the letters. From these we get a simple numerical description of the functions $\mathcal{E}$ and $\mathcal{D}$:

$$\mathcal{E}(x) \equiv (x + 3) \pmod{26}, \quad \mathcal{D}(x) \equiv (x - 3) \pmod{26}$$

These formulas suggest how the process of encrypting and decrypting can be implemented by a computer program.

The Caesar cipher is not particularly useful as it is easily broken and therefore offers little security. It does find use, however, as part of more complex ciphers such as the Vigenerère Cipher described in Section 25.7.

25.3 Zigzag Cipher

The *zigzag* or rail fence cipher is an example of a *transposition cipher*, an encryption scheme that takes the letters of the plaintext and rearranges them according to some well-defined rule. Decryption is then carried out by reversing the rule. We explain the cipher by an example.

Consider the message "Locate the file. They are here." rendered in the form

$$\text{LOCATETHEFILETHEYAREHERE}$$

Write this in the following zigzag pattern:

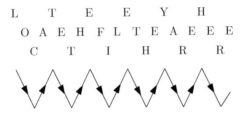

```
L     T     E     E     Y     H
  O  A  E  H  F  L  T  E  A  E  E  E
    C     T     I     H     R     R
```

FIGURE 25.3: Zigzag message.

Notice that there are 6 letters in the top and bottom rows and twice that many in the third row, 24 letters in all. Thus the ratios of the rows are 1:2:1. The encrypted message is then obtained by concatenating the rows (called "rails"):

$$\text{LTEEYHOAEHFLTEAEEECTIHRR}$$

The receiver notes the number of letters in the ciphertext, partitions them into rows with the ratios 1:2:1, and replicates the diagram in Figure 25.3, thus retrieving the original message.

The scheme works for any plaintext that can be rendered into rows with ratios 1:2:1. Thus if the length is L then there must be an integer n such that $L = n + 2n + n = 4n$. Otherwise the message must be padded with enough "dummy letters" to achieve this (X's are usually used for this purpose).

25.4 Spiral Cipher

The spiral cipher takes a message, writes it in a spiral starting from the center and moving outward and sends out the concatenated rows of the spiral as ciphertext. The message must be padded with X's so that the spiral is a complete square with an odd number of letters on each side. For example, consider the message "Rendezvous tonight at the same time and same place." Capitalizing and padding leads to the square in Figure 25.4, which we take as a matrix. Note that the first letter R of the message is the center of the matrix and the rest of the message spirals outward in a clockwise direction.

```
A  N  D  T  H  E  S
E  N  I  G  H  T  A
M  O  N  D  E  A  M
I  T  E  R  Z  T  E
T  S  U  O  V  T  P
E  M  A  S  E  H  L
X  X  X  X  E  C  A
```

FIGURE 25.4: Spiral cipher.

The X's pad the message so that each side of the square has an odd length and hence an exact center. The rows are then concatenated to form the ciphertext

ANDTHESENIGHTAMONDEAMITERZTETSUOVTPEMASEHLXXXXECA

Reversing the process yields the plaintext.

In the remainder off the section we develop encryption and decryption procedures in VBA. These use the following Public variables.

```
Public row As Integer, col As Integer, side As Integer
Public center As Integer, msgLen As Integer
```

The pair `row,col` are the current indices of the spiral matrix, `nRows` is the number of rows, and `center` is the row and column of the central entry of the matrix.

Encryption

The procedure `Encrypt` reads the message entered in B3 and pads it with X's, if necessary, to obtain a message whose spiral matrix is square with an odd number of rows and columns. The plaintext is then converted into the spiral matrix using the procedure `ptext2spiralMat`. The procedure `Spiral2ctext(spiralMat)` concatenates the rows of the matrix to form the ciphertext.

```
Sub Encrypt()
    Dim spiralMat() As String, ptext As String, ctext As String
    ptext = PadX(Range("B3").Value)
    msgLen = Len(ptext)
    nRows = msgLen ^ (1 / 2)
    center = Int(nRows / 2) + 1                    'nRows is odd
    spiralMat = ptext2spiralMat(ptext)
    ctext = Spiral2ctext(spiralMat)                'convert matrix to ctext
    Range("B5").Value = ctext
End Sub
```

The function `PadX` finds the smallest perfect square k^2 greater than or equal to the length of the plaintext. If k is not odd, the function increases k by one. The plaintext is then padded with $k^2 - n$ X's.

```
Function PadX(ptext As String) As String
    Dim k As Integer, n As Integer
    n = Len(ptext)
    For k = 1 To n
        If k ^ 2 >= n Then Exit For
    Next k
    If k ^ 2 Mod 2 = 0 Then k = k + 1        'nRows must have odd length
    PadX = ptext & String(k ^ 2 - n, "X")    'fill in with X's
End Function
```

The function `ptext2spiralMat` takes the padded plaintext and places its letters in a matrix in spiral form. The method used here is similar to that of the program `GrowingSpiral` of Section 7.8. It uses offsets 0, ± 1 to move from one leg of the spiral to the next, the legs increasing in length, causing the spiral to grow.

```
Function ptext2spiralMat(ptext As String) As String()
    Dim k As Integer, dupcount As Integer, spiralMat() As String
    dupcount = 1: k = 1
    ReDim spiralMat(1 To nRows, 1 To nRows)
    row = center: col = center                  'start at center of matrix
    spiralMat(row, col) = Mid(ptext, 1, 1)
    Do While k <= msgLen
        'go left for dupcount steps:
        Call MakeLeg(spiralMat, ptext, 0, -1, k, dupcount)
        'go up for dupcount steps:
        Call MakeLeg(spiralMat, ptext, -1, 0, k, dupcount)
        dupcount = dupcount + 1                  'increase leg length by one
        'go right for dupcount steps:
        Call MakeLeg(spiralMat, ptext, 0, 1, k, dupcount)
        'go down for dupcount steps:
        Call MakeLeg(spiralMat, ptext, 1, 0, k, dupcount)
        dupcount = dupcount + 1
    Loop
    ptext2spiralMat = spiralMat
End Function

Sub MakeLeg(spiralMat() As String, ptext As String, roff As Integer, _
            coff As Integer, k As Integer, dupcount As Integer)
    Dim j As Integer
    For j = 1 To dupcount                                'leg length
        row = row + roff                    'next position in spiral array
        col = col + coff
        k = k + 1: If k > Len(ptext) Then Exit Sub
        spiralMat(row, col) = Mid(ptext, k, 1)
    Next j
End Sub
```

The procedure `Spiral2ctext` forms the ciphertext by concatenating the rows of the spiral matrix.

```
Function Spiral2ctext(spiralMat() As String) As String
    Dim i As Integer, j As Integer, ctext As String
    For i = 1 To nRows                      'concatenate rows of spiral
        For j = 1 To nRows
            ctext = ctext & spiralMat(i, j)
        Next j
    Next i
    Spiral2ctext = ctext
End Function
```

Decryption

The decryption process is similar to encryption. It starts with the ciphertext, creates a matrix by splitting the text into rows, and then unwinds the matrix, viewed as a spiral.

```
Sub Decrypt()
    Dim spiralMat() As String, ctext As String
    ctext = Range("B5").Value                    'read the ciphertext
    msgLen = Len(ctext)
    nRows = msgLen ^ (1 / 2)
    center = Int(nRows / 2)
    spiralMat = ctext2Spiral(ctext)   'create matrix by row concatenation
    Range("B6").Value = Spiral2ptext(spiralMat)
End Sub
```

Here is the concatenation process:

```
Function ctext2Spiral(ctext As String) As String()
    Dim i As Integer, j As Integer, k As Integer, spiralMat() As String
    ReDim spiralMat(1 To nRows, 1 To nRows)
    k = 1
    For i = 1 To nRows
        For j = 1 To nRows
            spiralMat(i, j) = Mid(ctext, k, 1): k = k + 1
        Next j
    Next i
    ctext2Spiral = spiralMat
End Function
```

The procedure `Spiral2ptext` unwinds the spiral matrix to form the original plaintext. It is entirely similar to `ptext2spiralMat` except that instead of writing a character into the matrix the procedure reads it.

```
Function Spiral2ptext(spiralMat() As String) As String
    Dim i As Integer, j As Integer, ptext As String
    Dim k As Integer, dupcount As Integer
```

```
      dupcount = 1: k = 1
      row = center: col = center
      ptext = spiralMat(row, col)
      Do While k <= msgLen
                      'go left for dupcount steps
        Call GetLeg(spiralMat, ptext, 0, -1, k, dupcount)
                        'go up for dupcount steps
        Call GetLeg(spiralMat, ptext, -1, 0, k, dupcount)
        dupcount = dupcount + 1              'increase leg length by one
                      'go right for dupcount steps
        Call GetLeg(spiralMat, ptext, 0, 1, k, dupcount)
                      'go down for dupcount steps
        Call GetLeg(spiralMat, ptext, 1, 0, k, dupcount)
        dupcount = dupcount + 1
      Loop
      Spiral2ptext = ptext
    End Function

  Sub GetLeg(spiralMat() As String, ptext As String, roff As Integer, _
            coff As Integer, k As Integer, dupcount As Integer)
      Dim j As Integer
      For j = 1 To dupcount                          'leg length
          row = row + roff                'next position in spiral array
          col = col + coff
          k = k + 1: If k > msgLen Then Exit Sub
          ptext = ptext & spiralMat(row, col)
      Next j
    End Sub
```

25.5 Affine Cipher

This cipher converts a plaintext letter with alpha value p into the cipher text letter with alpha value $c \equiv (ap + b) \pmod{26}$, where a and b are positive integers such that a and 26 are relatively prime. The latter requirement allows one to solve for p. To see this, note by hypotheses that $ap = c - b + 26k$ for some k. Also, by Bézout's equation (Section 23.4), $1 = ax + 26y$ for some integers x and y that may be found by the extended gcd program in the same section. Multiplying the last equation by p yields

$$p = apx + 26py = (c - b + 26k)x \equiv (c - b)x \pmod{26}$$

Since the right side is known one can find p and therefore recover the message.

Implementation with VBA

The implementation uses both `Alph2Val` and `Val2Alph` of Section 25.1 The arguments in these function must be changed to Long to be compatible with the procedures `GCD` and `Rmdr` used below.

```
Function AffEncrypt(plaintext As String, a As Long, b As Long) As String
    Dim k As Integer, c As Long, ciphertext As String
    For k = 1 To Len(plaintext)
        p = Alph2Val(Mid(plaintext, k, 1))
        c = Rmdr(a * p + b, 26)
        ciphertext = ciphertext & Val2Alph(c)
    Next k
    AffEncrypt = ciphertext
End Function


Function AffDecrypt(ciphertext As String, a As Long, b As Long) As String
    Dim plaintext As String, k As Integer, x As Long, y As Long
    Dim p As Long, g As Long
    Call GCD(a, 26, x, y)
    For k = 1 To Len(ciphertext)
        c = Alph2Val(Mid(ciphertext, k, 1))
        p = Rmdr((c - b) * x, 26)
        plaintext = plaintext & Val2Alph(p)
    Next k
AffDecrypt = plaintext
End Function


Private Sub CommandButton1_Click()
    Dim plaintext As String, a As Long, b As Long
    a = Range("B3").Value
    b = Range("B4").Value
    plaintext = Range("B5").Value
    Range("B6").Value = AffEncrypt(plaintext, a, b)
End Sub


Sub CommandButton2_Click()
    Dim ciphertext As String, a As Long, b As Long
    a = Range("B3").Value
    b = Range("B4").Value
    ciphertext = Range("B6").Value
    Range("B7").Value = AffDecrypt(ciphertext, a, b)
End Sub
```

25.6   Permutation Cipher

The *permutation cipher* is the most general form of the *transposition cipher*. The function $\mathcal{E}$ for this cipher is a permutation $n_1, n_2, \ldots, n_p$ of the integers

$1, 2, \ldots, p$ for some p. A message is broken up into blocks or *message units* of length p, and the letters of each block are permuted by $\mathcal{E}$. The last block may have to be padded to reach length p.

For an example, consider the message "Able was I ere I saw Elba". Let the function $\mathcal{E}$ be based on the permutation

$$\begin{pmatrix} 1 & 2 & 3 & 4 & 5 \\ 3 & 5 & 4 & 2 & 1 \end{pmatrix},$$

which we abbreviate by the bottom row 3,5,4,2,1 (see Section 19.9). Divide the compressed and capitalized message ABLEWASIEREISAWELBA into message units of five letters each:

$$\text{ABLEW, \quad ASIER, \quad EISAW, \quad ELBAX,} \tag{25.1}$$

where we have filled out the last block with the letter X. The sender applies the permutation 3,5,4,2,1 to each block: letter 3 is written first, then letter 5, etc. This produces the blocks

$$\text{LWEBA \quad IRESA, \quad SWAIE, \quad BXALE} \tag{25.2}$$

The concatenation LWEBAIRESASWAIEBXALE is sent to the receiver.

The receiver is in possession of the key 3,5,4,2,1 and therefore can find $\mathcal{D}$, which is based on the inverse permutation

$$\begin{pmatrix} 1 & 2 & 3 & 4 & 5 \\ 5 & 4 & 1 & 3 & 2 \end{pmatrix} = 5, 4, 1, 3, 2.$$

The receiver breaks up the enciphered message into blocks and applies the permutation to each, obtaining the message units (25.1). Concatenating these and dropping the X produces the original compressed message.

Encryption with VBA

The main procedure of the encryption module is passed a message in compressed, capitalized form and a permutation and returns the encrypted form with the necessary padding in the manner of the above example.

```
Function PEncrypt(plaintext As String, perm As String) As String
    Dim blocksize  As Integer, pArr() As Integer
    pArr = ConvertPermutation(perm)      'convert string to integer array
    blocksize = Ubound(pArr)
    plaintext = PadX(plaintext, blocksize)      'pad last block with X's
    PEncrypt = ConvertText(plaintext, pArr)
End Function
```

The function `ConvertPermutation` takes a permutation string and converts it into an integer array starting at index 1. For example, the string "3,5,4,2,1" is returned as the array with entries `pArr(1)` = 3, `pArr(2)` = 5, etc.

```
Function ConvertPermutation(perm As String) As Integer()
    Dim k As Integer, temp() As String, pArr() As Integer, U As Integer
    temp = Split(perm, ",")          'split permutation into string array
    U = UBound(temp): ReDim pArr(1 To U)
    For k = 1 To U                              'convert to integer array
        pArr(k) = CInt(temp(k - 1))             'array starts at index 1
    Next k
    ConvertPermutation = pArr                              'return array
End Function
```

The function `PadX` takes `plaintext` and `blocksize` as arguments and returns the padded message. It first determines the number of X's to add to the incomplete block, if any, by finding the remainder r when the message length is divided by the block size. As r is the size of the last block, the number of X's to be added is the block size minus the remainder. The procedure uses the VBA function `String` to make the required string of X's. For example, `PadX` takes the string ABLEWASIEREISAWELBA and returns ABLEWASIEREISAWELBAX. In this example, the message length is 19, the block size is 5, and the remainder is 4, so exactly one X must be added.

```
Function PadX(plaintext As String, blocksize As Integer) As String
    Dim i As Integer, Xstr As String
    r = Len(plaintext) Mod blocksize        'size of incomplete block
    If r <> 0 Then
        Xstr = String(blocksize - r, "X")                      'X's
    End If
    PadX = plaintext & Xstr                  'return padded message
End Function
```

The function `ConvertText` takes as arguments a permutation array and text and returns the text with block members rearranged by the permutation. A For Next loop extracts each block and the function `PermuteBlock` permutes it. For example, `ConvertText` takes the string whose blocks are given by (25.1), and the permutation array $\{3, 5, 4, 2, 1\}$ and returns the string whose blocks are given by (25.2).

```
Function ConvertText(intext As String, pArr() As Integer) As String
    Dim numblocks As Integer, block As String, outtext As String
    Dim blocksize As Integer, i As Integer
    blocksize = Ubound(pArr): numblocks = Len(intext) / blocksize
    For i = 0 To numblocks - 1                      'convert each block
        block = Mid(intext, i * blocksize + 1, blocksize)    'get block
        outtext = outtext & PermuteBlock(block, pArr)      'convert it
    Next i
    ConvertText = outtext     'returned the concatenated, permuted blocks
End Function
```

The function `PermuteBlock` takes as arguments a block and the permutation array and returns the permuted block. The function `Mid` extracts the letters in the order established by the permutation array.

```
Function PermuteBlock(block As String, pArr() As Integer) As String
    Dim k As Integer, pblock As String
    For k = 1 To UBound(pArr)          'extract letters in permuted order
        pblock = pblock & Mid(block, pArr(k), 1)
    Next k
    PermuteBlock = pblock
End Function
```

Decryption with VBA

The decryption procedure `PDecrypt` is passed the ciphertext generated by `PEncrypt`, as well as the original permutation, and returns the plaintext. It is similar to `PEncrypt` except that it needs to invert the permutation. This is done using the function `InvPerm` (Section 19.9).

```
Function PDecrypt(ciphertext As String, perm As String) As String
    Dim blocksize As Integer, pArr() As Integer, iperm() As Integer
    pArr = ConvertPermutation(perm)     'convert perm into integer array
    iperm = InvPerm(pArr)
    PDecrypt = ConvertText(ciphertext, iperm)
End Function
```

Combining Encryption and Decryption

The above procedures may be incorporated into a single module `PermCipher` using command buttons as indicated in Figure 25.5 (code omitted). The sender enters the permutation and the message into cells B3 and B5, respectively, and then presses the command button encrypt . The encrypted message appears in cell B6 and is sent to the receiver. The recipient, who has the identical program, enters the permutation and encrypted message into cells B3 and B6, respectively, and presses the command button decrypt . The decrypted message then appears in B7. Figure 25.5 depicts the spreadsheet after both buttons were pressed.

	A	B
3	permutation	3,5,4,2,1
5	encrypt	ABLEWASIEREISAWELBA
6	decrypt	LWEBAIRESASWAIEBXALE
7		ABLEWASIEREISAWELBA

FIGURE 25.5: Input-output for `PermCipher`.

25.7   Vigenère Cipher

This cipher is based on the table of shifted alphabets in Figure 25.6. It is the best known example of a *polyalphabetic cipher*, that is, a cipher based on substitution using several alphabets. To encrypt a message the sender chooses a *text key*. The letters of the key determine which shifted alphabets to use to encode the message. Thus the cipher consists of a series of interlaced Caesar ciphers selected by the key.

1: message letters 2: key letters

	A	B	C	D	E	F	G	H	I	J	K	L	M	N	O	P	Q	R	S	T	U	V	W	X	Y	Z
A	A	B	C	D	E	F	G	H	I	J	K	L	M	N	O	P	Q	R	S	T	U	V	W	X	Y	Z
B	B	C	D	E	F	G	H	I	J	K	L	M	N	O	P	Q	R	S	T	U	V	W	X	Y	Z	A
C	C	D	E	F	G	H	I	J	K	L	M	N	O	P	Q	R	S	T	U	V	W	X	Y	Z	A	B
D	D	E	F	G	H	I	J	K	L	M	N	O	P	Q	R	S	T	U	V	W	X	Y	Z	A	B	C
E	E	F	G	H	I	J	K	L	M	N	O	P	Q	R	S	T	U	V	W	X	Y	Z	A	B	C	D
F	F	G	H	I	J	K	L	M	N	O	P	Q	R	S	T	U	V	W	X	Y	Z	A	B	C	D	E
G	G	H	I	J	K	L	M	N	O	P	Q	R	S	T	U	V	W	X	Y	Z	A	B	C	D	E	F
H	H	I	J	K	L	M	N	O	P	Q	R	S	T	U	V	W	X	Y	Z	A	B	C	D	E	F	G
I	I	J	K	L	M	N	O	P	Q	R	S	T	U	V	W	X	Y	Z	A	B	C	D	E	F	G	H
J	J	K	L	M	N	O	P	Q	R	S	T	U	V	W	X	Y	Z	A	B	C	D	E	F	G	H	I
K	K	L	M	N	O	P	Q	R	S	T	U	V	W	X	Y	Z	A	B	C	D	E	F	G	H	I	J
L	L	M	N	O	P	Q	R	S	T	U	V	W	X	Y	Z	A	B	C	D	E	F	G	H	I	J	K
M	M	N	O	P	Q	R	S	T	U	V	W	X	Y	Z	A	B	C	D	E	F	G	H	I	J	K	L
N	N	O	P	Q	R	S	T	U	V	W	X	Y	Z	A	B	C	D	E	F	G	H	I	J	K	L	M
O	O	P	Q	R	S	T	U	V	W	X	Y	Z	A	B	C	D	E	F	G	H	I	J	K	L	M	N
P	P	Q	R	S	T	U	V	W	X	Y	Z	A	B	C	D	E	F	G	H	I	J	K	L	M	N	O
Q	Q	R	S	T	U	V	W	X	Y	Z	A	B	C	D	E	F	G	H	I	J	K	L	M	N	O	P
R	R	S	T	U	V	W	X	Y	Z	A	B	C	D	E	F	G	H	I	J	K	L	M	N	O	P	Q
S	S	T	U	V	W	X	Y	Z	A	B	C	D	E	F	G	H	I	J	K	L	M	N	O	P	Q	R
T	T	U	V	W	X	Y	Z	A	B	C	D	E	F	G	H	I	J	K	L	M	N	O	P	Q	R	S
U	U	V	W	X	Y	Z	A	B	C	D	E	F	G	H	I	J	K	L	M	N	O	P	Q	R	S	T
V	V	W	X	Y	Z	A	B	C	D	E	F	G	H	I	J	K	L	M	N	O	P	Q	R	S	T	U
W	W	X	Y	Z	A	B	C	D	E	F	G	H	I	J	K	L	M	N	O	P	Q	R	S	T	U	V
X	X	Y	Z	A	B	C	D	E	F	G	H	I	J	K	L	M	N	O	P	Q	R	S	T	U	V	W
Y	Y	Z	A	B	C	D	E	F	G	H	I	J	K	L	M	N	O	P	Q	R	S	T	U	V	W	X
Z	Z	A	B	C	D	E	F	G	H	I	J	K	L	M	N	O	P	Q	R	S	T	U	V	W	X	Y

FIGURE 25.6: Table for Vigener cipher.

For an example, suppose the message is MOVEINATDAWN and the text key is FIDDLESTICKS. The first letter F in the key directs the sender to row F of the table. The first letter M of the message is then encrypted by the letter appearing in row F and column M, namely the letter R. These are underlined in the table for the purposes of illustration. Similarly, the second letter O of the message is encrypted by the letter appearing in row I, the second letter of the key, and column O, the second letter of the message, namely W. If one views the table as a String array VIG(0 to 25,0 to 25), then one can write the relationships in code as

```
R = VIG(Alph2Val(F),Alph2Val(M)), W = VIG(Alph2Val(I),Alph2Val(O)).
```

Continuing in this manner we obtain the encrypted message RWYHTRSMLCGF.

To decrypt the message the process is reversed: The recipient looks at the rows corresponding to the letters of the key FIDDLESTICKS, finds the ciphertext letter, and records the letter of its column. For example, to decode the third letter Y in the ciphertext RWYHTRSMLCGF, the recipient goes to row D, the third letter of the key, finds Y and decodes it with the letter of its column, namely V.

It is not necessary for the length of the keyword be the same as that of the message. Indeed, if the keyword is longer than the message, encrypting and decrypting stops when all the letters have been processed, whether or not there are leftover letters in the keyword. If the keyword is shorter, then it can be recycled. For example, if the keyword in the above example is simply FIDDLE, then after MOVEIN has been encrypted by FIDDLE as RWYHTR, as before, the remaining letters ATDAWN are encrypted using FIDDLE again.

Encryption with VBA

The encryption function VEncrypt uses a computational approach to encrypt a letter. First, the letters are converted to their alpha numbers $0, 1, 2, \cdots, 25$. If Mletter is the alpha number of the current letter in the message and Kletter the alpha number of the corresponding letter in the keyword, then the alpha number of the desired ciphertext letter Cletter in the table is given by

$$\text{Alph2Val(Cletter)} \equiv_{26} \text{Alph2Val(Mletter)} + \text{Alph2Val(Kletter)}$$

and the letter itself is Chr(Alph2Val(Cletter)) (see Figure 25.7). The formula is the basis of both the encryption and decryption process.

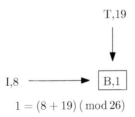

$$1 = (8 + 19) \,(\text{mod } 26)$$

FIGURE 25.7: Encrypting the letter T as B using the keyword letter I.

The function VEncrypt is passed the plaintext and keyword and returns the ciphertext. A For Next loop uses the string function Mid to retrieve the message and keyword characters. These are converted to their alpha values m and k, respectively. If the word is shorter than the message, then the keyword characters are recycled. For example, suppose that the message is MOVEINATDAWN and the keyword is FIDDLE. When the index i equals 7, the seventh letter of the message is retrieved. Since FIDDLE has only six letters, its first letter F is reused. This is done by the formula ((i-1) Mod LenK) + 1, where

LenK= 6, the length of the keyword. As i runs from 1 to LenM $= 12$, the message length, the formula takes on the values $1, 2, 3, 4, 5, 6, 1, 2, 3, 4, 5, 6$.

```
Function VEncrypt(plaintext As String, keyword As String) As String
    Dim i As Integer, LenM As Integer, LenK As Integer, ctext As String
    Dim m As Integer, k As Integer, Mletter As String, Kletter As String
    LenM = Len(plaintext)                        'length of message
    LenK = Len(keyword)                          'length of keyword
    For i = 1 To LenM
        Mletter = Mid(plaintext, i, 1)         'extract plaintext letter
        Kletter = Mid(keyword, (i - 1) Mod LenK + 1, 1)   'key letter
        m = Alph2Val(Mletter): k = Alph2Val(Kletter)      'alpha values
        ctext = ctext & Chr(65 + (m + k) Mod 26)   'append to ciphertext
    Next i
    VEncrypt = ctext                             'return ciphertext
End Function1
```

Decryption with VBA

The decryption function VDecrypt is passed the ciphertext and keyword and returns the plaintext. A For Next loop uses the string function Mid to retrieve the ciphertext and keyword characters. These are converted to alpha numbers c and k, respectively. The function Lookup finds the alpha number of the decrypted letter in the top row of the letter table.

```
Function VDecrypt(ciphertext As String, keyword As String) As String
    Dim plaintext As String, Cletter As String, Kletter As String
    Dim i As Integer, LenC As Integer, LenK As Integer
    Dim c As Integer, m As Integer, k As Integer
    LenC = Len(ciphertext): LenK = Len(keyword)
    For i = 1 To LenC                    'run through letters of ciphertext
        Cletter = Mid(ciphertext, i, 1)        'get current letter
        Kletter = Mid(keyword, (i - 1) Mod LenK + 1, 1)       'ditto
        c = Alph2Val(Cletter): k = Alph2Val(Kletter)     'alpha values
        m = Lookup(c, k)          'get alpha value of decrypted letter
        message = message & Chr(65 + m)           'attach the letter
    Next i
    VDecrypt = message
End Function
```

The function Lookup is passed the alpha numbers of the current ciphertext letter and keyword letter and returns the alpha number of the corresponding message letter found in the top row of the table.

```
Function Lookup(c As Integer, k As Integer) As Integer
    Dim m As Integer
    For m = 0 To 25          'run through the header row of table to find
        If (m + k) Mod 26 = c Then Exit For  'the cipher letter in table
    Next m
    Lookup = m                            'got the message letter
End Function
```

Combining Encryption and Decryption

The above procedures may be incorporated into a single module `Vcipher` using command buttons, as indicated in Figure 25.8.

```
Sub CommandButton1_Click()
    Dim keyword As String, message As String
    keyword = Range("B2").Value: message = Range("B4").Value
    Range("B6").Value = VEncrypt(message, keyword)  'print ciphertext
End Sub

Sub CommandButton2_Click()
    Dim keyword As String, ciphertext As String
    keyword = Range("B2").Value ciphertext = Range("B6").Value
    Range("B7").Value = VDecrypt(ciphertext, keyword)  'print message
End Sub
```

The sender enters the keyword and message into cells B2 and B4, respectively, and then presses the command button ⎸encrypt⎸. The encrypted message appears in cell B6 and is sent to the receiver. The recipient, who has the identical program, enters the keyword and encrypted message into cells B2 and B6, respectively, and presses the command button ⎸decrypt⎸. The decrypted message then appears in B7. The figure depicts the spreadsheet after both buttons were pressed.

	A	B
2	keyword	FIDDLE
3		
4	encrypt	MOVEINATDAWN
5		
6	decrypt	RWYHTRFBGDHR
7		MOVEINATDAWN

FIGURE 25.8: Input-output for `VCipher`.

25.8   Four Square Cipher

This cipher is based on four 5 by 5 tables of letters, as shown in Figure 25.9. The upper left and lower right tables (Tables 1 and 4) contain the alphabet with the infrequent letter Q removed. These are the tables that contain the plaintext letters. The upper right and lower left tables (Tables 2 and 3) contain keywords. These are the tables from which the ciphertext is formed. The message and the key words should not contain a Q. (Alternately, one could merge the infrequent letter J with I.)

A	B	C	D	E		H	O	W	C	A
F	G	H	I	K		L	M	Y	D	E
L	M	N	O	P		S	T	I	V	B
Q	R	S	T	U		R	N	K	G	F
V	W	X	Y	Z		J	P	U	X	Z
W	I	T	H	O		A	B	C	D	E
U	A	C	R	Y		F	G	H	I	K
P	E	N	B	L		L	M	N	O	P
F	D	S	G	J		Q	R	S	T	U
K	M	V	X	Z		V	W	X	Y	Z

$$GO \longrightarrow DE$$

FIGURE 25.9: Four square cipher.

The figure, which actually depicts part of the spreadsheet for the VBA implementation of the cipher, illustrates the case for the keywords

$$HOWCALMYDESTIVBRNKGFJPUXZ$$

in Table 2 and

$$WITHOUACRYPENBLFDSGJKMVXZ$$

in Table 3. The tables were derived, respectively, from the following lines from the first verse of a poem by Tennessee Williams in his play *The Night of the Iguana.*

"How calmly does the olive branch observe the sky begin to blanch"

and

"Without a cry, without a prayer with no betrayal of despair"

A message is processed by first breaking it up into letter pairs. This requires that the message have an even number of letters, which can be achieved by padding with an X, if necessary. For example, the message GONOW is rendered as GONOWX. Each pair of letters is then converted to ciphertext by locating the first letter of the pair in table 1 and the second letter of the pair in table 4. The letters of the pair are at opposite ends of the diagonal of a unique rectangle. The corresponding ciphertext letters are the letters at the opposite ends of the other diagonal. The figure illustrates how the pair GO (in the green cells of the figure) is encrypted to DE (in the red cells). The order of the letters in each pair is from the diagonal top to the diagonal bottom.

In the remainder of the section we develop a module `FourSquareCipher` that implements the cipher. Figure 25.10 shows an example of the spreadsheet accompanying the module. The sender enters the message in cell B2 and the keywords in cells B4 and B5, all in capital letters. Pressing | encrypt | generates

	A	B
2	encrypt	GONOWANDTAKETHEFILESWITHYOU
4	table 2 key	HOWCALMLYDOESTHEOLIVEBRANCHOBSERVE. . .
6	table 3 key	WITHOUTACRYWITHOUTAPRAYERWITHNO. . .
8	decrypt	EEBBXDETJHFIANVCTR
10		GONOWANDTAKETHEFILESWITHYOU

FIGURE 25.10: Four square cipher spreadsheet.

the tables in Figure 25.9 with spreadsheet the letter A of table 1 appearing in cell C10. The encryption is generated from the tables. This is sent to the receiver, who enters the keywords in cells B4 and B5 and encrypted message in cell B8. Pressing | decrypt | generates the tables and the decryption, which is placed in cell B10. For the sake of illustration the program assumes the part of both the receiver and the sender.

The module contains a number of Public variables. The first group gives the positions of the row and column of upper left entries of tables 1–4. For example, the letter A in table 1 has row-column position `Top1,Left1`.

```
Top1 As Integer, Left1 As Integer, Top2 As Integer, Left2 As Integer
Top3 As Integer, Left3 As Integer, Top4 As Integer, Left4 As Integer
```

The second group are string variables that hold the keys:

```
Table2Key As String, Table3Key As String
```

The third group are string variables that hold the plaintext and ciphertext:

```
ptext As String, ctext As String
```

There is also a Public integer variable `rate` which determines how fast letter pairs are deciphered. A high value slows down the procedure for illustrative purposes.

Encryption

Here is the procedure that is activated by the encrypt command button. After the grid for the tables is created, the procedure sets the positions and reads the message and keywords. The procedure then calls `ModifyKey` to reduce the keys as explained earlier. The alphabet tables 1 and 4 and key tables 2 and 3 are then created and the ciphertext derived from them.

```
Sub Encrypt()
    Dim Grid As TGrid
    Grid = MakeGrid(4, 4, 30, 30, 3, 13, -1)
    Top1 = 10: Left1 = 3                    'set the positions of the upper
    Top2 = Top1: Left2 = Left1 + 6            'left corners of the tables
    Top3 = Top1 + 6: Left3 = Left1
    Top4 = Top3: Left4 = Left3 + 6
    ptext = Range("B2").Value                     'read the message
    rate = Range("B11").Value          'rate letter pairs are deciphered
    Table2Key = Range("B4").Value         'get user-entered table keys
    Table3Key = Range("B6").Value
    Call ModifyKey(Table2Key)            'removing duplicates and attach
    Call ModifyKey(Table3Key)           'letters from rest of the alphabet
    Call MakeAlphabetTables             'make tables 1 and 4 from alphabet
    Call MakeKeyTables                   'make tables 2 and 3 from keys
    Call MakeCtext                      'use tables to make ciphertext
End Sub
```

The procedure `ModifyKey` takes a keyword, removes Q's if any, and then runs through the alphabet to see if a letter is in the key. If not it attaches the letter to the key. The process continues until the reduced key reaches 25 letters.

```
Sub ModifyKey(key As String)
    Dim i As Integer, j As Integer, LetterInKey As Boolean
    key = RemoveDupChar(key): key = Replace(key, "Q", "")
    Do While Len(key) < 25              'stop when key has 25 characters
        For i = 65 To 90                    'go through alphabet(caps)
            If Chr(i) = "Q" Then GoTo continue         'omit Q
            LetterInKey = False                    'initialize for each i
            For j = 1 To Len(key)                   'go through key letters
                If Mid(key, j, 1) = Chr(i) Then  'if key letter = current
                    LetterInKey = True           'alphabet letter then reset
                    Exit For    'key letter = Chr(i) so can't append Chr(i)
                End If
            Next j
            If Not LetterInKey And Len(key) < 25 Then 'if alphabet letter
                key = key & Chr(i)              'not in key then attach to key
            End If
        continue:
            Next i
    Loop
End Sub
```

Here are the procedures that make the tables.

```
Sub MakeAlphabetTables()
    Dim i As Integer, j As Integer, k As Integer
    k = 65 'start with ASCII A
    For i = 0 To 4                    'populate spreadsheet with alphabet
        For j = 0 To 4                   'letters at table 1,4 positions
```

```
            If k = Asc("Q") Then k = k + 1
            Cells(i + Top1, j + Left1) = Chr(k)
            Cells(i + Top4, j + Left4) = Chr(k)
            k = k + 1
        Next j
    Next i
    End Sub


Sub MakeKeyTables()
    Dim i As Integer, j As Integer, k As Integer
    k = 1
    For i = 0 To 4                  'populate spreadsheet with letters from
        For j = 0 To 4                   'Table2Key in table 2 positions
            Cells(i + Top2, j + Left2) = Mid(Table2Key, k, 1)
            k = k + 1
        Next j
    Next i
    k = 1
    For i = 0 To 4                  'populate spreadsheet with letters from
        For j = 0 To 4                   'Table3Key in table 2 positions
            Cells(i + Top3, j + Left3) = Mid(Table3Key, k, 1)
            k = k + 1
        Next j
    Next i
End Sub
```

The procedure `MakeCtext` locates the current pair of ptext letters located in
tables 1,4, finds their row column positions with the function `LetterToPosition`,
the positions being the endpoints of the diagonal of a rectangle as described
earlier, and then computes the positions of the ctext pair located at the ends of
the opposite diagonal. These are concatenated to form the ctext. The procedure
uses the green-red scheme illustrated in Figure 25.9 so the user can watch how
the ctext is created

```
Sub MakeCtext()
    Dim pPos1 As TPos, pPos4 As TPos         'letter positions in tables 1,4
    Dim cPos2 As TPos, cPos3 As TPos         'letter positions in tables 2,3
    Dim pLetter1 As String, pLetter4 As String 'letter pair in tables 1,4
    Dim cLetter2 As String, cLetter3 As String 'letter pair in tables 2,3
    Dim i As Integer
    Range("B8").Value = ""    'remove old ctext
    If Len(ptext) Mod 2 <> 0 Then
        ptext = ptext & "X" 'append if odd length
    End If
    For i = 1 To Len(ptext) - 1 Step 2            'run through ptext pairs
        pLetter1 = Mid(ptext, i, 1)               'extract a pair of letters
        pLetter4 = Mid(ptext, i + 1, 1)
        pPos1 = LetterToPosition(pLetter1, Top1, Left1)  'find positions
        pPos4 = LetterToPosition(pLetter4, Top4, Left4)  'in tables 1, 4
        Cells(pPos1.row, pPos1.col).Interior.ColorIndex=4'green cells for
        Cells(pPos4.row, pPos4.col).Interior.ColorIndex=4   'ptext pair
```

```
      cPos2.row = pPos1.row: cPos2.col = pPos4.col        'ctext pair on
      cPos3.row = pPos4.row: cPos3.col = pPos1.col      'opposite diagonal
      cLetter2 = Cells(cPos2.row, cPos2.col).Value      'retrieve letters
      cLetter3 = Cells(cPos3.row, cPos3.col)
      Cells(cPos2.row, cPos2.col).Interior.ColorIndex=3  'red cells for
      Cells(cPos3.row, cPos3.col).Interior.ColorIndex=3   'ctext pair
      Call Delay(rate)
      Cells(pPos1.row, pPos1.col).Interior.ColorIndex=0  'remove color
      Cells(pPos4.row, pPos4.col).Interior.ColorIndex=0   'after delay
      Cells(cPos2.row, cPos2.col).Interior.ColorIndex=0
      Cells(cPos3.row, cPos3.col).Interior.ColorIndex=0
      Range("B8").Value = _
      Range("B8").Value & cLetter2 & cLetter3    'display evolving ctext
   Next i
End Sub
```

The function `LetterToPosition` returns the position of a given letter in a table.

```
Private Function LetterToPosition(Letter As String, Top As Integer,
      Left As Integer) As TPos
   Dim i As Integer, j As Integer, pos As TPos
   For i = 0 To 4                 'run through the table to find an entry
      For j = 0 To 4                         'that matches Letter
         If Letter = Cells(i + Top, j + Left) Then
            pos.row = i + Top: pos.col = j + Left
         End If
      Next j
   Next i
   LetterToPosition = pos
End Function
```

Decryption

The process of decryption is entirely similar to that of encryption. The second command button on the spreadsheet invokes the procedure `Decrypt`. Which is the same as `Encrypt` except that the last line of code `MakeCtext` is replaced by the code `MakePtext`. The latter procedure differs from the former in that the roles of Tables 1,4 and Tables 2,3 are switched in calculating letter pair locations. The code for the decryption process is omitted.

25.9 Matrix Cipher

The *matrix* (or *Hill*) *cipher* is a generalization of the permutation cipher. As in the latter, the message is broken into blocks, the last block with possible padding. The blocks are written as column vectors of alpha numbers. These are multiplied by a matrix (the key) whose inverse has integer entries.[1] The resulting column vectors are sent to the receiver in letter form. The receiver knows the matrix key and so can retrieve the message by multiplying by the inverse of the matrix. We give the details in the following subsections and implement the process with VBA.

Encryption with VBA

Consider the message "Rendezvous same time same place". Assume the key is the 5×5 matrix

$$\text{KetMat} = \begin{bmatrix} 1 & 4 & 2 & 3 & 1 \\ 7 & 14 & 9 & 10 & 3 \\ 10 & 23 & 11 & 15 & 5 \\ 5 & 18 & 8 & 12 & 4 \\ 2 & 9 & 4 & 6 & 2 \end{bmatrix}$$

The message is capitalized and partitioned into size 5 blocks:

RENDE ZVOUS SAMET IMESA MEPLA CEXXX,

the last one padded with X's to fill out the block. The blocks are converted into columns of alpha numbers and assembled into a 5×6 matrix:

$$\text{BlockMat} = \begin{bmatrix} 17 & 25 & 18 & 8 & 12 & 2 \\ 4 & 21 & 0 & 12 & 4 & 4 \\ 13 & 14 & 12 & 4 & 15 & 23 \\ 3 & 20 & 4 & 18 & 11 & 23 \\ 4 & 18 & 19 & 0 & 0 & 23 \end{bmatrix}. \tag{25.3}$$

The matrix BlockMat is now multiplied on the left by KeyMat producing the matrix

$$\text{KeyMat} \cdot \text{BlockMat} = \begin{bmatrix} 72 & 215 & 73 & 118 & 91 & 156 \\ 334 & 849 & 331 & 440 & 385 & 576 \\ 470 & 1277 & 476 & 670 & 542 & 825 \\ 313 & 927 & 310 & 504 & 384 & 634 \\ 148 & 451 & 146 & 248 & 186 & 316 \end{bmatrix}.$$

The entries are then reduced modulo 26 and converted to letters:

$$\begin{bmatrix} 20 & 7 & 21 & 14 & 13 & 0 \\ 22 & 17 & 19 & 24 & 21 & 4 \\ 2 & 3 & 25 & 20 & 22 & 19 \\ 1 & 17 & 24 & 10 & 20 & 10 \\ 18 & 9 & 16 & 14 & 4 & 4 \end{bmatrix} \rightarrow \begin{bmatrix} U & H & V & O & N & A \\ W & R & T & Y & V & E \\ C & D & Z & U & W & T \\ B & R & Y & K & U & K \\ S & J & Q & O & E & E \end{bmatrix} \tag{25.4}$$

---

[1] One way to obtain such a matrix is to start with the identity matrix and apply random row operations.

The columns are the ciphertext blocks

The process of encryption is automated by function MEncrypt, which takes as input plaintext and the matrix key and returns the ciphertext.

```
Function MEncrypt(ptext As String, KeyMat() As Double) As String
    Dim  paddedmsg As String, BlockMat() As Double
    Dim numblocks As Integer, blocksize As Integer
    blocksize = Ubound(KeyMat,2)                'no. rows of BlockMat
    paddedmsg  = PadX(ptext, blocksize)          'pad ptext if needed
    numblocks = Len(paddedmsg ) / blocksize      'no. cols of BlockMat
    BlockMat = MakeBlockMatrix(paddedmsg , blocksize, numblocks)
    MEncrypt = MakeText(KeyMat, BlockMat)             'return ciphertext
End Function
```

The procedure MakeBlockMatrix is passed the padded message produced by the procedure PadX in Section 25.6, the block size and the number of blocks, these being, respectively, the number of rows and columns of the return matrix BlockMat.

```
Function MakeBlockMatrix(paddedmsg As String, blocksize As Integer, _
                  numblocks As Integer) As Double()
    Dim i As Integer, j As Integer, ch As String, BlockMat() As Double
    ReDim BlockMat(1 To blocksize, 1 To numblocks)
    For j = 1 To numblocks
        For i = 1 To blocksize
            ch = Mid(paddedmsg, i + blocksize * (j - 1), 1)
            BlockMat(i, j) = Asc(ch) - 65
        Next i
    Next j
    MakeBlockMatrix  = BlockMat
End Function
```

The procedure MakeText is passed the matrices KeyMat and BlockMat in that order and returns the ciphertext.

```
Function MakeText(KMat() As Double, BMat() As Double) As String
    Dim i As Integer, j As Integer, blocksize As Integer, Mat() As Double
    Dim numblocks As Integer, text As String, r As Double
    blocksize = UBound(KMat, 1): numblocks = UBound(BMat, 2)
    ReDim Mat(1 To blocksize, 1 To numblocks)
    Mat =  MultMat(KMat, BMat)                  'multiply the matrices
    For j = 1 To numblocks
        For i = 1 To blocksize
            r = Rmdr(CInt(Mat(i, j)), 26)          'reduce to alpha value
            text = text & Chr(r + 65)              'attach the letter
        Next i
    Next j
    MakeText = text
End Function
```

Decryption with VBA

Having been sent the ciphertext, the receiver goes through a process analogous to encryption but using the inverse of the key matrix, found to be

$$\text{KeyMatInv} = \begin{bmatrix} 0 & 0 & 0 & 1 & -2 \\ -2 & 0 & 0 & 0 & 1 \\ 1 & 0 & 1 & -5 & 7 \\ -2 & 1 & -3 & 11 & -15 \\ 13 & -3 & 7 & -24 & 29 \end{bmatrix}.$$

The ciphertext blocks UWCBS HRDRJ VTZYQ OYUKO NVWUE AETKE are placed into columns of a matrix and converted to alpha values:

$$\begin{bmatrix} U & H & V & O & N & A \\ W & R & T & Y & V & E \\ C & D & Z & U & W & T \\ B & R & Y & K & U & K \\ S & J & Q & O & E & E \end{bmatrix} \rightarrow \begin{bmatrix} 20 & 7 & 21 & 14 & 13 & 0 \\ 22 & 17 & 19 & 24 & 21 & 4 \\ 2 & 3 & 25 & 20 & 22 & 10 \\ 1 & 17 & 24 & 10 & 20 & 10 \\ 18 & 9 & 16 & 14 & 4 & 4 \end{bmatrix}.$$

The last matrix is multiplied on the left by KeyMatInverse to recover the plaintext. The function MDecrypt carries out the process in a manner analogous to MEncrypt.

```
Function MDecrypt(ctext As String, KeyMat() As Double) As String
    Dim pad As String, BlockMat() As Double, KeyMatInv() As Double
    Dim numblocks As Integer, blocksize As Integer, err As Boolean
    blocksize = Ubound(KeyMat,2)                    'no. rows of BlockMat
    numblocks = Len(ctext) / blocksize              'no. cols of BlockMat
    BlockMat = MakeBlockMatrix(ctext, blocksize, numblocks)
    KeyMatInv = InvertMat(KeyMat,err)                    'invert key
    If err Then Exit Function
    MDecrypt = MakeText(KeyMatInv, BlockMat)          'return plaintext
End Function
```

Combining Encryption and Decryption

The above procedures may be incorporated into a single module MatrixCipher by adding the command buttons as illustrated in Figure 25.11. The command button procedures retrieve the matrix key and the text to be encrypted or decrypted.

The sender and the receiver have copies of the module MatrixCipher. Each person enters the matrix key with first entry in cell C7 (not shown). The sender enters the message (plaintext) into B3; pressing the command button encrypt prints the encrypted message (ciphertext) in cell B5, which is sent to the receiver. The receiver enters the ciphertext in B5 and presses the command button decrypt , which prints the decrypted message in cell B6. Figure 25.11 depicts the state of the spreadsheet after both buttons were depressed. Here is the code for the command buttons.

	A	B
3	encrypt	RENDEZVOUSSAMETIMESAMEPLACE
4		
5	decrypt	UWCBSHRDRJVTZYQOYUKONVWUEAETKE
6		RENDEZVOUSSAMETIMESAMEPLACE

FIGURE 25.11: Input–output for `MatrixCipher`.

```
Private Sub CommandButton1_Click()
    Dim nrows As Integer, ncols As Integer
    Dim ptext As String, KeyMat() As Double
    ptext = Range("B3").Value              'read plaintext
    KeyMat = MatrixIn(7, 3)
    Range("B5").Value = MEncrypt(ptext, KeyMat)   'print ciphertext
End Sub

Private Sub CommandButton2_Click()
    Dim nrows As Integer, ncols As Integer
    Dim ctext As String, KeyMat() As Double
    ctext = Range("B5").Value              'read ciphertext
    KeyMat = MatrixIn(7, 3)
    Range("B6").Value = MDecrypt(ctext, KeyMat)   'print plaintext
End Sub
```

25.10   The RSA Algorithm

The preceding methods of encryption all use a single (private) key. By contrast, the encryption method in the current section, invented by R. Rivest, A. Shamir, and L. Adleman, uses two keys, one private, the other public. The public key is used to encrypt messages; it is available to anyone. The private key is used to decrypt messages; it is known only to the recipient of the message.

A public key/private key system is based on what is called a *trapdoor function*, that is, a function that is easy to calculate in one direction but very difficult to calculate in the reverse direction without additional information. For example, a combination lock may be viewed as a trapdoor: it is easy to close, but without knowing the combination the lock cannot be easily opened (unless one has a great deal of time to try all the permutations). A simple mathematical example of a trapdoor is the product of two very large prime numbers p and q: Given the number $n = pq$ (the public key) and nothing else, it would be difficult to calculate the primes p and q. However, if you are given one of the primes, then you can get the other by division.

The RSA algorithm described in this section is a more sophisticated version of the last example. Each person in a group wishing to send or receive private messages secretly chooses two large primes p and q and forms the product $n = pq$. The person also chooses an integer e which satisfies the conditions

$$0 < e < p - 1, \quad 0 < e < q - 1, \quad \gcd(e, p - 1) = \gcd(e, q - 1) = 1.$$

The program GCD in Section 23.4 may be used for this. Note that the conditions imply that

$$\gcd\left(e, (p - 1)(q - 1)\right) = 1$$

Indeed, by Bézout's equation, $1 = ue + v(p - 1)$, and $1 = we + x(q - 1)$ for some integers u, v, w, x; multiplying the two equations yields an equation of the form $1 = ye + z(p - 1)(q - 1)$, from which the assertion follows.

The pair (n, e) is the RSA public key for that person; it is available to everyone in the group. The person also has a private key d that satisfies the equation

$$ed \equiv 1 \ (\text{mod} \ (p - 1)(q - 1)) . \tag{25.5}$$

This key is known to no one else. Its value may be found with the following "brute force" function, which takes the primes p and q and the value e, and returns d. (We use the type Double to accommodate larger numbers.)

```
Function GetPrivateKey(p As Double, q As Double, e As Double) As Double
    Dim d As Double, modulus As Double, number As Double
    modulus = (p - 1) * (q - 1)
    For d = 0 To p * q
        number = e * d - 1
        If number = (Int(number / modulus) * modulus) Then Exit For
    Next d
    GetPrivateKey = d
End Function
```

Encryption with VBA

Suppose Aaron wishes to send to Betty the message MEETMETONIGHT. Aaron looks up Betty's public key, say $(n, e) = (589597, 7)$. Only Betty knows that n is the product of the primes $p = 727$ and $q = 811$.[2] Betty has chosen e to satisfy the requirements

$$\gcd(e, p - 1) = \gcd(7, 726) = 1, \quad \text{and} \quad \gcd(e, q - 1) = \gcd(7, 810) = 1.$$

Aaron now converts his message into a string of double digit alpha values

$$\text{MEETMETONIGHT} \rightarrow 12\ 04\ 04\ 19\ 12\ 04\ 19\ 14\ 13\ 08\ 06\ 07\ 19 \tag{25.6}$$

(without the spaces). The function TextToDigits accomplishes this:

---

[2] The primes were chosen for illustration. In practice much larger primes are used.

```
Function TextToDigits(text As String) As String
    Dim letter As String, letterval As Integer
    Dim i As Integer, digitstr As String
    For i = 1 To Len(text)
        letter = Mid(text, i, 1)                    'extract letter
        letterval = Asc(letter) - 65           'convert to alpha number
        If letterval < 10 Then digitstr = digitstr & "0"    '2nd digit
        digitstr = digitstr & CStr(letterval)         'build string
    Next i
    TextToDigits = digitstr            'return string of double digits
End Function
```

The digit string in (25.6) is too long to work with mathematically, so Aaron divides it into blocks. For the algorithm to work properly, each block must have the property that the integer it represents is less than n, the product of primes Betty has selected. This will be the case if the block size is one less than the number of digits in n. Applying this rule (assumed to be universally followed) Aaron chooses the block size to be $6 - 1 = 5$ and groups the digits in (25.6) accordingly:

$$12040\ 41912\ 04191\ 41308\ 06071\ 9. \tag{25.7}$$

As the last block is incomplete, Aaron appends the digits 2323 (two X's) to the digital message to obtain the full blocks

$$12040\ 41912\ 04191\ 41308\ 06071\ 92323. \tag{25.8}$$

Notice that the number of digits in the last block of (25.7), namely 1, is the remainder when the total number of digits in the string is divided by the number of complete blocks. The function MakeBlocks carries out this operation and returns the padded string as in (25.8) (with a single separating space between blocks).

```
Function MakeBlocks(digitstr As String, n As Double) As String
    Dim blksize As Integer, padsize As Integer, outblks As String
    Dim numblks As Integer, numdigits As Integer, i As Integer
    blksize = Len(CStr(n)) - 1                    'apply universal rule
    numdigits = Len(digitstr)
    numblks = numdigits \ blksize              'number of full blocks
    padsize = blksize - numdigits Mod numblks       'how much to pad
    If padsize <> 0 Then                    'pad with X's if needed
       For i = 1 To padsize
          digitstr = digitstr & "23"
       Next i
       numblks = numblks + 1             'update number of full blocks
    End If
    digitstr = Mid(digitstr, 1, numblks * blksize)     'chop any excess
    For i = 0 To numblks - 1             'break padded digit into blocks
       outblks = outblks & " " & Mid(digitstr, i*blksize+1, blksize)
    Next i
    MakeBlocks = Mid(outblks, 2, Len(outblks) - 1)     'remove last space
End Function
```

Each block of the padded digit string is now converted into an integer P and raised to the power e using the formula

$$C \equiv P^e \pmod{n} , \quad \text{where } n = 587597 \text{ and } e = 7.$$

Notice that converting a block with leading zeros, say 000001 (AAB), into P removes the leading zeros. These must be restored during the decryption process. The numbers C are then converted into blocks. The following function does this. It takes as arguments the modulus n, an exponent x ($= e$ in this subsection), and the string of blocks, as in (25.8). The function first splits the input string into an array Blocks and then runs a For Next loop, which employs PowerRmdr of Section 24.5 to raise each block to the power x modulo n.

```
Function BlockPowers(blks As String, x As Double, n As Double) As String
    Dim i As Integer, blksize As Integer, outblks As String
    Dim Blocks() As String, power As Double
    blksize = Len(CStr(n)) - 1              'general rule for blockize
    Blocks = Split(blks, " ")              'put input blocks into an array
    For i = 0 To UBound(Blocks)            'generate string of blocks^x
        power = PowerRmdr(CDbl(Blocks(i)), x, n)    'raise block to x
        outblks = outblks & " " & CStr(power)    'build string of powers
    Next i
    BlockPowers = Mid(outblks,2,Len(outblks) - 1)    'remove last space
End Function
```

For example, applying the function BlockPowers to (25.8) produces the cipher block string

$$540639 \ 280186 \ 286584 \ 433967 \ 326148 \ 42852, \tag{25.9}$$

(with single space separation), which Aaron sends to Betty.

Decryption

The decryption process relies on the formulas

$$ed \equiv 1 \pmod{(p-1)(q-1)} , \quad \gcd(e, p-1) = \gcd(e, q-1) = 1 \tag{25.10}$$

and the formula
$$C \equiv P^e \pmod{n} ,$$

which was used to convert the plaintext numbers P to the cipher numbers C. From these equations one may show that

$$P \equiv C^d \pmod{n} . \tag{25.11}$$

Having received the string of ciphertext numbers C in (25.9) sent by Aaron, Betty uses (25.11) to convert the numbers into plaintext numbers P. Only she can do this, since only she knows the components $p-1$ and $q-1$ in (25.10).

Betty uses the function `BlockPowers` to make the conversion producing (25.8), but without the leading zeros lost during encryption. Betty then uses the following function to add back the missing leading zeros and condense the string.

```
Function InsertZeros(blockstr As String, n As Double) As String
    Dim textdigits, i As Integer, blocksize As Integer
    Dim numzeros As Integer, Blocks() As String
    blocksize = Len(CStr(n)) - 1                    'apply universal rule
    Blocks = Split(blockstr, " ")            'split blocks into an array
    For i = 0 To UBound(Blocks)
        numzeros = blocksize - Len(Blocks(i))      'add this many zeros
        textdigits = textdigits & String(numzeros, "0") & Blocks(i)
    Next i
    InsertZeros = textdigits
End Function
```

Finally, Betty must take the string of digits (no spaces) returned by `InsertZeros` and convert the alpha number pairs to letters, yielding the original message. The following function does the work.

```
Function DigitsToText(digits As String) As String
    Dim alphanum As String, ptext As String, i As Integer
    For i = 1 To Len(digits) Step 2
        alphanum = Mid(digits, i, 2)           'two digit alpha no. string
        ptext = ptext & Chr(CInt(alphanum)+65)   'converted and attached
    Next i
    DigitsToText = ptext
End Function
```

The encryption and decryption procedures, together with the following command button procedures, may be assembled into a module `RSACipher`.

```
Sub CommandButton1_Click()
    Dim ptext As String, n As Double, e As Double
    Dim digits As String, digitblks As String
    n = Range("B4").Value: e = Range("B5").Value        'receiver's key
    ptext = Range("B6").Value               'plaintext entered by sender
    digits = TextToDigits(ptext)               'convert to alpha numbers
    digitblks = MakeBlocks(digits, n)
    Range("B12").Value = BlockPowers(digitblks, e, n)      'encryption
End Sub
```

```
Sub CommandButton2_Click()
    Dim cdigits As String, pdigits As String, n As Double
    Dim e As Double, d As Double, p As Double, q As Double
    n = Range("B4").Value:  e = Range("B5").Value
    p = Range("B10").Value: q = Range("B11").Value
    d = GetPrivateKey(p, q, e)
    cdigits = Range("B12").Value          'ciphertext entered by receiver
    pdigits = BlockPowers(cdigits, d, n)
```

```
        pdigits = InsertZeros(pdigits, n)
        Range("B13").Value = DigitsToText(pdigits)        'plaintext output
End Sub
```

Figure 25.12 shows the spreadsheet after Aaron enters Betty's key n, e and his message and depresses ⎡encrypt⎤; and Betty enters her primes p, q and Aaron's message and depresses ⎡decrypt⎤.

	A	B
3	receiver's key	
4	$n =$	589597
5	$e =$	7
6	encrypt	MEETMEATMIDNIGHT
8	encryption	540639 280186 286584 433967 326148 42852
9	receiver's primes	
10	$p =$	727
11	$q =$	811
12	decrypt	540639 280186 286584 433967 326148 42852
13	decryption	MEETMEATMIDNIGHTXX

FIGURE 25.12: Input-output for RSA spreadsheet.

25.11   Exercises

1. Encode by hand the message "Remove the troops now" using

   (a) the permutation cipher with the permutation 4,2,5,1,3

   (b) the Vigener cipher with keyword BANANASTAND

   (c) the Hill cipher with matrix $\begin{bmatrix} 5 & -1 & 2 \\ 12 & -3 & 4 \\ 4 & -1 & 1 \end{bmatrix}$.

   Check your answers with the programs.

2. Implement the ZigZag cipher in VBA.

3. (Polybius square cipher) This cipher is based on a 5 by 5 alphabet table used to translate letters into pairs of numbers, where I and J are

combined to fit the scheme:

	1	2	3	4	5
1	A	B	C	D	E
2	F	G	H	I	K
3	L	M	N	O	P
4	Q	R	S	T	U
5	V	W	X	Y	Z

Each letter is assigned its row an column so that, for example, the plaintext RETREAT becomes 42 15 44 42 15 11 44. To decipher a message simply find the letter in the specified row and column. Write a VBA implementation `PBSquareCipher`.

4. (Columnar transposition cipher). In this cipher a key word or phrase is used to permute the columns of a message. For example, consider the message "Execute plan B of the evacuation" with the key word RANDOM. Break up the message into stacked blocks of length 6 (the number of letters in the key), to form the *ptext matrix*

E	X	E	C	U	T
E	P	L	A	N	B
O	F	T	H	E	E
V	A	C	U	A	T
I	O	N	X	X	X

(The matrix has been suitably padded with X's.) The idea is to rearrange the columns as follows: Number the letters of the key RANDOM by 123456. Then rearrange the letters alphabetically to get ADMNOR and the corresponding permutation 246351. The latter gives the column rearrangement: column 2 is chosen first, then column 4, then column 6, etc., yielding the *ctext matrix*

X	C	T	E	U	E
P	A	B	L	N	E
F	H	E	T	E	O
A	U	T	C	A	V
O	X	X	N	X	I

Concatenating the columns yields the desired ciphertext: XPFAOC-AHUXTBETXELTCNUNEAXEEOVI. Write a VBA implementation `ColumnarCipher`.

Suggestion: Form a column array `col` which gives the order of the rearranged columns. For the example

$$\text{col}(1) = 2, \text{col}(2) = 4, \text{col}(3) = 6, \text{col}(4) = 3, \text{col}(5) = 5, \text{col}(6) = 1$$

Use this to rearrange the ptext matrix into the ctext matrix.

5. (Playfair cipher). The Playfair cipher is a substitution cipher developed to ensure telegraph secrecy. It was used by the British in World War I and by the Australians in World War II. The cipher was invented in 1854 by Sir Charles Wheatstone, who named it after Lyon Playfair.

Encryption and decryption is based on a 5 × 5 table of alphabet letters. The letters can be in any order, but since there are 26 letters in the alphabet, one letter has to be sacrificed. A common choice is the letter J, since it appears infrequently in the English language. The rules for encryption are as follows: First, any double letters in the plaintext are separated with the letter x. Second, the message is split into blocks of 2 letters each, padding the last block with an x if needed. Third, each pair of letters is replaced by another pair obtained from the table according the following rules:

(a) If the pair is in the same row, replace each by the letter to the immediate right, going back to the beginning of the row if necessary.

(b) If the pair is in the same column, replace each letter by the letter immediately below, going back to the beginning of the column if necessary.

(c) Neither (a) nor (b) holds. Then the letters are in positions that form the diagonal of a rectangle. The pair is then replaced by the pair at the ends of the other diagonal, preserving the up-down direction. This simply means that the columns of the letters are switched.

In the decryption process the rules are applied in reverse. Figure 25.13 illustrates the rules for the first eight letters of the message STORMY-TOMORROW. Write a program `PFCipher` that encrypts and decrypts

A	B	C	D	E
F	G	H	I	K
L	M	N	O	P
Q	R	S	T	U
V	W	X	Y	Z

ST → TU

A	B	C	D	E
F	G	H	I	K
L	M	N	O	P
Q	R	S	T	U
V	W	X	Y	Z

OR → MT

A	B	C	D	E
F	G	H	I	K
L	M	N	O	P
Q	R	S	T	U
V	W	X	Y	Z

MY → OW

A	B	C	D	E
F	G	H	I	K
L	M	N	O	P
Q	R	S	T	U
V	W	X	Y	Z

TO → YT

FIGURE 25.13: Playfair encryption rules.

messages using the above algorithm.

6. Add a space and a period to some of the ciphers in the chapter using modulus 28 instead of modulus 26.

Chapter 26

The Enigma Machine

The Enigma machine, invented by German engineer Arthur Scherbius, is an electromechanical device used in the early to mid-20th century to encrypt commercial, diplomatic, and military communication. It was employed extensively by the German military during World War II. In this chapter we describe the machine and develop a program that emulates its operation.

26.1 Description

The machine has three rotating disks called *rotors*, each labelled with the 26 letters of the alphabet. Each side of a rotor has 26 electrical contacts. The contacts are wired internally so that current can pass from one side of the rotor to the other, effectively converting one letter to another and therefore acting as a permutation of the alphabet. There is also a *reflector disk* that

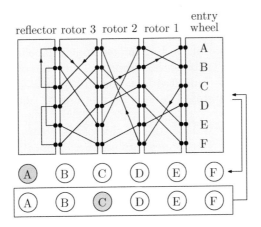

FIGURE 26.1: C encrypts to A.

sends the current back through the rotors along a different route. Letters to be encrypted or decrypted are typed into the machine with a keyboard. An entry wheel connects a keyboard to the rotors. When a key is pressed, current

DOI: 10.1201/9781003351689-26

is sent to the entry wheel and passes through the rotors from right to left. The reflector causes the current to flow back through the rotors left to right, eventually lighting a letter above the keyboard.[1] For example, by following the wiring in Figure 26.1 one sees that the depressed letter C on the keyboard undergoes several disk permutations and is ultimately transformed into the lighted letter A. The contacts are indicated by the black dots and the internal wiring by black lines.[2]

The complicating factor in all of this is that, after a key press, rotor 1 moves one position. After 26 key presses, rotor 1 returns to its initial position and rotor 2 moves one position, etc. Figure 26.2 shows the state of the machine after the initial key press C. [3]Pressing C again results in E being lighted. Thus the text CC encrypts to AE.

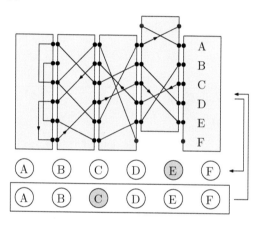

FIGURE 26.2: C encrypts to E.

The rim of each rotor of the Enigma machine is labelled with the capital letters of the alphabet, enabling the operator to view the position of the rotors through small windows (indicated by yellow in Figure 26.3). The setting may be changed by the operator. The same setting used to encrypt must be used to decrypt.

In Figure 26.4 we have labeled each side of the rotors with the letters to show how one letter is transformed into another. From this we see that each rotor acts as a pair of permutations, one going from right to left and the other, its inverse, going from left to right. We denote the permutations for the rotors going right to left by R_1, R_2, and R_3, the permutation for the reflector by R_4,

---

[1]The reflector allows the operator to encrypt and decrypt using the same procedure, thus eliminating the need for an additional mechanism to switch modes. For example, in Figure 26.1 not only does pressing C on the keyboard light A, but also pressing A lights C.

[2]For the purposes of illustration, we have used an abbreviated alphabet. In the module we use the standard 26 capital letters.

[3]Here, rotation of the disk is indicated by an upward shift of the rectangle that depicts rotor 1. Thus the red dots are connected to each other as are the blue dots.

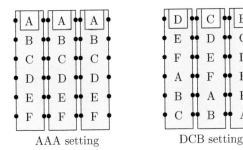

AAA setting   DCB setting

FIGURE 26.3: Rotor windows.

and the permutations for the rotors going left to right by R_5, R_6, and R_7, the inverses of the permutations R_3, R_2, R_1, respectively. Thus if the machine is in

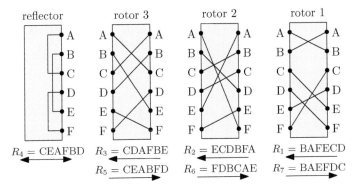

FIGURE 26.4: Disks as permutations.

state AAA, as is the case in Figure 26.4, then its internal operation is described mathematically by the permutation product $R_7R_6R_5R_4R_3R_2R_1$. We return to this idea later.

26.2   Implementation in VBA

In this section we use VBA simulate the electromechanical processes the machine undergoes in converting input text. The first step is to "wire" the rotors and the reflector. This is done in Sheet2 either by hand or with the procedure WireDisks. Figure 26.5 shows the permutation data entered by hand for the wiring of Figure 26.1.

The following procedure uses RndPerm of Section 10.1 to generate the rotor permutations in the right to left direction. The permutation inversion function

	A	B	C	D	E	F	G
2			Wire disks				
3	*R1*	*R2*	*R3*	*R4*	*R5*	*R6*	*R7*
4	B	E	C	C	C	F	B
5	A	C	D	E	E	D	A
6	F	D	A	A	A	B	E
7	E	B	F	F	B	C	F
8	C	F	B	B	F	A	D
9	D	A	E	D	D	E	C

FIGURE 26.5: Permutation data in Sheet2.

`InvPerm` of Section 19.9 is then used to generate the rotor permutations in the left to right direction. The function `TwoCycles` is used to wire the rotor. The functions work on the integers 1–26, which represent the alphabet—a deviation from the alpha number system used in the previous chapter.

```
Sub WireDisks()
    Dim R1() As Integer, R2() As Integer, R3() As Integer
    Dim R4() As Integer, R5() As Integer, R6() As Integer
    Dim R7() As Integer, i As Integer,
    R1 = RndPerm(26): R2 = RndPerm(26): R3 = RndPerm(26)   'right to left
    R4 = TwoCycles()                                        'reflector
    R5 = InvPerm(R3): R6 = InvPerm(R2): R7 = InvPerm(R1)   'left to right
    For i =1 To 26                      'print the permutations as letters
        Sheet2.Cells(i + 3, 1).Value = Chr(R1(i) + 64)
        Sheet2.Cells(i + 3, 2).Value = Chr(R2(i) + 64)
        Sheet2.Cells(i + 3, 3).Value = Chr(R3(i) + 64)
        Sheet2.Cells(i + 3, 4).Value = Chr(R4(i) + 64)
        Sheet2.Cells(i + 3, 5).Value = Chr(R5(i) + 64)
        Sheet2.Cells(i + 3, 6).Value = Chr(R6(i) + 64)
        Sheet2.Cells(i + 3, 7).Value = Chr(R7(i) + 64)
    Next i
End Sub
```

The function `TwoCycles` wires together pairs of contacts for the reflector. It does so by first generating a permutation `perm` of the numbers 1–26 and then associating adjacent numbers to produce a new permutation `outperm`. For example if in our abbreviated alphabet

$$\texttt{perm} = \begin{pmatrix} 1 & 2 & 3 & 4 & 5 & 6 \\ 5 & 3 & 6 & 1 & 4 & 2 \end{pmatrix} = 5, 3, 6, 1, 4, 2,$$

then we wire 5 to 3, 6 to 1, and 4 to 2 to produce the permutation

$$\texttt{outperm} = \begin{pmatrix} 1 & 2 & 3 & 4 & 5 & 6 \\ 6 & 4 & 5 & 2 & 3 & 1 \end{pmatrix}$$

```
Function TwoCycles() As Integer()
    Dim perm() As Integer, outperm() As Integer, i As Integer
    perm = RndPerm(26)                          'get a permutation
    ReDim outperm(1 To 26)
    For i = 1 To 13 'wire entry at odd position to that at adjacent even
        outperm(perm(2 * i - 1)) = perm(2 * i)
        outperm(perm(2 * i)) = perm(2 * i - 1)
    Next i
    TwoCycles = outperm
End Function
```

The remaining activity of the module takes place in Sheet1. There are
two ways to process text: letters may be entered one at a time using a
simulated keyboard, the input and output letters appearing in cells B3 and B4,
respectively, or to speed things up the text may be entered all at once in cell
B3 and processed with the command button $\boxed{\text{In}}$. In either case, the user may
observe the (simulated) path of the current through the wheels. Figure 26.6
depicts the spreadsheet with the original full German keyboard but with only
A–F active because of space limitations. The code in `CommandButton1_Click`,

	A	B	D	E	G	H	J	K	M	N
3	In	C	ref	rotor 3		rotor 2		rotor 1		ent
4	Out		A	B	B	C	C	E	E	A
5			B	C	C	D	D	F	F	B
6	Set at	BCE	C	D	D	E	E	A	A	C
7			D	E	E	F	F	B	B	D
8	current	BCE	E	F	F	A	A	C	C	E
9	speed	20	F	A	A	B	B	D	D	F

```
 Q   W   E   R   T   Z   U   I   O
   A   S   D   F   G   H   J   K
   P   Y   X   C   V   B   N   M   L
```

FIGURE 26.6: EnigmaCipher spreadsheet.

activated by pressing $\boxed{\text{In}}$, passes each letter of text in cell B3 to the function
`Convert`. The latter function changes the letter in accordance with the disk
positions and wiring.

```
Sub CommandButton1_Click()                      'command button In
    Dim inletter As String, outletter As String
    Dim text As String, i As Integer
    text = Range("B3").Value                     'get the entered text
    For i = 1 To Len(text)
        inletter = Mid(text, i, 1)               'extract a letter
```

```
        outletter = Convert(InLetter)          'send it through the disks
        Range("B4").Value = Range("B4").Value & outletter        'print
    Next i
End Sub
```

The function `Convert` runs a letter through the rotors right to left, then through the reflector, and finally back through the rotors left to right, producing an encrypted or decrypted letter. It uses the function `Disk`, which takes the row of an input letter in the column representing the disk and returns the row of the output letter. The column numbers are stored in the array `DiskCol`. The columns are indicated by the narrow double lines in Figure 26.6. There are nine entries since the current travels through the entry and exit disk (which are same), the reflector, and the three rotors right to left and left to right. The function also colors the path through the disks so that the user may follow the current through the wiring. (See Figure 26.7.)

```
Function Convert(InLetter As String) As String
    Dim rate As Integer, DiskCol(0 To 8) As Integer
    Dim InRow As Integer, OutRow As Integer
    DiskCol(0) = 14: DiskCol(1) = 12: DiskCol(2) = 9     'column numbers
    DiskCol(3) = 6:  DiskCol(4) = 3:  DiskCol(5) = 6      'of the disks
    DiskCol(6) = 9:  DiskCol(7) = 12: DiskCol(8) = 14
    Range(Cells(4, 3), Cells(29, 14)).Interior.ColorIndex = 0  'old path
    InRow = 4 + Alph2Val(InLetter)              'entry row of InLetter
    For i = 0 To 8
        OutRow = Disk(InRow, DiskCol(i), i): InRow = OutRow
    Next i
    Call MoveRotors(DiskCol)        'shift rotor column letters up by one
    Convert = Cells(OutRow, DiskCol(0)).Value    'return converted letter
End Function
```

The function `Disk` takes as arguments the row number `InRow` of the input letter to the disk, the column number of the disk, and the disk number. It outputs the row `OutRow` of the output letter of the disk. The output letter is found in Sheet2, which contains the permutations representing the disk wiring. The row of the output letter of the disk in Sheet1 is found by `LookUpRow`, which scans a disk column for the specified output letter and returns its row.

```
Function Disk(InRow As Integer, dcol As Integer, dnum As Integer) _
    As Integer
    Dim OutRow As Integer, OutCol As Integer, InCol As Integer
    Dim InLetter As String, OutLetter As String, color As Integer
    If dnum = 0 Or dnum = 8 Then    'if entry or exit, color red or green
        Cells(InRow, dcol).Interior.ColorIndex = 4 - dnum / 8
        Disk = InRow                        'output row same as input row
        Exit Function
    End If
    If 1 <= dnum And dnum <= 3 Then                'rotors right to left
        InCol = dcol + 1: OutCol = dcol - 1: color = 4
    ElseIf dnum = 4 Then                              'reflector
```

```
        InCol = dcol + 1: OutCol = dcol + 1: color = 24
    ElseIf 5 <= dnum And dnum <= 7 Then          'rotors left to right
        InCol = dcol - 1: OutCol = dcol + 1: color = 3
    End If
    InLetter = Cells(InRow, InCol).Value
    OutLetter = Sheet2.Cells(4 + Alph2Val(InLetter), dnum).Value
    OutRow = LookUpRow(OutLetter, OutCol)        'find row of OutLetter
    Call MakePath(InRow, InCol, OutRow, OutCol, dcol, color)
    Disk = OutRow
End Function


Function LookUpRow(letter As String, col As Integer) As Integer
    Dim row As Integer
    For row = 4 To 29
        If Cells(row, col).Value = letter Then Exit For
    Next row
    LookUpRow = row
End Function
```

The procedure `MoveRotors` shifts rotor 1 letters up after each input letter is processed. If rotor 1 returns to its original setting, then the letters of rotor 2 are shifted up. If rotor 2 returns to its original setting, letters of rotor 3 are shifted up.

```
Sub MoveRotors(DiskCol() As Integer)
    Dim setting As String, S1 As String, S2 As String
    setting = Range("B6").Value
    S1 = Mid(setting, 3)
    S2 = Mid(setting, 2)                              'original settings
    Call Shift(DiskCol(1))                    'shift rotor 1 each keystroke
    If Cells(4, DiskCol(1) + i).Value = S1 Then    'if rotor1 back to S1,
        Call Shift(DiskCol(2))                        'then shift rotor 2
    End If
    If Cells(4, DiskCol(2) + i).Value = S2 Then    'if rotor2 back to S2
        Call Shift(DiskCol(3))                        'then shift rotor 3
    End If
    Call DisplayPositions(DiskCol)        'show new positions of rotors
End Sub


Sub Shift(col As Integer)
    Dim i As Integer, temp1 As String, temp2 As String
    temp1 = Cells(4, col + 1).Value              'save current setting
    temp2 = Cells(4, col - 1).Value
    For i = 0 To 24                                  'move letters up
        Cells(4 + i, col + 1).Value = Cells(5 + i, col + 1).Value
        Cells(4 + i, col - 1).Value = Cells(5 + i, col - 1).Value
    Next i
    Cells(29, col + 1).Value = temp1        'put current setting at bottom
    Cells(29, col - 1).Value = temp2
End Sub
```

The procedure `DisplayPositions` simply lets the user know the top three letters in the rotor columns (which, can also be read directly from the columns).

```
Sub DisplayPositions(DiskCol() As Integer)
    Dim display As String
    For i = 1 to 3
        display = display & Cells(4, DiskCol(i) + 1).Value
    Next i
    Range("B8").Value = display
End Sub
```

The procedure `MakePath` connects the letter at `InRow` of a disk to the letter at `OutRow` with a colored path. A delay is added to slow down the "current" for visual purposes.

```
Sub MakePath(InRow As Integer, InCol As Integer, OutRow As Integer, _
            OutCol As Integer, dcol As Integer, color As Integer)
    Dim rate As Integer
    rate = Range("B9").Value
    Call Delay(rate)
    Cells(InRow, InCol).Interior.ColorIndex = color
    Call Delay(rate)
    Range(Cells(InRow, dcol), Cells(OutRow, dcol)) _
            .Interior.ColorIndex = color
    Call Delay(rate)
    Cells(OutRow, OutCol).Interiors.ColorIndex = color
    Call Delay(rate)
End Sub
```

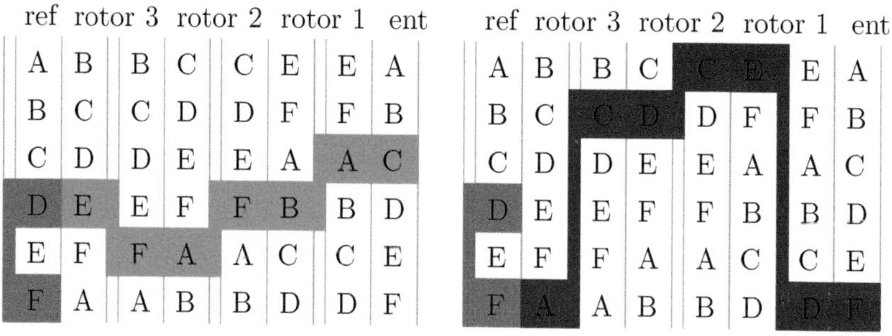

FIGURE 26.7: Green: right to left; violet: reflector; red: left to right.

The following command button procedure, activated by pressing [Set at], sets the rotors to the desired position `setting` in cell B6. For example, if `setting=` XYZ, then the rotors are set so that the first letter of the two rotor-3 columns is X, the first letter of the two rotor-2 columns is Y, and the first letter of the two rotor-1 columns is Z. The command button is used in the decryption process to return the setting to that used during encryption.

```
Sub CommandButton2_Click()
    Dim DiskCol(0 To 4) As Integer, i As Integer, setting As String
    Dim R1 As Integer, R2 As Integer, R3 As Integer
    DiskCol(0) = 14: DiskCol(1) = 12: DiskCol(2) = 9
    DiskCol(3) = 6:  DiskCol(4) = 3
    setting = Range("B6").Value
    R3 = Alph2Val(Mid(setting, 1, 1))
    R2 = Alph2Val(Mid(setting, 2, 1))
    R1 = Alph2Val(Mid(setting, 3, 1))
    For i = 0 To 25
        Cells(4 + i, DiskCol(3) + 1).Value = Chr((i + R3) Mod 26 + 65)
        Cells(4 + i, DiskCol(3) - 1).Value = Chr((i + R3) Mod 26 + 65)
        Cells(4 + i, DiskCol(2) + 1).Value = Chr((i + R2) Mod 26 + 65)
        Cells(4 + i, DiskCol(2) - 1).Value = Chr((i + R2) Mod 26 + 65)
        Cells(4 + i, DiskCol(1) + 1).Value = Chr((i + R1) Mod 26 + 65)
        Cells(4 + i, DiskCol(1) - 1).Value = Chr((i + R1) Mod 26 + 65)
    Next i
    Call DisplayPositions(DiskCol)
End Sub
```

26.3 A Closer Look at the Mathematics

For the purposes of this section we return to our abbreviated alphabet and write the permutations used in above examples in the standard form

$$R_1 = \begin{pmatrix} A & B & C & D & E & F \\ B & A & F & E & C & D \end{pmatrix} \quad R_2 = \begin{pmatrix} A & B & C & D & E & F \\ E & C & D & B & F & A \end{pmatrix} \quad R_3 = \begin{pmatrix} A & B & C & D & E & F \\ C & D & A & F & B & E \end{pmatrix}$$

$$R_4 = \begin{pmatrix} A & B & C & D & E & F \\ C & E & A & F & B & D \end{pmatrix} \quad R_5 = \begin{pmatrix} A & B & C & D & E & F \\ C & D & A & B & F & E \end{pmatrix} \quad R_6 = \begin{pmatrix} A & B & C & D & E & F \\ F & D & B & C & A & E \end{pmatrix}$$

$$R_7 = \begin{pmatrix} A & B & C & D & E & F \\ B & A & E & F & D & C \end{pmatrix}$$

Note that in state AAA the bottom row of R_1 is the input to the second rotor, etc. Moreover, the machine in this state is described by the permutation product $R_7 R_6 R_5 R_4 R_3 R_2 R_1$. To incorporate shifting of the rotors into the mathematical model we use the *left* and *right shift permutations* defined, respectively, by

$$\sigma = \begin{pmatrix} A & B & C & D & E & F \\ B & C & D & E & F & A \end{pmatrix}, \quad \sigma^{-1} = \begin{pmatrix} A & B & C & D & E & F \\ F & A & B & C & D & E \end{pmatrix}$$

Now suppose key D is pressed in state AAA. Current from letter D enters rotor 1 at contact D, travels to contact E, and enters rotor 2 at contact E. The machine is now in state AAB, rotor 1 having shifted up as in Figure 26.8.

Pressing key D again now causes current from letter D to enter rotor 1 at contact $E = \sigma(D)$, travel to contact C in rotor 1, and enter rotor 2 at contact $B = \sigma^{-1}(C)$. The net effect is given by the permutation product $\sigma^{-1}R_1\sigma$.

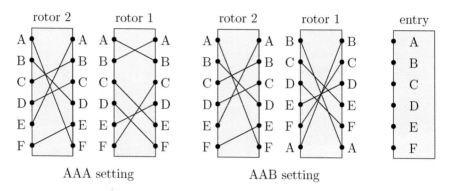

FIGURE 26.8: AAA and AAB states.

In summary, the machine in state AAB is described by the permutation product $\sigma R_7 \sigma^{-1} R_6 R_5 R_4 R_3 R_2 \sigma^{-1} R_1 \sigma$. In general if rotor 1 is moved a positions, rotor 2 is moved b positions, and rotor 3 is moved c positions, then the output of the machine in this state is given by the permutation product

$$(\sigma^a R_7 \sigma^{-a})(\sigma^b R_6 \sigma^{-b})(\sigma^c R_5 \sigma^{-c}) R_4 (\sigma^{-c} R_3 \sigma^c)(\sigma^{-b} R_2 \sigma^b)(\sigma^{-a} R_1 \sigma^a)$$
$$= \sigma^a R_7 \sigma^{b-a} R_6 \sigma^{c-b} R_5 \sigma^{-c} R_4 \sigma^{-c} R_3 \sigma^{c-b} R_2 \sigma^{b-a} R_1 \sigma^a,$$

where σ^a is σ applied a times.

This purely mathematical description may be used to emulate the enigma machine directly with a program that calculates permutation products and inverses, obviating the need for intricate electromechanical rotors. [4]Of course, electronic computers were unknown at the time of the machine's invention.

---

[4]We have actually incorporated this idea into the **Enigma** code available on the author's blog.

Chapter 27

Large Numbers

Arithmetic in VBA Excel is naturally limited by the size of Long and Double data types. However, by using strings instead of numbers, one can perform arithmetic on numbers of virtually any size. In this chapter we develop procedures that add, subtract, multiply, and divide integers and decimals whose size is limited only by the constraints of the spreadsheet. The results of the chapter suggest several interesting applications to cryptography. The reader is encouraged to devise and implement some of these.

27.1   Arithmetic of Large Positive Integers

There are a number of steps involved in performing arithmetic operations on digit strings. We begin with some formatting procedures.

Formatting

Since addition and subtraction of two numbers are performed digit by digit, it is convenient that the numbers be of equal length. The first procedure below takes a pair of digit strings and attaches leading zeros to whichever number is shorter. The second procedure removes leading zeros that may have been attached during an operation.

```
Sub AddLeadingZeros(a As String, b As String)
    Dim L As Integer
    L = Len(a) - Len(b)
    If L > 0 Then b = String(L, "0") & b      'attach L leading 0's to b
    If L < 0 Then a = String(-L, "0") & a     'attach -L leading 0's to a
End Sub

Sub RemoveLeadingZeros(a As String)
    Dim attachneg As Boolean, i As Integer: i = 1
    If IsZero(a) Then a = "0": Exit Sub
    If Mid(a, 1, 1) = "-" Then
        i = i + 1: attachneg = True   'skip over minus sign; attach later
    End If
    If Mid(a, i, 1) <> "0" Then Exit Sub
```

DOI: 10.1201/9781003351689-27

```
      Do While Mid(a, i, 1) = "0"
         i = i + 1
      Loop
      a = Mid(a, i)                                  'remove the zeros
      If attachneg Then a = "-" & a
   End Sub


   Function IsZero(str As String) As Boolean
      Dim i As Integer, zero As Boolean: zero = True
      For i = 1 To Len(str)
         If Mid(str, i, 1) Like "[1-9]" Then zero = False
      Next i
      IsZero = zero
   End Function
```

Addition of Positive Integers

The function AddPosInt(a,b) returns the sum of positive digit strings a,b. For ease of coding, the procedure first calls AddLeadingZeros to append zeros to the beginning of the numbers if necessary so that the numbers have the same length. The digits of a are then added to those of b starting from the right and moving left. At each stage, the procedure allows for a possible carry. (See Figure 27.1.)

$$\frac{\begin{array}{r}888\\888\end{array}}{} \rightarrow \frac{\begin{array}{r}1\\888\\888\end{array}}{6} \rightarrow \frac{\begin{array}{r}89\\88\end{array}}{6} \rightarrow \frac{\begin{array}{r}89\\88\end{array}}{76} \rightarrow \frac{\begin{array}{r}1\\89\\88\end{array}}{76} \rightarrow \frac{\begin{array}{r}9\\8\end{array}}{76} \rightarrow \frac{\begin{array}{r}9\\8\end{array}}{1776}$$

FIGURE 27.1: AddPosInt steps.

```
   Function AddPosInt(ByVal a As String, ByVal b As String) As String
      Dim sum As String, L As Integer, i As Integer
      Dim n As Integer, carry As Integer          'initialized to 0
      Call AddLeadingZeros(a, b)                   'give numbers same length
      L = Len(a)
      For i = L To 1 Step -1                       'add the digits in a and b
         n = CInt(Mid(a, i, 1)) + CInt(Mid(b, i, 1)) + carry
         carry = 0                                 'reset carry
         If n > 9 Then
            carry = 1                              'new carry for next addition
            n = n - 10                             'make into a digit
         End If
         sum = n & sum                             'append new digit to sum
      Next i
      If carry = 1 Then sum = 1 & sum              'if last carry=1, append to sum
      Call RemoveLeadingZeros(sum)                 'clean up
      AddPosInt = sum
   End Function
```

Subtraction of Positive Integers

The process of subtraction is somewhat more involved than addition owing to the fact that borrowing is more complicated than carrying. The function SubPosInt(a,b) takes positive numbers a,b and returns their difference a-b. The function first compares the numbers, switching them if a < b. In this case a negative sign must be attached to the result of the subtraction. The digits of the numbers are then subtracted from right to left, borrowing if needed. Figure 27.2 shows an example of the steps taken by the program.

$$
\begin{array}{ccccccc}
 & & 1 & & 1 & & \\
1022 & & 1012 & & 101 & & 9 \\
\underline{899} & \rightarrow & \underline{899} & \rightarrow & \underline{89} & \rightarrow & \underline{8} \\
 & & 3 & & 23 & & 123
\end{array}
$$

FIGURE 27.2: SubtractPosInt steps.

```
Function SubtractPosInt(ByVal a As String, ByVal b As String) As String
    Dim diff As String, temp As String, i As Integer, k As Integer
    Dim attachneg As Boolean, NeedToBorrow As Boolean
    attachneg = ComparePosInt(a, b) < 0        'attach minus sign later?
    If attachneg Then                          'if a < b then switch a and b
        temp = a: a = b: b = temp
    End If                                      'now a >= b
    Call AddLeadingZeros(a, b)                  'give numbers same length
    For i = Len(a) To 1 Step -1                 'start from right
        NeedToBorrow = False                    'default
        k = CInt(Mid(a, i, 1)) - CInt(Mid(b, i, 1))   'subtract digits
        If k < 0 Then k = k + 10: NeedToBorrow = True
        diff = k & diff                         'append new digit to diff
        If i > 1 Then
            a = Mid(a, 1, i - 1)                'remove last digits; they've done
            b = Mid(b, 1, i - 1)                           'their work
            If NeedToBorrow Then a = Borrow(a)  'e.g. 2300 --> 2299
        End If
    Next i
    Call RemoveLeadingZeros(diff)                          'clean up
    If attachneg Then diff = "-" & diff
    SubtractPosInt = diff
End Function
```

The function Borrow(num) subtracts 1 from num. It uses the VBA function String to attach 9's if needed, so, for example, Borrow(123500) = 123499. Since num could be large, ordinary subtraction may not work here.

```
Function Borrow(num As String) As String
    Dim L As Integer, n As Integer, i As Integer
    L = Len(num)
    If n <> 0 Then num = Mid(num, 1, L - 1) & n - 1: GoTo lastline
    For i = L To 1 Step -1
```

```
            n = CInt(Mid(num, i, 1))
        If n <> 0 Then Exit For 'found a nonzero digit n (= 1st digit<>0
    Next i                      'from the right; at position i from left) (n = 5)
    If n > 0 Then
        num = Mid(num, 1, i - 1)        '123500 becomes 1235 (L = 6, i = 4)
        num = num & n - 1                   '1235 becomes 1234
        num = num & String(L - i, "9")          '1234 becomes 123499
    End If
lastline:
    Borrow = num
End Function
```

The function `ComparePosInt` compares positive integers digit by digit, starting from the left.

```
Function ComparePosInt(a As String, b As String) As Integer
    Dim n1 As Integer, n2 As Integer, i As Integer: i = 1
    Call RemoveLeadingZeros(a)
    Call RemoveLeadingZeros(b)
    n1 = Len(a): n2 = Len(b)
    If n1 = n2 Then                             'if same length then
        Do                              'extract digits from left to right
            n1 = CInt(Mid(a, i, 1)): n2 = CInt(Mid(b, i, 1))
            i = i + 1
        Loop While n1 = n2 And i <= Len(a)          'stop when unequal
    End If
    ComparePosInt = n1 - n2                      'if > 0 then a > b, etc.
End Function
```

Multiplication of Positive Integers

The function `MultPosInt(a,b)` returns the product of the positive integer strings `a,b`. Most of the work is done by the special case `MultPosDigit`, which multiplies a positive integer by a single digit. The steps in the program are illustrated by the following example, the usual way one multiplies by hand.

$$
\begin{array}{ccccc}
 & 2 & 1 & 1 & \\
1234 & 1234 & 1234 & 1234 & 1234 \\
\underline{5} & \underline{5} & \underline{5} & \underline{5} & \underline{5} \\
 & 0 & 70 & 170 & 6170
\end{array}
$$

$1234 \xrightarrow{} 1234 \xrightarrow{} 1234 \xrightarrow{} 1234 \xrightarrow{} 1234$

FIGURE 27.3: Steps for MultByDigit.

```
Function MultDigit(num As String, digit As String) As String
    Dim p As Integer, c As Integer, prod As String
    Dim i As Integer, n As Integer, m As Integer
    n = CInt(digit):  c = 0                          'initialize
    For i = Len(num) To 1 Step -1                'go from right to left
```

```
m = CInt(Mid(num, i, 1))        'value of ith digit from right
p = m * n + c                   'product of digits + previous carry
If p > 9 Then                    'get carry and reduce p
    c = Int(p / 10)              'e.g., for p = 32: c = 3
    p = p - 10 * c              'new digit p = 32-30 = 2
Else
    c = 0                        'no carry
End If
prod = p & prod                 'adjoin new digit
Next i
If c > 0 Then prod = c & prod   'incorporate last carry
MultDigit = prod
End Function
```

The function `MultPosInt(a,b)` uses `MultPosDigit` to multiply the positive integer a by each digit of positive integer b from right to left, appending zeros on the right of each product and then adding, which is the usual way one multiplies.

$$
\begin{array}{r} 1234 \\ 78 \\ \hline 9872 \end{array}
\longrightarrow
\begin{array}{r} 1234 \\ 78 \\ \hline 9872 \\ 86380 \end{array}
\longrightarrow
\begin{array}{r} 1234 \\ 78 \\ \hline 9872 \\ 86380 \\ \hline 96252 \end{array}
$$

FIGURE 27.4: Steps for MultPosInt.

```
Function MultPosInt(a As String, b As String) As String
    Dim prod As String, d As String, digprod As String, i As Integer
    For i = Len(b) To 1 step -1
        d = Mid(b, i, 1)                          'get digit
        digprod = MultDigit(a, d)                 'multiply a by digit
        digprod = digprod & String(Len(b)-i, "0") 'attach Len(b)-i zeros
        prod = AddPosInt(prod, digprod)
    Next i
    MultPosInt = prod
End Function
```

Division Algorithm for Positive Integers

The procedure `DivAlgPos(a,b,q,r)` implements the algorithm for digit strings a,b, returning the quotient q and remainder r. The special cases a=b and a<b are treated first. The remaining case a>b is explained with the help of the long division example below, which yields $45678 = \underline{371} \cdot 123 + \underline{45}$: The procedure first extracts part of the dividend of length equal to that of the divisor 123, namely 456. The variable j is the position of the last digit in 456,

$$
\begin{array}{r}
371 \quad \text{quotient} \\
\text{divisor } 123 \, \overline{\big)\, 45678} \quad \text{dividend} \\
369 \\ \hline
877 \\
861 \\ \hline
168 \\
123 \\ \hline
45 \quad \text{remainder}
\end{array}
$$

namely 3. A For Next loop then multiplies 123 by digits $i = 0$ to 9 until the product exceeds 456, at which point $i = 4$. The values of the product and of i are then reduced to the preceding values 369 and 3, respectively. The digit 3 is appended to the quotient and the product 369 is subtracted from 456 to obtain a remainder of 87. The variable j is then incremented to 4 and the next digit in the dividend, 7, is appended to the remainder to obtain the new value str $= 877$. The process is repeated one more time ($j = 5$).

```
Sub LargeDivAlgPos(a As String, b As String, q As String, r As String)
    Dim str As String, p As String, i As Integer, j As Integer
    Dim val As Integer
    Call RemoveLeadingZeros(a): Call RemoveLeadingZeros(b)
    val = ComparePosInt(a, b)
    If val = 0 Then q = "1": r = "0": GoTo finish
    If val < 0 Then q = "0": r = a: GoTo finish
    j = Len(b)
    str = Mid(a, 1, j)                    'extract beginning part of number a
    q = ""
    Do While j <= Len(a)                             'keep increasing j
        For i = 0 To 9                   'multiply b by digits until exceeds str
            p = MultDigit(b, Chr(48 + i))
            If ComparePosInt(p, str) > 0 Then        'if p too large go back
                p = MultDigit(b, Chr(48 + i - 1))         'to last value
                Exit For
            End If
        Next i
        r = SubtractPosInt(str, p)        'subtract p from str for remainder
        q = q & i - 1                                 'update quotient
        If j = Len(a) Then Exit Do                    'if done exit
        j = j + 1
        nextdigit = Mid(a, j, 1)              'get next digit in dividend
        str = r & nextdigit                  'bring it down (attach to r)
    Loop
finish:
    Call RemoveLeadingZeros(q): Call RemoveLeadingZeros(r)     'clean up
End Sub
```

27.2 Arithmetic of Large Signed Integers

Negative Signs

The following functions are needed to reduce signed integer arithmetic to positive integer arithmetic.

```
Function IsNeg(x As String) As Boolean          'number negative?
    IsNeg = (Mid(x, 1, 1) = "-")
End Function

Function AbsVal(ByVal x As String) As String       'strip away any signs
    x = Replace(x, "-", ""): x = Replace(x, "+", "")
    AbsVal = x
End Function

Function Neg(a As String) As String     'return the negative of a number
    Dim x As String
    x = AbsVal(a)                                    'default
    If Not IsNeg(a) Then x = "-" & a          'attach neg. sign to a
    Neg = x
End Function
```

Addition and Subtraction of Integers

To find the sum of mixed sign integers a, b one need only consider the four cases determined by the signs of a and b and apply the functions `AddPosInt` and `SubtractPosInt`. Subtraction is then a special case of `AddInt`.

```
Function AddInt(a As String, b As String) As String
    Dim x As String
    If IsNeg(a) And IsNeg(b) Then
        x = "-" & AddPosInt(AbsVal(a), AbsVal(b))
    ElseIf IsNeg(a) Then
        x = SubtractPosInt(b, AbsVal(a))
    ElseIf IsNeg(b) Then
        x = SubtractPosInt(a, AbsVal(b))
    Else
        x = AddPosInt(a, b)
    End If
    AddInt = x
End Function

Function SubtractInt(a As String, b As String) As String
    SubtractInt = AddInt(a, Neg(b))
End Function
```

Multiplication of Integers

The product of signed integers is calculated by considering the signs of the numbers and applying `MultPosInt`.

```
Function MultInt(a As String, b As String) As String
    Dim x As String
    x = MultPosInt(AbsVal(a), AbsVal(b))
    If (IsNeg(a) And Not IsNeg(b)) Or (IsNeg(b) And Not IsNeg(a)) Then
        x = "-" & x
    EndIf
    MultInt = x
End Function
```

The function `PowerInt` takes an integer b and a nonnegative integer exponent e and calculates b^e.

```
Function PowerInt(b As String, exp As Integer) As String
    Dim i As Integer, attach_neg As Boolean, p As String: p = 1
    If exp = 0 Then GoTo lastline
    If exp = 1 Then p = b: GoTo lastline
    attach_neg = IsNeg(b) And Abs(exp) Mod 2 <> 0    'b < 0 and exp odd?
    For i = 1 To exp                          'repeat multiplication exp times
        p = MultInt(b, p)
    Next i
lastline: PowerInt = p
End Function
```

Division Algorithm for Integers

The general case of the division algorithm is analogous to the procedure of Section 23.1.

```
Sub LargeDivAlg(a As String, b As String, q As String, r As String)
    If IsZero(b) Then Exit Sub
    Call DivAlgPos(AbsVal(a), AbsVal(b), q, r)
    If Not IsNeg(a) And IsNeg(b) Then                    'a pos, b neg
        q = Neg(q)
    ElseIf IsNeg(a) And Not IsNeg(b) Then                'a neg, b pos
        If IsZero(r) Then q = Neg(q)
        If Not IsZero(r) Then
            q = Neg(AddInt(q, "1"))
            r = SubtractInt(b, r)
        End If
    ElseIf IsNeg(a) And IsNeg(b) Then                    'both neg
        Call LargeDivAlg(a, Neg(b), q, r): q = Neg(q)    'recurse!
    End If
End Sub
```

27.3 Arithmetic of Large Decimals

The idea behind decimal arithmetic is to remove the decimal points, do the calculations with the resulting integers, and then put back the decimal points. This allows the use of the procedures in Section 27.2.

Decimal Points

The following procedures facilitate the transition from signed integer arithmetic to decimal arithmetic. The first, SplitDec(a,s,w,f), takes a decimal string a and splits it into three strings: the sign s (either "-" or ""), the whole part w, and the fractional part f. For example, SplitDec("-123.45",s,w,f) returns "-", "123", and "45" in the variables s,w,f, respectively. For uniformity, the program returns f=0 if a is a whole number.

```
Sub SplitDec(a As String, s As String, w As String, f As String)
    Dim x As String, arr() As String, pos As Integer
    s = "": w = "0": f = "0": x = a                     'default
    If IsNeg(a) Then s = "-": x = AbsVal(a)
    pos = InStr(x, ".")                    'position of decimal point, if any
    If pos = 0 Then
        w = x                              'no decimal; return x = a
    ElseIf pos = 1 Then                    'decimal at beginning, no w
        f = Mid(x, pos + 1)
    ElseIf pos = Len(x) Then                   'decimal at end, no f
        w = Mid(x, 1, pos - 1)
    Else
        arr = Split(x, ".")               'decimal at intermediate position
        w = arr(0): f = arr(1)
    End If
End Sub
```

The procedure AddTrailingZeros(x,y,dright) adds trailing zeros to x,y, which are fractional parts of numbers, to ensure that these have the same length. The number dright of decimal places from the right is the largest of the lengths.

```
Sub AddTrailingZeros(x As String, y As String, dright As Integer)
    Dim numZeros As Integer
    dright = WorksheetFunction.Max(Len(x), Len(y))
    numZeros = Abs(Len(x) - Len(y))
    If Len(x) > Len(y) Then
        y = y & String(numZeros, "0")                 'append 0's to y
    ElseIf Len(x) < Len(y) Then
        x = x & String(numZeros, "0")                 'append 0's to x
    End If
End Sub
```

The function `RestoreDecPt(a,dright)` takes a whole number `a` and inserts a decimal `dright` positions from the right. For example, `RestoreDecPt(1234500,4)` returns `123.45`. Trailing zeros are deleted by `RemoveTrailingZeros`

```
Function RestoreDecPt(a As String, dright As Integer) As String
    Dim x As String, arr() As String
    x = a
    If IsZero(a) Then x = "0": GoTo finish
    If dright = 0 Then GoTo finish          'nothing to restore
    If v >= Len(x) Then
        x = "." & String(dright - Len(x), "0") & x
        GoTo finish
    End If
    x = InsertString(a, Len(a) - dright, ".")      'insert decimal pt.
    arr = Split(x, ".")                  'get whole and fractional parts
    If IsZero(arr(1)) Then                       'string of zeros?
        x = arr(0)                             'yes, so ignore frac part
    Else
        arr(1) = RemoveTrailingZeros(arr(1))    'otherwise remove 0 trail
        x = arr(0) & "." & arr(1)
    End If
finish:x = RemoveTrailingZeros(x)
    RestoreDecPt = x
End Function

Function RemoveTrailingZeros(a As String) As String
    Dim x As String, L As Integer
    L = Len(a): i = L: x = a
    Do While Midd(a, i, 1, "") = "0"
        i = i - 1
    Loop
    If i < L And InStr(a, ".") > 0 Then x = Mid(a, 1, i) 'remove the 0's
    RemoveTrailingZeros = x
End Function
```

Decimal Addition and Subtraction

The function `AddDec(a,b)` first splits `a` and `b` into their constituent parts, then adds trailing zeros to their fractional parts to balance length. The numbers are then reconstituted as integers and fed to `AddInt`

```
Function AddDec(a As String, b As String) As String
    Dim x As String, y As String, intsum As String
    Dim a_s As String, a_w As String, a_f As String
    Dim b_s As String, b_w As String, b_f As String
    Dim dright As Integer, decsum As String
    Call SplitDec(a, a_s, a_w, a_f)
    Call SplitDec(b, b_s, b_w, b_f)
    Call AddTrailingZeros(a_f, b_f, d_right)
    x = a_s & a_w & a_f: y = b_s & b_w & b_f
```

```
        intsum = AddInt(x, y)
        AddDec = RestoreDecPt(intsum, dright)
    End Function

    Function SubtractDec(a As String, b As String) As String
        SubtractDec = AddDec(a, Neg(b))
    End Function
```

Decimal Multiplication

The function MultDec(a,b) first splits a and b into their parts, adds trailing zero for balance, reassembles the parts into whole numbers and applies MultInt.

```
    Function MultDec(a As String, b As String) As String
        Dim x As String, y As String, prod As String, d_right As Integer
        Dim a_s As String, a_w As String, a_f As String
        Dim b_s As String, b_w As String, b_f As String
        Call SplitDec(a, a_s, a_w, a_f)
        Call SplitDec(b, b_s, b_w, b_f)
        Call AddTrailingZeros(a_f, b_f, d_right)
        x = a_s & a_w & a_f: y = b_s & b_w & b_f
        prod = RestoreDecPt(MultInt(x, y), 2 * d_right)
        MultDec = prod
    End Function
```

Decimal Division

The function DivDec(a,b,n) returns the quotient a/b carried out to n places. The procedure starts out as usual by splitting a and b into its constituent parts. Trailing zeros are added where needed and the parts reassembled to reduce the problem to division by integers. The numerator is then effectively multiplied by 10^n and DivAlg is called. The result is then effectively divided by 10^n, which achieves the desired accuracy. For example, for $n = 3$ we have

$$\frac{1234.56}{9.871} = \frac{1234560}{9871} = \frac{1}{10^3}\frac{1234560000}{9871} = \frac{1}{10^3}\frac{9871q + r}{9871} = \frac{q}{10^3} + \frac{r}{9871000}$$
$$= \frac{125069}{10^3} + \frac{r}{9871000} = 125.069 + \frac{r}{9871000}$$

where we have applied the division algorithm to the numerator 1234560000, yielding $q = 125069$ and $r = 3901$. From the inequality $0 \leq r < 9871$ we see that $0 \leq r/9871000 < .0001$, hence the approximation $q/10^3 = 125.069$ agrees with the actual value $1234.56/9.871 = 125.0693952$ in three decimal places.

```
    Function DivDec(a As String, b As String, divplaces As Integer) As String
        Dim x As String, y As String, q As String, r As String
        Dim a_s As String, a_w As String, a_f As String, d_right As Integer
        Dim b_s As String, b_w As String, b_f As String
        Call SplitDec(a, a_s, a_w, a_f)                          '1234, 56
```

```
    Call SplitDec(b, b_s, b_w, b_f)                         '9, 871
    Call AddTrailingZeros(a_f, b_f, d_right)               '560, 871
    x = a_s & a_w & a_f & String(divplaces, "0")         '1234560000
    y = b_s & b_w & b_f                                      '9871
    Call DivAlg(x, y, q, r)                      'q = 125069, r = 3901
    DivDec = RestoreDecPt(q, divplaces)              'q = 125.069
End Function
```

The function `LargeRound` rounds a decimal digit string to any desired number of places.

```
Function LargeRound(a As String, rndplaces As Integer) As String
    Dim a_s As String, a_w As String, a_f As String
    Dim b As String, c As String
    If InStr(a, ".") = 0 Then               'if no decimal point then
        LargeRound = a: Exit Function                'return a
    End If
    Call SplitDec(a, a_s, a_w, a_f)           'get decimal components
    If rndplaces >= Len(a_f) Then            'if too many places then
        LargeRound = a
        Exit Function                                'return a
    End If
    If rndplaces = 0 Then                       'if zero places then
        LargeRound = a_w
        Exit Function                       'return whole part of a
    End If
    b = a_w & "." & Mid(a_f, 1, rndplaces)    'chop fractional part of a
    If CInt(Mid(a_f, rndplaces + 1, 1)) >= 5 Then  'if next place >= 5
        c = "0." & String(rndplaces - 1, "0") & 1
        b = AddDec(b, c)                            'then round up
    End If
    LargeRound = b
End Function
```

A little experimentation might be required with the settings `divplaces` and `rndplaces` in calculations, especially with never-ending decimals such as $1/81 = .012345679012345679\ldots$. For example, using the preceding division, multiplications, and rounding functions with `divplaces`$-$ 8, one gets the correct result $(1/81) * 81 = 1$ only when `rndplaces`≤ 5.

Decimal Exponentiation

The function `PowerDec(b, e)` uses `MultDec` repeatedly to calculate b^e, where e is an integer.

```
Function PowerDec(base As String, exp As Integer)  As String
    Dim i As Integer, p As String, attach_neg As Boolean
    If exp = 0 Then p = 1: GoTo lastline
    If exp = 1 Then p = base: GoTo lastline
    If IsNeg(base) And Abs(exp) Mod 2 <> 0 Then    'is b < 0 and exp odd?
        attach_neg = True                        'yes: answer negative
```

```
      End If
      base = AbsVal(base)                          'remove sign
      If exp < 0 Then
          base = DivDec(1, base, n)             'replace base by 1/base
          exp = -exp                            'make power positive
      End If
      p = 1
      For i = 1 To exp
          p = MultDec(base, p)
      Next i
lastline: If attach_neg Then p = "-" & p
      PowerDec = p
End Function
```

*27.4 A Large Scale Calculator

In this section we develop a calculator that evaluates large arithmetic expressions using the procedures in the previous sections. The program takes as input in cell B6 a text expression such as 2.1*3.2^4 + 5.4*6/(7*8-.9) and returns the value in cell B3. The numbers are decimal strings of arbitrary length, allowing the calculator to be used for large scale arithmetic. The program obeys the usual natural hierarchy of operations.

```
Sub CommandButton1_Click()
    Dim expr As String, divplaces As Integer, roundplaces As Integer
    Dim error As Boolean, i As Integer, z As String
    Range("B3").Value = ""                       'erase previous answer
    divplaces = Range("B4").Value          'decimal places in division
    roundplaces = Range("B5").Value        'decimal places in rounding
    Do While Not IsEmpty(Cells(6 + i, 2).Value)     'read expr in col B
        expr = expr & Cells(6 + i, 2).Value            'concatenate rows
        i = i + 1
    Loop
    expr = RemoveWhiteSpace(expr)                          'compress
    z = LargeCalc(expr, 1, divplaces, 0, error)     'do the calculation
    If error Then z = "error"
    Range("B3").Value = LargeRound(z, roundplaces)
End Sub
```

The main procedure of the module is LargeCalc. It has essentially the same structure as ComplexEval of Section *8.4. The variable places determines the length of a division calculation. For example, if places = 5 then LargeCalc("1/3") returns .33333. The variable mode, initially set to zero, ensures the correct hierarchy operations. The procedure ErrorCheck is the same as the eponymous procedure in Section *8.4, except that there is no imaginary symbol i to worry about.

```
Function LargeCalc(expr As String, idx As Integer, mode As Integer, _
             places As Integer, error As Boolean) As String
   Dim z As String, w As String, char As String, LonelyOp As Boolean
   error = ErrorCheck(expr)
   Do While idx <= Len(expr) And error = False
      char = Mid(expr, idx, 1)                   'character at position idx
      If InStr("0123456789.", char) > 0 Then char = "#"        'decimal
      Select Case char
         Case Is = "#"
            z = GetDecimal(expr, idx)     'advances idx to after no.
         Case Is = "+"
            If mode > 0 Then Exit Do        'wait for higher mode ops
            LonelyOp = (idx = 1 Or Midd(expr, idx-1, 1, " ") = "(")
            idx = idx + 1
            w = LargeCalc(expr, idx, 0, places, error)
            If error Then Exit Do
            If LonelyOp Then z = "0"
            z = AddDec(z, w)
         Case Is = "-"
            If mode > 0 Then Exit Do        'wait for higher mode ops
            LonelyOp = (idx = 1 Or Midd(expr, idx - 1, 1, " ") = "(")
            idx = idx + 1
            w = LargeCalc(expr, idx, 1, places, error)
            If error Then Exit Do
            If LonelyOp Then z = "0"
            z = SubtractDec(z, w)
         Case Is = "*"
            If mode > 1 Then Exit Do        'wait for higher mode ops
            idx = idx + 1
            w = LargeCalc(expr, idx, 1, places, error)  'next factor
            If error Then Exit Do
            z = MultDec(z, w)
         Case Is = "/"
            If mode > 1 Then Exit Do        'wait for higher mode ops
            idx = idx + 1
            w = LargeCalc(expr, idx, 1, places, error)     'get denom.
            error = error Or (w = "0")
            If error Then Exit Do
            z = DivDec(z, w, places)
         Case Is = "^"                          'highest op--no wait
            idx = idx + 1                                  'skip "^"
            w = LargeCalc(expr, idx, 2, places, error)      'get exp
            error = error Or z = "0"
            If error Then Exit Do
            z = PowerDec(z, CInt(w), places)
         Case Is = "("
            idx = idx + 1                                  'skip "("
            z = LargeCalc(expr, idx, 0, places, error)
            If error Then Exit Do
            idx = idx + 1                                  'skip")"
         Case Is = ")": Exit Do
```

```
            Case Else: error = True: Exit Do
        End Select
    Loop
    Call RemoveLeadingZeros(z)
    LargeCalc = z
End Function
```

If the beginning of a number is found, then `GetDecimal` extracts the number string from `expr` and returns its decimal string. At this stage, `Idx` points to the symbol immediately after the number.

```
Function GetDecimal(expr As String, idx As Integer) As String
    Dim NumStr As String, m As Integer, char As String
    m = idx                    'remember position of beginning of number
    char = Midd(expr, idx, 1, " ")
    Do While InStr("0123456789.", char) > 0          'find end of number
        idx = idx + 1
        char = Midd(expr, idx, 1, " ")
    Loop                   'after loop idx points immediately after number
    NumStr = Mid(expr, m, idx - m)              'extract the number
    GetDecimal = NumStr
End Function
```

27.5 Exercises

1. Write a program `LargeFactorial` that calculates large factorials.

2. Write a program `LargeBinomial` that calculates large binomial coefficients.

3. Write a program `LargeSevenRule` that extends `SevenRule` of Exercise 2 in Chapter 23 to digit strings.

4. Write a program `LargeThirteenRule` that extends `ThirteenRule` of Exercise 3 in Chapter 23 to digit strings.

5. Write a program `LargeGCD` that extends `GCD` of Section 23.4 to digit strings.

6. Write a program `LargeSieve` that is the large string analog of the program `Sieve` of Chapter 7.

7. Write a function `LargeSymbolicAddFrac(frac1,frac2)` that returns the string obtained by adding the fractions. Each fraction should be a string of the form `num/den`, where `num,den` are digit strings. For example, `LargeAddFrac("1/2","3/4")` should return `"5/4"`. Write similar programs for subtraction, multiplication and division.

8. Write a procedure LargeBaseToBase that extends BaseToBase of Section 23.2. Use the procedure to devise a cipher that takes a message written in capitals as a base Z number and converts it into binary.

9. Write a function PowerSumDigits(n As String, e As Integer) where n is a possibly large digit string, that first calculates n^e then adds the digits of the resulting number, then adds the digits of that number, etc. until a one digit number is attained, which is returned by the function. For the case $e = 3$ that number will always be be either 1, 8, or 9. For example: $11^3 = 1331 \to 8$; $13^3 = 2197 \to 19 \to 10 \to 1$. Incorporate the function into a module that generates a column of random numbers n of random length and prints in a column the number and the result of applying the function. Experiment with various exponents e.

10. (Reverse and Add) Consider the following process: Take any positive integer n with at least two digits, reverse the digits, add the result to n, and repeat the operation with the new number. It is known that for many numbers n the process eventually results in a palindrome, that is a number which stays the same when the digits are reversed. For example, it takes 3 steps for the number 59 to become palindromic:

$$59 + 95 = 154, \quad 154 + 451 = 605, \quad 605 + 506 = 1111.$$

It takes 24 steps for the number 89 to become palindromic with final sum 8813200023188, and 55 steps for the number 10,911 to become palindromic with final sum 4668731596684224866951378664. Write a function ReverseAndAdd(n As String, steps As Long, maxsteps As Long) that takes a digit string n and returns the eventual palindromic sum of n and the number of steps required to achieve it. Use AddPosInt and the VBA function ReverseStr. (The upper limit maxsteps should guard against overflow.)

11. Write a program LargeDecimalSort that takes a column of positive decimal strings and sorts them from smallest to largest. Use the bubble sort method described in Section 5.5.

12. Make a digit string analog LargeDeterminant of the program DetRecursive (Section 14.2).

13. Make a digit string analog LargeCramer of the program CramersRule (Section 14.3).

Index

For Product Safety Concerns and Information please contact our
EU representative GPSR@taylorandfrancis.com Taylor & Francis
Verlag GmbH, Kaufingerstraße 24, 80331 München, Germany